Confidence interval for $\mu_1 - \mu_2$

$$\bar{y}_1 - \bar{y}_2 \pm t s_p \sqrt{\frac{1}{n_1} + \frac{1}{n_2}}$$

Statistical test of H_0: $\mu_1 - \mu_2 = 0$

$$t = \frac{\bar{y}_1 - \bar{y}_2}{s_p \sqrt{\frac{1}{n_1} + \frac{1}{n_2}}}$$

Confidence interval for p

$$\hat{p} \pm z\sigma_{\hat{p}}$$

Statistical test of H_0: $p = p_0$

$$z = \frac{\hat{p} - p_0}{\sigma_{\hat{p}}}$$

Confidence interval for $p_1 - p_2$

$$\hat{p}_1 - \hat{p}_2 \pm z\sigma_{\hat{p}_1 - \hat{p}_2}$$

Statistical test of H_0: $p_1 - p_2 = 0$

$$z = \frac{\hat{p}_1 - \hat{p}_2}{\sigma_{\hat{p}_1 - \hat{p}_2}}$$

Test of H_0: $\sigma^2 = \sigma_0^2$

$$\chi^2 = \frac{(n-1)s^2}{\sigma_0^2}$$

Test of H_0: $\sigma_1^2 = \sigma_2^2$

$$F = \frac{s_1^2}{s_2^2}$$

Least squares line

$$\hat{y} = \hat{\beta}_0 + \hat{\beta}_1 x$$

Slope of least squares line

$$\hat{\beta}_1 = \frac{S_{xy}}{S_{xx}}$$

Intercept of least squares line

$$\hat{\beta}_0 = \bar{y} - \hat{\beta}_1 \bar{x}$$

Test of H_0: $\beta_1 = 0$

$$t = \frac{\hat{\beta}_1}{\sqrt{s_\epsilon^2 / S_{xx}}}$$

Population correlation coefficient ρ

Sample correlation coefficient

$$\hat{\rho} = \frac{S_{xy}}{\sqrt{S_{xx}S_{yy}}}$$

Paired-difference test

$$t = \frac{\bar{d}}{s_d / \sqrt{n}}$$

Chi-square test of independence

$$\chi^2 = \sum \frac{(O - E)^2}{E}$$

Spearman's rank correlation coefficient

$$\hat{\rho}_s = 1 - \frac{6(\Sigma d^2)}{n(n^2 - 1)}$$

SIXTH EDITION

Understanding
Statistics

R. LYMAN OTT

Marion Merrell Dow, Inc.

WILLIAM MENDENHALL

Professor Emeritus, University of Florida

An Alexander Kugushev Book

Duxbury Press
An Imprint of Wadsworth Publishing Company
Belmont, California

Duxbury Press
An Imprint of Wadsworth Publishing Company
A division of Wadsworth, Inc.

Editorial Assistant   *Jennifer Burger*
Production   *The Book Company*
Design   *Cloyce Wall*
Print Buyer   *Barbara Britton*
Copy Editor   *Stephen Gray*
Technical Illustrator   *Lotus Art*
Cover   *Janet Bollow Associates*
Cover Photograph   *TIB/West Stanislaw Fernandez © 1993*
Compositor   *Interactive Composition Corporation*
Printer   *R. R. Donnelley and Sons*

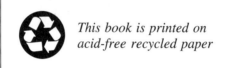

This book is printed on acid-free recycled paper

I(T)P ™

International Thomson Publishing
The trademark ITP is used under license

Printed in the United States of America
  2 3 4 5 6 7 8 9 10—98 97 96 95 94

Library of Congress Cataloging-in-Publication Data

Ott, Lyman.
    Understanding statistics. / Lyman Ott, William Mendenhall. — 6th ed.
       p. cm.
    "An Alexander Kugushev book."
    Includes bibliographical references and index.
    ISBN 0–534–20922–X
    1. Statistics. I. Mendenhall, William.   II. Title.
QA276, 12.087   1995
001.4'22—dc20   93–47902
                                                              CIP

4   Numerical Methods for Summarizing Data   94

Part 4   Background for Analyzing Data   147

5   Probability and Probability Distributions   148

Part 6   Step Three: Analyzing Data: Regression, Correlation, and Analysis of Variance   443

11   Regression and Correlation   444

12   Inferences Related to Linear Regression and Correlation   505

13 Analysis of Variance   558

Part 7   Step Four: Communicating Results   587

14 Communicating the Results of Analyses   588

T he sixth edition of *Understanding Statistics* offers a new and more modern approach to teaching an introductory statistics course with relevance to the real world. It is designed for a one-term course. Undergraduate majors in business or public administration, the arts and sciences, engineering, and education will find that this approach, with accompanying examples and exercises, will provide a basic understanding of statistical concepts. More importantly it will impart a skill—making sense of data—which is the central objective of this text.

This book emphasizes data. It is organized around an underlying pervasive theme: making sense of data. The four steps in making sense of data—data gathering, data summarization, data analysis, and the communication of results—are core segments of the text and are employed throughout. Most introductory texts focus on summarization and analysis. We emphasize that problems with any foundation in reality must deal with the collection, summarization, analysis, and communication of data. To ignore these steps is to take statistics out of the context of its applied use in business, in government, in research, and in all its other applications in the real world.

The theme of making sense of data is emphasized and sustained in a variety of pedagogical ways throughout the text:

- The table of contents is organized into parts reflecting the four steps: data collection, data summarization, data analysis, and communication of results.

- A section, "Using Computers to Help Make Sense of Data," concludes most chapters, to show how computers facilitate the process of making sense of data.

- A data disk is provided with ten real-life databases. It also includes data for most exercises in the text. The ten real-life databases are also printed in Appendix 1.

- An "Experiences with Real Data" section has been added to the exercises at the end of each chapter. The corresponding exercises reference data from the ten databases in Appendix 1.

- Graphical methods for summarization of data have been strengthened and modernized.

- The step of communication of results has been further developed and brought forward in the text.

- A discussion of quality improvement, as it relates to process changes, and of statistical tools useful in quality assessment has been added.

- An Instructor's Manual demonstrating how to teach an effective, modern course in statistics based on the theme of "making sense of data" is provided.

- The exercise sets have been strengthened throughout and organized into Basic Techniques, Applications, Supplementary Exercises, and Experiences with Real Data.

- The coverage has been reorganized into 14 tightly related chapters.

- Extensive error checking has been done to provide a virtually error-free text.

The making of this book through the years has been a team effort. For this edition, a special note of appreciation is extended to my administrative assistant, Jo Ann Spalding, who created polished word-processing documents in the revision cycle of this edition. We also want to thank Phyllis Barnidge who prepared the Solutions Manual and data disk.

Many students and professors who have used previous editions of this text have contacted us with suggestions and they are greatly appreciated. The authors especially want to acknowledge the reviewers for this edition: Dale O. Everson, University of Idaho; Alex S. Papadopoulos, University of North Carolina at Charlotte; Wesley W. Tom, Chaffey College; Kathie A. Yoder, Los Angeles Pierce College.

Finally, we want to acknowledge how grateful we are to be working again with our publisher, Alex Kugushev. He is the consummate professional and he has assembled a talented editorial and production team. We have been pleased as well with the ongoing encouragement and support of our families.

R. Lyman Ott

William Mendenhall

Statistics: Making Sense of Data

1

An Introduction to Making Sense of Data

1.1   What Is Statistics?

What is statistics? Is it the addition of numbers? Is it graphs, batting averages, percentages of passes completed, percentages of unemployment, and, in general, numerical descriptions of society and nature?

Statistics, as a subject, is the study of making sense of data. Almost everyone—including corporate presidents, marketing representatives, social scientists, chemists, and consumers—deals with data. These data could be in the form of quarterly sales figures, expenditures for goods and services, pulse rates for patients undergoing therapy, contamination levels in samples of surface water, or answers to census questions. In this text, we approach the study of statistics by considering the four steps involved in making sense of data: (1) gathering data, (2) summarizing data, (3) analyzing data, and (4) reporting the results of analyses. Following this introductory chapter, we begin addressing each of these four crucial steps in turn. Altogether, the text is divided into seven parts, most of which are centrally concerned with one or another of the four steps. Other parts are devoted to providing necessary background or connective material. The relationships among the seven parts of the textbook, the four steps involved in making sense of data, and the fourteen individual chapters are show in Table 1.1.

T A B L E 1.1   Organization of the Textbook

Part of the Textbook	Step in Making Sense of Data	Chapter of the Textbook	
1 Statistics: Making Sense of Data	—	1	An Introduction to Making Sense of Data
2 Step One: Gathering Data	1	2	Using Surveys and Experimental Studies to Gather Data
3 Step Two: Methods for Summarizing Data	2	3	Tabular and Graphical Methods for Summarizing Data
		4	Numerical Methods for Summarizing Data
4 Background for Analyzing Data	—	5	Probability and Probability Distributions
		6	Sampling Distributions
5 Step Three: Analyzing Data: Means, Variances, and Proportions	3	7	Inferences about μ
		8	Inferences About $\mu_1 - \mu_2$
		9	Inferences About Variances
		10	Analyzing Count Data
6 Step Three: Analyzing Data: Regression, Correlation, and Analysis of Variance	3	11	Regression and Correlation
		12	Inferences Related to Linear Regression and Correlation
		13	Analysis of Variance
7 Step Four: Communicating Results	4	14	Communicating the Results of Analyses

As you can see from this table, a great deal of time is spent discussing how to analyze data by using basic methods (for means, variances, and proportions), regression methods, and analysis of variance methods. However, you must remember that, for each data set requiring analysis, someone has developed a plan for gathering the data (step 1) and has prepared the data for analysis. Furthermore, following summarization (step 2) and analysis of the data (step 3), someone has to communicate the results of the analysis (step 4) verbally or in writing to the intended audience. All four steps are important in making sense of data; the analysis step, while time-consuming, is only one of these steps. Throughout the text, we will try to keep you focused on the bigger picture of making sense of data. Periodically, you may want to refer to this table as a reminder of where each chapter fits into the overall scheme of things.

Before jumping into the study of statistics, let's consider three instances in which the application of statistics could help to solve a practical problem.

1   Suppose that a manufacturer of light bulbs produces roughly a half-million bulbs per day. Because of some adverse customer reaction to its product, the firm wishes to determine the fraction of bulbs produced on a given day that are defective. It can solve the problem in two ways. The half-million bulbs could be inserted into sockets and tested, but the cost of this solution would be substantial and could greatly increase the price per bulb. A second method for determining the fraction of defective bulbs is to select 1000 bulbs from the half-million produced and test each of these. The fraction of bulbs defective in the 1000 tested could be used to estimate the fraction defective in the entire day's production. We will show in later chapters that the fraction defective in the bulbs tested will probably be quite close to the fraction defective for the entire half-million bulbs. We will also be able to tell you by how much you might expect this estimate to differ from the actual fraction of defective bulbs produced on any given day.

2   A similar application of statistics is brought to mind by the frequent use of the Gallup poll, the Harris poll, and other public opinion polls. How can these pollsters presume to know the opinions of more than 100 million Americans? They certainly cannot reach their conclusions by contacting every voter in the United States. Rather, as we have suggested in the light bulb example, they sample the opinions of a small number of voters, perhaps as few as 1500, to estimate the reaction of every voter in the country. The amazing result of this process is that the fraction of people contacted who hold a particular opinion will match very closely the fraction of voters holding that opinion in the total population at that point in time. Most students find this assertion difficult to believe, but we will supply convincing supportive evidence in subsequent chapters.

3   Another example of a statistical problem comes from the field of medicine. Suppose that a research physician wishes to investigate the stimulatory effect of a new drug on a patient's heart. The physician, who is interested in the effect of the drug on all future heart patients who might be treated with the drug, selects fifty heart patients and treats each with the drug. The increase in pulse rate is recorded for each patient over a period of time. After observing the effects of the drug on the fifty patients, the physician may infer that the drug will have similar effects on all heart patients in the future.

Each of these problems illustrates the four steps involved in making sense of data. First came a data-gathering stage, using sampling. A group (sample) of light bulbs was selected from the day's production; a sample of people was obtained from the entire voting population in the United States; and a sample of fifty heart patients was obtained from the population of individuals with heart problems. Then a measurement was obtained for each element (bulb, voter, or patient) in the sample. These measurements (data) are then used to solve the problem.

Next, in order to make sense of the data collected, someone must summarize and analyze them. In the light bulb example, the fraction of defective bulbs could be computed for the sample of bulbs tested. Based on this value and on the number of bulbs tested (1000), one could accurately predict the fraction of defective bulbs in the entire day's production of a half-million bulbs. Similarly, for the voter opinion poll, one could compute the fraction of sample voters who favored each of the candidates; then, based on the results for the 1500 sample voters, one could accurately predict the voting pattern for the entire voting public in the United States at that point in time. In the study of fifty heart patients, measurements of variables such as exercise capacity, oxygen consumption, and quality of life could be used to predict the effect (efficacy) of the experimental drug on other patients who might be candidates for similar treatment.

Finally, having collected, summarized, and analyzed the data, one must report the results in unambiguous terms to interested persons. For the light bulb example, management and technical staff need to know the quality of their production batches. Based on this information, they could determine whether adjustments in the process are necessary. The results of statistical analysis cannot be presented in ambiguous terms; decisions must be made from a well-defined knowledge base. The results of a voter opinion poll are of vital interest to political candidates, to campaign managers, and to potential campaign contributors, and they might lead to major shifts in campaign and funding strategies. Also, the results of a heart patient study are important to the physician treating the patients, to the company developing the medicinal compound, to the U.S. Food and Drug Administration, and to the medical community in general. The results must be presented clearly so that informed decisions can be made regarding the future development of the compound for treating heart patients.

DEFINITION   1.1   **Population**   A **population** is the set of all measurements of interest to the sample collector.   ▪

DEFINITION   1.2   **Sample**   A **sample** is any subset of measurements selected from the population.   ▪

FIGURE 1.1
Population and Sample

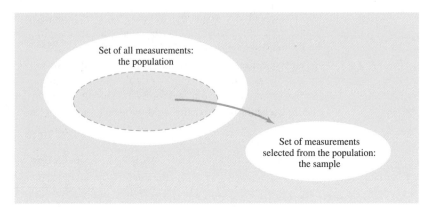

Set of all measurements:
the population

Set of measurements
selected from the population:
the sample

1.2 Why Study Statistics?

We can think of two good reasons for taking an introductory course in statistics. One reason is that you need to know how to evaluate published numerical facts. Every person is exposed to manufacturers' claims for products; to the results of sociological, consumer, and political polls; and to the published achievements of scientific research. Many of these results are inferences based on sampling. Some of the inferences are valid; others are invalid. Some are based on samples of adequate size; others are not. Yet all such published results seem to have the ring of truth. Some people say that statistics can be made to support almost anything (particularly statisticians). Others say that it is easy to lie with statistics. Both statements are true. It is easy, purposely or unwittingly, to distort the truth by using statistics when presenting the results of sampling to the uninformed. Therefore, to be an informed consumer and citizen, each of us needs to have a basic understanding of how to make sense of data. Indeed it's a virtual necessity if we are to function effectively in our society.

A second reason for studying statistics is that your profession or employment may require you to interpret the results of sampling (surveys or experimentation) or to employ statistical methods of analysis to make inferences in your work. For example, practicing physicians receive large amounts of advertising describing the benefits of new drugs. These advertisements frequently display the numerical results of experiments that compare a new drug with an older one. Do such data really imply that the new drug is more effective, or is the observed difference in results due simply to random variation in the experimental measurements?

In a similar way, the past decade has emphasized that making sense of data can be fundamental to good management and decision making. When deciding whether to change to a new, radically different menu for a fast-food chain, the accountable managers could follow their own tastes and hunches or they could use information about menu preferences obtained from a sample of customers. When a foundry's shipment of steel is rejected by an automobile manufacturer because the steel does not meet the latter's quality specifications, the foundry's design, engineering, purchasing, and production managers can argue about who is to blame or they can work as a team to obtain the data necessary for correcting the problem.

Recent trends in the conduct of court trials indicate an increasing use of probability and statistical inference in evaluating the quality of legal evidence. The use of statistics in the social, biological, and physical sciences is essential because all these sciences rely on observations of natural phenomena, through sample surveys or experimentation, to develop and test new theories. Statistical methods are employed in business settings when sample data are used to forecast sales and profit. They are used in engineering and manufacturing to monitor product quality. The sampling of accounts is a new and useful tool to assist accountants in conducting audits. Thus statistics plays an important role in almost all areas of science, business, and industry; persons employed in these areas need to know the basic concepts, strengths, and limitations of statistics.

In this textbook, we will focus on making sense of data: gathering data, summarizing data, analyzing data, and reporting the results of analyses. Computers will be used as an important aid in making sense of data.

1.3 Some Current Applications of Statistics

Acid Rain: A Threat to Our Environment

The accepted causes of acid rain are sulfuric and nitric acids; the primary sources of these acidic components of rain are hydrocarbon fuels, which spew sulfur and nitric oxide into the atmosphere when burned. The effects of acid rain are many, including the following:

- Acid rain, when present in spring snow melts, invades breeding areas for many fish and amphibians, preventing successful reproduction. Forms of life that depend on ponds and lakes now contaminated by acid rain begin to disappear.

- In areas surrounded by affected bodies of water, nutrients are leached from the soil, affecting the fertility of the soil, especially in forests.

- Man-made structures are also affected by acid rain. Experts from the United States estimate that acid rain has caused nearly $15 billion in damage to buildings and other structures thus far.

Solutions to the problems associated with acid rain will not be easy. The National Science Foundation (NSF) has recommended that we strive for a 50% reduction in sulfur oxide emissions. Perhaps that is easier said than done. High-sulfur coal is a major source of these emissions, but in states that depend on coal for energy a shift to lower-sulfur coal is not always possible. Rather, better scrubbers must be developed to remove these contaminating oxides from the burning process before they are released into the atmosphere. Fuels for internal combustion engines are also a major source of the nitric and sulfur oxides of acid rain. Clearly, better emission control is needed for automobiles and trucks.

Reducing the oxide emissions from coal-burning furnaces and motor vehicles will require greater use of existing scrubbers and emissions control devices, as well as development of new technology to allow us to use other available energy sources. Developing alternative, cleaner energy sources is also important if we are to meet NSF's goal. Statistics and statisticians will play a key role in monitoring atmospheric conditions, testing the effectiveness of proposed emissions control devices, and developing new control technology and alternative energy sources.

Determining the Effectiveness of a New Drug

The development and preliminary testing of the Salk vaccine for protection against poliomyelitis (polio) provide an excellent example of how statistics can be used in solving practical problems. Most parents and children growing up before 1954 can recall the panic brought on by the outbreak of polio cases during the summer months. Although relatively few children fell victim to the disease each year, the pattern of outbreak of polio was unpredictable and caused great concern because of the possibility of paralysis or death. The fact that very few of today's youth have even heard of polio demonstrates the great success of the vaccine and of the testing program that preceded its release on the market.

It is standard practice in establishing the effectiveness of a particular drug to conduct an experiment (often called a *clinical trial*) with human subjects. For some clinical trials, assignments to subjects are made at random, with half receiving the drug product and the other half receiving a solution or tablet (called a *placebo*) that contains no active medication. One statistical problem involves determining the total number of subjects to be included in the clinical trial. This problem was particularly important in the testing of the Salk vaccine, because data from previous years suggested that the incidence rate might be less than 50 cases for every 100,000 children. Hence a large number of subjects had to be included in the clinical trial to enable researchers to detect a difference in the incidence rates for those treated with the vaccine and those receiving the placebo.

With the assistance of statisticians, it was decided that a total of 400,000 children should be included in the Salk clinical trial begun in 1954, with half of them randomly assigned the vaccine and the remaining children assigned the placebo. No other clinical trial had ever been attempted on such a large group of subjects. Through a public school inoculation program, the 400,000 subjects were treated and then observed over the summer to determine the number of children contracting polio. Although fewer than 200 cases of polio were reported for the 400,000 subjects in the clinical trial, more than three times as many cases appeared in the group receiving the placebo than in the vaccinated group. These results, together with some statistical calculations, were sufficient to indicate the effectiveness of the Salk polio vaccine. However, these conclusions would not have been possible if the statisticians and scientists had not planned for and conducted such a large clinical trial.

The development of the Salk vaccine is not an isolated example of the use of statistics in the testing and developing of drug products. In recent years the Food and Drug Administration (FDA) has placed stringent requirements on pharmaceutical firms to establish the effectiveness of proposed new drug products. In fact, the use of statistics in partnership with the practice of medicine has played an important role in the development and testing of compounds for treating diabetes, heart disease, hypertension, epilepsy, depression, measles, various forms of cancer, AIDS, and most other diseases and ailments that affect human beings.

Applications in the Courts

Libel suits related to consumer products have touched each one of us; you may have been involved as a plaintiff or defendant in a suit, or you may know someone who was involved in such litigation. Certainly we all help to fund the costs of this litigation indirectly through increased insurance premiums and increased costs of goods. The testimony in libel suits concerning a particular product (automobile, drug product, and so on) frequently leans heavily on the interpretation of data from one or more scientific studies examining the product. This is how and why statistics and statisticians have been pulled into the courtroom.

For example, epidemiologists have used statistical concepts applied to data to determine whether a statistically significant "association" exists between a specific characteristic, such as the use of a brand-name tampon, and a disease condition, such as toxic shock syndrome. An epidemiologist who finds such an association in a study should try to determine whether it is due to random variation or whether it reflects an actual association between the characteristic and the disease. Arguments in courtrooms over how to interpret these types of associations involve data analyses using statistical concepts as well as clinical interpretations of the data.

The Energy Crisis: A Search for New Energy Sources and a Search for Oil

The OPEC oil crisis of 1973–1974 brought to Americans' attention a problem that is with us today and will continue to plague us for decades: a shortage of energy. The United States is confronted by staggering annual demands for energy and must depend on supplies that may not meet current and future demands, especially if a major supplier "interrupts" service. Such an interruption by OPEC in 1974 led to an energy panic, huge price increases, and subsequent supply problems.

Possible sources of energy needed to supply the present and future requirements of the United States include vast known coal and oil shale reserves, nuclear reactors, newly discovered oil and natural gas reserves, solar energy, and alternative new fuels. For example, methanol (wood alcohol) and ethanol (grain alcohol) may be major contributors to octane boost as leaded fuels are phased out. These alcohols are also likely candidates to reduce our dependence on foreign crude oil.

In which of these resources should we, the American public, invest the capital necessary for development? Which source will yield a given amount of energy at minimum cost? What unfavorable impact will each have on the environment or on our quality of life? Which might yield dangerous side effects? These questions and others must be answered by experimentation. Statisticians will assist in this effort by designing experiments and interpreting experimental data.

Opinion and Preference Polls

Public opinion, consumer preference, and election polls are commonly used to assess the opinions or preferences of a segment of the public with regard to issues, products, or candidates of interest. And we, the American public, are exposed to the results of these polls on a daily basis in newspapers, in magazines, on the radio, and on television. For example, the results of polls related to the following subjects were printed in local newspapers over a two-day period:

- Consumer confidence in relation to future expectations about the economy
- Preferences for candidates in upcoming elections and caucuses
- Attitudes toward cheating on federal income tax returns
- Preference polls related to specific products (for example, foreign vs. American cars, Coke vs. Pepsi, McDonald's vs. Wendy's)
- Reactions of North Carolina residents toward arguments about the morality of tobacco production, sales, and use
- Opinions of voters toward proposed tax increases and proposed changes in the Defense Department budget

A number of questions can be raised about these polls. How many people were polled? What questions were asked? Was each person asked the same question? How were people chosen or selected for participation in the poll? Can we believe the results of these polls? Do these results "represent" how the general public feels?

Opinion and preference polls are an important, visible application of statistics for the consumer. We will discuss this topic in more detail in Chapter 10. After studying this material, you should have a better understanding of how to interpret the results of these polls.

Quality and Process Improvement

You might wonder, at this stage, why we would bring up the subject of quality and process improvement in a statistics textbook. We do so to make you aware of some of the broader issues involved in making sense of data in the business and scientific communities.

The immediate post–World War II years saw U.S. business and the U.S. economy emerge to dominate world business, and this pattern lasted for about 30 years. During this time, little attempt was made to change the ways things were done; the major focus was on doing things on an increasingly grand scale, perfecting mass production. Beginning in the mid-1970s, however, many American industries have had to face fierce competition from their counterparts overseas.

quality

Quality, rather than *quantity*, has become the principal buying gauge used by consumers, and American industries have had a difficult time adjusting to this new emphasis. Unless drastic changes occur in the way many American industries ap-proach their businesses, the "quality" revolution will claim many more casualties.

Japanese businesses were the first to learn the lessons of quality. They adopted the statistical quality-control and process-control suggestions espoused by Deming (1981) and others and installed total quality programs. Throughout their organiz-ations—from top management down—they were committed to improving the quality of their products and the way they did things. They were never satisfied with the status quo and continually sought new and better ways of doing things.

quality improvement

A number of American companies have now begun the journey toward excellence by instituting a **quality improvement** process. Harrington (1987) identified ten basic requirements that would make a quality-improvement process successful:

Harrington's Fundamental Requirements
for a Successful Quality-Improvement Process

1 Focus on the customer as the most important part of the process

2 Long-term commitment by management to make the quality improvement process part of the management system

3 Belief that there is room to improve

4 Belief that preventing problems is better than reacting to problems

5 Management focus, leadership, and participation

6 A performance standard(goal) of zero errors

7 Participation by all employees, both as groups and as individuals

8 Improvement focus on the process, not the people

9 Belief that suppliers will work with you if they understand your needs

10 Recognition for success

reengineering

Embedded in a companywide quality-improvement process or running concurrently with such a process is the idea of improving or **reengineering** the work processes. For years, companies that wanted to boost and improve performance have tried to speed up their processes, usually with additional people or technology, without addressing possible deficiencies in the work processes themselves. As Michael Hammer (1990) states, "It is time to stop paving the cow paths." In many cases, businesses need to rethink completely the processes that have persisted despite the ever-changing business and technological environment. Fundamental to reengineering is the task of identifying and discarding outdated habits, rules, and assumptions. Dramatic, radical improvements in quality, efficiency, and effectiveness are possible with these reengineered efforts. Within a workplace, management can begin a reengineering effort by developing a case for action that focuses on the following five questions:

- What is important to our mission?

- How are we using people, processes, and technology to perform our mission?

- Are there opportunities to use technology or process change to improve our work processes significantly?

- What opportunities have the greatest potential, and what are the costs and payoffs of implementing these opportunities?

- How do we get started?

At the heart of any quality-improvement process or engineering effort lie the data that reflect the state of health of a process and the reports based on these data. Here's where statistics and statistical tools can help. A number of these tools and techniques are listed next.

Statistical Tools, Techniques, and Methods Used
in Quality Improvement and Reengineering

- Histograms
- Numerical descriptive measures (means, standard deviations, proportions, and so on)
- Scatterplots
- Line graphs (scatterplots with dots connected)
- Control charts: $\bar{y}$(sample mean), r(sample range), and s(sample standard deviation)
- Sampling schemes
- Experimental designs

The statistical tools and concepts listed here and discussed in this textbook represent only a small component of a total quality-improvement process.

Keep in mind situations where you think these tools and concepts may have application in a quality-improvement process or reengineering project as you encounter them in various parts of the text. Quality improvement and process reengineering are clearly the focus of American industry for the 1990s in world markets characterized by increased competition, more consolidation, and greater specialization. These shifts will have impacts on us all, either as consumers or as business participants, and it will be useful to know some of the statistical tools that are part of this revolution.

1.4   What Do Statisticians Do?

What do statisticians do? In the context of making sense of data, statisticians participate in all aspects of gathering, summarizing, and analyzing data, and reporting the results of their analyses. There are both good and bad ways to gather data. Statisticians apply their knowledge of existing survey techniques and scientific study designs, or they develop new techniques to provide a guide to good methods of data collection. We will explore these ideas further in Chapter 2.

Once the data are gathered, they must be summarized before any meaningful interpretation can be made. Statisticians can recommend and apply useful methods for summarizing data in graphical, tabular, and numerical forms. Intelligible graphs and tables are useful first steps in making sense of the data. Measures of the average (or typical) value and some measure of the range or spread of data also help in interpretation. These topics will be discussed in detail in Chapters 3 and 4.

The objective of statistics is to make an inference about a population of interest based on information obtained from a sample of measurements from that population. The analysis stage of making sense of data deals with making inferences. For example, a market research study reaches only a few of the potential buyers of a new product, so the probable reaction of the entire set of potential buyers (population) must be inferred from the reactions of the buyers included in the study (sample). If the market research study has been carefully planned and executed, the reactions of those included in the sample should agree reasonably well (but not necessarily exactly) with the reactions of the population as a whole. We can say this because the basic concepts of probability allow us to make an inference about a population of interest that includes our best guess plus a statement of the probable error in our best guess.

We will illustrate how inferences are made by way of an example. Suppose that an auditor samples 2000 financial accounts from a set of more than 25,000 accounts and finds that 84 (4.2%) are in error. What can be said about the set of 25,000 accounts? In other words, what inference can we make about the percentage of accounts in error for the population of 25,000 accounts based on the information we obtained from the sample of 2000 accounts? We will show (in Chapter 10) that our best guess (inference) about the percentage of accounts in error for the population is 4.2%, and that this best guess should be within ±.9% of the actual unknown percentage of accounts in error in the population. The plus-or-minus factor is called the probable error of our inference. Anyone can make a guess about the percentage of accounts in error; concepts of probability allow us to define the (probable) error of our guess.

In dealing with analyses of data, statisticians can apply existing methods for making inferences; some theoretical statisticians engage in developing new methods with more advanced mathematics and probability theory. Our study of the methods for analyzing sample data will begin in Chapter 7, after we discuss the basic concepts of probability and sampling distributions in Chapters 5 and 6.

Finally, statisticians help communicate the results of their analyses as the final stage in making sense of data. The form of the communication varies from an informal conversation to a formal report. The advantage of a more formal verbal presentation with visual aids or a study report is that the communication can make use of graphical, tabular, and numerical displays as well as of the analyses done on the data to help convey the "sense" found in the data. Too often this is lost in an informal conversation. The report or communication should convey to the intended audience what can be gleaned from the sample data, and it should be conveyed as nontechnically as possible to avoid any confusion as to what is being inferred. More information about the communication of results is presented in Chapter 14.

1.5 Using Computers to Help Make Sense of Data

computer software

Making sense of data has been made easier with the increasing availability and accessibility of **computer software** (programs or program systems) that perform the calculations and construct the graphs required to summarize and analyze sample data from surveys or experimental studies. Thus computer technology in the form of computers and computer software is an "enabler" of our ultimate objective—making sense of data. It is not an end in itself but will be used throughout the text to bypass various tedious calculations, allowing us to spend more time dealing with the issues of summarizing, analyzing, and reporting results.

Most chapters in the text include a section called "Using Computers to Help Make Sense of Data." In this section, we will apply the methods and techniques of the chapter to real-life situations, using computer software to help wherever possible. Emphasis will be placed on how computers can facilitate the process of generating useful data and on where the work of the chapter fits into the larger process of making sense of data. In addition, the Supplementary Exercises at the end of each chapter contain an expanded exercise set called "Experiences with Real Data." Here you will be asked to apply what you have learned in the chapter to several real-life situations. Access to and familiarity with one of the many statistical software packages available on a mainframe or personal computer/workstation will be a great help to you in doing the computations required for these exercises; when data are referred to in the exercises, the data are available on the data disk supplied with the text.

software systems

What statistical **software systems** or packages can be used to help you make sense of data? Literally hundreds of computer software packages are available to do the computations required for summarizing and analyzing data. Some of these, such as SAS, Minitab, SPSS, and BMDP, were originally developed for mainframe computers and later adapted for use with personal computers. Others, such as Execustat and Systat, were developed specifically for the personal computer market.

You need not know how to program a computer in order to use these packages, and they are fairly easy to use. Similarly, you need not learn everything about the package in order to use it effectively. What you must do is learn the steps needed to run the particular analysis of interest to you. Typically, you will have to enter the data (or access it from a data disk), assign labels so that the output from your analysis is understandable, select the variables of interest, and choose the program or programs in the package needed to do the required summarization or analysis. The results (output) can then be viewed on your monitor or sent to a line printer for a hardcopy version of the information.

We will use output from several widely available packages throughout the text, especially in the "Using Computers to Help Make Sense of Data" section of each chapter. Most of the calculations we will need for summarizing and analyzing data from a survey or experimental study can by done with any one of these software systems. As stated previously, there are too many packages available and they change too rapidly to permit us to describe them all or to focus on only one. Instead, we will show how several of these packages perform the various computations, allowing us to focus on making sense of data. When output is given, look for the specific results you need and don't worry about the rest. Some software systems include almost everything you could conceivably want, making much of it irrelevant to your immediate needs. As you learn more about statistics and about making sense of the data, however, you will discover that more of the output from these packages is relevant. In the meantime, try to focus on what you need.

Finally, we will emphasize how computers and software systems facilitate the summarization and analysis steps of making sense of data. Perhaps of equal importance is the role computers and microprocessors (small computers) play in data gathering. Laboratories, assembly lines, and offices of large corporations provide for the direct acquisition of relevant data and signals using computers or microprocessors. Think about how many appliances and how many components of automobiles, airplanes, and spacecraft are "run" by or outfitted with computers and microprocessors. We are all aware of some of the problems arising from a computer malfunction. But on the positive side, as these become more reliable, the direct signal (or data) acquisition they perform eliminates many of the transcription and recording errors that occur when humans gather data manually.

End of paper trail: Offices do everything via computer

Quick access to records seen as improvement for customer service.

By ROBERT A. CRONKLETON

Three months ago, paperwork littered Juanita K. Craft's desk at Twentieth Century Investors Inc.

Investor's files crammed her desk drawers. Finished work tumbled over the edge of her "out" basket.

Today her desk is pristine — hardly a wrinkled memo or folder in sight. In turn, Craft has become an enthusiastic occupier of the "paperless" office.

"I would never want to go back," said the account service representative.

The paperless office, for years touted as just around the corner for every big American business, is at last reaching Kansas City. Technology called "imaging" revolutionizes how companies handle information. Each day, millions of paper documents are scanned by machines that convert the information into electronic files in huge computer databases. Ultimately, it's designed to help the customer.

Many experts doubt corporate America will ever become totally paperless. Even those with paperless systems have some paper floating around.

But the trend is powered by enhanced computer capability. And corporations, inspired by competition and the recession, are intent on improving customer service, getting better use of their work force, and pushing decision-making to the lower ranks. Imaging can help accomplish all three.

Kansas City executives point to these specifics:

• Workers can respond to customer inquiries quicker and cut response time to minutes compared with days or even weeks. Managers can get up-to-the-minute data on how much work is processed and any delays in processing customer requests.

• Managers can gauge the work flow of each employee and reroute work if necessary to lessen the burden on overworked staff members.

• Clerical workers can spend less time filing records, which usually gives them a more meaningful job and more time to work on customers' needs.

A storage sensation

Imaging, the converting of paper documents into electronic images, essentially is what a facsimile machine does. it takes a picture of your document, changes it to an electronic image, sends the image along phone lines, then reconstitutes it as a document.

But paperless companies are mainly interested in the first two steps, using high-speed scanners to store these electronic images into computers.

These images usually are permanently fixed onto 12-inch optical disks, similar to compact discs that play music. Each disk can store 60,000 multipage documents – enough to fill at least five filing cabinets.

Imaging systems can also categorize and route work to individual employees, giving managers a better way to monitor and control progress. Statistical reports can be made on individual worker performance. And employee "fingerprints" can be tracked to learn which employees handled which documents when.

But the biggest advantage is the ability to store and have access to a large number of documents in a few minutes – or seconds.

Despite the advantages, experts point to some hurdles.

The switch to imaging is time consuming, said Norman G. Tsiguloff, Twentieth Century's vice president and director of account services. Every aspect of the company's operations must be evaluated, he said. It took Twentieth Century four years.

And the systems aren't cheap. Desktop versions run as low as $10,000, but large systems can cost as much as $18 million.

Source: Kansas City *Star*, January 25, 1993. Reprinted courtesy of the Kansas City *Star*.

No Trash Cans Needed

The paperless office, long touted among technologists as the next wave of office management, has arrived at a few Kansas City area companies. Here's a look at how the process works:

STEP ONE

Incoming mail is opened and sorted by employees, then prepared for scanning.

STEP TWO

The material is fed through a high-speed computer scanner that essentially takes a picture of the documents and transforms the information into electronic images.

STEP THREE

A computer program electronically routes the electronic images to the appropriate queues and sets priorities for each document.

STEP FOUR

An employee then calls up the electronic "paper work" based on its priority and processes that account. When finished, the electronic images are cleared from the screen and all transactions are saved to a central computer database.

STEP FIVE

Three months later, the information is saved to 10-to-12-inch optical disks and placed in a machine commonly referred to as a "juke box." When information from a disk is needed, the machine selects the appropriate disk and places it on a optical reader like a juke box of the 1950s would place an album on a turntable.

When data are gathered electronically via a computer or microprocessor, the user must provide an electronic interface or data transfer between the computer used for data acquisition and the one used for data summarization and analysis. In effect, the accompanying data disk for many of the data sets used throughout this text serves as an electronic interface between your computer and the computer that was used for data gathering. Not all data gathering situations can be supported with direct, electronic access of the data via a computer. But as you will no doubt notice, more and more of the technical and business data we deal with daily are recorded and summarized using computers. Where possible, we will try to draw this phenomenon to your attention.

1.6 A Note to the Student

We think with words and concepts. A study of the discipline of statistics requires the memorization of new terms and concepts (as does the study of a foreign language). You must commit these definitions, theorems, and concepts to memory.

Remember to focus on the broader concept of making sense of data. Do not let details obscure these broader characteristics of the subject. The teaching objective of this text is to identify and amplify these broader concepts of statistics.

Summary

The discipline of statistics and those who apply the tools of that discipline deal with making sense of data. As such, statisticians are involved with methods of data collection, data summarization, and data analysis, as well as with communicating the results of its analyses. Computers and software systems are not an end in themselves; rather, they facilitate the process of making sense of data.

Key Terms

population

sample

quality

quality improvement

reengineering

computer software

software systems

Supplementary Exercises

1.1   Selecting the proper diet for shrimp or other marine animals is an important aspect of sea farming. A researcher wishes to estimate the mean weight of shrimp maintained on a specific diet for a period of six months. One hundred shrimp are randomly selected from an artificial pond, and each is weighed.

a   Identify the population of measurements that is of interest to the researcher.

b   Identify the sample.

c   What characteristics of the population are of interest to the researcher?

d   If the sample measurements are used to make inferences about certain characteristics of the population, why would a measure of the reliability of the inferences be important?

1.2   Radioactive waste disposal and the production of radioactive material in some mining operations are creating a serious pollution problem in some areas of the United States. State health officials decided to investigate the radioactivity levels in one suspect area. Two hundred geographical points were randomly selected in the area, and the level of radioactivity was measured at each point. Answer questions a, b, c, and d from Exercise 1.1 for this sampling situation.

1.3   A social researcher in a particular city wishes to obtain information on the number of children in households that receive welfare support. A random sample of 400 households was selected from the welfare rolls of the city. A check of welfare recipient data identified the number of children in each household. Answer questions a, b, c, and d from Exercise 1.1 for this sample survey.

1.4   Search copies of your local newspaper to find the results of a recent Harris or Gallup survey.

    a   Identify the items that were observed in obtaining the sample measurements.

    b   Identify the measurement made on each item.

    c   Clearly identify the population associated with the survey.

    d   What characteristic(s) of the population is (are) of interest to the pollster?

    e   Does the article explain how the sample was selected?

    f   Does the article include the number of measurements in the sample?

    g   What type of inference is made concerning the population characteristics?

    h   Does the article tell you how much numerical faith you can place in the inference about the population characteristic?

Step One: Gathering Data

2

Using Surveys and Experimental Studies to Gather Data

2.1 Introduction

As we've mentioned previously, the first step in making sense of data is to gather data on one or more variables of interest. But intelligent data gathering doesn't just happen; it takes a conscious, concerted effort focused on the following steps:

- Specify the objective of the data-gathering exercise.
- Identify the variable(s) of interest.
- Choose an appropriate design for the survey or scientific study.
- Collect the data.

To specify the objective of the data-gathering exercise, you must understand the problem under study. For example, if the management of a large manufacturing company is considering whether to institute a new incentive pay plan for its production workers, it might want to determine the attitudes of the production supervisors toward the proposed plan. This, then, could be the objective of a data-gathering exercise.

To identify the variable(s) of interest, you must examine the objective of the data-gathering exercise. For the production incentive plan, the variable of interest would be the attitude of the production supervisors. Measurements would consist of preferences (favor, oppose, indifferent) accompanied by comments as to why a production supervisor may favor or oppose the new incentive plan.

Once the objective has been determined and the variable(s) of interest specified, you must choose how to collect the data. In statistics, data can be gathered by way of a survey, a study, or a combination of the two. Survey theory and the theory of experimental study design provide good methods for collecting data. Usually surveys are passive, and the aim is to gather (survey) data on existing conditions, attitudes, or behaviors. Thus the management of the manufacturing company would use a survey to sample the opinions of the production supervisors on the merits of introducing a new incentive plan. Experimental studies, on the other had, tend to be more active: the person conducting the study may deliberately vary certain conditions in order to reach a conclusion. For example, a plant manager interested in the effects of noise level on productivity in a manufacturing plant could vary the noise level and certain other controlled conditions in order to measure directly the gains or losses in productivity.

In this chapter, we will consider a number of survey methods and designs for experimental studies. We will also make a distinction between an experimental study and an observational study.

2.2 Surveys

Information from surveys affects almost every facet of our daily lives. Surveys determine such government policies as the control of the economy and the promotion of social programs. Opinion polls are the basis of much of the news reported by the various news media. Ratings of television shows determine which shows are made available for viewing in the future.

One usually thinks of the U.S. Census Bureau as contacting every household in the country. Actually, in the 1980 census only 14 questions were asked of all households. Information on an additional 42 questions was obtained from only a sample of households. The resulting information is used by many agencies and individuals for a multitude of purposes. For example, the federal government uses it to determine allocations of funds to states and cities; businesses use it to forecast sales, to manage personnel, and to establish future site locations; urban and regional planners use it to plan land use, transportation networks, and energy consumption; and social scientists use it to study economic conditions, racial balance, and other aspects of the quality of life.

The U.S. Bureau of Labor Statistics (BLS) routinely conducts over twenty surveys at varying intervals. Among the best known and most widely used of these are the surveys that establish the consumer price index (CPI), which measures price change for a fixed market basket of goods and services over time. The CPI is used as a measure of inflation and serves as an economic indicator for guiding government policies. Many businesses tie wage rates and pension plans to the CPI. Federal health and welfare programs, as well as many state and local programs, define their bases of eligibility in terms of the CPI. Escalator clauses in rents and mortgages are also based on the CPI. Clearly this one index, determined on the basis of sample surveys, plays a fundamental role in our society.

Many other surveys from the BLS are crucial to society, too. The monthly Current Population Survey compiles basic information on the labor force, employment, and unemployment. Consumer expenditure surveys collect data on family expenditures for goods and services used in day-to-day living. The Establishment Survey collects information on employment hours and earnings for nonagricultural business establishments. A survey on occupational outlook provides predictive information on future employment opportunities for various occupations, projecting conditions approximately ten years into the future. Other activities of the BLS are addressed in the *BLS Handbook of Methods* (1982).

Opinion polls are constantly in the news, and the names of Gallup and Harris have become well known to everyone. These polls, or sample surveys, report the attitudes and opinions of citizens on everything from politics and religion to sports and entertainment. The Nielsen ratings determine the success or failure of TV shows.

Businesses conduct sample surveys for their internal operations, in addition to relying on government surveys to make crucial management decisions. Auditors estimate account balances and confirm a company's compliance with operating rules by sampling accounts. Quality control of manufacturing processes relies heavily on sampling techniques.

One particular area of business activity that depends on detailed sampling activities is marketing. Decisions about which products to market, where to market them, and how to advertise them are often made on the basis of sample survey data. The data may come from surveys conducted by the firm that manufactures the product, or they may be purchased from survey firms that specialize in marketing data.

The activities of three such firms are discussed here. The Nielsen retail index is less famous than the Nielsen television ratings, but it is very important to firms that market products for retail sale. This index furnishes continuous sales data on foods, cosmetics, pharmaceuticals, beverages, and many other classes of products. It can provide estimates of total sales for a product class, of sales for a client's particular brand, of sales for a competing brand, of retail and wholesale prices, and of the percentage of stores stocking a particular item. The data come from audits of inventories and sales in 1600 stores across the United States conducted every 60 days.

Selling Areas–Marketing, Inc. (SAMI), collects information on the movement of products from warehouses and wholesalers. Data are obtained in 36 major television market areas, which account for 4% of national food sales, and cover 425 product categories.

The Marketing Research Corporation of America provides many types of marketing data gathered through the use of surveys; some of the more interesting results come from its National Menu Census, which samples families and observes their eating patterns for two weeks. As many as 4,000 families may participate during a year. Data are obtained on the number of times a particular food item is served, how it is served, how many persons eat the item, and many other details, including what happens to the leftovers. Such details are important for product development and advertising.

Many interesting examples of the practical uses of statistics in general and of sampling in particular can be found in *Statistics: A Guide to the Unknown* (see the references in the Appendix). In this book you might want to look at some of the methods and uses of opinion polling discussed in the articles "Opinion Polling in a Democracy" by George Gallup and "Election Night on Television" by R. F. Link. If you are interested in wildlife ecology, you should read "The Plight of the Whales" by D. G. Chapman. Find out how sampling allows interrailroad and interairline billing to be handled economically by reading "How Accountants Save Money by Sampling" by John Neter.

Sampling Techniques

simple random sampling

The basic design (**simple random sampling**) consists of selecting a group of n sampling units in such a way that each sample of size n has the same chance of being selected. Thus we can obtain a random sample of n eligible voters in a bond-issue poll by drawing names from the list of registered voters in such a way that each sample of size n has the same probability of selection. (The details of simple random sampling can be found in Chapter 4 of Scheaffer et al. 1990.) At this point we merely state that a simple random sample will contain as much information on community preferences as any other sample survey design, provided that all voters in the community have similar socioeconomic backgrounds.

Suppose, however, that the community consists of people in two distinct income brackets: high and low. Voters in the high bracket may have opinions on the bond issue that differ significantly from the opinions of voters in the low bracket. Therefore, to obtain accurate information about the population, we must sample voters from each bracket. We can divide the population elements into two groups, or strata, according to income and then select a simple random sample from each group. The resulting

stratified random sample

sample is called a **stratified random sample**. (See Chapter 5 of Scheaffer et al. 1990.)

Note that stratification is accomplished by using knowledge of an auxiliary variable—namely, personal income. By stratifying on high and low values of income,

ratio estimation

we increase the accuracy of our estimator. **Ratio estimation** is a second method for using the information contained in an auxiliary variable. Ratio estimators not only use measurements on the response of interest, they incorporate measurements on an auxiliary variable. Ratio estimation can also be used with stratified random sampling.

Although surveys ultimately seek to uncover individual preferences, a more economical procedure—especially in urban areas—may be to sample specific families, apartment buildings, or city blocks rather than individual voters. Individual preferences can then be obtained from each eligible voter within the unit sampled. This technique is called **cluster sampling**. Although both cluster sampling and stratified random sampling divide the population into groups, the techniques differ. In stratified random sampling we take a simple random sample within each group, whereas in cluster sampling we take a simple random sample of groups and then sample all items within the selected groups (cluster). (See Chapters 8 and 9 of Scheaffer et al. 1990 for details.)

cluster sampling

Sometimes, the names of persons in the population of interest are available in a list, such as a registration list, or on file cards stored in a drawer. For this situation an economical sampling technique is to draw the sample by selecting one name near the beginning of the list and then selecting every tenth or fifteenth name thereafter. If sampling is conducted in this manner, we obtain a **systematic sample**. As you might expect, systematic sampling offers a convenient means of obtaining sample information; unfortunately, it does not necessarily provide the most information for a specified amount of money. (Details are given in Chapter 7 of Scheaffer et al. 1990.)

systematic sample

The important point to understand is that different kinds of surveys can be used to collect sample data. For the surveys discussed in this text, we will be dealing with simple random sampling and with methods for summarizing and analyzing data collected in such a manner. More complicated surveys lead to even more complicated problems at the summarization and analysis stages of statistics.

Data Collection Techniques

Having chosen a particular sample survey, how does one actually collect the data? The most commonly used methods of data collection in sample surveys are personal interviews and telephone interviews. These methods, with appropriately trained interviewers and carefully planned callbacks, commonly achieve response rates of 60% to 75% and sometimes even higher. A mailed questionnaire sent to a specific group of interested persons can achieve good results, but generally the response rates for this type of data collection are so low that all reported results are suspect. Frequently, objective information can be found from direct observation rather than from an interview or mailed questionnaire.

Data are frequently obtained by **personal interviews**. For example, we can use personal interviews with eligible voters to obtain a sample of the public sentiments toward a community bond issue. The procedure usually requires the interviewer to ask prepared questions and to record each respondent's answers. The primary advantage of these interviews is that people will usually respond when confronted in person. In addition, the interviewer can note specific reactions and eliminate misunderstandings about the questions asked. The major limitations of the personal interview (aside from the cost involved) relate to the interviewers. If they are not thoroughly trained, they may deviate from the required protocol, thus introducing bias into the sample data. Any movement, facial expression, or statement by the interviewer can affect the

personal interviews

response obtained. For example, a leading question such as "Are you also in favor of the bond issue?" may tend to elicit a positive response. Finally, errors in recording the responses can yield erroneous results.

Information can also be obtained from persons in the sample through **telephone interviews**. With the advent of wide-area telephone service lines (WATS lines), an interviewer can place any number of calls to specified areas of the country for a fixed monthly rate. Surveys conducted through telephone interviews are frequently less expensive than personal interviews, owing to the elimination of travel expenses. The interviews can also be monitored to confirm that the specified interview procedure is being followed.

A major problem with telephone surveys involves establishing a sample frame that closely corresponds to the population. Telephone directories have many numbers that do not belong to households, and many households have unlisted numbers. A few households have no phone service, although lack of phone service is now only a minor problem for most surveys in the United States. A technique that avoids the problem of unlisted numbers is random-digit dialing. In this method a telephone exchange number (the first three digits of the seven-digit number) is selected, and then the last four digits are dialed randomly until a fixed number of households of a specified type are reached. This technique seems to produce unbiased samples of households in selected target populations and avoids many of the problems inherent in sampling based on a telephone directory.

With random-digit dialing in a residential survey, only about 20% of the numbers will lie within the population of interest. Most of the remaining 80% will be unused numbers or numbers belonging to businesses and institutions. The rate of usable numbers can be improved by using random-digit dialing to locate clusters (blocks of numbers). Once a residential number is identified, more residences can be selected from the same cluster by repeating the first eight digits and randomizing only the last two. This improves the proportion of usable responses, because telephone companies assign numbers in blocks.

A study of post-election attitudes and voting behavior reported by Bergstein in "Some Methodological Results from Four Statewide Telephone Surveys Using Random-Digit Dialing," in *Proceedings of the Section on Survey Research Methods* of the American Statistical Association (1979), used this clustered technique. It found that only about 23% of first-stage calls resulted in usable residential numbers; however, with the clustering technique, the percentage of usable residential numbers rose to approximately 57%. This technique, therefore, holds out the prospect of big savings in time and money. Incidentally, in this same study, trained interviewers (those with over six months of interviewing experience) produced a 77% response rate, while those with less training produced only a 67% response rate.

Telephone interviews generally must be kept shorter than personal interviews because respondents tend to get impatient more quickly when talking over the telephone. Nonetheless, with appropriately designed questionnaires and trained interviewers, telephone interviews can be as successful as personal interviews. (See Schuman & Presser (1981) for more details.)

Another useful method of data collection is the **self-administered questionnaire**, to be completed by the respondent. These questionnaires usually are mailed to the individuals included in the sample, although other distribution methods can be used.

The questionnaire must be carefully constructed if it is to encourage participation by the respondents.

The self-administered questionnaire does not require interviewers, and thus it costs less to conduct. However, the savings are usually obtained at the expense of a lower response rate. Nonresponse can be a problem in any form of data collection, but since we have the least contact with respondents in a mailed questionnaire, we frequently have the lowest rate of response from it. The low response rate can introduce a bias into the sample, because the people who answer questionnaires may not be representative of the population of interest. To eliminate some of the bias, investigators frequently attempt to contact the nonrespondents through follow-up letters, telephone interviews, or personal interviews.

direct observation

The fourth method for collecting data is **direct observation**. For example, if we are interested in estimating the number of trucks that use a particular road during the 4–6 P.M. rush hours, we can assign a person to count the number of trucks passing a specified point during this period. Possibly, electronic counting equipment can also be used. The disadvantage of using an observer is the possibility of errors in observation.

Direct observation is used in many surveys that do not involve measurements on people. The U.S. Department of Agriculture, for instance, measures certain variables affecting crops in sections of fields in order to produce estimates of crop yields. Wildlife biologists may count animals, animal tracks, eggs, or nests in order to estimate the size of animal populations.

A notion closely related to direct observation is that of getting data from objective sources that are not affected by the respondents themselves. For example, health information can sometimes be obtained from hospital records, and income information from employers' records (especially for state and federal government workers). This approach may take more time, but it can yield large rewards in important surveys.

Exercises

Basic Techniques

2.1 An experimenter wants to estimate the average water consumption per family in a city. Discuss the relative merits of choosing individual families, dwelling units (single-family houses, apartment buildings, etc.), and city blocks as sampling units.

2.2 A forester wants to estimate the total number of trees on a tree farm that possess diameters exceeding 12 inches. A map of the farm is available. Discuss the problem of choosing what to sample and how to select the sample.

2.3 A safety expert is interested in estimating the proportion of automobile tires with unsafe treads. Should he use individual cars or collections of cars, such as those in parking lots, in his sample?

2.4 An industry is composed of many small plants located throughout the United States. An executive wants to survey the opinions of the employees on the vacation policy of the industry. What should she sample?

2.5 A state department of agriculture wishes to estimate the number of acres planted in corn within the state. How might it conduct such a survey?

2.6 A political scientist wants to estimate the proportion of adult residents of a state who favor a unicameral legislature. What could be sampled? Discuss the relative merits of personal interviews, telephone interviews, and mailed questionnaires as methods of data collection.

2.7 Discuss the relative merits of personal interviews, telephone interviews, and mailed questionnaires as methods of data collection for each of the following situations:

    a A television executive wants to estimate the proportion of viewers in the country who are watching her network at a certain hour.

    b A newspaper editor wants to survey the attitudes of the public toward the type of news coverage offered by his paper.

    c A city commissioner is interested in determining how homeowners feel about a proposed zoning change.

    d A county health department wants to estimate the proportion of dogs in the county that have had rabies shots within the last year.

2.8 A Yankelovich, Skelly, and White poll taken in the fall of 1984 reported that one-fifth of the 2207 people surveyed admitted to having cheated on their federal income taxes. Do you think that this fraction is close to the actual proportion who cheated? Why? (Discuss the difficulties of obtaining accurate information on a question of this type.)

2.3 Some Hints on Designing a Questionnaire (Optional)

One objective of any survey is to minimize the errors that may occur due to faulty design of the questionnaire. Some major concerns in questionnaire design are outlined here. More details appear in Schuman & Presser (1981).

Question Ordering

Respondents to questionnaires generally try to be consistent in their responses to questions. For this reason, the ordering of the questions on the questionnaire may affect the responses, sometimes in ways that seem unpredictable to the inexperienced investigator. An example discussed in Schuman & Presser (1981) illustrates this point.

An experiment was conducted with the following two questions:

A. Do you think the United States should let Communist newspaper reporters from other countries come in here and send back to their papers the news as they see it?

B. Do you think a Communist country like Russia should let American newspaper reporters come in and send back to America the news as they see it?

For surveys in 1980 in which the questions appeared in the order (A, B), 54.7% of the respondents answered yes to A and 63.7% answered yes to B. For surveys in which the questions appeared in the order (B, A), 74.6% answered yes to A and 81.9% answered yes to B. The evidence suggests that asking question B first put respondents in a more libertarian frame of mind toward the question of allowing Communist reporters into the United States. Conversely, asking question A first appears to have encouraged respondents to adopt a more restrictive view of the propriety of giving free editorial rein to American reporters in Communist countries—again because of a presumed wish to appear consistent. In other words, a significant number of those who answered yes to B when it was asked first tried to be consistent by answering yes to a similar question A; and some of those who answered no to A when it was asked first appear to have felt constrained to answer no to B as well. Thus the context in which a question is asked is very important and should be understood and explained in the analysis of questionnaire data.

Order is also an important consideration in the positioning of specific versus general questions. For example, respondents may be asked the following questions:

A. Will you support an increase in state taxes for education?

B. Will you support an increase in state taxes?

It would not be surprising to find more people answering yes to B if the questions are asked in the order (B, A) than if they are asked in the order (A, B). If question A is asked first, persons who support taxes for education and answer A affirmatively may think that B implies an increase in taxes *not* necessarily going to education, and they may then say no to this question. If B is presented first, the same people who support more taxes for education may answer affirmatively, since they have not yet seen a specific question on taxes for education.

A respondent's attitude toward a question in a survey is very often set or changed by preceding questions that bear on the same topic. Reportedly (Schuman & Presser (1981)), more crime victimization was reported by respondents when the question on victimization occurred *after* a series of questions on crime had helped the respondent remember small incidents when he or she was a victim of crime—incidents that might otherwise have been forgotten. Likewise, a respondent's attitude toward government can be quite negative after the respondent has answered a series of questions emphasizing government waste and inefficiency, but they can be much more positive after the respondent has answered a series of questions emphasizing the necessary and timely functions government performs.

In a series of questions that ask for ratings, the first question is often considered in a different light from those that follow, and it tends to receive more extreme ratings. For example, suppose that a person is asked to rate a number of possible vacation sites, using a numerical scale that ranges from 1 to 10 (10 being very good). If the first site is viewed favorably by the respondent, it will tend to receive a rating close to 10 and subsequent sites will tend to be rated lower. If the first site looks unattractive to the respondent, it will tend to be rated close to 1 and subsequent sites will tend to be rated higher. Thus, among the group of good sites, each will tend to receive its highest rating when it appears first on the list. Evidently, the first item on the list is used as a reference point, and other items are rated up or down as a group relative to the first item.

For many survey questions, the order of the possible responses (or choices) to a particular question is as important as the position of the question on the questionnaire. If the person being interviewed is presented with a long list of possible choices, or if each possible choice is wordy or difficult to interpret, the person is likely to select the most recent choice (the last one on the list). If the respondent must choose items from a long written list, the items appearing toward the top of the list have a selection advantage. For example, consider the election of candidates for office from a long slate: Those toward the top of the list tend to get elected. Even in a relatively short list of simple choices, such as strongly agree, agree, disagree, and strongly disagree in an attitude survey, alternatives tend to receive their highest frequency of response when listed first. That is, the proportion who strongly agree will tend to be higher when that option is offered as the first choice rather than as the fourth choice.

Researchers attempting to design a questionnaire should be aware of the common ordering problems for question and response. They should attempt to counter potential difficulties by considering the following techniques:

1   Print questionnaires with different orderings for different subsets of the sample.

2   Use show cards or repeat the question as often as necessary in an interview so that the question and possible answers are clearly understood.

3   Carefully explain the context in which a question was asked in the analysis of the survey data.

Open Versus Closed Questions

closed questions

Since questionnaires today are often designed to be scored electronically after completion, with the data recorded in a form convenient for computer handling, most questions are **closed questions**: each question has either a single numerical answer (like the age of the respondent) or an answer selected from among a fixed number of predetermined choices.

open questions

Even though closed questions allow for easy data coding and analysis, some thought should be given to asking **open questions**, in which the respondent is allowed to state freely an unstructured answer. Open questions allow the respondent to impart some depth and shades of meaning to the answers. But they can also cause great difficulties in analysis, because individual answers may not be easily quantified and answers may be nearly impossible to compare across questionnaires. In contrast, closed questions may not always provide appropriate alternatives, and the alternatives listed may themselves influence the opinion of the person responding. Once a closed-question questionnaire is completed, however, the data handling is fairly routine, and valid statistical summaries of reported answers are easily constructed.

A typical open question, similar to ones actually used in Gallup polls, is as follows:

What is the most important problem facing the United States today?

This question can elicit meaningful results, since many people will choose one of a handful of problems as being most important. However, their choices can be forced into predetermined categories by rephrasing the open question as a closed question:

The most important problem facing the United States today is (check one):

A.   national security.

B.   crime.

C.   inflation.

D.   unemployment.

E.   budget deficits.

Clearly, any closed form of this question limits the alternatives and may force a respondent into an answer that would not necessarily be a first choice.

A good plan for designing a closed question with appropriate alternatives is to use a similar open question on a pretest; then choose as the closed question's alternative answers those that were expressed most frequently in the open answers. Coming up with a short list of alternatives from all the open-ended answers received is not always easy, but this approach will provide more realistic alternatives than could be obtained on the basis of mere speculation.

Response Options

In response to almost any question that can be posed, some interviewees will want to say that they don't know or that they have no opinion. Since such responses give no useful information about the question and essentially reduce the sample size, typical survey practice is to avoid using these options. Instead, the respondent is forced to make a choice from among the listed informative answers, unless the interviewer decides that such a choice simply cannot be made.

However, forcing people to take positions on questions they know nothing about is inappropriate. Thus a good questionnaire will provide screening questions to determine whether the respondent has enough information to form an opinion on a certain issue. If so, the main question is asked without a "no opinion" option. If not, the question may be skipped.

In other words, questions about which nearly everyone has enough information to form some opinion, such as questions on stricter enforcement of speed limit laws for drivers, should be stated without a "no opinion" option. Questions of a specific, narrow, or detailed nature, such as questions on a recently passed city ordinance, should be prefaced with screening questions that enable the interviewer to determine whether the respondent has any information on the subject.

Even after the "no opinion" option has been eliminated from a question, one must decide how many options to allow. Frequently, questionnaires attempt to polarize opinion on one side or the other, as in the following question:

> Do you think the enforcement of traffic laws in our city is too strict or too lenient?

Here no middle ground is offered. One reason for not allowing a middle choice, such as "just right the way things are," is that respondents may take this choice far too often as an easy way to avoid choosing between the two extreme positions. The two-choice option forces the respondent to think about the preferential direction of the response, but the interviewer should explain that the two extremes comprehend many different degrees of support—whether or not these degrees are spelled out in the reported results. "Which pole am I closer to?" is the point that the respondent should be urged to consider. Of course, if one actually wants to categorize the degrees of support respondents bring to this question, more than two options can be presented. However, questionnaire designers usually wish to keep the number of options as small as possible.

Wording of Questions

Even in the case of questions for which the number of response options is clearly determined, the designer should be concerned about the phrasing of the main body of the question. Yes–no questions like

> Do you favor the use of capital punishment?

should be asked in a more balanced form, such as

> Do you favor or oppose the use of capital punishment?

Some questions have strong arguments and counterarguments woven into them. The study by Schuman & Presser (1981, p. 186) shows results for a comparison of the following questions:

A. If there is a union at a particular company or business, do you think that all the workers there should be required to be union members, or are you opposed to this?

B. If there is a union at a particular company or business, do you think that all the workers there should be required to be union members, or should it be left to the individual to decide whether or not he wants to be in the union?

Among persons presented with question A, 32.1% responded that workers should be required to be union members; but among those presented with question B, only 23.0% responded in this way. Question B incorporates a stronger counterargument in the second phase of the question. People with no strong feelings either way are particularly susceptible to strong arguments or counterarguments framed within the body of the question. Again, questions should be asked in a balanced form, with little argument or counterargument within the text of the question. The question

> Do you agree that courts are too lenient with criminals?

will receive many more yes responses than it should, simply because that response seems to agree with the interviewer's notion of the correct response. Leading questions should be rephrased in a balanced form, as discussed earlier in this subsection.

Responses to many questions can be drastically altered just by introducing an appropriate, or inappropriate, choice of words. The study by Schuman & Presser (1981, p. 277) reports on studies of the following questions:

A. Do you think the United States should forbid public speeches against democracy?

B. Do you think the United States should allow public speeches against democracy?

In one study, 21.4% of those presented with question A gave *yes* responses, while 47.8% of those presented with question B gave *no* responses. People are somewhat reluctant to *forbid* public speeches against democracy, but they are much more willing *not to allow* such speeches. *Forbid* is a strong, prohibitory word that elicits a negative feeling that many cannot favor. *Allow* is a milder, permissive word whose denial doesn't elicit such strong feelings. The important point to remember is that the tone of the question, set by the words employed, can have a significant impact on the responses.

Questions also must be stated in clearly defined terms, in order to minimize response errors. A question like

How much water do you drink?

is unnecessarily vague. It may be reworded as follows:

Here is an 8-ounce glass. [Hold one up.] How many 8-ounce glasses of water do you drink each day?

If total water intake is important, the interviewer must remind the person that coffee, tea, and other drinks are mostly water.

Similarly, a question like

How many children are in your family?

is too ambiguous. It may be restated as follows:

How many persons under the age of 21 live in your household and receive more than one-half of their financial support from you?

Again, the question must be specific, with all components well defined.

In designing a questionnaire, we must always remember that people do not remember factual information very well. An interesting study in this area is reported by Bradburn et al., "Answering Autobiographical Questions: The Impact of Memory and Inference on Surveys," *Science*, April 10, 1987, pp. 157–61. Three main points emerge from the article:

1 Do not count on people to know or remember even the simplest facts. One study reported that only 31% of respondents correctly identified their savings account balance, and only 47% got it correct *when allowed to consult their records.*

2 People do not generally determine frequencies of events by simple counting. If asked "How many times have you visited a doctor in the past year?" they will tend to establish a rate for a shorter period of time and then multiply. For example, a certain respondent may think she visits a doctor about once a month, and then multiply this by 12 to get an annual figure. If asked "How many times have you eaten at a restaurant in the past month?" an interviewee may decompose the event into breakfast, lunch, and dinner, and approximate an answer for each meal before adding them back together.

3 People tend to telescope events that they remember well into a shorter time frame. Thus, an automobile accident or a reward on the job may seem to have occurred more recently than it actually did. Similarly, events that are not recalled easily may seem to have occurred longer ago than they actually did.

Knowledge of these facets of human behavior can help you in designing a good questionnaire. For example, consider the following guidelines:

1 Ask questions about facts in more than one way, seek out more than one source, and use direct observation as much as possible.

2 Help with the decomposition process by decomposing the questions you ask (such as asking about water, soft drink, beer, and coffee consumption rather than simply asking about drink consumption).

3   Tie questions about events in a relationship to important milestones in life (such as "Was the hospital visit before or after you moved to this address?"; "Was it before or after your daughter left for college?") to compensate for telescoping.

Responses will always contain some errors, but careful questioning can reduce these errors to a point at which the results are still useful.

Many more items could be discussed on the topic of questionnaire construction. But the items presented here are the most important ones, and each should be considered very carefully before sampling is begun.

Exercises

2.9   Discuss problems associated with question ordering. List two or three questions for which you think order is important, and explain why.

2.10   Discuss the use of open versus closed questions. Give an example of an appropriate open question. Give an example of how a similar question could be closed. What are the advantages of closed questions?

2.11   Give an example of a question that contains a weak counterargument. Give an example of a question that contains a strong counterargument.

2.12   Discuss the pros and cons of including a no-opinion option in a closed question.

2.13   Give an example of a question that could force a response in a certain direction because of its strong wording.

2.4   Experimental Studies

The subject of designs for experimental studies cannot be given much justice in the beginning of an introductory course in statistics, since entire courses at the undergraduate and graduate levels are needed to get a comprehensive understanding of the methods and concepts of designs. Even so, we will attempt to give you a brief overview of the subject because many data requiring summarization and analysis arise from experimental studies involving one of a number of designs. We will work by way of examples.

A multinational oil company has been developing an unleaded gasoline that can be competitively priced while delivering higher gasoline mileage. A number of gasoline blends have been proposed and undergone initial testing. One blend in particular appears to yield good gasoline mileage and can be produced economically. An additional test is planned to obtain an accurate estimate of the gasoline mileage (miles per gallon) for the blend under normal road conditions. In the test, a standard car model is chosen; then each of ten cars of this model type is drive over a predetermined course, and the miles per gallon are recorded. (Later, in Chapter 7, we will show how these data could be summarized to provide an estimate of the miles per gallon for the blend under these fixed road conditions.)

To change the problem slightly, suppose that the oil company had three different gasoline blends it wanted to test in the gasoline mileage test conducted under normal road conditions. For this study, the company could take nine standard model cars, randomly assign three cars to each gasoline blend, and test the cars under the road conditions dictated by the experiment. There would be a recorded gasoline mileage for each car, and three cars per gasoline blend. The methods presented in Chapters 3 and 13 could be used to summarize and analyze the sample mileage data in order to make comparisons (inferences) among the three gasoline blends. One inference of interest might involve identifying the best gasoline blend from a gasoline mileage standpoint. Which blend performed best? Can the best-performing blend in the sample data be expected to provide superior gasoline mileage a second time, if the same study is repeated?

Experimental Designs

completely randomized design

The experimental design for this scientific study is called a **completely randomized design**. Table 2.1 displays a completely randomized design for the gasoline blend study.

TABLE 2.1
Completely Randomized Design
of Gasoline Blends

Blend 1	Blend 2	Blend 3
3 cars	3 cars	3 cars

In general, a completely randomized design is used when one is interested in comparing t "treatments" (in our $t = 3$ case, the treatments were gasoline blends). For each of the treatments, we obtain a sample of observations, and the samples are not necessarily of the same size for the different treatments. The sample of observations from a treatment is assumed to be equivalent to a simple random sample of observations from the hypothetical population of possible values that could have been obtained for that treatment. In our example, the sample of three gasoline mileages obtained from blend 1 was considered to be the outcome for a simple random sample of three observations selected from the hypothetical population of possible mileages for standard model cars using gasoline blend 1. The same reasoning applies to the samples from blends 2 and 3.

This experimental design could be changed to accommodate the study of the same three blends using each of, say, three different drivers. Because of individual driving habits, not all drivers get the same gasoline mileage for the same blend of gasoline; so it would be desirable to have each of the drivers test each of the blends over the specified road course. Here we avoid having the comparison of blends distorted by differences among drivers. The experimental design is called a **randomized block design** because we have "blocked" out any differences among drivers in order to get a precise comparison of the three blends (see Table 2.2). Note that each driver tests each blend, and the order of testing the blends is randomized for each driver (i.e., driver 2 will test blend 2 first, then blend 3, and finally blend 1).

randomized block design

What happens if the order of testing influences a driver's performance and the first blend tested generally receives a higher mileage rating than the ones tested second or third? Then blend 1 could (possibly) look better than blends 2 and 3 simply because it was tested first by two of the three drivers (see Table 2.2).

TABLE 2.2
Randomized Block Design of
Gasoline Blends

Driver 1	Driver 2	Driver 3
Blend 1	Blend 2	Blend 1
Blend 3	Blend 3	Blend 2
Blend 2	Blend 1	Blend 3

Latin square design

A variation on the randomized block design, called a **Latin square design**, eliminates the order of testing as a factor affecting the comparison of treatments (blends). A Latin square design for our example is shown in Table 2.3. With this design, each blend is tested first once, second once, and third once, *and* each driver tests all blends.

TABLE 2.3
Latin Square Design of Gasoline
Blends

Order of Testing	Driver 1	Driver 2	Driver 3
1st	Blend 1	Blend 3	Blend 2
2nd	Blend 2	Blend 1	Blend 3
3rd	Blend 3	Blend 2	Blend 1

The randomized block and Latin square designs are both extensions of the completely randomized design, where the objective is to compare t treatments. The analysis of data collected according to a completely randomized design and the inferences made from such analyses are discussed further in Chapter 13. A special case of the randomized block design is presented in Chapter 8, where the number of treatments is $t = 2$, and again the analysis of data and the inferences from the analyses are discussed. We will not cover the summarization, analysis, and reporting of data from randomized block designs and Latin square designs. See Ott (1993) for details.

Factorial Experiments

Suppose that we want to examine the effects of two or more variables (factors) on a response. For example, suppose that an experimenter is interested in examining the effects of two independent variables—nitrogen and phosphorus—on the yield of a crop. For simplicity we will assume that two input levels have been selected for the study of each factor: 40 and 60 pounds per plot for nitrogen, and 10 and 20 pounds per plot for phosphorus. For this study the experimental units are small, relatively homogeneous plots that have been partitioned from the acreage of a farm.

one-at-a-time approach

One approach for examining the effects of two or more factors on a response is called the **one-at-a-time approach**. To examine the effect of a single variable, an experimenter varies the levels of this variable while holding the levels of the other independent variables fixed. This process is continued until the effect of each variable on the response has been examined (see Table 2.4).

Hypothetical yields corresponding to the three factor-level combinations of our experiment are given in Table 2.5. Suppose that the experimenter is interested in using

TABLE 2.4
Factor-Level Combination for a
One-at-a-Time Approach

Combination	Nitrogen	Phosphorus
1	60	10
2	40	10
3	40	20

the sample information to determine the factor-level combination that will give the maximum yield. From the table we see that crop yield increases when the nitrogen application is increased from 40 to 60 (holding phosphorus at 10). Yield also increases when the phosphorus setting is changed from 10 to 20 (at a fixed nitrogen setting of 40). Thus it might seem logical to predict that increasing both the nitrogen and the phosphorus applications to the soil will result in a larger crop yield. The fallacy in this argument is that our prediction is based on the unconfirmed assumption that the effect of one factor is the same for both levels of the other factor.

TABLE 2.5
Yields for the Three Factor-Level
Combinations

Observation (Yield)	Nitrogen	Phosphorus
145	60	10
125	40	10
160	40	20
?	60	20

From our investigation we know what happens to yield when the nitrogen application is increased from 40 to 60 for a phosphorus setting of 10. But will the yield also increase by approximately 20 units when the nitrogen application is changed from 40 to 60 at a setting of 20 for phosphorus?

To answer this question we could apply the factor-level combination of 60 nitrogen–20 phosphorus to another experimental plot and observe the crop yield. If the yield is 180, then the information obtained from the three factor-level combinations would be correct and useful in predicting the factor-level combination that produces the greatest yield. But what if the yield obtained from the high settings of nitrogen and phosphorus turns out to be 110? If this happens, the two factors nitrogen

interaction

and phosphorus are said to **interact**. That is, the effect of one factor on the response does not remain the same for different levels of the second factor, and the information obtained from the one-at-a-time approach would lead to a faulty prediction.

The two outcomes just discussed for the crop yield at a 60–20 setting is displayed in Figure 2.1, along with the yields at the initial factor-level settings. Figure 2.1 (a) illustrates a situation in which no interaction occurs between the two factors. The effect of nitrogen on yield is the same for both levels of phosphorus. In contrast, Figure 2.1 (b) illustrates a case in which the two factors nitrogen and phosphorus do interact.

We have seen that the one-at-a-time approach to investigating the effect of two factors on a response is suitable only for situations in which the two factors do not interact. Although this was illustrated for the simple case in which two factors were

FIGURE 2.1   Recorded Yields of the Three Design Points and Two Possible Yields at a Fourth Design Point (N = 60 and P = 20)

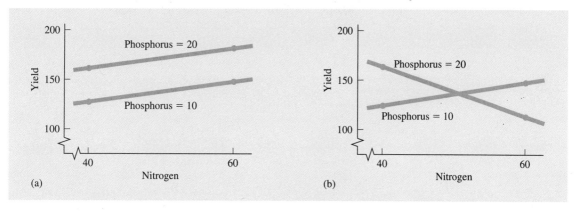

(a)

(b)

to be investigated at each of two levels, the inadequacies of a one-at-a-time approach are even more salient when one is trying to investigate the effects of more than two factors on a response.

Factorial experiments are useful for examining the effects of two or more factors on a response, whether or not interaction exists. As before, the choice of the number of levels of each variable to use and the actual settings of these variables are important.

When the factor-level combinations are assigned to experimental units at random, we have a completely randomized design, with treatments being the factor-level combinations.

DEFINITION   2.1

Factorial Experiment   A **factorial experiment** is an experiment in which the response is observed at all factor-level combinations of the independent variables. ∎

Using our previous example, if we are interested in examining the effect of two levels of nitrogen at 40 and 60 pounds per plot and two levels of phosphorus at 10 and 20 pounds per plot on the yield of a crop, we can use a completely randomized design where the four factor-level combinations (treatments) of Table 2.6 are assigned at random to the experimental units.

TABLE 2.6
2 × 2 Factorial Experiment for Crop Yield

Factor-Level Combinations	
Nitrogen	**Phosphorus**
40	10
40	20
60	10
60	20

Similarly, if we wish to examine nitrogen at two levels (40 and 60) and phosphorus at three levels (10, 15, and 20), we can use the six factor-level combinations of Table 2.7 as treatments in a completely randomized design.

TABLE 2.7
2 × 3 Factorial Experiment for Crop Yield

Factor-Level Combinations	
Nitrogen	**Phosphorus**
40	10
40	15
40	20
60	10
60	15
60	20

The examples of factorial experiments presented in this section deal with the effects of two variables (factors) on a response. However, the procedure applies equally well to any number of factors and any number of levels per factor. Thus, if we have four different factors at two, three, three, and four levels, respectively, we could formulate a $2 \times 3 \times 3 \times 4$ factorial experiment by considering all $2 \cdot 3 \cdot 3 \cdot 4 = 72$ factor-level combinations. Analyses and inferences of data obtained from factorial experiments in various designs are discussed in Ott (1993), Chapters 15, 16, and 17.

More Complicated Designs

Sometimes the objectives of a study direct us to investigate the effects of certain factors on a response while blocking out certain extraneous sources of variability. Such situations require a block design with treatments from a factorial experiment. This approach can be illustrated with the following example.

An investigator wants to examine the effects of two factors (factors A and B, each measured at three levels) on a response y. Two observations are desired at each factor-level combination, but only nine observations can be done each day. To satisfy these constraints, the investigator can run a complete replication of the 3×3 factorial experiment on two different days to get the desired number of observations. The design is shown in Table 2.8.

TABLE 2.8
Block Design Combined with Factorial Experiment

Day 1				Day 2			
Factor B				**Factor B**			
Factor A	1	2	3	**Factor A**	1	2	3
1				1			
2				2			
3				3			

Notice that this design is really a randomized block design in which the blocks are days and the treatments are the nine factor-level combinations of the 3×3 factorial experiment. Other, more complicated combinations of block designs and factorial experiments are possible. As we did in the case of sample surveys, though, we will deal here only with the simplest experimental designs. The point we want to make is that many different designs can be used in experimental studies for designating the collection of sample data. Each has certain advantages and disadvantages. A more extensive discussion of experimental designs appears in Ott (1993), where emphasis is placed on the summarization, analysis, and reporting from experimental studies using these designs.

2.5 Observational Studies

observational study

Before leaving the subject of sample data collection, we will draw a distinction between an **observational study** and a scientific study. In experimental designs for scientific studies, the observation conditions are fixed or controlled. For example, with a factorial experiment laid off in a completely randomized design, an observation is made at each factor-level combination. Similarly, with a randomized block design, an observation is obtained on each treatment in every block. These "controlled" conditions are very different from the conditions that prevail in certain observational studies where it is not feasible to do a proper scientific study. This can be illustrated by way of an example.

Much research and public interest centers on the effect of cigarette smoking on lung cancer and cardiovascular disease. One possible experimental design would be to randomize a fixed number of individuals (say 1000) to each of two groups: one group would be required to smoke cigarettes for the duration of the study (say, 10 years), while those in the second group would not be allowed to smoke throughout the study. At the end of the study, the two groups would be compared for incidence of lung cancer and cardiovascular disease. Even if we ignore ethical questions, this type of study would be impossible to do. Because of the long duration, it would be difficult to monitor all participants and make certain that they follow the study plan. And it would be difficult to find nonsmoking individuals willing to take the chance of being assigned to the smoking group.

Another possible study would be to sample a fixed number of smokers and a fixed number of nonsmokers to compare the groups for lung cancer and for cardiovascular disease. Assuming that one could obtain willing groups of participants, this study could be done in a *much shorter* period of time.

What has been sacrificed? Well, the fundamental difference between an observational study and a scientific study lies in the inference(s) that can be drawn. For a scientific study comparing smokers to nonsmokers, assuming that the two groups of individuals followed the study plan, the observed differences between the smoking and nonsmoking groups could be attributed to the effects of cigarette smoking because individuals were randomized to the two groups; hence, the groups were assumed to be comparable at the outset.

This type of reasoning does not apply to the observational study of cigarette smoking. Differences between the two groups in the observation are not necessarily attributable to the effects of cigarette smoking, because (for example) there may be hereditary factors that predispose people to smoking and cancer of the lungs and/or cardiovascular disease. Thus, differences between the groups might be due to hereditary factors, smoking, or a combination of the two. Typically the results of an observational study are reported by way of a statement of association. For our example, if the observational study showed a higher frequency of lung cancer and cardiovascular disease for smokers than for nonsmokers, it would be stated that this study showed that cigarette smoking was *associated* with an increased frequency of lung cancer and cardiovascular disease. This careful rewording is done so as not to claim that cigarette smoking causes lung cancer and cardiovascular disease.

Many times, however, an observational study is the only type of study that can be run. Our job is to make certain that we understand the type of study run and hence understand how the data were collected. Then we can critique inferences drawn from an analysis of the study data.

2.6   Data Management: Preparing Data for Summarization and Analysis

Making sense of data involves four basic steps: data collection, data summarization, data analysis, and communication of results. We have already dealt with data collection methods (step 1) for some surveys and experimental studies. Before one can begin summarizing and analyzing the study or survey data (steps 2 and 3), some data management "chores" must be performed; these will provide a means for storing and readily retrieving important study or survey information at a later date.

We begin with a discussion of some of these data management activities. In practice, these activities may consume 75% of the researcher's total effort from receipt of the raw data to presentation of the results of the analysis. What is involved in the management of data? Why are these activities so important, and why are they so time-consuming?

To answer these questions, let's first list the major data-processing activities we must perform before we can proceed with steps 2 and 3 of making sense of data. Then we will discuss each activity separately.

Data Management Activities Needed for Summarization and Analysis of Data (Steps 2 and 3 of Making Sense of Data)

1   Receive and retain the raw data source.

2   Create the database from the raw data source.

3   Edit the database.

4   Correct and clarify the raw data source.

5   Finalize the database.

6   Create workfiles from the database.

1. Receive and retain the raw data source.

For each study or survey that is to be summarized and analyzed, the data arrive in some form, which we will refer to as the **raw data source**. In a clinical trial, the raw data source is usually case report forms—sheets of $8\frac{1}{2} \times 11$-inch paper that have been used to record study data for each patient included in the study. In other types of experimental studies, the raw data source may be sheets of paper from a laboratory notebook, a data file on a disk (or diskette) obtained from a microprocessor or computer, hand-tabulations, or the like. In a survey, the raw data source could be completed questionnaires on predesigned forms or a data file from the electronic polling equipment used.

It is important to retain the raw data source, since it is the beginning of the **data trail**, which leads from the raw data to the conclusions drawn from a study or survey. Many consulting operations involved in the analysis and summarization of numerous studies and surveys keep a log that contains vital information related to each study and its raw data source. In a regulated environment such as the pharmaceutical industry, one may have to reproduce data, data summarizations, and data analyses based on previous work. Other situations outside the pharmaceutical industry may also require a retrospective review of what was done in the analysis of a study or survey. In these situations, the study or survey can be an invaluable source of information. The types of general information contained in a log are listed next.

Log for Study or Survey Data

1   Date received, and from whom
2   Name of study investigator or survey coordinator
3   Names of project teams members
4   Brief description of the study or survey, and identification of the design used
5   Treatments (compounds, preparations, etc.) studied or population surveyed
6   Raw data source
7   Response(s) measured, and how measured
8   Reference number for the study or survey
9   Estimated (actual) completion date
10 Other pertinent information

Later, when the study or survey has been analyzed and the results have been communicated, additional information can be added to the log describing how the results were communicated, where the results are recorded, what data files have been saved, and where these files are stored.

2. Create the database from the raw data source.

Computers can be of enormous help in doing the calculations required to summarize and analyze data, even for studies or surveys involving small amounts of data. For most studies or surveys, a computerized database is created from the raw data source, using available hardware and software. The steps taken to create the database and the eventual form of the database vary from one situation to another, depending on the computer software and hardware available for summarizing and analyzing the data and on the form of the raw data source. However, we can offer a few guidelines.

If the raw data source is a hard copy rather than an electronic file, you must create a database. If the data are to be entered at a terminal, PC, or workstation, the raw data must first be checked for legibility. Any illegible numbers or letters or other problems should be brought to the attention of the study or survey coordinator. Then a coding guide that assigns column numbers and variable names to the data should be filled out. Certain codes for missing values (for example, "not available") are also defined here, and a brief description of each variable is given. The data file keyed in **machine-readable** is referred to as the **machine-readable database**. A listing (printout) of the database **database** should be obtained and checked carefully against the raw data source. Any errors should be corrected at the terminal and verified by checking the marked copy against an updated listing.

Sometimes data are received in machine-readable form. In these situations, the disk file is considered to be the database. You must, however, have a coding guide to "read" the database. Using the coding guide, obtain a listing of the database and check it *carefully* to confirm that all numbers and characters look reasonable and that proper formats were used to create the file. Any problems that arise must be resolved before you proceed any further.

Some data sets are so small that it is not necessary to create a machine-readable data file from the raw data source. Instead, calculations are performed by hand, or the data are entered into an electronic calculator. In such situations check all calculations to see that they make sense. Don't simply accept everything you see; redoing the calculations is not a bad idea.

3. Edit the database.

The types of edits you do and the completeness of your editing process in a particular case depend on the type of study and on how concerned you are about the accuracy and completeness of your data prior to the analysis. For example, in using a statistical software package, it is wise to examine the minimum, the maximum, and the frequency distribution for each variable to make certain that nothing looks unreasonable.

Certain other checks should be made as well. Plot the data and look for problems. (Some of the graphical methods presented in Chapter 3 will be helpful). In addition, **logic checks** certain **logic checks** should be done, depending on the structure of the data. If, for example, data are recorded for patients on several different visits, data recorded for visit 2 can't be earlier than the data for visit 1; similarly, if a patient is lost to follow-up after visit 2, data for that patient on later visits can't exist.

For small data sets, we can do the data edits by hand; but for large data sets, the job may be too time-consuming and tedious. If machine-editing is required, look for a software system that allows you to specify certain data edits. Even so, more complicated edits and logic checks may require you to have a customized edit program written in order to machine-edit the data. This programming chore can be time-consuming; plan for this well in advance of receiving the data.

4. Correct and clarify the raw data source.

Questions may (and frequently do) arise with regard to the legibility or accuracy of the raw data during any step of the process, from receipt of the raw data to communication of the results of the statistical analysis. We have found it helpful to keep a list of these problems or discrepancies in order to define the data trail for a study. If a correction (or clarification) to the raw data source is required, this should be indicated on the form and the appropriate change should then be made to the raw data source. If no correction is required, this should be indicated on the form as well. Keep in mind that the machine-readable database must be amended to reflect any changes made to the raw data source.

5. Finalize the database.

You may have been led to believe that all data for a study arrive at one time. This, of course, is not always the case. For example, with a marketing survey, different geographic locations may be surveyed at different times and, hence, those responsible for data processing do not receive all the data at one time. All these subsets of data, however, must be processed through the cycles required to create, edit, and correct the database. Eventually the study is declared complete and the data is processed into the database. At this time, the database should be reviewed again and final corrections made before beginning the summarization and analysis of the survey or study data. In actuality, once some of the data have been entered into the database, summarization and analysis of the data can begin, and because of deadlines and curiosity concerning the study or survey results, this is usually what happens. However, the final summarizations and analysis which will be used to officially convey the results of the survey or study should (must) be done on the finalized database.

6. Create workfiles from the database.

workfiles

Generally these are data files created from the machine-readable database. These files, called **workfiles**, are created from the database and are designed to facilitate the analysis; they may not reflect the original structure of the database. For example, the workfiles may include only a selection of important variables, or they may create or add new variables by insertion, computation, or transformation. A listing of the workfiles should be checked against that of the database, to ensure proper restructuring and variable selection. Computed and transformed variables are checked by hand calculations to verify the program code. You should also use the labeling and documentation features of your software system to document what you have in the workfiles. Label the variables and give names to files for future reference.

You are now ready to proceed to the next stage of making sense of data. You've created, checked, and documented the database and data files to be used in the summarization and analysis of the study or survey data. How much data management is needed? Clearly, the conclusions drawn from the study or survey are only as good as the data on which they are based. So you be the judge. The amount of time you should spend on these data-processing chores before summarization and analysis ultimately depends on the nature of the study or survey, the quality of the raw data source, and how confident you want to be about the data's completeness and accuracy.

Summary

The first step in making sense of data consists of gathering data intelligently. This involves specifying the objectives of the data-gathering exercise, identifying the variables of interest, and choosing an appropriate design for the survey of scientific study. In this chapter we discussed various designs for surveys and for experimental studies. We also discussed questionnaire design. Armed with a basic understanding of some major design considerations for conducting surveys or experimental studies, one can address how data are to be collected on the variables of interest in order to address the stated objectives of the data-gathering exercise.

We also drew a distinction between observational and experimental studies based on the inferences (conclusions) that can be drawn from the sample data. Differences found between treatment groups in an observational study are said to be *associated with* the use of the treatments; on the other hand, differences found between treatments in an experimental study are said to be *due to* the treatments. In Chapters 3 and 4, we will examine methods of summarizing data, the second step in making sense of data.

Key Terms

simple random sampling

stratified random sampling

ratio estimation

cluster sampling

systematic sample

personal interviews

telephone interviews

self-administered questionnaire

direct observation

closed questions

open questions

completely randomized design

randomized block design

Latin square design

one-at-a-time approach

interaction

factorial experiment

observational study

raw data source

data trail

machine-readable database

logic checks

workfiles

Supplementary Exercises

2.14   A pharmaceutical company runs a clinical study to test three new compounds against a currently used drug for the treatment of congestive heart failure. Twelve different patients are assigned at random to each of the four compounds, treated, and measured for the effectiveness of the compound.

a   Is this a survey or an experimental study? Give reasons for your choice.

b   Identify the design used.

2.15   A direct-mail retailer tested three different ways of incorporating order forms into its catalog. The Type I catalogs had the order form at the end of the catalog; Type II had the form in the middle; and Type III had order forms both in the middle and at the end. Each type of catalog was sent to a random sample of 1000 potential customers, none of whom had previously bought from the retailer. The number of orders placed by recipients of each type of catalog was recorded.

a   Is this a survey or an experimental study? Why?

b   What design was used?

2.16   Respondents commonly receive telephone calls from people taking surveys during the evening dinner hour. The survey planners probably think that many potential respondents will be home at that time. Discuss the pros and cons of this approach.

2.17   You are hired to estimate the proportion of registered Republicans in your county who favor a decrease in the number of nuclear weapons maintained by the United States. How would you plan the survey?

2.18   In a Gallup youth survey, 414 high school juniors and seniors were asked the following question:

> What course or subject that you have studied in high school has been the best for preparing you for your future education or career?

In their responses to the question, 25% of the students chose mathematics and 25% chose English. Do you think that this question is a good question, with informative results? Explain.

2.19   A survey by Group Attitudes, Inc., attempted to measure American attitudes toward college. The polling firm mailed questionnaires to 4200 people across the United States and received 1188 responses. About 55% of those polled said that they had major concerns about being able to pay for their child's college education. Would you regard this figure as highly reliable and representative of the true proportion of Americans with this concern? What groups of people are most likely to respond to such a question?

3

Step Two: Methods for Summarizing Data

Tabular and Graphical Methods for Summarizing Data

3.1   Introduction

In Chapter 2 we dealt with gathering data—the first step in making sense of data. The second step in this process is to summarize the data that have been gathered so that meaningful interpretations can be made. Imagine having to describe annual incomes for all families registered in the last census or having to describe sales data by quarter for the steel industry, broken up by domestic consumption, exports, inventory change, and imports. Because it is almost impossible to understand and describe such detailed data, it is first necessary to summarize the "raw" data.

The first step in summarizing data is to graph the data. Take a "look" at the data to see if you can get a better understanding of what the data say. Are there any trends? Do most of the values fall near some central value? Do the data have considerable variability? Sometimes answers to these (and other) questions will be obvious from the data plots (graphs) that you construct. In this chapter we will examine how to use graphical displays to summarize data. In Chapter 4 we will discuss a second way to summarize data, focusing on the average or typical value in a data set and the spread or variability of the measurements.

3.2   Types of Data

In Chapter 2, we focused on gathering data and planning various designs for gathering data from surveys or experimental studies. We did not discuss the different types of data that we might gather in these surveys or experimental studies. Four basic data types will be discussed and illustrated in this section. Succeeding sections will provide tabular and graphical ways to summarize data.

Nominal Data

nominal data

Nominal data categorize qualitative objects by name. The categories established are assigned identifying symbols such as numbers or letters. For example, in categorizing the political party affiliation of potential voters in the 1992 presidential elections, we could use four categories: "Democrat," "Republican," "Independent," and "Other." In this way, every potential voter could be assigned to one of the categories. For purposes of identification, we could assign the number 1 to a Democrat, 2 to a Republican, 3 to an Independent, and 4 to a person with any other political affiliation. (We could just as easily have assigned letters to denote the categories.) Data gathered on the political party affiliation of potential voters would thus consist of a collection of 1s, 2s, 3s, and 4s corresponding to the political affiliation of the potential voters sampled. You must remember, however, that, although we have used numbers to indicate categories for the nominal data on the variable "political party affiliation," we cannot assign an order to the categories. For example, we cannot say that the category Democrat is "higher" or "lower" than the category Republican.

Ordinal Data

ordinal data

Ordinal data incorporate the features of nominal data but also allow us to establish the order (or ranking) of the categories. For example, we could use the categories "professional," "white collar," and "blue collar" to (crudely) classify occupations according to prestige. Here we could order (or rank) occupations from low to high, using these categories; note, however, that we could not determine a distance between the categories, because the categories have no inherent quantitative value. We could also assign a rank of 1 to occupations assigned to the professional category, a 2 to those assigned to the white-collar category, and a 3 to those occupations in the blue-collar category. Then the data we subsequently gathered on the prestige ranking of occupations for a sample of workers would yield a set of 1s, 2s, and 3s.

Another example of ordinal data would be used in a clinical evaluation of the severity of side effects observed for patients undergoing chemotherapy, categorized as none, mild, moderate, or severe.

Discrete Data

discrete data

Discrete data enable us to observe the magnitude as well as the order of data values. For nominal and ordinal data, categories were identified by a name, symbol, or number, but the categories did not represent measurements with any numerical value. Discrete data can assume any of various specified numerical values. (These possible values are usually represented by integers or counts.) For example, data on the number of homicides recorded for each county in the state of New York in 1993 would correspond to discrete data. The possible values for each county include the numbers 0, 1, 2, 3, and so on. Note, however, that intermediate values such as 2.5 for a given county are not possible. A set of data gathered on the numbers of homicides in each county in New York during 1993 would consist of the values corresponding to the observed number of homicides for each county.

Other examples of discrete data include the number of abortions done each week at a federally funded clinic, the number of voters favoring a lowering of the capital gains tax, and the number of new AIDS cases recorded each month in the United States over the past 5 years.

Continuous Data

continuous data

When the possible values a measurement can assume or correspond to the countless number of values on a number line, the data are termed **continuous**. For example, the daily recorded maximum temperature (°F) in the United States on a specific day could be any one of a countless number of possible values, limited only by the accuracy of the measuring equipment. The recorded value could thus be any possible integer or fractional value on the entire Fahrenheit scale. Other typical examples of continuous data include times, weights, heights, distances, and concentrations. The only limiting factor for continuous data is the degree of accuracy of the measuring equipment or (from a practical standpoint) the accuracy of measurement required. In many situations it is sufficient to have time data recorded to the nearest second, weights recorded to the nearest pound (or kilogram), heights recorded to the nearest inch (or centimeter), and so on.

What type of data should we gather when trying to conduct a survey or experimental study? Sometimes the choice is obvious from past surveys or studies, and sometimes the decision has already been made for us so that we are merely following standard practice.

In these situations, when trying to make sense of the data, we may require a lesser degree of data quantification than was obtained in the original data-gathering step. For example, in a study investigating the association of low initial birth weight with infant mortality, the birth weight data in a sample of infants may have been recorded to the nearest tenth of a pound; but summarization of the data may call for classifying them according to one of several ordinal categories such as extremely low (less than 1 pound), below average (1–4 pounds), or average or above (greater than 4 pounds).

As we shall see in this chapter and later, the summarization, analysis, and interpretation of ordinal data usually are easier than the corresponding actions for discrete or continuous data. And, since we are focused on making sense of data, we must consider the type of data to be gathered, summarized, and analyzed in order to effectively communicate our conclusions. Another fact to be considered in the choice of data relates to the ability of the intended audience to understand the results of the data summarization and analysis. Thus, making sense of data implicates not only our ability to "understand" the data, but perhaps more importantly our ability to communicate our understanding of the data effectively. The choice of data type and the summarization and analysis tools available for that data type can greatly affect the ability of the intended audience to understand our communication of results.

3.3 Pie Charts

pie chart

The simplest graphical procedure for displaying data is the **pie chart**. The pie chart is especially appropriate for nominal and ordinal data because it can be used to display the percentage of the total number of measurements that fall into each of the data categories; this is accomplished by partitioning a circle (much as one might slice a pie) into sectors (wedges) whose size reflects their proportion of the whole.

The data of Table 3.1 represent a summary of information from a study to determine paths to authority for individuals occupying top positions of responsibility in key public-interest organizations. Using biographical information, each of 1345 individuals was classified according to how she or he was recruited for the current elite position.

TABLE 3.1
Recruitment to Top
Public-interest Positions †

Source of Recruitment (Sector)	Number	Percentage
Corporate	501	37.2
Public-interest	683	50.8
Government	94	7.0
Other	67	5.0

†Includes trustees of private colleges and universities, directors of large private foundations, senior partners of top law firms, and directors of certain large cultural and civic organizations.

Source: Thomas R. Dye and L. Harmon Zeigler, *The Irony of Democracy,* 5th ed. (Monterey, Calif.: Duxbury Press, 1981), p. 130.

Although you can develop some idea of how these numbers compare by scanning the data in Table 3.1, the results are easier to interpret by using a pie chart. From Figure 3.1 we can quickly make certain inferences about channels to positions of authority. For example, we can see that more people were recruited for elite positions from public-interest organizations (approximately 51%) than from all other types of organizations.

Other variations of the pie chart are shown in Figures 3.2 and 3.3. Clearly, from Figure 3.2, cola soft drinks have gained popularity during the period from 1980 to 1990 at the expense of some other types of soft drinks. It's also evident, from Figure 3.3, that the loss of a major fast-food chain account affected fountain sales for PepsiCo, Inc.

F I G U R E **3.1**
Pie Chart for the Data of
Table 3.1

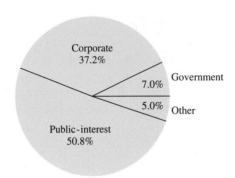

F I G U R E **3.2**
Approximate Market Share of
Soft Drinks, by Type, in 1980
and 1990

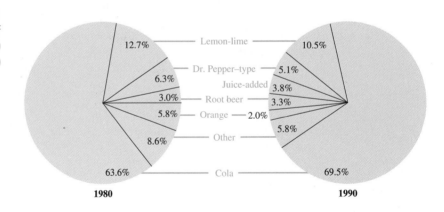

F I G U R E **3.3**
Estimated U.S. Market Share
Before and After Switch in
Accounts[†]

[†]A major fast-food chain switched its account from Pepsi to Coca-Cola for fountain sales.

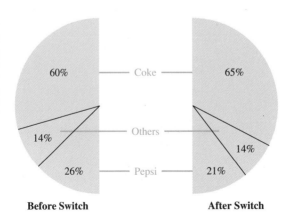

In summary, the pie chart can be used to display percentages associated with each category of the variable. The following guidelines should help you to obtain clarity of presentation when you use pie charts.

Guidelines for Constructing Pie Charts

1   Choose a small number of categories for the variable—preferably a maximum of five or six. Too many categories make the pie chart difficult to interpret.

2   Whenever possible, construct the pie chart so that percentages are arranged in either ascending or descending order.

3.4   Bar Charts

bar chart

A second graphical technique for presenting data is the **bar chart** or bar graph. As with the pie chart, this graphical method is especially appropriate for nominal and ordinal data. Figure 3.4 displays the number of workers in the greater Cincinnati, Ohio, area employed by companies from each of the five largest foreign investor nations. The estimated total work force is 680,000. The bar chart has many variations, some of which you may have seen in printed form. Sometimes the bars are displayed horizontally, as shown in Figure 3.5. They can also be used to display data across time, as shown in Figure 3.6. Sometimes the bar chart displays the percentage rather than the number of measurements in each category.

FIGURE **3.4**
Number of Workers Employed by Companies from Major Foreign Investor Nations

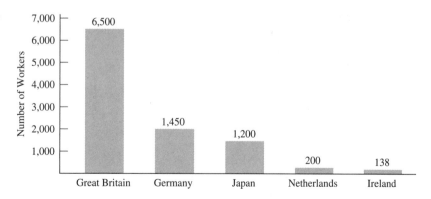

FIGURE 3.5
Greatest per Capita
Consumption, by Country

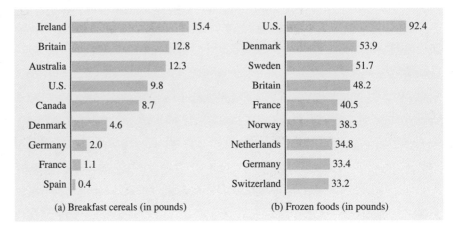

(a) Breakfast cereals (in pounds)   (b) Frozen foods (in pounds)

FIGURE 3.6
Estimated Direct and Indirect
Costs for Developing a New Drug
in Selected Years

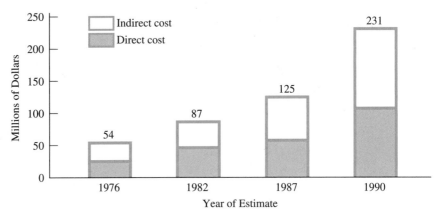

Bar charts are relatively easy to construct if you use the guidelines given here.

Guidelines for Constructing Bar Charts

1 Label frequencies (or percentages) along one axis, and label categories of the variable along the other axis.

2 Construct a rectangle at each category of the variable, with a height equal to the frequency (or percentage) of observations in the category.

3 Leave a space after each category to connote distinct, separate categories and to clarify the presentation.

Exercises

Basic Techniques

 3.1   The accompanying table of data gives a breakdown of total U.S. oil consumption, in percentages, categorized by the various purposes for which the oil is used. Present these data in a pie chart.

Use	Percentage of Total
Gasoline	43.1
Industrial fuel oil	12.0
Heating oil	17.0
Jet fuel	6.9
Diesel fuel	5.4
Petrochemical	3.6
Other	12.0
Total	100.0

3.2   Use the consumption figures in Exercise 3.1 to construct a bar graph. Which presentation, the bar graph or pie chart, seems to provide a clearer display of the data?

3.3   The U.S. Bureau of the Census publishes the *Statistical Abstract of the United States.* This reference was used to obtain the data listing in the accompanying table. Use these data to construct a pie chart for the percentage of males, categorized by years of school completed.

| | Percentage of Persons 65 or Older in 1980 | |
Years of Schooling Completed	Male	Female
8 years or less	45.3	41.6
1–3 years of high school	15.5	16.7
4 years of high school	21.4	25.8
1–3 years of college	7.5	8.6
4 years or more of college	10.3	7.4
	100.0	100.1[†]

[†]This column sums to 100.1 rather than 100.0 due to rounding errors.

 3.4   Refer to Exercise 3.3. Construct a bar chart for the female population, categorized by years of schooling.

 3.5   Can we combine the male and female data for each group in the table accompanying Exercise 3.3 and construct a bar chart for the combined population?

 3.6   Can we combine the male and female data for each group in the table accompanying Exercise 3.3 and construct a pie chart for the combined population?

 3.7   A large study of employment trends from 1990 to 1995, based on a survey of 45,000 businesses, was conducted by a large state university. Assuming an unemployment rate of 8% or less, it is predicted that 2.1 million job openings will be created between 1990 and 1995. This employment growth is shown in the accompanying table, categorized by major industry groups.

Industry Group	Percentage of Job Openings, 1990–1995
Service	33.2%
Manufacturing	25.0%
Retail trade	17.9%
Finance, insurance, and real estate	6.6%
Wholesale trade	4.8%
Construction	4.6%
Transportation	3.9%
Government	2.7%
Other	1.3%

Construct a pie chart to display these data.

3.8 From the same study described in Exercise 3.7, the following data were obtained on the job openings between 1990 and 1995, categorized by major occupational groups. Use the data to construct a bar chart.

Occupational Group	Percentage of Job Openings, 1990–1995
Clerical workers	20.9%
Sales	7.3%
Managers	9.5%
Professional and technical	16.3%
Laborers	3.7%
Service workers	18.1%
Operatives	13.1%
Craft and kindred workers	11.1%

3.5 Frequency Histograms

The next graphical technique we will consider is the frequency histogram, which is constructed from a frequency table. The frequency table and the frequency histogram are applicable only to discrete and continuous types of data. Consider the following continuous data: weight gains (in grams) were obtained for each of 100 baby chicks fed a new diet and observed over an 8-week period. These data are recorded in Table 3.2.

TABLE 3.2
Weight Gains for Chicks (grams)

3.7	4.2	4.4	4.4	4.3	4.2	4.4	4.8	4.9	4.4
4.2	3.8	4.2	4.4	4.6	3.9	4.3	4.5	4.8	3.9
4.7	4.2	4.2	4.8	4.5	3.6	4.1	4.3	3.9	4.2
4.0	4.2	4.0	4.5	4.4	4.1	4.0	4.0	3.8	4.6
4.9	3.8	4.3	4.3	3.9	3.8	4.7	3.9	4.0	4.2
4.3	4.7	4.1	4.0	4.6	4.4	4.6	4.4	4.9	4.4
4.0	3.9	4.5	4.3	3.8	4.1	4.3	4.2	4.5	4.4
4.2	4.7	3.8	4.5	4.0	4.2	4.1	4.0	4.7	4.1
4.7	4.1	4.8	4.1	4.3	4.7	4.2	4.1	4.4	4.8
4.1	4.9	4.3	4.4	4.4	4.3	4.6	4.5	4.6	4.0

frequency table

class interval

In trying to summarize the set of measurements recorded in Table 3.2, we know that the largest weight gain is 4.9 grams and the smallest is 3.6 grams. But even if we examine the table very closely, it is difficult for us to describe how the measurements are situated along the interval from 3.6 to 4.9. Are most of the measurements near 3.6, are they near 4.9, or are they evenly distributed along the interval? To answer the questions, we summarize the data in a **frequency table**.

To construct a frequency table, we first divide the range from 3.6 to 4.9 into an arbitrary number of subintervals called **class intervals**. The number of subintervals we choose depends on the number of measurements in the set; normally, the appropriate number of class intervals ranges from 5 to 20. The more data we have, the larger the number of classes we tend to use. The guidelines given here can be used for constructing the appropriate class intervals.

range

Guidelines for Constructing Class Intervals

1 Divide the **range** of the measurements (the difference between the largest measurement and the smallest measurement) by the approximate number of class intervals desired. Generally, we should try to establish from 5 to 20 class intervals.

2 After dividing the range by the desired number of subintervals, round the resulting number to a convenient (easy to work with) unit. This unit represents a common width for the class intervals.

3 Choose the first class interval so that it contains the smallest measurement. It is also advisable to choose a starting point for the first interval in such a way that no measurement falls on a point of division between two subintervals. This eliminates any ambiguity in deciding where individual measurements belong in the set of class intervals. (One way to do this is to choose boundaries that extend to one more decimal place than the data do.)

For the data in Table 3.2,

$$\text{Range} = 4.9 - 3.6 = 1.3.$$

Let's assume that we want approximately 10 subintervals. Dividing the range by 10 and rounding to a convenient unit, we obtain $1.3/10 = .13 \approx .1$. Thus, the appropriate class interval width is .1.

It is convenient to choose the first interval to be 3.55–3.65, the second to be 3.65–3.75, and so on. Notice that the smallest measurement, 3.6, falls in the first interval and that no measurement falls on the endpoint of any class interval (see Table 3.3).

Having determined the class interval, we can construct a frequency table for the data. The first column in Table 3.3 labels the classes by number, and the second column indicates the class intervals. We then examine the 100 measurements of Table 3.2, keeping a tally of the number of measurements that fall into each interval. The number of measurements that fall into a given class interval is called the **class**

TABLE 3.3
Frequency Table for the Chick Data

Class	Class Interval	Frequency f_i	Relative Frequency $f_i n$
1	3.55–3.65	1	.01
2	3.65–3.75	1	.01
3	3.75–3.85	6	.06
4	3.85–3.95	6	.06
5	3.95–4.05	10	.10
6	4.05–4.15	10	.10
7	4.15–4.25	13	.13
8	4.25–4.35	11	.11
9	4.35–4.45	13	.13
10	4.45–4.55	7	.07
11	4.55–4.65	6	.06
12	4.65–4.75	7	.07
13	4.75–4.85	5	.05
14	4.85–4.95	4	.04
Totals		$n = 100$	1.00

class frequency

relative frequency

frequency histogram

frequency. These data are recorded in the third column of the frequency table (see Table 3.3).

The **relative frequency** of a class is the frequency of the class divided by the total number of measurements in the set (total frequency). Thus, if we let f_i denote the frequency for class i and we let n denote the total number of measurements in the set, the relative frequency for class i is f_i/n. The relative frequencies for all the classes are listed in the fourth column of Table 3.3.

Now that the data of Table 3.2 have been organized into a frequency table, this can be used to construct a *frequency histogram* or a *relative frequency histogram*. To construct a **frequency histogram**, draw two axes: a horizontal axis labeled with the class intervals, and a vertical axis labeled with the frequencies. Then construct a rectangle over each class interval, with a height equal to the number of measurements falling in the given interval. The frequency histogram for the data of Table 3.1 is shown in Figure 3.7.

The relative frequency histogram for these data is constructed in much the same way as is the frequency histogram. In the relative frequency histogram, however, the vertical axis is labeled as relative frequency, and the height of the rectangle constructed over each class interval is equal to the class interval's relative frequency (the fourth column of Table 3.3). The relative frequency histogram for the data of Table 3.1 is shown in Figure 3.8. Clearly, the two histograms in Figures 3.7 and 3.8 have the same shape and would be identical if their vertical axes were equivalent. We will frequently refer to either type of graph simply as a histogram.

Several comments should be made concerning histograms. First, the distinction between bar charts and histograms is based on the distinction between the types of data they present. Bar charts are used to display nominal and ordinal data; histograms display discrete and continuous data recorded for quantitative variables.

Second, the histogram is the most important graphical technique we will present, because of the role it plays in statistical inference—a subject we will discuss in

F I G U R E **3.7**
Frequency Histogram for the
Chick Data of Table 3.2

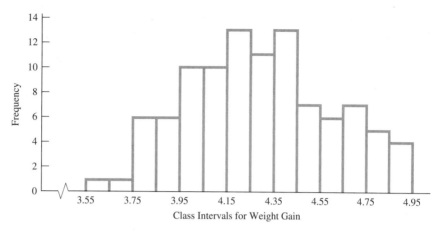

F I G U R E **3.7**
Frequency Histogram for the
Chick Data of Table 3.2

F I G U R E **3.8**
Relative Frequency Histogram
for the Chick Data of Table 3.2

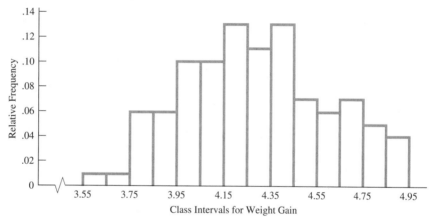

later chapters. Third, if we had an extremely large set of measurements, and if we constructed a histogram that used many class intervals, each with a very narrow width, the histogram for the set of measurements would, for all practical purposes, form a smooth curve. Fourth, the fraction of the total number of measurements accounted for in an interval is equal to the fraction of the total area under the histogram that lies within the rectangle (or rectangles) over the interval. For example, if we consider the chick data interval 3.75 to 4.35 in Table 3.3, we see that exactly 56 of the 100 measurements lie in that interval. Thus, .56, the fraction of the total number of sample measurements falling into that interval, is equal to the fraction of the total area under the histogram that lies within the rectangles over that interval, as indicated in Figure 3.9.

Fifth, if a single measurement is selected at random from the set of sample measurements, the chance, or **probability**, that it lies in a particular interval is equal to the fraction of the total number of sample measurements that fall into that interval. This same fraction is used to estimate the probability that a measurement randomly selected from the population lies in the interval of interest. For example, from the sample data of Table 3.2, we can estimate that the chance or probability of selecting a baby chicken whose weight gain fell into the interval 3.75 to 4.35 is .56. This value,

probability

FIGURE 3.9
Fraction of Measurements in the
Interval 3.75 to 4.35 for the
Chick Data

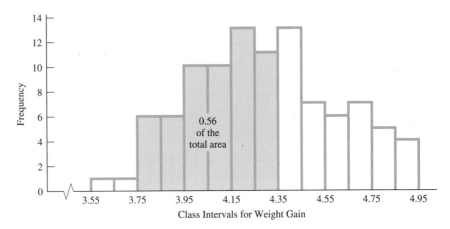

.56, is an approximation of the probability of selecting a measurement in the interval 3.75 to 4.35 from the population of all weight gains recorded for baby chicks.

Finally, since we use proportions rather than frequencies in a relative frequency histogram, we can compare two different samples (or populations) by examining their relative frequency histograms, even if the samples (populations) are of different sizes.

Exercises

Basic Techniques

 3.9 During a crackdown, the number of speeding citations delivered by each of the 30 members of a police department was recorded. These data are summarized in the following frequency table.

Number of Citations	Frequency
0.5–4.5	2
4.5–9.5	3
9.5–14.5	5
14.5–19.5	8
19.5–24.5	9
24.5–29.5	2
29.5–34.5	1

Construct a frequency histogram for these data.

 3.10 In a survey of prices of a loaf of white bread at regional stores, we obtained the following data.

Cost (cents)	Frequency
89.5–94.5	5
94.5–99.5	10
99.5–104.5	14
104.5–109.5	13
109.5–114.5	15
114.5–119.5	21
119.5–124.5	16
124.5–129.5	7
129.5–134.5	2

a   Use the results of this survey to compute the relative frequency for each class interval.

b   Construct a relative frequency histogram for these data.

Applications

3.11   The degree of job satisfaction among employees in any job classification is difficult to quantify. Attempts have been made to develop questionnaires (sometimes called *instruments*) composed of specific questions related to a variable of interest. Each respondent is asked to quantify his or her answer to each question on a scale (perhaps ranging from 0 to 5). In this way a respondent obtains a total score (often called *index score*) for the entire questionnaire. Designers of such questionnaires hope that different numerical scores will differentiate the variable of interest, such as the degree of job satisfaction among employees. One such instrument was used to measure the degree of job satisfaction among a sample of 219 nurses. (A high index score indicates a high degree of job satisfaction.) The resulting index data are summarized in the accompanying frequency table. Construct a frequency histogram for these data.

Degree of job satisfaction among 219 nurses:

Index Score	f	Index Score	f
19.5–23.5	2	47.5–51.5	43
23.5–27.5	1	51.5–55.5	38
27.5–31.5	1	55.5–59.5	43
31.5–35.5	3	59.5–63.5	24
35.5–39.5	7	63.5–67.5	13
39.5–43.5	7	67.5–71.5	4
43.5–47.5	33		
		Total	219

Source: Marshall, unpublished paper.

3.12   Refer to the data of Exercise 3.11. Compute the relative frequency for each class and construct a relative frequency histogram. (The shape of the relative frequency histogram should be the same as that of the frequency histogram you constructed for Exercise 3.11.)

3.13   Many metropolitan areas throughout the country have experienced staggering population increases over the past decade. It has been estimated that nearly 70% of the entire United States population now lives in 264 metropolitan areas; and by the year 2000 this percentage could reach 83%. In these 264 metropolitan areas, the average number of people per square mile is 400. The accompanying table lists the 50 most crowded and the 50 least crowded metropolitan areas from the original list of 264.

Most Crowded		Least Crowded	
Metropolitan Area	**People per Square Mile**	**Metropolitan Area**	**People per Square Mile**
1. Jersey City	12,963	1. Reno	19
2. New York City	7,206	2. Laredo	22
3. Paterson–Clifton–Passaic	2,400	3. Richland–Kennewick, Wash.	31
4. Boston	2,351	4. Great Falls, Mont.	31
5. Meriden, Conn.	2,332	5. Billings	33
6. Nassau–Suffolk, N.Y.	2,096	6. Yakima	34
7. Newark, N.J.	2,039	7. Las Vegas	35
8. Bridgeport	2,029	8. Duluth–Superior	36
9. Chicago	1,877	9. Tucson	38
10. New Brunswick–Perth Amboy–Sayreville, N.J.	1,871	10. Bakersfield, Calif.	40
11. Anaheim–Santa Ana–Garden Grove	1,816	11. Riverside–San Bernardino–Ontario	42
12. Los Angeles–Long Beach	1,728	12. Fargo–Moorhead	43
13. Stamford	1,706	13. Abilene, Tex.	45
14. New Britain	1,670	14. Eugene–Springfield, Ore.	47
15. Norwalk	1,449	15. Fort Smith	47
16. Cleveland	1,359	16. San Angelo, Tex.	47
17. Philadelphia	1,356	17. Pueblo, Colo.	49
18. Trenton	1,333	18. Texarkana	56
19. San Francisco–Oakland	1,254	19. St. Cloud, Minn.	62
20. Lowell, Mass.	1,219	20. Alexandria, La.	66
21. Providence–Warwick–Pawtucket	1,212	21. Albuquerque	68
22. Detroit	1,132	22. Provo–Orem	68
23. New Haven–West Haven	1,109	23. Fresno	69
24. Brockton, Mass.	1,098	24. Midland, Tex.	70
25. Honolulu	1,056	25. Fayetteville–Springdale, Ark.	71
26. Washington	1,034	26. Salinas–Seaside–Monterey, Calif.	75
27. Long Branch–Asbury Park, N.J.	965	27. Killeen–Temple, Tex.	76
28. Milwaukee	964	28. Wichita Falls, Tex.	76
29. Baltimore	917	29. Amarillo	80
30. Bristol, Conn.	885	30. Salt Lake City–Ogden	82
31. Springfield–Chicopee–Holyoke	856	31. Tallahassee	86
32. Buffalo	849	32. Tuscaloosa	87
33. Lawrence–Haverhill, Mass.	848	33. Colorado Springs	88
34. Waterbury, Conn.	844	34. Bloomington–Normal, Ill.	89
35. San Jose	819	35. Sherman–Denison, Tex.	89
36. Fall River, Mass.	807	36. Williamsport, Pa.	93
37. Pittsburgh	788	37. Florence, Ala.	94
38. New Bedford, Mass.	783	38. Tulsa	97
39. Akron	752	39. Santa Barbara–Santa Maria–Lompoc	97
40. Hartford	698	40. Lynchburg, Va.	97
41. Gary–Hammond–East Chicago	675	41. Salem, Ore.	98
42. Worcester, Mass.	667	42. Pine Bluff, Ark.	98
43. Cincinnati	644	43. Bryan–College Station, Tex.	99
44. Louisville	623	44. Lawton, Okla.	100
45. Miami	621	45. Odessa, Tex.	101
46. Lewiston–Auburn, Maine	604	46. Topeka	102
47. Fitchburg–Leominster, Mass.	581	47. Sioux City	103
48. Nashua, N.H.	560	48. Wilmington, N.C.	103
49. Norfolk–Virginia Beach–Portsmouth	548	49. Tyler, Tex.	104
50. New Orleans	532	50. Phoenix	106

Notice first that Jersey City and New York City have population densities (numbers of people per square mile) that far exceed the densities of the remaining 48 cities in the list of the 50 most crowded metropolitan areas. Since it would be difficult to include Jersey City and New York on the same graph with the remaining 48, because of the extremely high densities in these two cities, graph only the remaining 48. Use a frequency histogram to describe the population densities for the 48 cities from Paterson–Clifton–Passaic to New Orleans. Begin the first interval at 531.5, and construct each interval with a width of 190.

3.14   Refer to Exercise 3.13. Construct a frequency histogram for the 50 least crowded cities among the original 264 cities. Use approximately 10 intervals, with an interval width of 9. Begin the first interval at 18.5.

3.15   Refer to Exercise 3.13. Construct a relative frequency histogram for the most crowded metropolitan area data (excluding Jersey City and New York City). Use the same intervals as in Exercise 3.13.

3.16   Refer to Exercise 3.13. Construct a relative frequency histogram for the 50 least crowded metropolitan areas, using the same intervals as in Exercise 3.14. Notice that the shape of the relative frequency histogram is identical to that of the frequency histogram of Exercise 3.14.

3.17   The length of time an outpatient must wait for treatment is a variable that plays an important role in the design of outpatient clinics. The waiting times (in minutes) for 50 patients at a pediatric clinic are as follows:

35	22	63	6	49	19	15	83	46	19
16	31	24	29	36	68	42	57	64	8
23	47	21	51	7	40	19	46	16	32
108	33	55	32	22	36	25	27	37	58
39	10	42	28	72	13	51	45	77	16

Construct a relative frequency histogram for these data.

3.6 Stem-and-Leaf Plots

exploratory data analysis (EDA)

The next graphical technique we will discuss in this section is a display technique taken from an area of statistics called **exploratory data analysis (EDA)**. Professor John Tukey (1977) has been the leading proponent of this practical philosophy of data analysis, which is aimed at exploring and understanding data.

stem-and-leaf plot

The **stem-and-leaf plot** is a clever, simple device for constructing a histogram-like picture of a frequency distribution. It is used with discrete and continuous data. It allows us to use the information contained in a frequency distribution to show the range of scores, where the scores are concentrated, the shape of the distribution, whether there are any specific values or scores not represented, and whether there are any stray or extreme scores. The stem-and-leaf plot does not follow the organizing principles stated earlier for histograms. We will use the data shown in Table 3.4 to illustrate how to construct a stem-and-leaf plot. The original scores in Table 3.4 are either three- or four-digit numbers. We will use the first (or **leading**) **digit** of each

leading digit
trailing digits

score as the stem (see Figure 3.10) and the **trailing digits** as the leaf. For example, the violent crime rate in Albany is 876. The leading digit is 8 and the trailing digits are 76. In the case of Fresno, the leading digits are 10 and the trailing digits are 20. If our data consisted of six-digit numbers such as 104,328, we might use the first two

TABLE 3.4
Violent Crime Rates for 90 Standard Metropolitan Statistical Areas Selected from the South, North, and West

South	Rate[†]	North	Rate	West	Rate
Albany, GA	876	Allentown, PA	189	Abilene, TX	570
Anderson, SC	578	Battle Creek, MI	661	Albuquerque, NM	928
Anniston, AL	718	Benton Harbor, MI	877	Anchorage, AK	516
Athens, GA	388	Bridgeport, CT	563	Bakersfield, CA	885
Augusta, GA	562	Buffalo, NY	647	Brownsville, TX	751
Baton Rouge, LA	971	Canton, OH	447	Denver, CO	561
Charleston, SC	698	Cincinnati, OH	336	Fresno, CA	1020
Charlottesville, VA	298	Cleveland, OH	526	Galveston, TX	592
Chattanooga, TN	673	Columbus, OH	624	Houston, TX	814
Columbus, GA	537	Dayton, OH	605	Kansas City, MO	843
Dothan, AL	642	Des Moines, IA	496	Lawton, OK	466
Florence, SC	856	Dubuque, IA	296	Lubbock, TX	498
Fort Smith, AR	376	Gary, IN	628	Merced, CA	562
Gadsden, AL	508	Grand Rapids, MI	481	Modesto, CA	739
Greensboro, NC	529	Janesville, WI	224	Oklahoma City, OK	562
Hickory, NC	393	Kalamazoo, MI	868	Reno, NV	817
Knoxville, TN	354	Lima, OH	804	Sacramento, CA	690
Lake Charles, LA	735	Madison, WI	210	St. Louis, MO	720
Little Rock, AR	811	Milwaukee, WI	421	Salinas, CA	758
Macon, GA	504	Minneapolis, MN	435	San Diego, CA	731
Monroe, LA	807	Nassau, NY	291	Santa Ana, CA	480
Nashville, TN	719	New Britain, CT	393	Seattle, WA	559
Norfolk, VA	464	Philadelphia, PA	605	Sioux City, IA	505
Raleigh, NC	410	Pittsburgh, PA	341	Stockton, CA	703
Richmond, VA	491	Portland, ME	352	Tacoma, WA	809
Savannah, GA	557	Racine, WI	374	Tucson, AZ	706
Shreveport, LA	771	Reading, PA	267	Victoria, TX	626
Washington, DC	685	Saginaw, MI	684	Waco, TX	631
Wilmington, DE	448	Syracuse, NY	685	Wichita Falls, TX	639
Wilmington, NC	571	Worcester, MA	460	Yakima, WA	585

[†] Rates represent the number of violent crimes (murder, forcible rape, robbery, and aggravated assault) per 100,000 inhabitants, rounded to the nearest whole number.

Source: Department of Justice, *Uniform Crime Reports for the United States*, 1990.

FIGURE 3.10
Stem-and-Leaf Plot for the Violent Crime Rates of Table 3.4

1	89
2	98 96 24 10 91 67
3	88 76 93 54 36 93 41 52 74
4	64 10 91 48 47 96 81 21 35 60 66 98 80
5	78 62 37 08 29 04 57 71 63 26 70 16 61 92 62 62 59 05 85
6	98 73 42 85 61 47 24 05 28 05 84 85 90 31 26 39
7	18 35 19 71 51 39 20 58 31 03 06
8	76 56 11 07 77 68 04 85 14 43 17 09
9	71 28
10	20

digits as stem numbers, use the second two digits as leaf numbers, and ignore the last two digits.

For the data on violent crime, the smallest rate is 189, the largest is 1020, and the leading digits are 1, 2, 3, ..., 10. In the same way that a class interval determines where a measurement is placed in a frequency table, the leading digit (stem of a score) determines the row in which a score is placed in a stem-and-leaf plot. The trailing digits for the score are then written in the appropriate row. In this way, each score is listed in the stem-and-leaf plot. This has been done in Figure 3.10 for the violent crime data.

As you can see, each stem defines a class interval, and the limits of each interval are the largest and smallest possible scores for the class. The values represented by each leaf must fall between the lower and upper limits of the interval.

Graphically, a stem-and-leaf plot looks much like a histogram turned sideways, as in Figure 3.10. The plot can be made a bit more useful by ordering the data (leaves) within a row (stem) from lowest to highest (see Figure 3.11). The advantage of such a graph over a histogram is that it reflects not only the frequencies, the concentration(s) of scores, and the shapes of the distribution, but also the actual scores.

FIGURE 3.11
Stem-and-Leaf Plot with Ordered Leaves

```
 1 | 89
 2 | 10  24  67  91  96  98
 3 | 36  41  52  54  74  76  88  93  93
 4 | 10  21  35  47  48  60  64  66  80  81  91  96  98
 5 | 04  05  08  16  26  29  37  57  59  61  62  62  62  63  70  71  78  85  92
 6 | 05  05  24  26  28  31  39  42  47  61  73  84  85  85  90  98
 7 | 03  06  18  19  20  31  35  39  51  58  71
 8 | 04  07  09  11  14  17  43  56  68  76  77  85
 9 | 28  71
10 | 20
```

Guidelines for Constructing Stem-and-Leaf Plots

1 Split each score or value into two sets of digits. The first or leading set of digits is the stem, and the second or trailing set of digits is the leaf.

2 List all possible stem digits, from lowest to highest.

3 For each score in the mass of data, write down the leaf values on the line labeled by the appropriate stem number.

4 If the display looks too cramped and narrow, stretch the display by using two lines per stem—so that, for example, leaf digits 0, 1, 2, 3, and 4 are placed on the first line of the stem, and leaf digits 5, 6, 7, 8, and 9 are placed on the second line.

5 If too many digits are present, such as in a six- or seven-digit score, drop the rightmost trailing digit(s) to maximize the clarity of the display.

6   The rules for developing a stem-and-leaf plot differ somewhat from
the rules for establishing class intervals for the traditional frequency
distribution and for various other procedures that we will consider in later
sections of the text. Class intervals for stem-and-leaf plots are thus, in a
sense, atypical.

Exercises

Basic Techniques

3.18   Final examination scores in a mathematics course are listed for 50 students.

43	51	53	55	57	58	58	59	61	61
61	62	63	64	65	65	66	66	67	68
68	69	69	69	69	70	70	70	71	71
72	73	73	74	74	75	76	76	77	78
79	79	81	82	82	85	87	89	91	96

Construct a stem-and-leaf plot for these data.

3.19   Refer to the data from Exercise 3.18. Construct a histogram for these data. Compare this
histogram to the stem-and-leaf plot you constructed in Exercise 3.18

3.7 Time Series Plots

The last graphical technique we will discuss in this chapter deals with how certain
discrete and continuous variables change over time. For both macroeconomic data
(such as disposable income) and microeconomic data (such as weekly sales data of
one particular product at one particular store), plots of data over time are fundamental
to business management. Similarly, social researchers are often interested in showing
how variables change over time. They might be interested in changes with time in
attitudes toward various racial and ethnic groups, or changes in the rate of savings in
the United States, or changes in crime rates for various cities. A pictorial method of
presenting changes in a variable over time is called a **time series**. Figure 3.12 shows a
time series tracing the percentage of white women aged 30 to 34 in the United States
who have not had any children. This trend is presented for the years from 1970 to
1986.

time series

Usually, time points are labeled chronologically across the horizontal axis (ab-
scissa), and numerical values (frequencies, percentages, rates, etc.) of the variable
of interest are labeled along the vertical axis (ordinate). Time can be measured in
days, months, years, or whatever other unit is most appropriate. As a rule of thumb, a
time series should consist of no fewer than four time points; typically, the time points
represented are equally spaced. Many more than four time points are desirable, in
order to present a more complete picture of changes in a variable over time.

How one displays the time axis in a time series frequently depends on the time
intervals at which data are available. For example, the U.S. Census Bureau reports

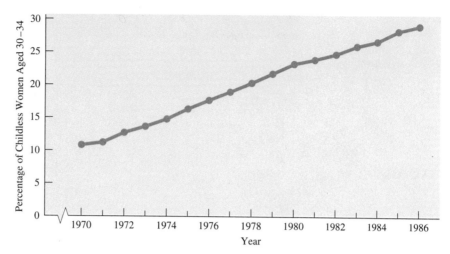

FIGURE 3.12
Percentage of Childless Women
Aged 30 to 34, 1970–1986

average family income in the United States only on a yearly basis. When information about a variable of interest is available in different units of time, one must decide which unit or units are most appropriate for the research. In an election year, a political scientist would most likely examine weekly or monthly changes in candidate preferences among registered voters. On the other hand, a manufacturer of machine-tool equipment might keep track of sales (in dollars and number of units) on a monthly, quarterly, and yearly basis. Figure 3.13 shows the quarterly sales (in thousands of units) of a machine-tool product over the past three years. From this time series, it is clear that the company has experienced a gradual but steady growth in number of units sold over the past 3 years.

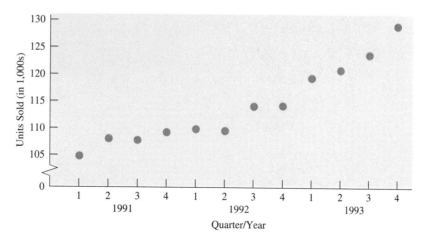

FIGURE 3.13
Quarterly Sales (in thousands)

Time series plots are useful for examining general trends and seasonal or cyclic patterns. For example, the "Money and Investing" section of the *Wall Street Journal* gives the daily workday values for the Dow Jones Industrial, Transportation, and Utilities Averages for the preceding 6-month period. These are displayed in Figure 3.14

F I G U R E 3.14
Time Series Plots for the Dow
Jones Industrial, Utilities, and
Transportation Averages

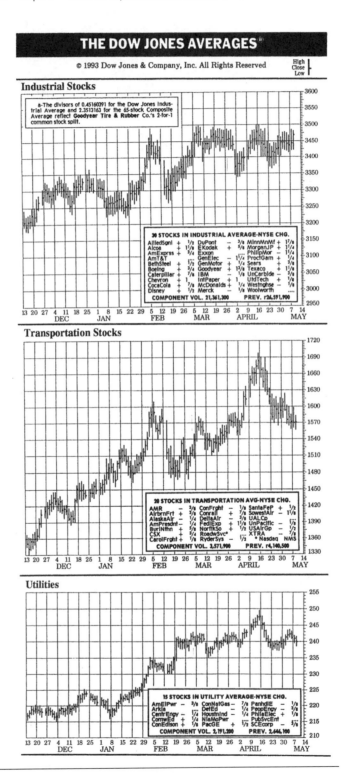

Source: Wall Street Journal, 12 May 1993.

for a typical period. A quick glance at these plots reveals general trends downward in the Industrial and Transportation indices over the July through September period of 1993. Seasonal or cyclical patterns might be detectable if we had weekly (or perhaps monthly) data for several years.

E X A M P L E **3.1**   Monthly data on the number of paid admissions (in thousands) to an indoor sports and entertainment arena for 72 months are given in the following table and are plotted in Figure 3.15. Identify any apparent trends, seasonal effects, and cyclic effects.

J	F	M	A	M	J	J	A	S	O	N	D
89	101	116	111	94	59	44	44	73	78	99	93
96	110	118	116	107	69	52	54	85	92	113	109
120	136	155	155	129	98	69	78	116	143	154	166
183	199	227	219	198	148	94	108	160	201	215	246
242	264	359	308	265	193	150	146	243	260	332	293
323	377	470	422	345	239	176	182	288	342	380	379

Solution   The upward trend in admissions is obvious in the figure. The trend is evidently not linear, however. Rather, it involves an initial period of slow growth, then a period of more rapid growth, and then a tapering-off. (Of course, the trend upward cannot increase forever. Paid admissions to the arena are limited by the available number of dates and seats.) The data plotted reveal a strong seasonal effect: admissions drop dramatically in the summer months. If there's any cyclic component, it's not apparent in the figure.   ∎

F I G U R E **3.15**
Admissions to Arena;
Example 3.1

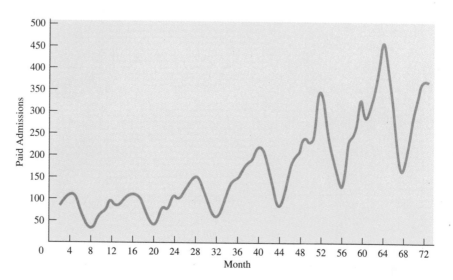

Sometimes it is important to compare trends over time in a variable for two or more groups. For example, Figure 3.16 reports the values of two ratios over the period from 1976 to 1988: the ratio of median family income of African-Americans to median family income of Anglo-Americans, and the ratio of the median family income of Hispanics to median family income of Anglo-Americans.

F I G U R E 3.16
Ratio of African-American and Hispanic Median Family Income to Anglo-American Median Family Income, 1976–1988

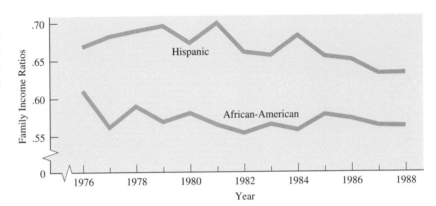

Median family income is the income amount that divides family incomes exactly in half, with half receiving a higher income and half receiving a lower income. In 1987, the median family income for African-Americans was $18,098, meaning that 50% of all African-American families had incomes above $18,098, and 50% had incomes below $18,098. The median, one of several measures of central tendency, is discussed more fully later in this chapter.

Figure 3.16 shows that the ratios of African-American and Hispanic to Anglo-American family income fluctuated between 1976 and 1988, but overall they both declined over this time. A social researcher would interpret these trends to mean that the income of African-American and Hispanic families generally declined relative to the income of Anglo-American families.

Sometimes information is not available in equal time intervals. For example, polling organizations such as Gallup or the National Opinion Research Center do not necessarily ask the American public the same questions about their attitudes or behavior every year. Sometimes there may be a time gap of more than 2 years before a question is asked again.

When information is not available in equal time intervals, it is important for the interval width between time points (the horizontal axis) to reflect this fact. If, for example, a social researcher is plotting values of a variable for 1985, 1986, 1987, and 1990, the interval width between 1987 and 1990 on the horizontal axis should be three times the interval width between the other years. If all of the interval widths were spaced evenly, the resulting trend line could be seriously misleading. Other examples of graphical distortions are discussed in Chapter 14.

Figure 3.17 presents the trend in church attendance among American Catholics and Protestants from 1958 to 1988. As the width of the intervals between time points reflects, Catholics were not asked about their church attendance every year. Before

FIGURE **3.17**
Church Attendance of American
Protestants and Catholics in a
Typical Week, 1954–1988

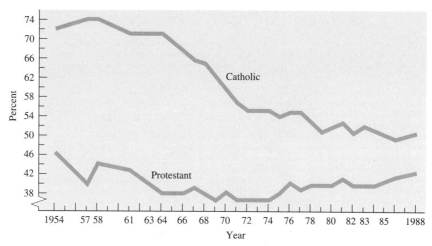

leaving graphical methods for describing data, let's review several general guidelines that can be helpful in developing effective graphs. These guidelines focus on design and presentation techniques and should help you 876 make better, more informative graphs.

General Guidelines for Successful Graphics

1   Before constructing a graph, set your priorities. What messages should the viewer get?

2   Choose the type of graph (pie chart, bar graph, histogram, and so on).

3   Pay attention to the title. One of the most important aspects of a graph is its title. The title should immediately inform the viewer of the point of the graph and should draw the eye toward the most important elements of the graph.

4   Fight the urge to use many type sizes, styles, and color changes. The indiscriminate and possibly excessive use of different type sizes and styles and of numerous colors will confuse the viewer. Generally, we recommend using only two typefaces; color changes and italics should be used in only one or two places.

5   Convey the tone of your graph by using colors and patterns. The more intense, warm colors (yellows, oranges, reds) are more dramatic than the blues and purples and, hence, help to stimulate enthusiasm in the viewer. On the other hand, pastels (particularly grays) convey a conservative business-like tone. Similarly, the simple patterns convey a conservative tone, whereas the busier patterns help to stimulate more excitement.

6   Don't underestimate the effectiveness of a simple, straightforward graph. Practice drawing graphs frequently. As with almost anything, practice improves skill.

3.8   Using Computers to Help Make Sense of Data

This section presents several examples of computer-generated graphs from MINITAB, using data from this chapter. First we give a MINITAB program that can be used to generate a histogram and stem-and-leaf plot for the data of Table 3.4. Compare the output shown here to the plots we did by hand earlier in the chapter. The histogram displayed in the output is fairly straightforward. The middle of each interval and the number of observations that fall into each interval are given. The histogram is displayed horizontally rather than vertically, as we're used to seeing it. Each asterisk (*) in the histogram represents an observation.

The stem-and-leaf plot in the MINITAB output is not as clearly labeled as the histogram. Ignore the left-hand column for now; it is used to locate the central measurement (median), which is discussed in Chapter 4. The middle column of the display gives the stems of the stem-and-leaf plot. The last column of the output gives the leaf portion of the stem-and-leaf plot. As indicated by the note at the top of the output, the leaf digit unit is 10.0. The first row of our stem-and-leaf plot in Figure 3.11 is 1 | 89. If we truncate the leaf digit 89 to the next lowest 10, we have the number 80. This digit gives rise to the first row of the MINITAB stem-and-leaf plot, 1 | 8. Similarly, the second row in Figure 3.11 is 2 | 10 24 67 91 96 98. Truncating each of the leaf digits 10, 24, 67, 91, 96, and 98 to the next lowest 10, we get the second row of the MINITAB stem-and-leaf plot: 2 | 1 2 6 9 9 9. Armed with this basic understanding, don't worry about trying to reconstruct all the leaves in a MINITAB stem-and-leaf plot; you should, however, be able to describe a set of measurements from a MINITAB stem-and-leaf plot; that is, from a MINITAB plot you should be able to determine (approximately) the concentrations of scores, the shape of the distribution, the range of scores, low and high values, and whether or not there are extreme values.

Next we have a SAS program that shows how to use PROC CHART to generate a histogram for the data of Table 3.4 and how to use PROC UNIVARIATE with the PLOT option to construct a stem-and-leaf plot for the same data. Following the listing of data in the output, we see a histogram (or as SAS calls it, a *frequency bar chart*) of the crime rate data. The midpoints of the intervals are given along the horizontal axis, and class frequency is recorded along the vertical axis. Each line of five asterisks in a given interval represents one observation.

The stem-and-leaf plot for SAS (as with MINITAB) is a bit more difficult to read than a SAS histogram; and the one displayed in the output looks different from the plot that we constructed in Figure 3.11. The first column of the plot gives the stems. The leaf portions of the plot are listed in the second column. According to the note at the bottom of the plot, we must multiply each number represented in the form of a stem and a leaf by $10^2 = 100$. The first line of the output 10 | 2 is taken to be 10.2. If we multiply 10.2 by 100, we get 1020. The highest crime rate is 1020. In the second line 9|37 represents 9.3 and 9.7, which become 930 and 970 when multiplied by 100. These are rounded versions of the original measurements 928 and 971, respectively. The bottom line of the stem-and-leaf plot is 1|9 (or 1.9), which becomes 190 and corresponds to the original crime rate 189. Note that SAS, unlike MINITAB, rounds to the nearest unit, not to the nearest lower unit. Thus 189 is rounded to 190 in SAS but is rounded to 180 in MINITAB.

Again, don't worry too much about how the leaves of a SAS stem-and-leaf plot are constructed. Just be able to describe a set of measurements from the plot. You can ignore the other sections of the output from this program until we cover these topics later in the text.

EXECUSTAT was used as well to construct a histogram and stem-and-leaf plot for the data of Table 3.4. This output is shown next.

The histogram for EXECUSTAT is well labeled, and the stem-and-leaf plot is identical to the one obtained by using MINITAB.

The MINITAB, SAS, and EXECUSTAT programs presented in this section can be used as models for similar analyses to be performed on other data sets, particularly for histograms and stem-and-leaf plots. Additional details about MINITAB statements, SAS procedures, and EXECUSTAT programs, are available in the user's manuals for these software systems (see references at the end of this text). Exercises 3.41, 3.49, and 3.50 offer other examples of the use of MINITAB to graph and plot data.

MINITAB Output

```
MTB > SET INTO C1
DATA> 876   578   718   388   562   971   698   298   673   537
DATA> 642   856   376   508   529   393   354   735   811   504
DATA> 807   719   464   410   491   557   771   685   448   571
DATA> 189   661   877   563   647   447   336   526   624   605
DATA> 496   296   628   481   224   868   804   210   421   435
DATA> 291   393   605   341   352   374   267   684   685   460
DATA> 570   928   516   885   751   561  1020   592   814   843
DATA> 466   498   562   739   562   817   690   720   758   731
DATA> 480   559   505   703   809   706   631   626   639   585
DATA> END

MTB > PRINT C1

C1
   876    578    718    388    562    971    698    298    673    537    642
   856    376    508    529    393    354    735    811    504    807    719
   464    410    491    557    771    685    448    571    189    661    877
   563    647    447    336    526    624    605    496    296    628    481
   224    868    804    210    421    435    291    393    605    341    352
   374    267    684    685    460    570    928    516    885    751    561
  1020    592    814    843    466    498    562    739    562    817    690
   720    758    731    480    559    505    703    809    706    631    626
   639    585

MTB > HISTOGRAM C1

Histogram of C1   N = 90

Midpoint    Count
     200        3   ***
     300        6   ******
     400       12   ************
     500       15   ***************
     600       21   *********************
     700       15   ***************
     800       10   **********
     900        6   ******
    1000        2   **
```

MINITAB Output (continued)

```
MTB > STEM-AND-LEAF C1

Stem-and-leaf of C1          N  = 90
Leaf Unit = 10

     1     1 8
     7     2 126999
    16     3 345577899
    29     4 1234466688999
   (19)    5 0001223556666677789
    42     6 0022233446788899
    26     7 00112333557
    15     8 000111456778
     3     9 27
     1    10 2


MTB > STOP
```

SAS Output

```
OPTIONS NODATE NONUMBER PS=60 LS=78;
TITLE1 'VIOLENT CRIME RATES FOR 90 SMSAS SELECTED';
TITLE2 'FROM THE SOUTH, NORTH, AND WEST';

DATA CRIME;
INPUT RATE @@;
CARDS;
    876  578  718  388  562  971  698  298  673  537
    642  856  376  508  529  393  354  735  811  504
    807  719  464  410  491  557  771  685  448  571
    189  661  877  563  647  447  336  526  624  605
    496  296  628  481  224  868  804  210  421  435
    291  393  605  341  352  374  267  684  685  460
    570  928  516  885  751  561 1020  592  814  843
    466  498  562  739  562  817  690  720  758  731
    480  559  505  703  809  706  631  626  639  585
;
PROC PRINT N;
TITLE3 'LISTING OF THE DATA';

PROC CHART DATA=CRIME;
  VBAR RATE;
TITLE3 'EXAMPLE OF PROC CHART';
TITLE4 'FREQUENCY DISTRIBUTION OF THE VARIABLE CRIME RATE';

PROC UNIVARIATE PLOT DATA=CRIME;
  VAR RATE;
TITLE3 'EXAMPLE OF PROC UNIVARIATE WITH THE PLOT OPTION';
RUN;
```

SAS Output (continued)

VIOLENT CRIME RATES FOR 90 SMSAS SELECTED
FROM THE SOUTH, NORTH, AND WEST
LISTING OF THE DATA

OBS	RATE
1	876
2	578
3	718
4	388
5	562
6	971
7	698
8	298
9	673
10	537
11	642
12	856
13	376
14	508
15	529
16	393
17	354
18	735
19	811
20	504
21	807
22	719
23	464
24	410
25	491
26	557
27	771
28	685
29	448
30	571
31	189
32	661
33	877
34	563
35	647
36	447
37	336
38	526
39	624
40	605
41	496
42	296
43	628
44	481
45	224
46	868
47	804
48	210

VIOLENT CRIME RATES FOR 90 SMSAS SELECTED
FROM THE SOUTH, NORTH, AND WEST
LISTING OF THE DATA

OBS	RATE
49	421
50	435
51	291
52	393
53	605
54	341
55	352
56	374
57	267
58	684
59	685
60	460
61	570
62	928
63	516
64	885
65	751
66	561
67	1020
68	592
69	814
70	843
71	466
72	498
73	562
74	739
75	562
76	817
77	690
78	720
79	758
80	731
81	480
82	559
83	505
84	703
85	809
86	706
87	631
88	626
89	639
90	585

N = 90

SAS Output (continued)

VIOLENT CRIME RATES FOR 90 SMSAS SELECTED
FROM THE SOUTH, NORTH, AND WEST
EXAMPLE OF PROC CHART
FREQUENCY DISTRIBUTION OF THE VARIABLE CRIME RATE

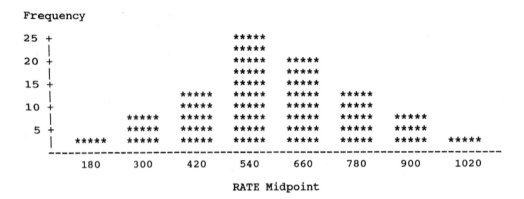

```
Frequency

 25 +                          *****
    |                          *****
 20 +                          *****     *****
    |                          *****     *****
 15 +                          *****     *****
    |                 *****    *****     *****     *****
 10 +                 *****    *****     *****     *****
    |        *****    *****    *****     *****     *****            *****
  5 +        *****    *****    *****     *****     *****     *****   *****
    | *****  *****    *****    *****     *****     *****     *****   *****   *****
      ------------------------------------------------------------------------
        180    300      420      540       660       780      900    1020

                              RATE Midpoint
```

VIOLENT CRIME RATES FOR 90 SMSAS SELECTED
FROM THE SOUTH, NORTH, AND WEST
EXAMPLE OF PROC UNIVARIATE WITH THE PLOT OPTION

Univariate Procedure

Variable=RATE

Moments

N	90	Sum Wgts	90
Mean	588.7333	Sum	52986
Std Dev	184.3092	Variance	33969.88
Skewness	0.004993	Kurtosis	-0.49396
USS	34217944	CSS	3023320
CV	31.30606	Std Mean	19.4279
T:Mean=0	30.3035	Pr>|T|	0.0001
Num ^= 0	90	Num > 0	90
M(Sign)	45	Pr>=|M|	0.0001
Sgn Rank	2047.5	Pr>=|S|	0.0001

Quantiles(Def=5)

100% Max	1020	99%	1020
75% Q3	719	95%	877
50% Med	574.5	90%	830
25% Q1	464	10%	346.5
0% Min	189	5%	291
		1%	189

Range	831
Q3-Q1	255
Mode	562

SAS Output (continued)

```
                              Extremes

               Lowest    Obs      Highest    Obs
                  189(    31)        877(     33)
                  210(    48)        885(     64)
                  224(    45)        928(     62)
                  267(    57)        971(      6)
                  291(    51)       1020(     67)
```

```
              VIOLENT CRIME RATES FOR 90 SMSAS SELECTED
                 FROM THE SOUTH, NORTH, AND WEST
             EXAMPLE OF PROC UNIVARIATE WITH THE PLOT OPTION


                      Univariate Procedure

Variable=RATE


       Stem Leaf                           #        Boxplot
         10 2                              1
          9 37                             2           |
          8 011112467888                  12           |
          7 001222344567                  12        +-----+
          6 002333445678889               15        |     |
          5 000012334666666677889         21        *--+--*
          4 12455667889                   11        +-----+
          3 00445578999                   11           |
          2 1279                          4            |
          1 9                             1            |
            ----+----+----+----+-
       Multiply Stem.Leaf by 10**+2
```

Variable=RATE

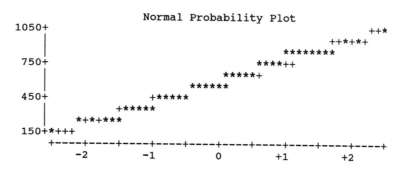

EXECUSTAT Output

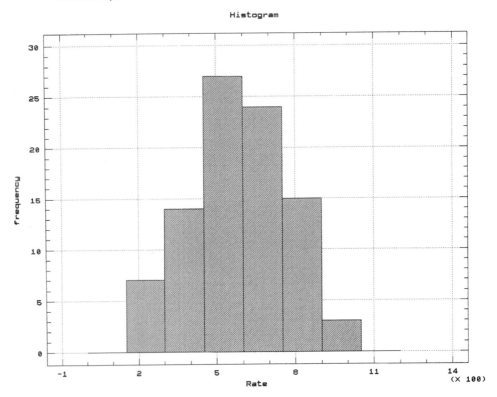

Stem-and-leaf display for Rate: unit = 10   1|2 represents 120

```
   1      1 | 8
   7      2 | 126999
  16      3 | 345577899
  29      4 | 1234466688999
 (19)     5 | 0001223556666677789
  42      6 | 0022233446788899
  26      7 | 00112333557
  15      8 | 000111456778
   3      9 | 27
   1     10 | 2
```

Summary

In Chapter 2, we discussed how to gather data—the first step in making sense of data. Chapter 3 is one of two chapters dealing with summarizing data—the second step in making sense of data. We began by considering the different types of data that might be gathered using the methods of Chapter 2. Nominal data have values that fall into unordered categories; ordinal data have values that can be categorized and ordered; discrete data can assume any of a specified number of measurable values, usually represented by counts; and continuous data can assume any of a countless number of measurable values on a line interval, such as one with measurements of time, weight, and height.

Graphical methods for summarizing data were then discussed. Pie charts and bar graphs are appropriate for nominal and ordinal data. Frequency histograms, relative frequency histograms, and the stem-and-leaf plot are used to graph discrete and continuous data; this involves dividing the possible values into a number of contiguous intervals and then plotting the frequency (or relative frequency) of values falling into the intervals. Finally, we discussed several versions of a graphical technique (the time series plot) for graphically representing discrete or continuous data across time.

Each of the graphical methods presented is designed to help us "see" the data in a summarized format. Many statistical software packages can be used to do the actual graphing of the study or survey data; this is especially helpful for large data sets, allowing us more time to explore various different graphs rather than choosing one specific graph for summarizing the data. The ultimate goal is to facilitate our efforts to make sense of the survey or experimental study data.

Key Terms

nominal data	class frequency
ordinal data	relative frequency
discrete data	frequency histogram
continuous data	probability
pie chart	exploratory data analysis (EDA)
bar chart	stem-and-leaf plot
frequency table	leading digit
class interval	trailing digits
range	time series

Supplementary Exercises

3.20 University officials periodically review the distribution of undergraduate majors within the colleges of the university to help determine a fair allocation of resources to the various departments within the colleges. During one such review, the accompanying data were obtained.

College	Number of Majors
Agriculture	1,500
Arts and Sciences	11,000
Business Administration	7,000
Education	2,000
Engineering	5,000

    a   Construct a pie chart for these data.

    b   Use the same data to construct a bar graph.

3.21   Because of the difficult times many basic industries have endured in recent years, financial analysts have begun monitoring the influx of foreign materials. The following data represent steel industry imports (in thousands of tons) for the years 1980 to 1990.

Year	1980	1981	1982	1983	1984	1985	1986	1987	1988	1989	1990
Import	15,491	19,898	16,663	17,061	26,171	23,650	19,650	18,982	17,675	17,050	16,493

    a   Would a pie chart be an appropriate graphical method for describing these data? Explain.

    b   Construct a bar graph for the data.

3.22   Graph the data shown here on the allocation of our food dollars to the categories listed in the table. Try a pie chart and a bar graph. Which seems better?

Where Our Food Dollars Go	Percent
Dairy products	13.4
Cereal and baked goods	12.6
Nonalcoholic beverages	8.9
Poultry and seafood	7.5
Fruit and vegetables	15.6
Meat	24.5
Other foods	17.5

3.23   The regulations of the Board of Health in a particular state specify that the fluoride level in the drinking water must not exceed 1.5 parts per million (ppm). The 25 measurements given here represent the fluoride levels for a sample of 25 days. Although fluoride levels are measured more than once per day, these data represent the early morning readings for the 25 days sampled.

.75	.86	.84	.85	.97
.94	.89	.84	.83	.89
.88	.78	.77	.76	.82
.72	.92	1.05	.94	.83
.81	.85	.97	.93	.79

    a   Determine the range of the measurements.

    b   Dividing the range by 7 (the number of subintervals selected) and rounding, we have a class interval width of .05. Using .705 as the lower limit of the first interval, construct a frequency histogram.

    c   Compute relative frequencies for each class interval, and construct a relative frequency histogram. Notice that the frequency and relative frequency histograms for these data have the same shape.

    d   If one of these 25 days were selected at random, what is the chance (probability) that the fluoride reading would be greater than .90 ppm? Guess (predict) what proportion of days in the coming year will have a fluoride reading of greater than .90 ppm.

3.24   The National Highway Traffic Safety Administration has studied the use of rear-seat automobile lap and shoulder seat belts. The numbers of lives potentially saved through the use of lap and shoulder seat belts are shown for various percentages of use.

Percentage of Use	Lives Saved Wearing	
	Lap Belt Only	Lap and Shoulder Belt
100	529	678
80	423	543
60	318	407
40	212	271
20	106	136
10	85	108

Suggest several different ways to graph these data. Which one seems most appropriate, and why?

3.25 Construct a frequency histogram for the data of Table 3.4. Compare the histogram to the stem-and-leaf plot of Figure 3.11. Which is more informative?

3.26 Construct a relative frequency histogram for the data in the accompanying table.

Per-Capita Public Welfare Expenses, by Number of States

Dollars	Number of States
50–74	3
75–99	6
100–124	14
125–149	11
150–174	2
175–199	5
200–224	2
225–249	5
250–274	1
275–299	1 1
Total	50

3.27 Construct a frequency table with suitable class intervals for the following data.

32.3	22.8	30.5
31.3	31.3	30.0
29.4	31.1	29.4
31.6	27.6	29.7
30.4	28.8	31.6
31.2	30.7	32.5
29.8	30.3	29.2

3.28 Construct a relative frequency histogram for the data of Exercise 3.27.

3.29 Construct a frequency histogram for the following data.

2.9	3.0	4.4
0.8	2.7	1.6
3.5	3.6	1.2
1.9	3.8	2.2
2.6	3.9	1.5
2.8	4.4	0.9
2.5	4.1	2.3
4.5	3.5	2.5

3.30   Construct a stem-and-leaf diagram for the data of Exercise 3.29. Which plot seems to be more informative for these data?

3.31   Survival times (in months) are shown for patients with severe chronic left-ventricular heart failure. Construct a stem-and-leaf diagram for these data.

4	15	24	10
1	27	31	14
2	16	32	7
13	36	29	6
12	18	14	15
18	6	13	21
20	8	3	24

3.32   Use the data of Exercise 3.31 to construct a frequency histogram. Which plot (the stem-and-leaf diagram or the frequency histogram) describes these data better? Why?

3.33   Data from SAT exams are given for selected years. Plot these time series data, and interpret the data.

	Year				
Gender, Type	1967	1970	1975	1980	1990
Male, math	514	509	495	491	493
Female, math	467	465	449	443	445
Male, verbal	463	459	437	428	430
Female, verbal	468	461	431	420	420

Source: College Entrance Examination Board.

3.34   Using the data shown in the accompanying table, construct a time series plot, taking into account the unequal time points.

Year	Percentage of All Families Headed by a Single Woman
1960	5.3
1965	6.6
1970	9.1
1975	13.4
1980	16.7
1983	18.2
1985	20.0
1987	21.7
1988	22.9
1989	25.4

3.35   Using data from the table in Exercise 3.34, construct a time series plot in which the time points along the horizontal axis are evenly spaced.

3.36   Compare the time series you obtained in Exercises 3.34 and 3.37. Discuss how a time series can be misleading if the interval width between time points does not correspond to the actual length of time between data observation points.

3.37   Construct a frequency histogram plot for the telephone data in the accompanying table (telephones per 1000 people).

State	Telephones	State	Telephones
Alabama	500	Montana	540
Alaska	350	Nebraska	590
Arizona	550	Nevada	720
Arkansas	480	New Hampshire	590
California	610	New Jersey	650
Colorado	570	New Mexico	470
Connecticut	620	New York	530
Delaware	630	North Carolina	530
Florida	620	North Dakota	560
Georgia	570	Ohio	550
Hawaii	480	Oklahoma	580
Idaho	550	Oregon	560
Illinois	650	Pennsylvania	610
Indiana	580	Rhode Island	560
Iowa	570	South Carolina	510
Kansas	600	South Dakota	540
Kentucky	480	Tennessee	540
Louisiana	520	Texas	570
Maine	540	Utah	560
Maryland	610	Vermont	520
Massachusetts	570	Virginia	530
Michigan	580	Washington	570
Minnesota	560	West Virginia	450
Mississippi	470	Wisconsin	540
Missouri	570	Wyoming	580

3.38 Construct a stem-and-leaf plot for the data of Exercise 3.37. Interpret the data display.

3.39 Computer output is shown here for the data of Exercise 3.37. Compare the stem-and-leaf plot in the output to the one you constructed in Exercise 3.38.

```
MTB > name c1 'phones'
MTB > set c1
MTB > end

MTB > STEM-AND-LEAF 'PHONES'

Stem-and-leaf of phones    N  = 50
Leaf Unit = 10

     1      3 5
     1      4
     7      4 577888
    19      5 012233344444
   (21)     5 5556666677777777888899
    10      6 0111223
     3      6 55
     1      7 2

MTB > STOP
```

3.40 A supplier of high-quality audio equipment for automobiles accumulates monthly sales data on speakers and receiver/amplifier units for five years. The data (in thousands of units per month) are shown in the accompanying table. Plot the sales data. Do you see any overall trend in the data? Do there seem to be any cyclic or seasonal effects?

Year	J	F	M	A	M	J	J	A	S	O	N	D
1	101.9	93.0	93.5	93.9	104.9	94.6	105.9	116.7	128.4	118.2	107.3	108.6
2	109.0	98.4	99.1	110.7	100.2	112.1	123.8	135.8	124.8	114.1	114.9	112.9
3	115.5	104.5	105.1	105.4	117.5	106.4	118.6	130.9	143.7	132.2	120.8	121.3
4	122.0	110.4	110.8	111.2	124.4	112.4	124.9	138.0	151.5	139.5	127.7	128.0
5	128.1	115.8	116.0	117.2	130.7	117.5	131.8	145.5	159.3	146.5	134.0	134.2

3.41 A machine-tool firm that produces a variety of products for manufacturers has quarterly records of total activity for the previous 8 years. The data reflect activity rather than price, so inflation is irrelevant. The data are shown in the following table.

	Quarter			
Year	1	2	3	4
1	97.2	100.2	102.8	102.6
2	106.1	107.8	110.5	110.6
3	116.5	117.3	119.9	119.3
4	126.1	125.7	128.3	132.1
5	133.2	133.8	141.1	142.1
6	144.2	146.1	151.6	154.0
7	155.8	158.6	165.8	167.0
8	171.1	172.6	176.5	179.7

a Plot the data against time (quarters 1–32).

b Does there appear to be a trend? If so, what form of trend might be appropriate?

c Can you detect cyclic or seasonal variation? Explain.

3.42 An investigator was interested in studying the sedative effect on rats of different doses of a drug. A small cage was constructed, with several electric eyes focused at different angles. Attached to each electric eye was a counter that monitored the number of times a rat broke any of the light beams in a 15-minute period. Twenty-five rats were injected with a specified dose of the drug, and each one was observed in the cage for a 15-minute period. These data are recorded in the accompanying table. Construct a frequency histogram for these data, using five or more class intervals.

Rat	Number of Times a Light Beam Was Broken in the 15-minute Period	Rat	Number of Times a Light Beam Was Broken in the 15-minute Period
1	107	14	128
2	99	15	106
3	171	16	177
4	116	17	144
5	101	18	102
6	109	19	196
7	199	20	191
8	142	21	169
9	118	22	182
10	173	23	148
11	155	24	130
12	184	25	159
13	132		

3.43   A questionnaire circulated last year asked 94 economic forecasters to estimate the probability of an increase in the gross national product from the third quarter to the fourth quarter of the year. The results of the survey are summarized in the accompanying table. Construct a relative frequency histogram for these results.

Estimated Probability of an Increase in GNP	Frequency
.01–.10	38
.11–.20	18
.21–.30	15
.31–.40	15
.41–.50	8

3.44   A study was conducted among smokers to examine symptoms of breathlessness and wheeze. A total of 1827 subjects who exhibited breathlessness and wheeze were classified by age. Use the data in the accompanying table to construct a relative frequency histogram.

Age	Frequency	Age	Frequency
20–24	9	45–49	269
25–29	23	50–54	404
30–34	54	55–59	406
35–39	121	60–64	372
40–44	169		

3.45   Summarize the data from Exercise 3.44 into the following age categories: 20–29, 30–39, 40–49, and 50 and over. Construct a pie chart depicting the percentage of subjects in each of these categories.

3.46   The accompanying table gives the age at inauguration and at death for 35 American presidents.

President	Age at Inauguration	Age at Death	President	Age at Inauguration	Age At Death
Washington	57	67	Hayes	54	70
J. Adams	61	90	Garfield	49	49
Jefferson	57	83	Arthur	50	56
Madison	57	85	Cleveland	47	71
Monroe	58	73	B. Harrison	55	67
J. Q. Adams	57	80	McKinley	54	58
Jackson	61	78	T. Roosevelt	42	60
Van Buren	54	79	Taft	51	72
W. H. Harrison	68	68	Wilson	56	67
Tyler	51	71	Harding	55	57
Polk	49	53	Coolidge	51	60
Taylor	64	65	Hoover	54	90
Fillmore	50	74	F. Roosevelt	51	63
Pierce	48	64	Truman	60	88
Buchanan	65	77	Eisenhower	62	78
Lincoln	52	56	Kennedy	43	46
A. Johnson	56	66	L. Johnson	55	64
Grant	46	63			

a   Construct a frequency table for the age-at-inauguration data.

b   Use a frequency histogram to graph the age-at-inauguration data.

3.47   Refer to Exercise 3.46. Use the following computer-generated stem-and-leaf plot to describe the age-at-death data.

```
MTB > SET INTO C1
DATA>   67    90    83    85    73    80    78
DATA>   79    68    71    53    65    74    64
DATA>   77    56    66    63    70    49    56
DATA>   71    67    58    60    72    67    57
DATA>   60    90    63    88    78    46    64

MTB > PRINT C1

C1
    67      90      83      85      73      80      78      79      68      71      53      65      74
    64      77      56      66      63      70      49      56      71      67      58      60      72
    67      57      60      90      63      88      78      46      64

MTB > STEM-AND-LEAF C1

Stem-and-leaf of C1       N  = 35
Leaf Unit = 1.0

     2      4 69
     3      5 3
     7      5 6678
    13      6 003344
    (6)     6 567778
    16      7 011234
    10      7 7889
     6      8 03
     4      8 58
     2      9 00

MTB > STOP
```

    3.48   Refer to Exercise 3.46. MINITAB was used to compute the years lived after inauguration for each president and to graph these data using a stem-and-leaf plot. Describe the distribution of years lived following inauguration.

```
MTB > READ INTO C1 C2          MTB > LET C3 = C2 - C1
DATA>    57     67
DATA>    61     90             MTB > PRINT C1 C2 C3
DATA>    57     83
DATA>    57     85             ROW        C1     C2     C3
DATA>    58     73
DATA>    57     80              1         57     67     10
DATA>    61     78              2         61     90     29
DATA>    54     79              3         57     83     26
DATA>    68     68              4         57     85     28
DATA>    51     71              5         58     73     15
DATA>    49     53              6         57     80     23
DATA>    64     65              7         61     78     17
DATA>    50     74              8         54     79     25
DATA>    48     64              9         68     68      0
DATA>    65     77             10         51     71     20
DATA>    52     56             11         49     53      4
DATA>    56     66             12         64     65      1
DATA>    46     63             13         50     74     24
DATA>    54     70             14         48     64     16
DATA>    49     49             15         65     77     12
DATA>    50     56             16         52     56      4
DATA>    47     71             17         56     66     10
DATA>    55     67             18         46     63     17
DATA>    54     58             19         54     70     16
DATA>    42     60             20         49     49      0
DATA>    51     72             21         50     56      6
DATA>    56     67             22         47     71     24
DATA>    55     57             23         55     67     12
DATA>    51     60             24         54     58      4
DATA>    54     90             25         42     60     18
DATA>    51     63             26         51     72     21
DATA>    60     88             27         56     67     11
DATA>    62     78             28         55     57      2
DATA>    43     46             29         51     60      9
DATA>    55     64             30         54     90     36
DATA> END                     31         51     63     12
                              32         60     88     28
                              33         62     78     16
                              34         43     46      3
                              35         55     64      9
```

```
MTB > STEM-AND-LEAF C3

Stem-and-leaf of C3          N  = 35
Leaf Unit = 1.0

     8     0 00123444
    11     0 699
    17     1 001222
   (7)     1 5666778
    11     2 01344
     6     2 56889
     1     3
     1     3 6


MTB > STOP
```

3.49   Average salaries for classroom teachers were compiled for school districts in Hamilton County, Ohio. Determine the graphical technique that you think will best describe the data.

School District	Average Salary ($1000s)
Cincinnati	21.6
Deer Park	18.6
Forest Hills	22.5
Greenhills	22.9
Indian Hills	24.9
Loveland	20.0
Madeira	22.5
Mariemont	19.9
Princeton	25.2
Sycamore	24.7
Wyoming	23.5

3.50   After the merger of two pharmaceutical companies, one of the two main offices was closed and the affected employees were offered an opportunity to relocate to the other office, 1000 miles away. The decisions employees reached are summarized here.

Decision	Number
Accept company offer to relocate	160
Reject company offer to relocate because:	
• Better job offer elsewhere	110
• Spouse's job precludes relocation; will seek other employment	70
• Retirement	42
• Other	28

a   What kind of data were collected and tabulated here?

b   Calculate the percentages for each category of response, and use an appropriate graph to summarize the data.

c   Summarize (in plain English) what the affected employees' response to the relocation offer was.

Experiences with Real Data

3.51   Refer to the Clinical Trials database on your data disk (or in Appendix 1).

The data presented here are drawn from a clinical trial that was conducted to compare the safety and efficacy of three different active compounds (A, B, and C) and a placebo (D) in the treatment of patients who exhibited characteristic signs and symptoms of depression. Certain predrug (baseline) determinations were made on each of the 40 patients to determine their suitability for the study. Each patient who qualified for the study was assigned at random to one of the four treatment groups and was dispensed medication for the duration of the study. Neither the investigator nor the patient knew which medication had been assigned.

At the end of the study, scores on numerous anxiety and depression scales were made.

a   Use the data from this study to form separate stem-and-leaf plots of the HAM-D anxiety scores for the four treatment groups. Discuss your findings.

b   Do the same for the HAM-D retardation, sleep disturbance, and total scores.

    c   Based on your plots in parts a and b, are the results for the four treatment groups comparable at the end of the study? Why might they be different?

3.52   Refer to the Clinical Trials database discussed in Exercise 3.51.

    a   Suggest some graphical technique for comparing the data on daily consumption of tobacco for the four treatment groups prior to their receiving the assigned medication. Graph the data and discuss your findings.

    b   Refer to part a. Do the same for the history of alcohol use prior to medication. Graph the data and discuss your findings.

    c   Why is it important to have comparable groups before starting the study?

3.53   Refer to the Insurance Claims database on your data disk (or in Appendix 1). A large insurance company conducted a study prior to reaching a decision about whether to expand its car insurance sales effort at its midwest regional office. The damage claims (in $1000s) for all automobile insurance policies recorded at the regional office over the past year are shown here.

    a   Construct a stem-and-leaf plot for these data.

    b   Assume that you work for the insurance company. Describe the summarized data to your boss.

4

Numerical Methods For Summarizing Data

4.1 Introduction

So far we have discussed how the process of making sense of data involves four steps: gathering data, summarizing data, analyzing data, and communicating results. The first step, *gathering data*, was presented in Chapter 2. The second step, *summarizing data*, was introduced in Chapter 3 as it related to using graphical methods. This chapter expands the discussion of summarizing data to cover numerical methods.

Measures of central tendency (modes, median, and averages) and their definitions, interpretations, and applications are presented in Section 4.2. Measures of variability, their calculation, and their interpretation are discussed in Section 4.3. An additional graphical method that incorporates measures of central tendency and variability is presented in Section 4.4, while Section 4.5 introduces the subject of summarizing data from more than one variable. Finally Section 4.6 emphasizes how computers and computer software can help in summarizing data.

4.2 Measures of Central Tendency

mode

The first measure of central tendency we consider is the **mode**.

DEFINITION 4.1

Mode   The **mode** of a set of measurements is the measurement that occurs most often (with the highest frequency).   ▪

We illustrate how to determine the mode in an example.

EXAMPLE 4.1

Slaughter weights (in pounds) for a sample of 15 Herefords, each with a frame size of 3 (on a 1–7 scale), are shown here.

962	1005	1033
980	965	1030
975	989	955
1015	1000	970
1042	1005	955

Determine the modal slaughter weight.

Solution

For these data, the weight 1005 occurs twice, and all the others occur once. Hence, the mode is 1005.   ▪

Identifying the mode in Example 4.1 was quite easy because we were able to count the number of times each measurement occurred. When dealing with grouped data—data presented in the form of a frequency table—we can define the modal interval to be the class interval with the highest frequency. However, since we do not know the actual measurements but only how many measurements fall into each interval, we identify the mode as the midpoint of the modal interval; it is thus an approximation to the mode of the actual sample measurements.

The mode is also commonly used as a measure of popularity reflecting central tendency or opinion. For example, we might talk about the most preferred stock, a most preferred model of washing machine, or the most popular candidate for political office. In each case, we would be referring to the mode of the distribution.

Some distributions have more than one measurement that occurs with the highest frequency. Thus, we sometimes encounter bimodal, trimodal, and so on, distributions.

median

The second measure of central tendency we consider is the **median**.

DEFINITION 4.2

Median   The **median** of a set of measurements is the middle value when the measurements are arranged from lowest to highest.   ▪

The median is most often used to measure the midpoint of a large set of measurements. For example, we may read about the median wage increase won by union members, the median age of persons receiving social security benefits, and the median weight of cattle prior to slaughter during a given month. Each of these situations involves a large set of measurements, and the median reflects the central value of the data.

However, we may use the definition of median for small sets of measurements by using the following convention. The median for an even number of measurements is the average of the two middle values when the measurements are arranged from lowest to highest. When there is an odd number of measurements, the median is still the middle value. Thus, whether there is an even or odd number of measurements, there is an equal number of measurements above and below the median.

EXAMPLE   4.2   Each of 10 children in the second grade was given a reading aptitude test. The scores were as follows:

95   86   78   90   62   73   89   92   84   76

Determine the median test score.

Solution   We must first arrange the scores in order of magnitude:

62   73   76   78   84   86   89   90   92   95

Since the number of measurements is even, the median is the average of the two midpoint scores.

$$\text{Median} = \frac{84 + 86}{2} = 85.$$

EXAMPLE   4.3   An experiment was conducted to measure the effectiveness of a new procedure for pruning grapes. Each of 13 workers was assigned the task of pruning grapes. The following productivity, measured in worker-hours/acre, was recorded for each person:

4.4   4.9   4.2   4.4   4.8   4.9   4.8   4.5   4.3   4.8   4.7   4.4   4.2

Determine the mode and the median productivity for the group.

Solution   First arrange the measurements in order of magnitude:

4.2   4.2   4.3   4.4   4.4   4.4   4.5   4.7   4.8   4.8   4.8   4.9   4.9

In these data, two measurements appear three times each. Hence, the data are bimodal, with modes of 4.4 and 4.8. The median for the odd number of measurements is the middle score, 4.5.

grouped data median

The **median for grouped data** is slightly more difficult to compute. Since the actual values of the measurements are unknown, we can only say for certain that the median occurs in a particular class interval; we do not know where to locate the median within the interval. If we assume that the measurements are spread evenly throughout the interval, we get the following result. Let

L = Lower class limit of the interval that contains the median

n = Total frequency

cf_b = Sum of frequencies (cumulative frequency) for all classes before the median class

f_m = Frequency of the class interval containing the median

w = Interval width

Then, for grouped data,

$$\text{Median} = L + \frac{w}{f_m}(.5n - cf_b).$$

The next example illustrates how to find the median for grouped data.

E X A M P L E   4.4   Table 4.1 is the frequency table for the chick data of Table 3.3. Compute the median weight gain for these data.

T A B L E 4.1
Frequency Table for the Chick
Data of Table 3.3

Class Interval	f_i	Cumulative f_i	f_i/n	Cumulative f_i/n
3.55–3.65	1	1	.01	.01
3.65–3.75	1	2	.01	.02
3.75–3.85	6	8	.06	.08
3.85–3.95	6	14	.06	.14
3.95–4.05	10	24	.10	.24
4.05–4.15	10	34	.10	.34
4.15–4.25	13	47	.13	.47
4.25–4.35	11	58	.11	.58
4.35–4.45	13	71	.13	.71
4.45–4.55	7	78	.07	.78
4.55–4.65	6	84	.06	.84
4.65–4.75	7	91	.07	.91
4.75–4.85	5	96	.05	.96
4.85–4.95	4	100	.04	1.00
Totals	$n = 100$		1.00	

Solution   Let the cumulative relative frequency for class j equal the sum of the relative frequencies for class 1 through class j. To determine the interval that contains the median, we must find the first interval for which the cumulative relative frequency exceeds .50. This interval is the one that contains the median. For these data, the interval from 4.25 to 4.35 is the first interval for which the cumulative relative frequency exceeds .50, as shown in Table 4.1, column 5. So this interval contains the median. Then

$$L = 4.25, \quad f_m = 11, \quad n = 100, \quad w = .1, \quad cf_b = 47,$$

and

$$\text{Median} = L + \frac{w}{f_m}(.5n - cf_b) = 4.25 + \frac{.1}{11}(50 - 47) = 4.28. \quad \blacksquare$$

The third measure of central tendency we will discuss in this text is the arithmetic mean, known simply as the **mean**.

mean

DEFINITION   4.3   **Arithmetic Mean (Mean)**   The **arithmetic mean**, or **mean**, of a set of measurements is the sum of the measurements divided by the total number of measurements. $\quad \blacksquare$

μ
$\bar{y}$

When people talk about an "average," they quite often are referring to the mean. Because of the important role that the mean plays in statistical inference (discussed in later chapters), we give special symbols to the population mean and to the sample mean. The *population mean* is denoted by the Greek letter μ (read "mu"), and the *sample mean* is denoted by the symbol $\bar{y}$ (read "y-bar"). As indicated in Chapter 1, a population of measurements is the complete set of measurements of interest to us; a sample of measurements is a subset of measurements selected from the population of interest. If we let $y_1, y_2, \ldots, y_n$ denote the measurements observed in a sample of size n, then the sample mean $\bar{y}$ can be written as

$$\bar{y} = \frac{\sum_i y_i}{n},$$

where the symbol appearing in the numerator, $\sum_i y_i$, is the notation used to designate a sum of n measurements, y_i:

$$\sum_i y_i = y_1 + y_2 + \cdots + y_n.$$

The corresponding population mean is μ.

In most situations, we will not know the population means; consequently the sample mean will be used to make inferences about the corresponding (but unknown) population mean. Details about how this is done and what inferences we can make will be discussed in Chapter 7.

EXAMPLE   4.5   A sample of $n = 15$ overdue accounts in a large department store yields the following amounts due:

$55.20	$ 4.88	$271.95
18.06	180.29	365.29
28.16	399.11	807.80
44.14	97.47	9.98
61.61	56.89	82.73

a   Determine the mean amount due for the 15 accounts sampled.

b Assuming that there are a total of 150 overdue accounts, use the sample mean to predict the total amount overdue for all 150 accounts.

Solution   **a** The sample mean is computed as follows:

$$\bar{y} = \frac{\sum_i y_i}{15} = \frac{55.20 + 18.06 + \cdots + 82.73}{15} = \frac{2483.56}{15} = \$165.57.$$

b In part a we found that the 15 accounts sampled averaged $165.57 overdue. Using this information, we would predict, or estimate, the total amount overdue for the 150 accounts to be $150(165.57) = \$24,835.50.$ ▪

grouped data mean   The formula for the sample **mean for grouped data** is only slightly more complicated than the formula just presented for ungrouped data. Usually, we do not know the individual sample measurements, but only the interval to which a measurement is assigned, so this formula will be an approximation to the actual sample mean. Hence, when the sample measurements are known, the formula for ungrouped data should be used. We will use the same symbol $\bar{y}$ to designate the sample mean for grouped data. If there are k class intervals and

$$y_i = \text{Midpoint of the } i\text{th class interval}$$
$$f_i = \text{Frequency associated with the } i\text{th class interval}$$
$$n = \text{Total number of measurements}$$

then

$$\bar{y} = \frac{\sum_i f_i y_i}{n}.$$

E X A M P L E   4.6   The data of Example 4.4 are reproduced in Table 4.2, along with two additional columns, y_i and $f_i y_i$, that will be helpful in computing the mean. Compute the sample mean for this set of grouped data.

Solution   Adding the entries in the $f_i y_i$ column and substituting this value into the formula, we find the sample mean to be

$$\bar{y} = \frac{\sum_i f_i y_i}{100} = \frac{429.2}{100} = 4.29.$$ ▪

The mean is a useful measure of the central value of a set of measurements, but it is subject to distortion due to the presence of one or more extreme values in the set. **outliers** In these situations, the extreme values (called **outliers**) pull the mean in the direction of the outliers, thus distorting the mean as a measure of central value. A variation of

TABLE 4.2
Chick Data for Example 4.6

Class Interval	f_i	y_i	$f_i y_i$
3.55–3.65	1	3.6	3.6
3.65–3.75	1	3.7	3.7
3.75–3.85	6	3.8	22.8
3.85–3.95	6	3.9	23.4
3.95–4.05	10	4.0	40.0
4.05–4.15	10	4.1	41.0
4.15–4.25	13	4.2	54.6
4.25–4.35	11	4.3	47.3
4.35–4.45	13	4.4	57.2
4.45–4.55	7	4.5	31.5
4.55–4.65	6	4.6	27.6
4.65–4.75	7	4.7	32.9
4.75–4.85	5	4.8	24.0
4.85–4.95	4	4.9	19.6
Totals	100		429.2

trimmed mean

the mean, called a **trimmed mean**, drops the highest and lowest extreme values and averages the rest. For example, a 20% trimmed mean drops the highest 20% and the lowest 20% of the measurements and averages the rest. Similarly, a 10% trimmed mean drops the highest and the lowest 10% of the measurements and averages the rest. By trimming the data, we are able to reduce the impact of very large (or very small) values on the mean and thus get a more reliable measure of the central value of the set. This becomes particularly important when the sample mean is used to predict the corresponding population central value.

skewness

In this section, we have discussed the mode, the median, the mean, and the trimmed mean. How are these measures of central tendency related for a given set of measurements? The answer depends on the **skewness** of the data. If the distribution is mound-shaped and symmetrical about a single peak, the mode (M_o), median (M_d), mean (μ), and trimmed mean (*TM*) will all be the same. This is shown by using a smooth curve and population quantities in Figure 4.1(a). If the distribution is skewed, so that it has a long tail in one direction and a single peak, the mean is pulled in the direction of the tail; the median falls between the mode and the mean; and (depending on the degree of trimming) the trimmed mean usually falls between the median and the mean. Figure 4.1(b) and (c) illustrates this for distributions skewed to the left and to the right.

The important thing to remember is that we are not restricted to using only one measure of central tendency. For some data sets, it will be necessary to use more than one of these measures to provide an accurate descriptive summary of central tendency for the data.

FIGURE 4.1   Relation Among the Mean μ, the Trimmed Mean TM, the Median M_d, and the Mode M_o

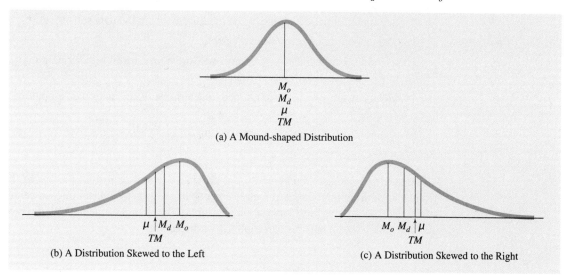

(a) A Mound-shaped Distribution

(b) A Distribution Skewed to the Left

(c) A Distribution Skewed to the Right

Major Characteristics of Each Measure of Central Tendency

Mode

1 It is the most frequent or probable measurement in the data set.

2 There can be more than one mode for a data set.

3 It is not influenced by extreme measurements.

4 Modes of subsets cannot be combined to determine the mode of the complete data set.

5 For grouped data, its values can change depending on the categories used.

6 It is applicable to both qualitative and quantitative data.

Median

1 It is the central value; 50% of the measurements in the data set lie above it, and 50% fall below it.

2 There is only one median for a data set.

3 It is not influenced by extreme measurements.

4 Medians of subsets cannot be combined to determine the median of the complete data set.

5 For grouped data, its value is rather stable, even when the data are organized into different categories.

6 It is applicable to quantitative data only.

Mean

1   It is the arithmetic average of the measurements in the data set.

2   There is only one mean for a data set.

3   Its value is influenced by extreme measurements; trimming can help to reduce the degree of outlier influence.

4   Means of subsets can be combined to determine the mean of the complete data set.

5   It is applicable to quantitative data only.

Measures of central tendency do not provide a complete mental picture of the frequency distribution of a set of measurements. In addition to determining the center of the distribution, we must have some measure of the spread of the data. In the next section, we discuss measures of variability or dispersion.

Exercises

Basic Techniques

4.1   Compute the mean, the median, and the mode for the following data:

11	17	18	10	22	23	15	17
14	13	10	12	18	18	11	14

4.2   Refer to the data in Exercise 4.1, but with the measurements 22 and 23 replaced by 42 and 43. Recompute the mean, the median, and the mode. Discuss the impact of these extreme measurements on the three measures of central tendency.

4.3   Refer to Exercises 4.1 and 4.2. Compute a 10% trimmed mean for both data sets. Do the extreme values affect the 10% trimmed mean? Would a 5% trimmed mean be affected?

4.4   Determine the mode, the median, and the mean for the following measurements:

10	2	1	5
1	5	7	10
3	4	8	12
5	6	8	9

4.5   Determine the mean, the median, and the mode for the data presented in the following frequency table:

Class Interval	Frequency
0–2	1
3–5	3
6–8	5
9–11	4
12–14	2

Applications

4.6   Salaries for 40 recent M.B.A. graduates from a major university are summarized in the accompanying frequency table (in thousands of dollars). Determine the mode, the median, and the mean for the data shown. What do the relationships among the three measures indicate about the shape of the histogram for these data?

Interval	Frequency
24.9–29.9	6
29.9–24.9	10
34.9–39.9	15
39.9–44.9	7
44.9–49.9	2

4.7   Exercise capacity (in seconds) was determined for each of 11 patients who were being treated for chronic heart failure. Determine the median and the mean.

906	1320
711	1170
684	1200
837	1056
897	882
1008	

4.8   Daily crude oil output (in millions of barrels) is shown in the accompanying table for the years 1971 to 1990. Compute the mean and the median daily output for these years.

Year	Output
1971	9.45
1972	9.40
1973	9.25
1974	8.75
1975	8.30
1976	8.10
1977	8.25
1978	8.70
1979	8.55
1980	8.60
1981	8.55
1982	8.65
1983	8.70
1984	8.70
1985	8.91
1986	8.60
1987	8.20
1988	7.70
1989	7.20
1990	6.75

4.9   Given the frequency distribution contained in the data in the accompanying table, what is the class interval width?

Normal Daily Mean
Temperatures, Annual Average

Temperature	Frequency
39–41	3
42–44	2
45–47	8
48–50	10
51–53	9
54–56	10
57–59	8
60–62	7
63–65	3
66–68	3
69–71	2
72–74	0
75–77	2
Total	67

Determine the mode, the median, and the mean for these data. Would a trimmed mean better describe the center of the distribution than the mean does? Explain.

4.10 Nitrogen is a limiting factor in the yield of many different plants. In particular, the yield of apple trees is directly related to the nitrogen content of apple tree leaves, and this content must be carefully monitored to protect the trees in an orchard. Research has shown that the nitrogen content should be approximately 2.5% for best yield results. (Some researchers report their results in parts per million (ppm); hence, 1% would be equivalent to 10,000 ppm.)

To determine the nitrogen content of trees in an orchard, the growing tips of 150 leaves are clipped from trees throughout the orchard. These leaf tissues are ground to form one composite sample, which the researcher assays for percentage of nitrogen. Composite samples obtained from a random sample of 36 orchards throughout the state give the following percentages for nitrogen content:

2.0968	2.8220	2.1739	1.9928	2.2194	3.0926
2.4685	2.5198	2.7983	2.0961	2.9216	2.1997
1.7486	2.7741	2.8241	2.6691	3.0521	2.9263
2.9367	1.9762	2.3821	2.6456	2.7678	1.8488
1.6850	2.7043	2.6814	2.0596	2.3597	2.2783
2.7507	2.4259	2.3936	2.5464	1.8049	1.9629

a Round each of these measurements to the nearest hundredth. (Use the convention that 5 is rounded up.)

b Determine the sample mode for the rounded data.

c Determine the sample median for the rounded data.

d Determine the sample mean for the rounded data.

4.11 Refer to the data of Exercise 4.10, rounded to the nearest hundredth. Replace the fourth measurement (2.94) with the value 29.40. Compute the sample mean, median, and mode for these data. Compare these results to those you found in Exercise 4.10.

4.12 Refer to the data of Example 4.5. Since the sample mean is greater than 10 of the 15 observations, suggest a more appropriate measure of central tendency. Compute its value. How does the distribution of amounts for overdue accounts appear to be skewed?

4.13 The effective tax rate (per $100) on residential property for three groups of large cities, ranked by residential property tax rate, is shown here.

Group 1	Rate	Group 2	Rate	Group 3	Rate
Detroit, MI	4.10	Burlington, VT	1.76	Little Rock, AR	1.02
Milwaukee, WI	3.69	Manchester, NH	1.71	Albuquerque, NM	1.01
Newark, NJ	3.20	Fargo, ND	1.62	Denver, CO	.94
Portland, OR	3.10	Portland, ME	1.57	Las Vegas, NV	.88
Des Moines, IA	2.97	Indianapolis, IN	1.57	Oklahoma City, OK	.81
Baltimore, MD	2.64	Wilmington, DE	1.56	Casper, WY	.70
Sioux Falls, IA	2.47	Bridgeport, CT	1.55	Birmingham, AL	.70
Providence, RI	2.39	Chicago, IL	1.55	Phoenix, AZ	.68
Philadelphia, PA	2.38	Houston, TX	1.53	Los Angeles, CA	.64
Omaha, NE	2.29	Atlanta, GA	1.50	Honolulu, HI	.59

Source: Government of the District of Columbia, Department of Finance and Revenue, *Tax Rates and Tax Burdens in the District of Columbia: A Nationwide Comparison* (annual).

a Compute the mean, the median, and the mode separately for the three groups.

b Compute the mean, the median, and the mode for the complete set of 30 measurements.

c What measure or measures best summarize the center of these distributions? Explain.

4.14 Refer to Exercise 4.13. Average the three group means, the three group medians, and the three group modes, and compare your results to those of part b of Exercise 4.13. Comment on your findings.

4.3 Measures of Variability

variability

The need for some measure of variability is illustrated in the relative frequency histograms of Figure 4.2. All of these histograms have the same mean, but each has a different spread, or **variability**, about the mean. For purposes of illustration, we have shown the histograms as smooth curves.

FIGURE 4.2
Relative Frequency Histograms with Different Variabilities but the Same Mean

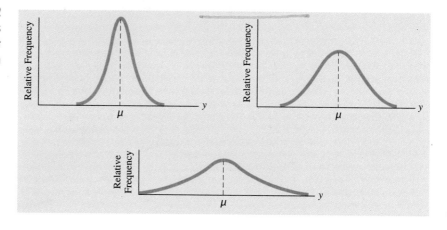

range

The simplest but least useful measure of data variation is the **range**. Recall that we alluded to the range in Section 3.2. We now present its definition.

DEFINITION   4.4

Range   The **range** of a set of measurements is the difference between the largest and the smallest measurements of the set.   ▪

EXAMPLE   4.7   Determine the range of the 15 overdue accounts of Example 4.5.

Solution   The smallest measurement is \$4.88, and the largest is \$807.80. Hence, the range is

$$807.80 - 4.88 = \$802.92. \quad \blacksquare$$

grouped data range

For **grouped data**, since we do not know the individual measurements, the **range** is taken to be the difference between the upper limit of the last interval and the lower limit of the first interval.

Although the range is easy to compute, it is sensitive to outliers, since it depends on the most extreme values. It does not give much information about the pattern of variability. In Figure 4.2, the middle and the bottom distributions have the same mean and the same range, yet they differ substantially in their variability about the mean. We seek a measure of variability that is more sensitive to the piling up of data about the mean.

percentile

A second measure of variability involves the use of **percentiles**.

DEFINITION   4.5

pth Percentile   The **pth percentile** of a set of n measurements arranged in order of magnitude is the value that has at most $p\%$ of the measurements below it and at most $(100 - p)\%$ of the measurements above it.   ▪

For example, Figure 4.3 illustrates the 60th percentile of a set of measurements.

FIGURE 4.3
The 60th Percentile of a Set of
Measurements

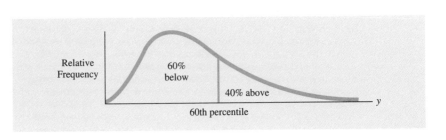

Percentiles are frequently used to describe the results of achievement test scores and the ranking of a person in comparison to all other people who took a particular examination. Specific percentiles of interest are 25th, 50th, and 75th percentiles, often called the *lower quartile*, the *middle quartile* (median), and the *upper quartile*, respectively (see Figure 4.4).

FIGURE 4.4
Quartiles of a Distribution

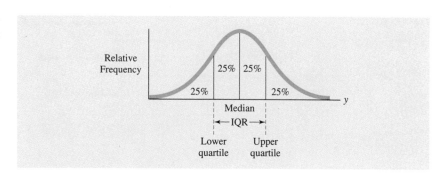

interquartile range

The second measure of variability, the **interquartile range**, can now be defined. A slightly different definition of the interquartile range is given in connection with the box plot (see Section 4.4).

DEFINITION   4.6

Interquartile Range (IQR)   The **interquartile range (IQR)** of a set of measurements is the difference between the upper and lower quartiles (see Figure 4.4). That is,

IQR = 75th percentile − 25th percentile.   ∎

The interquartile range, although more sensitive to data pileup about the midpoint than the range, is still not satisfactory for our purposes. In particular, the IQR can be used for comparing the variability of two sets of measurements; but not much useful information can be gained from the IQR for interpreting the variability of a single set of measurements.

We seek now a sensitive measure of variability, not only for comparing the variabilities of two sets of measurements, but also for interpreting the variability of

deviation

a single set of measurements. To do this, we work with the **deviation** $y - \bar{y}$ of a measurement y from the mean $\bar{y}$ of the set of measurements.

To illustrate, suppose that we have five sample measurements $y_1 = 68$, $y_2 = 67$, $y_3 = 66$, $y_4 = 63$, and $y_5 = 61$, representing the percentages of registered voters in five cities who voted at least once during the past year. These measurements are

dot diagram

shown in the **dot diagram** of Figure 4.5. Each measurement is identified by a dot above the horizontal axis of the diagram. We use the sample mean

$$\bar{y} = \frac{\sum_i y_i}{n} = \frac{325}{5} = 65$$

to locate the center of the set, and we construct horizontal lines in Figure 4.5 to represent the deviations of the sample measurements from their mean. The deviations of the measurements are computed by performing the subtraction $y_i - \bar{y}$. The five measurements and their deviations are shown in Figure 4.5.

FIGURE 4.5
Dot Diagram of the Percentages
of Registered Voters in Five Cities

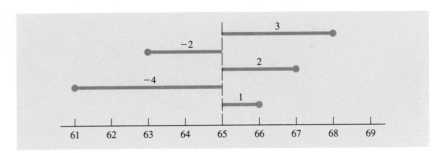

In a data set with very little variability, most measurements will be located near the center of the distribution. In a more variable set of measurements, deviations from the mean will be relatively large.

Many different measures of variability can be constructed on the basis of the deviations $y_i - \bar{y}$. A first thought would be to use the mean deviation, but this will always equal zero, as it does for our example. A more easily interpreted function of the deviations involves the sum of the squared deviations of the measurements from their mean. This measure is called the **variance**.

variance

DEFINITION   4.7

Variance   The **variance** of a set of n measurements $y_1, y_2, \ldots, y_n$ with mean $\bar{y}$ is the sum of the squared deviations divided by $n - 1$:

$$\frac{\sum_i (y_i - \bar{y})^2}{n - 1} \quad \blacksquare$$

As with sample and population means, we have special symbols to denote sample and population variances. The symbol s^2 represents the sample variance, and the corresponding population variance is denoted by the symbol σ^2. The definition of the variance of a set of measurements depends on whether the data are regarded as a sample or as a population of measurements. The definition we have given here assumes that we are working with a sample, since the population measurements usually are not available.

s^2

σ^2

standard deviation

Another useful measure of variability, the **standard deviation**, involves the square root of the variance.

DEFINITION   4.8

Standard Deviation   The **standard deviation** of a set of measurements is the positive square root of the variance. $\blacksquare$

s

σ

Thus, we have s denoting the sample standard deviation and σ denoting the corresponding population standard deviation.

EXAMPLE 4.8   The time lapse between occurrence of an electric light stimulus and the pressing of a bar to avoid a shock was noted for each of five conditioned rats. Use the following data to compute the sample variance and the sample standard deviation.

shock avoidance times (seconds): 5, 4, 3, 1, 3

Solution   The deviations and the squared deviations are shown in the accompanying table. The sample mean $\bar{y}$ is 3.2.

y_i	$y_i - \bar{y}$	$(y_i - \bar{y})^2$
5	1.8	3.24
4	.8	.64
3	-.2	.04
1	-2.2	4.84
3	-.2	.04

Totals	16	0	8.80

Using the total from the squared deviations column, we find that the sample variance is

$$s^2 = \frac{\sum_i (y_i - \bar{y})^2}{4} = \frac{8.80}{4} = 2.2. \quad \blacksquare$$

Computations of the quantities s^2 and s are sometimes simplified by using the following algebraic identity:

$$\sum_i (y_i - \bar{y})^2 = \sum_i y_i^2 - \frac{(\sum_i y_i)^2}{n}.$$

Hence, we have the shortcut formula for s^2 (and s) given next.

Shortcut Formula for s^2 and s

$$s^2 = \frac{1}{n-1} \left[\sum_i y_i^2 - \frac{(\sum_i y_i)^2}{n} \right] \quad \text{and} \quad s = \sqrt{s^2}$$

E X A M P L E   **4.9**   Use the data of Example 4.8 to compute the sample variance, using the shortcut formula.

Solution   It is convenient to construct the following table to perform the calculations.

y_i	y_i^2
5	25
4	16
3	9
1	1
3	9
Totals 16	60

Using the totals from the table, we have

$$s^2 = \frac{1}{4}\left[60 - \frac{(16)^2}{5}\right] = \frac{1}{4}[60 - 51.2] = 2.2,$$

which is exactly the result we obtained in Example 4.8.   ▪

We can make a simple modification of our shortcut formula to approximate the sample variance if only grouped data are available. Recall that, in approximating the sample mean for grouped data, we let y_i and f_i denote the midpoint and the frequency, respectively, for the ith class interval. With this notation, the sample variance for grouped data is

$$s^2 = \frac{1}{n-1}\left[\sum_i f_i y_i^2 - \frac{\left(\sum_i f_i y_i\right)^2}{n}\right] \quad \text{or} \quad \frac{\sum_i f_i (y_i - \bar{y})^2}{n-1}$$

The sample standard deviation is $\sqrt{s^2}$.

E X A M P L E   **4.10**   Refer to the chick data from Table 4.2 of Example 4.6. Calculate the sample variance and standard deviation for these data.

Solution   In addition to the calculations in Table 4.2, we also need the calculations for $f_i y_i^2$. These calculations, formed by multiplying corresponding elements in the y_i and $f_i y_i$ columns, are shown in the following listing.

$f_i y_i^2$
12.96
13.69
86.64
91.26
160.00
168.10
229.32
203.39
251.68
141.75
126.96
154.63
115.20
96.04

The sum of the $f_i y_i^2$ calculations is 1851.62. Using this total in conjunction with the total for $f_i y_i$ in Table 4.2, we can determine s^2 and s:

$$s^2 = \frac{1}{n-1}\left[\sum_i f_i y_i^2 - \frac{(\sum_i f_i y_i)^2}{n}\right]$$

$$= \frac{1}{99}\left[1851.62 - \frac{(429.2)^2}{100}\right] = \frac{9.49}{99} = .10$$

$$s = \sqrt{.10} = .32. \quad \blacksquare$$

We have now discussed several measures of variability, each of which can be used to compare the variabilities of two or more sets of measurements. The standard deviation is particularly appealing for two reasons: using it, we can compare the variabilities of *two or more* sets of data; and we can also use the results of the following rule to interpret the standard deviation of a single set of measurements. This rule applies to data sets with roughly a "mound-shaped" histogram—that is, a histogram that has a single peak, is symmetrical, and tapers off gradually in the tails. Since so many data sets can be classified as mound-shaped, the rule has wide applicability. For this reason, it is called the *Empirical Rule*.

Empirical Rule

Given a set of n measurements that possesses a mound-shaped histogram,

the interval $\bar{y} \pm s$ contains approximately 68% of the measurements;

the interval $\bar{y} \pm 2s$ contains approximately 95% of the measurements;

the interval $\bar{y} \pm 3s$ contains approximately all the measurements.

E X A M P L E   4.11   Data for a sample of 20 days throughout the previous year indicated that the average wholesale price per pound for steers at a particular stockyard was $.61, with a standard deviation of $.07. If the histogram for the measurements is mound-shaped, describe the variability of the data, using the Empirical Rule.

Solution   Applying the Empirical Rule, we see that the interval

$$.61 \pm .07 \quad \text{or} \quad \$.54 \text{ to } \$.68$$

contains approximately 68% of the measurements. Similarly, the interval

$$.61 \pm .14 \quad \text{or} \quad \$.47 \text{ to } \$.75$$

contains approximately 95% of the measurements, and the interval

$$.61 \pm .21 \quad \text{or} \quad \$.40 \text{ to } \$.82$$

contains approximately all the measurements.

In English, approximately $\frac{2}{3}$ of the steers sold for between $.54 and $.68 per pound; and 95% sold for between $.47 and $.75 per pound, with minimum and maximum prices per pound being approximately $.40 and $.82, respectively.   ▪

To increase our confidence in the Empirical Rule, let us see how well it describes the five frequency distributions of Figure 4.6. We calculated the mean and the standard deviation for each of the five data sets (not given), and the results of these calculations are shown next to each frequency distribution. Figure 4.6 (a) shows the frequency distribution for measurements made on a variable that can take values $y = 0, 1, 2, \ldots, 10$. The mean $\bar{y} = 5.50$ and the standard deviation $s = 1.49$ for this symmetric mound-shaped distribution were used to calculate the interval $\bar{y} \pm 2s$, which is marked below the horizontal axis of the graph. We found that 94% of the measurements fall in this interval—that is, lie within two standard deviations of the mean. Note that this percentage is very close to the 95% specified in the Empirical Rule. We also calculated the percentage of measurements that lie within one standard deviation of the mean. We found this percentage to be 60%, a figure not too far from the 68% specified by the Empirical Rule. Consequently, we conclude that the Empirical Rule provides an adequate description for Figure 4.6(a).

Figure 4.6(b) shows another mound-shaped frequency distribution, but one that is less sharply peaked than the distribution of Figure 4.6(a). The mean and the standard deviation for this distribution, shown to the right of the figure, are 5.50 and 2.07, respectively. The percentages of measurements that lie within one and two standard deviations of the mean are 64% and 96%, respectively. Once again, these percentages agree very well with the Empirical Rule.

Now let us look at three other distributions. The distribution in Figure 4.6(c) is perfectly flat, while the distributions in Figure 4.6(d) and (e) are nonsymmetric and skewed to the right. The percentages of measurements that lie within two standard

FIGURE 4.6
A Demonstration of the Utility of
the Empirical Rule

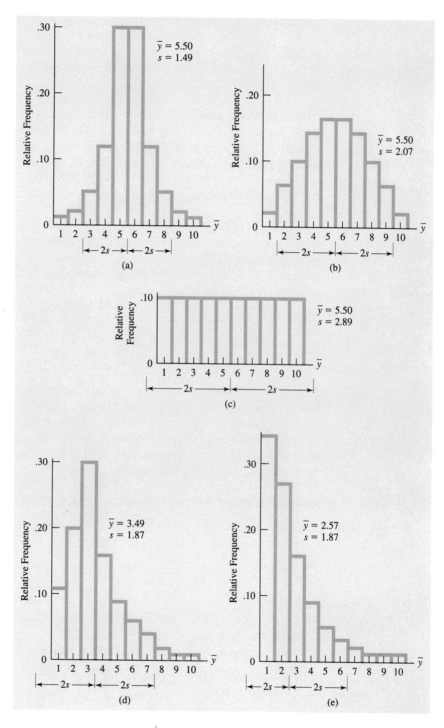

deviations of the mean are 100%, 96%, and 95%, respectively, for these three distributions. All of these percentages are reasonably close to the 95% specified by the Empirical Rule. The percentages that lie within one standard deviation of the mean (60%, 75%, and 87%, respectively) show some disagreement with the 68% of the Empirical Rule.

To summarize, you can see that the Empirical Rule accurately forecasts the percentage of measurements that fall within two standard deviations of the mean for all five distributions of Figure 4.6, even for the distributions that are flat, as in Figure 4.6(c), or highly skewed to the right, as in Figure 4.6(e). The Empirical Rule is less accurate in forecasting the percentage of measurements that fall within one standard deviation of the mean, but the forecast, 68%, compares reasonably well for the three distributions that might reasonably be called mound-shaped—Figure 4.6(a), (b), and (d).

The results of the Empirical Rule enable us to obtain a quick approximation to the sample standard deviation s. The Empirical Rule states that approximately 95% of the measurements lie in the interval $\bar{y} \pm 2s$. The length of this interval is, therefore, $4s$. Since the range of the measurements is approximately $4s$, we obtain an **approximate value for** s by dividing the range by 4.

approximate value for s

$$\text{Approximate value of } s = \frac{\text{Range}}{4}$$

You may wonder why we did not equate the range to $6s$, since the interval $\bar{y} \pm 3s$ should contain almost all the measurements. The reason is that this procedure would yield a smaller approximate value for s than the one obtained by the procedure above. If we are going to make an error (as we are bound to do with any approximation), it is better to overestimate the sample standard deviation so that we are not led to believe that there is less variability than may be the case.

E X A M P L E   4.12   The following data represent the percentages of family income allocated to groceries by each member of a sample of 30 shoppers:

26	28	30	37	33	30
29	39	49	31	38	36
33	24	34	40	29	41
40	29	35	44	32	45
35	26	42	36	37	35

For these data, $\sum y_i = 1043$ and $\sum y_i^2 = 37{,}331$.

Compute the mean, the variance, and the standard deviation of the percentage of income spent on food. Check your calculation of s.

Solution   The sample mean is

$$\bar{y} = \frac{\sum_i y_i}{30} = \frac{1043}{30} = 34.77.$$

The corresponding sample variance and standard deviation are

$$s^2 = \frac{1}{n-1}\left[\sum_i y_i^2 - \frac{(\sum_i y_i)^2}{n}\right]$$

$$= \frac{1}{29}[37,331 - 36,261.63] = \frac{1069.27}{29} = 36.87$$

$$s = \sqrt{36.87} = 6.07.$$

We can check our calculation of s by using the range approximation. The largest measurement is 49 and the smallest is 24. Hence an approximate value of s is

$$s \approx \frac{\text{Range}}{4} = \frac{49 - 24}{4} = 6.25.$$

Notice how close the approximation is to the computed value, 6.07.   ▪

Exercises

Basic Techniques

4.15   Calculate $\bar{y}$ and s for the following sample of five measurements: 3, 9, 4, 1, 3.

4.16   Given the measurements 4, 12, 9, 16, 7, 6, and 2, compute the range, the variance, and the standard deviation.

4.17   Refer to Exercise 4.16. Suppose that the last measurement is 30 rather than 2. How does this change the range, the variance, and the standard deviation?

4.18   For practice, consider a set of five measurements—say 5, 4, 1, 2, 3.

   a So that you can see the variation in the measurements, construct a dot diagram.

   b Use $\sum y$ and $\sum y^2$ to calculate $\sum(y - \bar{y})^2$.

   c Calculate s^2 and s.

   d Because the number of measurements in the sample is so small, the frequency distribution for the sample measurements is not mound-shaped. Nevertheless, notice that the interval $(\bar{y} \pm 2s)$ contains all the measurements. (Construct this interval on the dot diagram for the data so that you can see the location of the points within the interval.)

4.19   Repeat the instructions of Exercise 4.18 for the six measurements 1, 0, 3, 1, 2, 2.

4.20   Repeat the instructions of Exercise 4.18 for the ten measurements 4, 1, 3, 5, 2, 3, 1, 4, 0, 2.

Applications

4.21   The test scores reported for a nationally administered college achievement test have a mound-shaped distribution with a mean of 520 and a standard deviation of 110.

   a Use the Empirical Rule to assist you in describing this distribution of scores.

   b The median for the scores is reported to be 510, and the 98th percentile is 795. Interpret these two statistics.

   c The range of scores is 702. Interpret this statistic.

4.22 Students in a chemistry class were assigned the task of determining the purity of a chemical substance. Three hundred class members independently analyzed the substance. The mean and the standard deviation for the $n = 300$ measurements were 78.1% and 1.3%, respectively. Joe Smith missed his laboratory class but analyzed the substance the next day. His analysis gave a purity reading of 83.4%. Is there reason to suspect his analytical results? Why?

4.23 In the use of an oxygen tent it is very important that the actual percentage of oxygen generated at a particular time be close to the amount specified by a physician and the amount indicated on the tent's oxygen control valve. In an investigation of a particular manufacturer's oxygen tents, 50 tents were selected and their control valves were adjusted to the same oxygen input setting. Then the atmosphere within each tent was sampled, and the difference between the actual percentage of oxygen and the valve setting was recorded for each. If the mean and the standard deviation for the sample were 1.3% and .6% respectively, describe the distribution of the 50 readings.

4.24 The treatment times (in minutes) for patients at a health clinic are as follows:

21	20	31	24	15	21	24	18	33	8
26	17	27	29	24	14	29	41	15	11
13	28	22	16	12	15	11	16	18	17
29	16	24	21	19	7	16	12	45	24
21	12	10	13	20	35	32	22	12	10

Use the shortcut formula to calculate s^2 and s. You can verify that $\sum y = 1016$ and $\sum y^2 = 24,080$ for the 50 treatment times.

4.25 Refer to Exercise 4.24. To increase your confidence in the applicability of the Empirical Rule, construct the intervals $(\bar{y} \pm s)$, $(\bar{y} \pm 2s)$, and $(\bar{y} \pm 3s)$, and count the number of treatment times falling into each of the three intervals. From these frequencies, calculate the corresponding percentage of measurements falling into the three intervals. Does the Empirical Rule give a reasonable approximation to the relative frequencies you have observed?

4.26 To assist in estimating the amount of lumber in a tract of timber, an owner decided to count the number of trees whose diameters exceeded 12 inches in randomly selected 50 × 50-foot squares. Seventy 50 × 50-foot squares were randomly selected from the tract, and the number of trees with diameters in excess of 12 inches were counted for each. The data gathered were as follows:

7	8	6	4	9	11	9	9	9	10
9	8	11	5	8	5	8	8	7	8
3	5	8	7	10	7	8	9	8	11
10	8	9	8	9	9	7	8	13	8
9	6	7	9	9	7	9	5	6	5
6	9	8	8	4	4	7	7	8	9
10	2	7	10	8	10	6	7	7	8

a Construct a relative frequency histogram to describe these data.

b Calculate the sample mean $\bar{y}$ as an estimate of μ, the mean number of timber trees with diameters exceeding 12 inches for all 50 × 50-foot squares in the tract.

c Calculate s for the data. Construct the intervals $(\bar{y} \pm s)$, $(\bar{y} \pm 2s)$, and $(\bar{y} \pm 3s)$. Count the percentages of sample squares falling into each of the three intervals, and compare these percentages with the corresponding percentages given by the Empirical Rule.

4.4 The Box Plot

As mentioned in Chapter 3, a stem-and-leaf plot provides a graphical representation of a set of scores that can be used to examine the shape of the distribution, the range

box plot

of scores, and places where the scores are concentrated. The **box plot**, which builds on the information displayed in a stem-and-leaf plot, is more concerned with the symmetry of the distribution; it incorporates numerical measures of central tendency and location in order to permit users to study the variability of the scores and the concentration of scores in the tails of the distribution.

Before we show how to construct and interpret a box plot, we must introduce several new terms that are peculiar to the language of exploratory data analysis (EDA). You are familiar with the definitions for the first, second (median), and third quartiles of a distribution presented earlier in this chapter. The box plot uses the median and

hinges

the **hinges** of a distribution. Hinges are very similar to the quartiles of a distribution, but owing to the method by which they are computed for sample data, the lower and upper hinges of a distribution may differ very slightly from the first and third quartiles of a set of scores.

Having said this, and recognizing the slight distinction, we will compute hinges in this text but will refer to them as the lower and upper quartiles of the sample data.

We can now illustrate a *skeletal box plot* by way of an example.

E X A M P L E   **4.13**   Use the stem-and-leaf plot in Figure 4.7 (summarizing the 90 violent crime rates of Table 3.4) to construct a skeletal box plot.

Solution   When the scores are ordered from lowest to highest, the median score and the quartile scores are located as follows:

$$\text{Median location} = \frac{n+1}{2}$$

$$\text{Quartile location} = \frac{\textit{truncated median location} + 1}{2},$$

where the truncated median location is simply the median location with the decimal .5 omitted where present. For the distribution of $n = 90$ violent crime rates, we have

$$\text{Median location} = \frac{90+1}{2} = 45.5$$

$$\text{Truncated median location} = 45$$

$$\text{Quartile location} = \frac{45+1}{2} = 23.$$

F I G U R E   **4.7**
Stem-and-Leaf Plot

1	89
2	10 24 67 91 96 98
3	36 41 52 54 74 76 88 93 93
4	10 21 35 47 48 60 64 66 80 81 91 96 98
5	04 05 08 16 26 29 37 57 59 61 62 62 62 63 70 71 78 85 92
6	05 05 24 26 28 31 39 42 47 61 73 84 85 85 90 98
7	03 06 18 19 20 31 35 39 51 58 71
8	04 07 09 11 14 17 43 56 68 76 77 85
9	28 71
10	20

Since the median location is score 45.5 in the distribution, we average the 45th and 46th scores to compute the median. For these data, the 45th score (counting from the lowest to the highest in Figure 4.7) is 571, and the 46th score is 578. Hence, the median is

$$M = \frac{571 + 578}{2} = 574.5.$$

To find the lower and upper quartiles for this distribution of scores, we determine the 23rd scores, counting in from the low side of the distribution and counting in from the high side of the distribution. The 23rd-lowest and 23rd-highest scores are 464 and 719, respectively. Therefore,

Lower quartile, $Q_1 = 464$

Upper quartile, $Q_3 = 719$

skeletal box plot

These three descriptive measures—M, Q_1, and Q_3—and the smallest and largest values in a data set are used to construct a **skeletal box plot** (see Figure 4.8). The box plot is constructed by drawing a box between the lower and upper quartiles, with a solid line drawn across the box to locate the median. A straight line is then drawn to connect the box to the largest value; a second line is drawn from the box to the smallest value. These straight lines are sometimes called whiskers and the entire

box-and-whiskers plot

graph, a **box-and-whiskers plot**.   ■

With a quick glance at a skeletal box plot, it is easy to obtain an impression about the following aspects of the data:

1   The lower and upper quartiles, Q_1 and Q_3
2   The interquartile range (IQR)—the distance between the lower and upper quartiles
3   The most extreme (lowest and highest) values
4   The symmetry or asymmetry of the distribution of scores

If we had been presented with Figure 4.8 without having seen the original data, we would have observed the following:

$$Q_1 \approx 475$$
$$Q_3 \approx 725$$

FIGURE 4.8
Skeletal Box Plot for the Data of
Figure 4.7

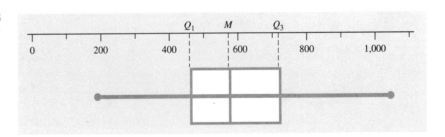

$$\text{IQR} \approx 725 - 475 = 250$$
$$M \approx 575$$

Most extreme values: 175 and 1025.

Because the median is closer to the lower quartile than it is to the upper quartile, and because the upper whisker is a little longer than the lower whisker, the distribution is slightly nonsymmetrical. To see that this conclusion is true, construct a frequency histogram for these data (or refer to your results in Exercise 3.27).

What information can be drawn from a box plot? First, the center of the distribution of scores is indicated by the median line in the box plot. Second, a measure of the variability of the scores is given by the interquartile range (the length of the box). Recall that the box is constructed between the lower and upper quartiles, so it contains the middle 50% of the scores in the distribution, with 25% on either side of the median line falling inside the box. Third, by examining the relative position of the median line, we can gauge the symmetry of the middle 50% of the scores. For example, if the median line is closer to the lower quartile than to the upper one, there is a greater concentration of scores on the lower side of the median within the box than on the upper side; a symmetric distribution of scores would cause the median line to be located in the center of the box. Fourth, additional information about skewness is obtained from the lengths of the whiskers; the longer one whisker is relative to the other one, the more skewness there is in the tail with the longer whisker.

Exercise

Application

4.27   The number of persons who volunteered to give a pint of blood at a central donor center was recorded for each of 20 successive Fridays. The data are shown here:

| 320 | 370 | 386 | 334 | 325 | 315 | 334 | 301 | 270 | 310 |
| 274 | 308 | 315 | 368 | 332 | 260 | 295 | 356 | 333 | 250 |

a   Construct a stem-and-leaf plot.

b   Construct a box plot, and interpret the results.

4.5 Summarizing Data from More Than One Variable

In the previous sections of this chapter and in Chapter 3, we discussed graphical methods and numerical descriptive methods for summarizing data from a single variable. Frequently, however, more than one variable is being studied at the same time; and although we might be interested in summarizing the data from each variable separately, we might also be interested in studying relations among the variables. For example, we might be interested in the prime interest rate and in the consumer price index, as well as in the relation between the two. In this section, we'll discuss a few techniques for summarizing data from two or more variables. Material in this section will provide a brief preview and introduction to chi-square methods (Chapter 10), analysis of variance (Chapter 13), and regression (Chapters 11 and 12).

Consider first the problem of summarizing data from two qualitative variables. Cross-tabulations can be constructed to form a **contingency table**. The rows of the table identify the categories of one variable, and the columns identify the categories of the other variable. The entries in the table identify the number of times each value of one variable occurred with each possible value of the other. For example, a television viewing survey was conducted on 1500 individuals. Each individual surveyed was asked to state his or her place of residence and network preference for national news. The results of the survey are shown in Table 4.3. As you can see, 144 urban residents preferred ABC, 135 urban residents preferred CBS, and so on.

contingency table

TABLE **4.3**
Data from a Survey of Television News Viewing

Network Preference	Residence			
	Urban	Suburban	Rural	Total
ABC	144	180	90	414
CBS	135	240	96	471
NBC	108	225	54	387
Other	63	105	60	228
Total	450	750	300	1500

The simplest method for looking at relationships between variables in a contingency table is to do a percentage comparison based on the row totals, the column totals, or the overall total. If we calculate percentages within each row of Table 4.3, we can compare the distribution of residences within each network preference. A percentage comparison such as this, based on the row totals, is shown in Table 4.4.

TABLE **4.4**
Percentage Comparison of Distribution of Residences for Each Network

Network Preference	Residence			
	Urban	Suburban	Rural	Total
ABC	34.8	43.5	21.7	$100(n = 414)$
CBS	28.7	50.9	20.4	$100(n = 471)$
NBC	27.9	58.1	14.0	$100(n = 387)$
Other	27.6	46.1	26.3	$100(n = 228)$

Except for ABC, which has the highest urban percentage among the networks, the differences among the residence distributions are in the suburban and rural categories. The percentage of suburban preferences rises from 43.5% for ABC to 58.1% for NBC. Inversely corresponding shifts downward occur in the rural category. In Chapter 10, we will use chi-square methods to explore further relations between two (or more) qualitative variables.

An extension of the bar graph provides a convenient method of summarizing joint data from a single qualitative and a single quantitative variable. We will discuss this method by way of an example. Suppose that a company wants to investigate the relative effects of three different employee-incentive systems on productivity. A total of 15 work teams are selected randomly. Of these teams, 7 participate in a released-time plan, by which teams that achieve certain goals are allowed to take extra time off, with pay; 5 participate in a bonus-pay plan; and 3 participate in a profit-sharing plan. The company has a standard productivity measure and calculates the increased productivity of each work team over a 3-month period. Suppose that the following results are obtained:

Released time, R:	16.2	15.6	19.4	18.8	16.9	15.9	17.6
Bonus pay, B:	12.4	15.8	14.0	9.8	10.0		
Profit sharing, P:	4.6	8.0	6.0				

Do the data indicate a strong relationship between plan and productivity gain?

The data summarized in Figure 4.9 clearly indicate that productivity gains are generally largest for the released-time plan, R; gains for the bonus-pay plan, B, are in the middle; and gains for the profit-sharing plan, P, are lowest. For now, we will use a plot like the one in Figure 4.9. Later, we will use analysis-of-variance methods (Chapter 13) to examine the relationships between a quantitative variable and one or more qualitative variables.

F I G U R E 4.9
Relationship Between
Productivity and the Incentive
Plans

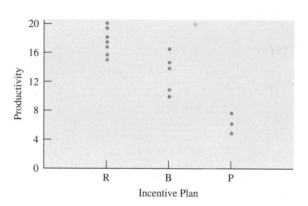

Finally, we can construct data plots for summarizing the relationships between two quantitative variables. Consider the following example. A manager of a small machine shop examined the starting hourly wage y offered to machinists with x years of previous experience. The data are shown here:

y(dollars):	8.90	8.70	9.10	9.00	9.79	9.45	10.00	10.65	11.10	11.05
x(years):	1.25	1.50	2.00	2.00	2.75	4.00	5.00	6.00	8.00	12.00

Is there a relationship between x and y?

scatterplot   One way to summarize these data is to use a **scatterplot**, as shown in Figure 4.10. Each point on the plot represents a machinist with a particular starting wage and number of years of experience. The point circled corresponds to $y = 9.45, x = 4.00$.

FIGURE **4.10**
Scatterplot of Starting Hourly
Wage and Years Experience

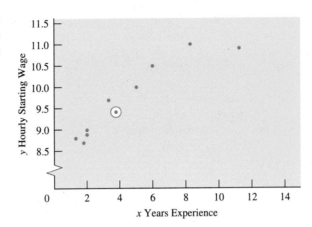

In general, the data displayed in Figure 4.10 indicate that, as the number of years of previous experience x increases, the hourly starting wage y for machinists increases. This basic idea of relating two quantitative variables is discussed and expanded in the chapters on regression (11 and 12).

Exercises

Basic Techniques

4.28   Refer to the television survey data of Table 4.3. Do a percentage comparison based on the column totals. Interpret the data.

Applications

4.29   Data on the age of employees at the time of a job turnover and on the reason for the job turnover are displayed here for 250 job changes in a large corporation.

Reason for Turnover	Age (Years)				Total
	≤ 29	30–39	40–49	≥ 50	
Resigned	30	6	4	20	60
Transferred	12	45	4	5	66
Retired/fired	8	9	52	55	124
Total	50	60	60	80	250

Do a percentage comparison based on the row totals, and use this to describe the data.

4.30   Refer to Exercise 4.29. What different summary would you get by performing a percentage comparison based on the column totals? Do this summary and describe your results.

4.31   The lengths of hospital stays were recorded for patients undergoing a particular surgical procedure at each of four hospitals. The resulting data are shown here.

Hospital	Length of Stay (days)							
A	18	20	22	22	24	26		
B	14	15	17	17	18	19	21	21
C	21	25	27	31				
D	27	33						

a   Compute the mean stay for each hospital.

b   Plot the sample data.

c   Use parts a and b to describe the data. Which hospital appears to have the shortest stays?

4.32   The federal government keeps a close watch on money growth versus targets that have been set for that growth. The accompanying table lists two measures of the money supply in the United States—M2 (private checking deposits, cash, and some savings) and M3 (M2 plus some investments)—for 20 consecutive months.

Month	Money Supply (trillions of dollars)	
	M2	M3
1	2.25	2.81
2	2.27	2.84
3	2.28	2.86
4	2.29	2.88
5	2.31	2.90
6	2.32	2.92
7	2.35	2.96
8	2.37	2.99
9	2.40	3.02
10	2.42	3.04
11	2.43	3.05
12	2.42	3.05
13	2.44	3.08
14	2.47	3.10
15	2.49	3.10
16	2.51	3.13
17	2.53	3.17
18	2.53	3.18
19	2.54	3.19
20	2.55	3.20

a   Would a scatterplot describe the relationship between M2 and M3?

b   Construct a scatterplot. Is there an obvious relation?

4.33   Refer to Exercise 4.32. What other plot might be used to summarize these data? Make the plot and interpret your results.

4.6 Using Computers to Help Make Sense of Data

MINITAB, SAS, and EXECUSTAT programs for generating numerical descriptive statistics and a box plot are displayed here for the data of Table 3.2. The DESCRIBE statement in MINITAB produces the sample size (n), the mean, the median, the trimmed mean (TRMEAN, which you can ignore), the standard deviation, the standard error of the mean, the maximum and minimum values, and the upper and lower quartiles (i.e., the 75th and 25th percentiles of the data set, shown as Q_3 and Q_1, respectively, on the printout). The BOXPLOT statement produces a box plot and also identifies the location of the more extreme values of the data set (by way of an asterisk or an "oh").

Refer to the following MINITAB output to see the box plot for the data in Table 3.2. Because various numerical descriptive statistics have been computed and then shown in the output, we know that the median is 4.3 and the lower and upper quartiles are 4.1 and 4.5, respectively. Notice that the box plot correctly locates these quartiles.

In summary, using the legend below the box plot in MINITAB, we can determine the scale in order to extract desired descriptive measures of the data set.

The PROC MEANS procedure in SAS computes the sample size, the mean, the standard deviation, the minimum, the maximum, the standard error of the mean, the sum, and the variance. The PROC UNIVARIATE procedure computes many of the same numerical descriptive measures (plus others); with the PLOT option it also gives a stem-and-leaf plot and a box plot.

MINITAB Output

```
MTB > SET INTO C1
DATA>   3.7   4.2   4.4   4.4   4.3   4.2   4.4   4.8   4.9   4.4
DATA>   4.2   3.8   4.2   4.4   4.6   3.9   4.3   4.5   4.8   3.9
DATA>   4.7   4.2   4.2   4.8   4.5   3.6   4.1   4.3   3.9   4.2
DATA>   4.0   4.2   4.0   4.5   4.4   4.1   4.0   4.0   3.8   4.6
DATA>   4.9   3.8   4.3   4.3   3.9   3.8   4.7   3.9   4.0   4.2
DATA>   4.3   4.7   4.1   4.0   4.6   4.4   4.6   4.4   4.9   4.4
DATA>   4.0   3.9   4.5   4.3   3.8   4.1   4.3   4.2   4.5   4.4
DATA>   4.2   4.7   3.8   4.5   4.0   4.2   4.1   4.0   4.7   4.1
DATA>   4.7   4.1   4.8   4.1   4.3   4.7   4.2   4.1   4.4   4.8
DATA>   4.1   4.9   4.3   4.4   4.4   4.3   4.6   4.5   4.6   4.0
DATA> END

MTB > PRINT 'WEIGHT'

WEIGHT
    3.7     4.2     4.4     4.4     4.3     4.2     4.4     4.8     4.9     4.4     4.2
    3.8     4.2     4.4     4.6     3.9     4.3     4.5     4.8     3.9     4.7     4.2
    4.2     4.8     4.5     3.6     4.1     4.3     3.9     4.2     4.0     4.2     4.0
    4.5     4.4     4.1     4.0     4.0     3.8     4.6     4.9     3.8     4.3     4.3
    3.9     3.8     4.7     3.9     4.0     4.2     4.3     4.7     4.1     4.0     4.6
    4.4     4.6     4.4     4.9     4.4     4.0     3.9     4.5     4.3     3.8     4.1
    4.3     4.2     4.5     4.4     4.2     4.7     3.8     4.5     4.0     4.2     4.1
    4.0     4.7     4.1     4.7     4.1     4.8     4.1     4.3     4.7     4.2     4.1
    4.4     4.8     4.1     4.9     4.3     4.4     4.4     4.3     4.6     4.5     4.6
    4.0
```

MINITAB Output (Continued)

```
MTB > DESCRIBE 'WEIGHT'

                  N      MEAN    MEDIAN    TRMEAN     STDEV    SEMEAN
WEIGHT          100    4.2920    4.3000    4.2900    0.3097    0.0310

                MIN       MAX        Q1        Q3
WEIGHT       3.6000    4.9000    4.1000    4.5000

MTB > STEM-AND-LEAF 'WEIGHT'

Stem-and-leaf of WEIGHT     N  = 100
Leaf Unit = 0.010

      1    36 0
      2    37 0
      8    38 000000
     14    39 000000
     24    40 0000000000
     34    41 0000000000
     47    42 00000000000000
    (11)   43 00000000000
     42    44 00000000000000
     29    45 0000000
     22    46 000000
     16    47 0000000
      9    48 00000
      4    49 0000

MTB > BOXPLOT 'WEIGHT'


                                    -----------------
              ---------------------I        +       I----------------
                                    -----------------
         +---------+---------+---------+---------+---------+------WEIGHT
       3.50      3.75      4.00      4.25      4.50      4.75

MTB > STOP
```

SAS Output

```
          OPTIONS NODATE NONUMBER LS=80 PS=60;
          DATA CHICK;
            INPUT WEIGHT @@;
            LABEL WEIGHT = 'WEIGHT GAINS FOR CHICKS';
            LIST;
            CARDS;
            3.7   4.2   4.4   4.4   4.3   4.2   4.4   4.8   4.9   4.4
            4.2   3.8   4.2   4.4   4.6   3.9   4.3   4.5   4.8   3.9
            4.7   4.2   4.2   4.8   4.5   3.6   4.1   4.3   3.9   4.2
            4.0   4.2   4.0   4.5   4.4   4.1   4.0   4.0   3.8   4.6
            4.9   3.8   4.3   4.3   3.9   3.8   4.7   3.9   4.0   4.2
            4.3   4.7   4.1   4.0   4.6   4.4   4.6   4.4   4.9   4.4
            4.0   3.9   4.5   4.3   3.8   4.1   4.3   4.2   4.5   4.4
            4.2   4.7   3.8   4.5   4.0   4.2   4.1   4.0   4.7   4.1
            4.7   4.1   4.8   4.1   4.3   4.7   4.2   4.1   4.4   4.8
            4.1   4.9   4.3   4.4   4.4   4.3   4.6   4.5   4.6   4.0
          ;
          PROC PRINT N;
          TITLE1 'WEIGHT GAINS FOR 100 BABY CHICKS OVER AN 8-WEEK PERIOD';
          TITLE2 'LISTING OF DATA';
          PROC MEANS DATA = CHICK N MEAN STD MIN MAX STDERR SUM VAR;
            VAR WEIGHT;
          TITLE2 'EXAMPLE OF PROC MEANS';
          PROC UNIVARIATE PLOT;
            VAR WEIGHT;
          TITLE2 'EXAMPLE OF PROC UNIVARIATE WITH THE PLOT OPTION';
          RUN;
```

SAS Output (Continued)

WEIGHT GAINS FOR 100 BABY CHICKS OVER AN 8-WEEK PERIOD
LISTING OF DATA

OBS	WEIGHT
1	3.7
2	4.2
3	4.4
4	4.4
5	4.3
6	4.2
7	4.4
8	4.8
9	4.9
10	4.4
11	4.2
12	3.8
13	4.2
14	4.4
15	4.6
16	3.9
17	4.3
18	4.5
19	4.8
20	3.9
21	4.7
22	4.2
23	4.2
24	4.8
25	4.5
26	3.6
27	4.1
28	4.3
29	3.9
30	4.2
31	4.0
32	4.2
33	4.0
34	4.5
35	4.4
36	4.1
37	4.0
38	4.0
39	3.8
40	4.6
41	4.9
42	3.8
43	4.3
44	4.3
45	3.9
46	3.8
47	4.7
48	3.9
49	4.0
50	4.2
51	4.3
52	4.7
53	4.1
54	4.0
55	4.6

WEIGHT GAINS FOR 100 BABY CHICKS OVER AN 8-WEEK PERIOD
LISTING OF DATA

OBS	WEIGHT
56	4.4
57	4.6
58	4.4
59	4.9
60	4.4
61	4.0
62	3.9
63	4.5
64	4.3
65	3.8
66	4.1
67	4.3
68	4.2
69	4.5
70	4.4
71	4.2
72	4.7
73	3.8
74	4.5
75	4.0
76	4.2
77	4.1
78	4.0
79	4.7
80	4.1
81	4.7
82	4.1
83	4.8
84	4.1
85	4.3
86	4.7
87	4.2
88	4.1
89	4.4
90	4.8
91	4.1
92	4.9
93	4.3
94	4.4
95	4.4
96	4.3
97	4.6
98	4.5
99	4.6
100	4.0

N = 100

WEIGHT GAINS FOR 100 BABY CHICKS OVER AN 8-WEEK PERIOD
EXAMPLE OF PROC MEANS

Analysis Variable : WEIGHT GAINS FOR CHICKS

N	Mean	Std Dev	Minimum	Maximum	Std Error
100	4.2920000	0.3096691	3.6000000	4.9000000	0.0309669

Sum	Variance
429.2000000	0.0958949

WEIGHT GAINS FOR 100 BABY CHICKS OVER AN 8-WEEK PERIOD
EXAMPLE OF PROC UNIVARIATE WITH THE PLOT OPTION

UNIVARIATE PROCEDURE

Variable=WEIGHT   WEIGHT GAINS FOR CHICKS

Moments

N	100	Sum Wgts	100
Mean	4.292	Sum	429.2
Std Dev	0.309669	Variance	0.095895
Skewness	0.117976	Kurtosis	-0.67139
USS	1851.62	CSS	9.4936
CV	7.21503	Std Mean	0.030967
T:Mean=0	138.5996	Prob>$|T|$	0.0001
Sgn Rank	2525	Prob>$|S|$	0.0001
Num ^= 0	100		

Quantiles(Def=5)

100% Max	4.9	99%	4.9
75% Q3	4.5	95%	4.8
50% Med	4.3	90%	4.7
25% Q1	4.1	10%	3.9
0% Min	3.6	5%	3.8
		1%	3.65
Range	1.3		
Q3-Q1	0.4		
Mode	4.2		

Extremes

Lowest	Obs	Highest	Obs
3.6(	26)	4.8(	90)
3.7(	1)	4.9(	9)
3.8(	73)	4.9(	41)
3.8(	65)	4.9(	59)
3.8(	46)	4.9(	92)

```
Stem Leaf                                     #        Boxplot
  49 0000                                     4           |
  48 00000                                    5           |
  47 0000000                                  7           |
  46 000000                                   6           |
  45 0000000                                  7        +-----+
  44 0000000000000                           13        |     |
  43 00000000000                             11        *-----*
  42 0000000000000                           13        |  +  |
  41 0000000000                              10        +-----+
  40 0000000000                              10           |
  39 000000                                   6           |
  38 000000                                   6           |
  37 0                                        1           |
  36 0                                        1           |
     ----+----+----+----+
Multiply Stem.Leaf by 10**-1
```

SAS Output (Continued)

**WEIGHT GAINS FOR 100 BABY CHICKS OVER AN 8-WEEK PERIOD
EXAMPLE OF PROC UNIVARIATE WITH THE PLOT OPTION**

UNIVARIATE PROCEDURE

Variable=WEIGHT   WEIGHT GAINS FOR CHICKS

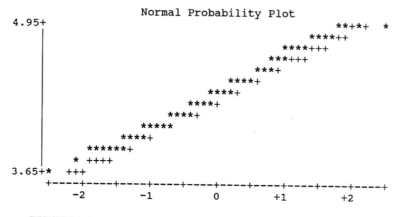

EXECUSTAT Output

Summary Statistics

Variable: WEIGHT

Sample size	100
Mean	4.292
Median	4.3
Variance	0.0958949
Std. deviation	0.309669
Std. error	0.0309669
Minimum	3.6
Maximum	4.9
Range	1.3
Lower quartile	4.1
Upper quartile	4.5
Interquartile range	0.4

Stem-and-leaf display for WEIGHT: unit = 0.01   1|2 represents 0.12

```
    1    36 | 0
    2    37 | 0
    8    38 | 000000
   14    39 | 000000
   24    40 | 0000000000
   34    41 | 0000000000
   47    42 | 0000000000000
  (11)   43 | 00000000000
   42    44 | 0000000000000
   29    45 | 0000000
   22    46 | 000000
   16    47 | 0000000
    9    48 | 00000
    4    49 | 0000
```

EXECUSTAT Output (Continued)

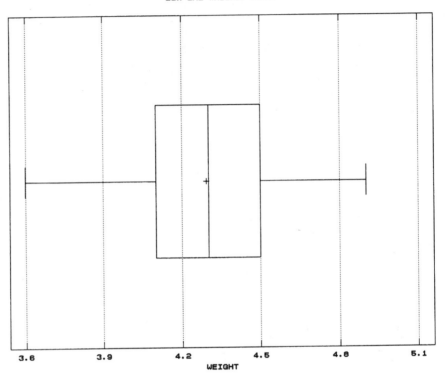

WEIGHT

The box plot for SAS is displayed in the output immediately to the right of the stem-and-leaf plot. Notice that no scale (other than the scale for the stems of the stem-and-leaf plot) is provided to enable you to identify the median, the lower and upper quartiles, and other characteristics of the data set. However, using the summary statistics labeled Quantiles, we can determine the median (4.3), the lower and upper quartiles (4.1 and 4.5), and the largest and smallest measurements (4.9 and 3.6).

The EXECUSTAT software system was used to provide descriptive statistics, a stem-and-leaf plot, and a box plot for the weight gain data of Table 3.2. This output is shown here as well. Notice that the descriptive statistics obtained from EXECUSTAT are the same as those obtained from MINITAB and SAS. The box plot for EXECUSTAT is much easier to read and to use in identifying appropriate values than are those for the other two software systems.

The programs illustrated in this section can be used to analyze other data sets in a similar manner. Additional details about the statements and procedures are available in the SAS, MINITAB, and EXECUSTAT user's manuals.

Summary

The objective of statistics is to make sense of data. After the data have been collected for a survey or for an experimental study, we can summarize the data graphically as one approach to understanding the data. (Several different graphical descriptive measures for summarizing data were discussed in Chapter 3.) A second way to summarize data involves computing one or more *numerical descriptive measures*— numbers that convey a mental picture of the frequency distribution for a set of measurements.

The most important of these descriptive measures are the *mean* and *standard deviation*, which measure the center and the spread, respectively, of a frequency distribution. The standard deviation of a set of measurements with a mound-shaped graphical distribution is a meaningful measure of variability when interpreted by the Empirical Rule. Numerical descriptive measures are suitable for developing both descriptions and inferences. Thus we can use a numerical descriptive measure of the sample—say, the *sample mean*—to estimate a parameter such as the *population mean*. The great advantage of numerical descriptive measures in making inferences is that they enable us to give a quantitative measure of the "goodness" of the inference. In particular, we can show that the sample mean will lie within a specified distance of the population mean with some predetermined probability.

Now that we have learned how to summarize data using graphical and numerical methods, we will take up the analysis of data. The results of a statistical analysis of sample data are expressed in terms of inferences about the population of measurements from which the sample data were obtained—all of which are subject to a degree of uncertainty. If we decide that a population possesses a certain characteristic, we may be correct, but there is always some element of doubt. Thus we find that probability, which is a measure of uncertainty, plays a major role both in making an inference and in measuring how good it is.

Probability and probability distributions (Chapter 5) and sampling distributions (Chapter 6) provide fundamental tools and concepts for proceeding with the third step in making sense of data—analyzing data (Chapters 7–13).

Key Terms

mode	mean
median	grouped data mean
grouped data median	outliers

trimmed mean	s^2
skewness	σ^2
variability	standard deviation
range	approximate value for s
grouped data range	box plot
percentile	hinges
interquartile range	skeletal box plot
deviation	box-and-whiskers plot
dot diagram	contingency table
variance	scatterplot

Key Formulas

1 Median, grouped data:

$$\text{Median} = L + \frac{w}{f_m}(.5n - cf_b)$$

2 Sample mean:

$$\bar{y} = \frac{\sum_i y_i}{n}$$

3 Sample mean, grouped data:

$$\bar{y} = \frac{\sum_i f_i y_i}{n}$$

4 Sample variance:

$$s^2 = \frac{1}{n-1}\left[\sum_i y_i^2 - \frac{\left(\sum_i y_i\right)^2}{n}\right]$$

5 Sample variance, grouped data:

$$s^2 = \frac{1}{n-1}\left[\sum_i f_i y_i^2 - \frac{\left(\sum_i f_i y_i\right)^2}{n}\right]$$

6 Sample standard deviation:

$$s = \sqrt{s^2}$$

Supplementary Exercises

4.34 The rounded nitrogen content (in percent) for the 36 composite apple leaf samples of Exercise 4.10 are as follows:

2.10	2.82	2.17	1.99	2.22	3.09
2.47	2.52	2.80	2.10	2.92	2.20
1.75	2.77	2.82	2.67	3.05	2.93
2.94	1.98	2.38	2.65	2.77	1.85
1.69	2.70	2.68	2.06	2.36	2.28
2.75	2.43	2.39	2.55	1.80	1.96

a Use the shortcut formula to compute s^2 and s. You can verify that

$$\sum_i y_i = 87.61 \quad \text{and} \quad \sum_i y_i^2 = 218.7297.$$

b Use the range approximation to check you calculation of s.

c To increase your confidence in the Empirical Rule, construct the intervals $\bar{y} \pm s$, $\bar{y} \pm 2s$, and $\bar{y} \pm 3s$. Count the number of rounded nitrogen content readings falling in each of the three intervals. Convert these numbers into percentages, and compare your results to those predicted by the Empirical Rule.

4.35 The College of Dentistry at the University of Florida has made a commitment to develop its entire curriculum around the use of self-paced instructional materials such as videotapes, slide tapes, and syllabi. It is hoped that each student will proceed at a pace commensurate with his or her ability and that the instructional staff will have more free time for personal consultation in student-faculty interaction. One such instructional module was developed and tested on the first 50 students proceeding through the curriculum. The following measurements represent the number of hours it took these students to complete the required modular material:

16	8	33	21	34	17	12	14	27	6
33	25	16	7	15	18	25	29	19	27
5	12	29	22	14	25	21	17	9	4
12	15	13	11	6	9	26	5	16	5
9	11	5	4	5	23	21	10	17	15

a Calculate the mode, the median, and the mean for these recorded completion times.

b Guess the value of s.

c Compute s using the shortcut formula, and compare your answers to the answer in part b.

d Would you expect the Empirical Rule to describe adequately the variability of these data? Explain.

4.36 A study was conducted to determine urine flow in sheep (in milliliters/minute) when sheep are infused intravenously with the antidiuretic hormone ADH. The recorded urine flows of 10 sheep are as follows:

0.7	0.5	0.5	0.6	0.5	0.4	0.3	0.9	1.2	0.9

a Determine the mean, the median, and the mode for these sample data.

b Suppose that the largest measurement is 6.8 rather than 1.2. How does this affect the mean, the median, and the mode?

4.37 Refer to Exercise 4.36.

a Compute the range and the sample standard deviation.

b Check you calculation of *s* by using the range approximation.

c How are the range and then standard deviation affected if the largest measurement is 6.8 rather than 1.2? What if the largest measurement is 68?

4.38 A stem-and-leaf plot is shown for the telephone data from Exercise 3.39. Compute the mean, the median, the mode, and the standard deviation for these data.

```
MTB > PRINT 'PHONES'

phones
    500    350    550    480    610    570    620    630    620    570    480
    550    650    580    570    600    480    520    540    610    570    580
    560    470    570    540    590    720    590    650    470    530    530
    560    550    580    560    610    560    510    540    540    570    560
    520    530    570    450    540    580

MTB > STEM-AND-LEAF 'PHONES'

Stem-and-leaf of phones     N  = 50
Leaf Unit = 10

      1     3 5
      1     4
      7     4 577888
     19     5 012233344444
    (21)    5 5556666677777777888899
     10     6 0111223
      3     6 55
      1     7 2

MTB > STOP
```

4.39 A box plot was constructed for the data Exercise 4.38, using MINITAB. Describe the data by reference to information conveyed by the box plot.

```
MTB > BOXPLOT 'PHONES'
```

```
MTB > STOP
```

 4.40 A random sample of 90 standard metropolitan statistical areas (SMSA) was studied to obtain information on murder rates. The murder rate (number of murders per 100,000 people) for each area was recorded; these data are summarized in the accompanying frequency table. Construct a relative frequency histogram for these data.

Class Interval	f_i	Class Interval	f_i
.5–1.5	2	13.5–15.5	9
1.5–3.5	18	15.5–17.5	4
3.5–5.5	15	17.5–19.5	2
5.5–7.5	13	19.5–21.5	1
7.5–9.5	9	21.5–23.5	1
9.5–11.5	8	23.5–25.5	1
11.5–13.5	7		

4.41 Refer to the data of Exercise 4.40.

a Compute the sample median and the mode.

b Compute the sample mean.

c Which measure of central tendency would you use to describe the center of the distribution of murder rates?

d Compute the sample standard deviation.

e Describe the data set by using the Empirical Rule.

4.42 Every 20 minutes, a sample of 10 transistors is drawn from the outgoing product on a production line and tested. The data are summarized below for the first 500 samples of 10 items each (y_i represents the number of defective items in each sample of 10 items).

y_i	0	1	2	3	4	5	6	7	8	9	10
f_i	170	185	75	25	15	10	8	5	4	2	1

Construct a relative frequency distribution depicting the interquartile range for these data.

4.43 Refer to Exercise 4.42.

a Determine the sample median and the mode.

b Calculate the sample mean.

c Based on the mean, the median, and the mode, how is the distribution skewed?

4.44 Per capita expenditure (dollars) for health and hospital services by state are shown here.

Dollars	f
45–59	1
60–74	4
75–89	9
90–104	9
105–119	12
120–134	6
135–149	4
150–164	1
165–179	3
180–194	0
195–209	1
Total	50

a Construct a relative frequency histogram.

b Compute approximate values for $\bar{y}$ and s from the grouped expenditure data.

4.45 The Insurance Institute for Highway Safety published data on the total damage suffered by compact automobiles in a series of controlled, low-speed collisions. The data, in dollars, with brand names removed, are as follows:

361	393	430	543	566	610	763	851	886	887	976	1,039
1,124	1,267	1,328	1,415	1,425	1,444	1,476	1,542	1,544	2,048	2,197	

a Draw a histogram of the data, using six or seven categories.

b On the basis of the histogram, what would you guess the mean to be?

c Calculate the median and the mean.

d What does the relationship between the mean and the median indicate about the shape of the data?

4.46 Production records for an automobile manufacturer show the following figures for production per shift (maximum production is 720 cars per shift):

688	711	625	701	688	667	694	630	547	703	688	697	703
656	677	700	702	688	691	664	688	679	708	699	667	703

a Would the mode be a useful summary statistic for these data?

b Find the median.

c Find the mean.

d What does the relationship between the mean and the median indicate about the shape of the data?

4.47 Draw a stem-and-leaf plot of the data in Exercise 4.46. The stems should include (from highest to lowest) 71, 70, 69, ... Does the shape of the stem-and-leaf display confirm your judgment in part d of Exercise 4.46?

4.48 Refer to Exercise 4.47.

a Find the median and the IQR.

b Draw a box plot of the data.

4.49 Data are collected on the weekly expenditures by each of a sample of urban households on food (including restaurant expenditures). The data, obtained from diaries kept by each household, are grouped by number of members of the household. The expenditures were as follows:

1 member:	67	62	168	128	131	118	80	53	99	68		
	76	55	84	77	70	140	84	65	67	183		
2 members:	129	116	122	70	141	102	120	75	114	81	106	95
	94	98	85	81	67	69	119	105	94	94	92	
3 members:	79	99	171	145	86	100	116	125				
	82	142	82	94	85	191	100	116				
4 members:	139	251	93	155	158	114	108					
	111	106	99	132	62	129	92					
5+ members:	121	128	129	140	206	111	104	109	135	136		

a Calculate the mean expenditure separately for households with each number of members.

b Calculate the median expenditure separately for households with each number of members.

4.50 Refer to the data in Exercise 4.49.

a Calculate the mean of the combined data, using the raw data.

b Can the combined mean be calculated from the means for households with each number of members?

c Calculate the median of the combined data, using the raw data.

d Can the combined median be calculated from the medians for households with each number of members?

4.51 A company revised a long-standing policy by eliminating the time clocks and cards for nonexempt employees. Along with this change, all employees (exempt and nonexempt) were expected to account for their own time on the job as well as for absences due to sickness, vacation, holidays, and so on. The previous policy of allocating a certain number of sick days was eliminated; if an employee was sick, he or she was given time off with pay; otherwise, he or she was expected to be working.

In order to see how well the new program was working, the records of a random sample of 15 employees were examined to determine the number of sick days this year (under the new plan) and the corresponding number for the preceding year. These data are shown in the accompanying table.

Employee	This Year (new policy)	Preceding Year (old policy)
1	0	2
2	0	2
3	0	3
4	0	4
5	2	5
6	1	2
7	1	6
8	3	8
9	1	5
10	0	4
11	5	5
12	6	12
13	1	3
14	2	4
15	12	4

a Obtain the mean and the standard deviation for each column.

b Based on the sample data, what might you conclude (infer) about the new policies? Explain your reason(s).

4.52 Refer to Exercise 4.51. What happens to $\bar{y}$ and s for each column if we eliminate the two 12s and substitute values of 7? Are the ranges for the old and new policies affected by these substitutions?

4.53 Federal authorities have destroyed considerable quantities of wild and cultivated marijuana plants. The following table shows the number of plants destroyed and the number of arrests made during a 12-month period in 15 states.

State	Plants	Arrests
1	110,010	280
2	256,000	460
3	665	6
4	367,000	66
5	4,700,000	15
6	4,500	8
7	247,000	36
8	300,200	300
9	3,100	9
10	1,250	4
11	3,900,200	14
12	68,100	185
13	450	5
14	2,600	4
15	205,844	33

a Discuss the appropriateness of using the sample mean to describe these two variables.

b Compute the sample mean, the 10% trimmed mean, and the 20% trimmed mean. Which trimmed mean seems more appropriate for each variable? Why?

4.54 Refer to Exercise 4.53. Does there appear to be a relation between the number of plants destroyed and the number of arrests? How might you examine this question? What other variable(s) might be related to the number of plants destroyed?

 4.55 Monthly readings for the FDC Index, a popular barometer of the health of the pharmaceutical industry, are shown here. As can be seen, the Index has several components—one for pharmaceutical companies, one for diversified companies, one for chain drugstores, and one for drug and medical supply wholesalers.

Month	Pharmaceuticals	Diversified	Chain	Wholesaler
January	123.1	154.6	393.3	475.5
February	122.4	146.0	407.6	504.1
March	125.2	169.2	405.0	476.6
April	136.1	156.7	415.1	513.3
May	149.3	177.0	418.9	543.5
June	145.7	158.1	443.2	552.6
July	162.4	156.6	419.1	526.2
August	168.0	178.6	404.0	516.3
September	155.6	170.4	391.8	482.1
October	177.0	162.9	410.9	484.0
November	196.6	182.4	459.8	522.6
December	195.2	195.4	431.9	536.8

a Plot these data on a single graph.

b Discuss trends within each component and any apparent relationships among the separate components of the FDC Index.

4.56 Refer to Exercise 4.55. Compute the percentage change for each month of each component of the Index. (Assume that the percentage changes in January were 12.3, −.7, 12.1, and 16.1, respectively, for the four components.) Plot these data. Are they more revealing than the original measurements were?

The Changing Dow

The mighty Dow Jones industrial average has hardly been dormant over the years. Only 10 companies have remained on the barometer since 1928, when the Dow was first made up of 30 companies.

Since then, there have been 39 changes. Even IBM was dropped and then added back years later.

Today's Dow lists many companies that are experiencing trouble. The highlighted companies have either suffered recent earnings problems or made big layoffs.

What it was on Oct. 1, 1928		What it is today	
†Allied Chemical & Dye	Atlantic Refining	**Allied-Signal**	Goodyear
Wright Aeronautical	Chrysler	Aluminum Co. of America	**IBM**
North American	Paramount Public	**American Express**	International Paper
Victor Talking Machine	General Railway Signal	AT&T	McDonald's
†Bethlehem Steel	Mack Trucks	Bethlehem Steel	Merck
International Nickel	American Smelting	**Boeing**	Minnesota Mining
International Harvester	American Can	Caterpillar	J.P. Morgan
Goodrich	Postum Inc.	Chevron	Philip Morris
Texas Gulf Sulphur	Nash Motors	Coca-Cola	Procter & Gamble
U.S. Steel	†Sears	Walt Disney	**Sears**
American Sugar	†Texas Corp.	Du Pont	Texaco
American Tobacco	†Union Carbide	**Eastman Kodak**	Union Carbide
†Standard Oil (N.J.)	Radio Corp.	Exxon	**United Technologies**
†General Electric	†Westinghouse Electric	General Electric	Westinghouse Electric
†General Motors	†Woolworth	**General Motors**	Woolworth

†Denotes a company that has remained in the average for the 64 years.

Source: Wall Street Journal

 4.57 The Dow Jones Industrial Average on a particular day is obtained by computing the average price of the 30 stocks that make up the index and then dividing this average by a "price divisor" that appears on pg C3 of the *Wall Street Journal*. This divisor changes some, but not drastically, from day to day. The closing prices on March 10, 1993 for the stocks included in the Dow Jones Industrial Average are listed in the accompanying table and on the data disk.

   a Compute the average closing price of the 30 stocks for March 10, 1993.

   b The price divisor listed in the *Wall Street Journal* for that data was 0.46268499. Compute the Dow Jones Industrial Average (DJIA) at the close of that day.

   c Compute the DJIA for a day last week, using the closing stock prices and the price divisor listed in the *Wall Street Journal* for that day. Check you result against the DJIA listed in the paper for that same day.

Stock	Abbreviation	Closing Price 3/10/93
Allied-Signal	AlliedSgnl	67.125
Aluminum Co of America	Alcoq	73.375
American Express	AmExprss	28.000
AT&T	AmT&T	58.625
Bethlehem Steel	BethSteel	18.500
Boeing	Boeing	34.125
Caterpillar	Caterpillar	59.625
Chevron	Chevron	77.375
Coca-Cola	CocaCola	43.125
Walt Disney	Disney	45.500
DuPont	DuPont	47.875
Eastman Kodak	EKodak	54.625
Exxon	Exxon	64.125
General Electric	GenElec	87.250
General Motor	GenMotor	40.000
Goodyear	Goodyear	74.625
IBM	IBM	56.625
International Paper	IntPaper	66.625
McDonalds	McDonalds	52.750
Merck	Merck	38.125
Minnesota Mining	MinnMnMf	109.375
J. P. Morgan	MorganJP	67.250
Philip Morris	PhilipMor	64.000
Procter & Gamble	ProctGam	53.500
Sears	Sears	53.500
Texaco	Texaco	62.750
Union Carbide	UnCarbide	17.125
United Technologies	Utd Tech	47.000
Westinghouse Electric	Westinghouse	13.500
Woolworth	Woolworth	30.500

4.58 The number of telephones (per 1000 people) is shown by state in the accompanying table. These data were plotted in Exercise 3.38.

State	Telephones	State	Telephones	State	Telephones
Alabama	500	Louisiana	520	Ohio	550
Alaska	350	Maine	540	Oklahoma	580
Arizona	550	Maryland	610	Oregon	560
ArkanSAS	480	Massachusetts	570	Pennsylvania	610
California	610	Michigan	580	Rhode Island	560
Colorado	570	Minnesota	560	South Carolina	510
Connecticut	620	Mississippi	470	South Dakota	540
Delaware	630	Missouri	570	Tennessee	540
Florida	620	Montana	540	Texas	570
Georgia	570	Nebraska	590	Utah	560
Hawaii	480	Nevada	720	Vermont	520
Idaho	550	New Hampshire	590	Virginia	530
Illinois	650	New Jersey	650	Washington	570
Indiana	580	New Mexico	470	West Virginia	450
Iowa	570	New York	530	Wisconsin	540
KanSAS	600	North Carolina	530	Wyoming	580
Kentucky	480	North Dakota	560		

a Might the Empirical Rule be used to describe the data?

b Compute $\bar{y}$ and s, and count the number (percentage) of measurements falling in the intervals $\bar{y} \pm s$, $\bar{y} \pm 2s$, $\bar{y} \pm 3s$.

4.59 Refer to Exercise 4.58. Are there many extreme values affecting $\bar{y}$? Should this have been anticipated based on the data plot in Exercise 4.58? Compute the 10% trimmed mean for these data.

4.60 As one part of a review of middle-manager selection procedures, a study was made of the relationship between hiring source (promoted from within, hired from related business, hired from unrelated business) and 3-year job history (additional promotion, same position, resigned, dismissed). The data for 120 middle managers follows.

	Source			
Job History	Within Firm	Related Business	Unrelated Business	Total
Promoted	13	4	10	27
Same position	32	8	18	58
Resigned	9	6	10	25
Dismissed	3	3	4	10
Total	57	21	42	120

a Calculate job-history percentages within each source.

b Would you say that there is a strong correlation between source and job history?

4.61 A survey was taken of 150 residents of major coal-producing states, 200 residents of major oil- and natural-gas-producing states, and 450 residents of other states. Each resident chose a most preferred national energy policy. The results are shown in the SAS printout on page 142.

a Interpret the values 62, 32.8, 41.3, and 7.8 in the upper left-hand cell of the cross-tabulation. Notice the labels COUNT, ROW PCT, COL PCT, and TOT PCT at the upper left-hand corner.

b Which of the percentage calculations seems most meaningful to you?

c According to the percentage calculations that you prefer, does there appear to be a strong association between state and opinion?

STATE

FREQUENCY PERCENT ROW PCT COL PCT	COAL	OIL AND GAS	OTHER	TOTAL
OPINION COAL ENCOURAGED	62 7.75 32.80 41.33	25 3.12 13.23 12.50	102 12.75 53.97 22.67	189 23.62
FUSION DEVELOP	3 0.38 7.32 2.00	12 1.50 29.27 6.00	26 3.25 63.41 5.78	41 5.12
NUCLEAR DEVELOP	8 1.00 22.22 5.33	6 0.75 16.67 3.00	22 2.75 61.11 4.89	36 4.50
OIL DEREGULATION	19 2.37 12.58 12.67	79 9.88 52.32 39.50	53 6.62 35.10 11.78	151 18.87
SOLAR DEVELOP	58 7.25 15.14 38.67	78 9.75 20.37 39.00	247 30.87 64.49 54.89	383 47.88
TOTAL	150 18.75	200 25.00	450 56.25	800 100.00

STATISTIC	DF	VALUE	PROB
CHI-SQUARE	8	106.194	0.000
LIKELIHOOD RATIO CHI SQUARE	8	97.258	0.000
MANTEL-HAENSZEL CHI-SQUARE	1	11.288	0.001
PHI COEFFICIENT		0.364	
CONTINGENCY COEFFICIENT		0.342	
CRAMER'S V		0.258	

 4.62   A municipal workers' union that represents sanitation workers in many small midwestern cities studied the contracts that had been signed in previous years. The contracts were subdivided into those settled by negotiation without a strike, those settled by arbitration without a strike, and those settled after a strike. For each contract, the first-year percentage wage increase was determined. Summary figures follow on page 143.

Does there appear to be a relationship between contract type and mean percentage wage increase? If you were management- rather than union-affiliated, which posture would you prefer to take in future contract negotiations?

Contract Type	Negotiation	Arbitration	Poststrike
Mean percentage wage increase	8.20	9.42	8.40
Variance	0.87	1.04	1.47
Standard deviation	0.93	1.02	1.21
Sample size	38	16	6

Experiences with Real Data

4.63   Refer to the Clinical Trials database on your data disk (or in Appendix 1). Compute the mean, the range, and the standard deviation of the HAM-D total score for the four treatment groups. Then graph the data sets. What do you observe?

4.64   Again refer to the Clinical Trials database. Complete a profile of the four treatment groups by computing the mean, the range, and the standard deviation for the anxiety, retardation, sleep disturbance, and total scores from the HAM-D scale. Are there any obvious differences among the groups following the treatment period?

4.65   Combine the OBRIST scores for the four treatment groups of the Clinical Trials database in Appendix 1.

    a   Generate a stem-and-leaf plot, and determine the mean and the median from the histogram.

    b   Generate the actual sample mean and median for the combined data, and compare these values to the approximations obtained from part a.

4.66   Refer to the Insurance Claims database on your data disk (or in Appendix 1).

    a   Generate an appropriate graph for these data or refer to the one done for Exercise 3.53.

    b   Generate a numerical descriptive measure to summarize the data.

    c   From parts a and b, describe the data; that is, make sense of the data from the graphical and numerical summaries you've generated.

4.67   Refer to Exercise 4.66.

    a   Notice that $\bar{y} - \bar{s}$ includes negative values—impossible for insurance claims. Do you think that the Empirical Rule adequately describes the data? Why or why not?

    b   Notice also that the median is less than the mean. What does this tell you, and why might this have happened?

4.68   Select an area of your undergraduate major that utilizes experimental data. Typical data sources might be chemistry, biology, psychology, geology, or physics laboratories. Or you might seek data contained in social science or business journals. Either by experimentation or by use of a professional journal, select a sample of at least $n = 25$ observations on some experimental variable.

    a   Define the population from which your sample was drawn.

    b   Construct a relative frequency histogram for the data.

    c   Calculate y and s for the data.

    d   Do the data appear to be mound-shaped and therefore to make the Empirical Rule applicable?

    e   How many of the observations (expressed as a fraction) lie within two standard deviations of y? Do these results agree with the Empirical Rule?

4.69   Suppose that you are an environmental scientist and must measure the dissolved oxygen content (which is a measure of pollution) at a particular point in a lake. Why would a single chemical determination be unsatisfactory?

    You can answer this question by recalling your own experiences. If you have ever constructed something—measuring and cutting pieces to be assembled—you may have noticed that they sometimes did not fit. This problem is often attributable to errors in measuring the components prior to assembly. Similar random-measurement errors occur in almost all experimentation (if the measuring instruments are accurate enough to detect the variation).

    Most errors caused by inaccurate measuring instruments can be reduced (not eliminated) by using not one measurement but the mean of several measurements to characterize the true value of the quantity being measured.

    To illustrate, have 10 people in your class measure some object in the room (for instance, the length of the room). Notice the variation in the recorded measurements. Calculate the mean for the 10 sample measurements ($n = 10$). Notice that the mean falls near the center of the set of measurements and that it tends to offset overly large measurements with overly small ones.

If we were to measure an object repeatedly, millions and millions of times, a population of measurements would be generated. It is likely that the mean of this population, μ, would coincide with the true length of the object. Viewed in this manner, the 10 classroom measurements represent a sample and the sample mean $\bar{y}$ estimates μ. In later chapters we will learn how to evaluate the error of this estimate—that is, the difference between the estimate $\bar{y}$ and the true mean μ.

4.70   Refer to the Patient Stay database on the data disk (or in Appendix 1). In order for a health clinic to be capable of handling the desired patient load, designers need to know something about the demand for services in the area where the clinic is to be located. This would include information on the patient arrival rate as well as on the length of time required to treat a patient. Both the arrival rate and the treatment time vary in a random manner. The arrival rate varies because of the random occurrence of outbreaks of flu and other common illnesses; the treatment time varies depending on the patient's particular illness. Thus the demand for physician and nurse time varies in a random manner. By studying the frequency distributions of patient arrival rates and treatment times, the designer can specify the numbers of doctors, nurses, technicians, orderlies, and items of physical equipment needed to meet the demand. These numbers will affect the length of time a patient must wait in the clinic before receiving attention.

To answer some important questions about clinic treatment times, a designer acquired data from an established clinic in a locale that possessed similar characteristics to those of the proposed new clinic location. The treatment times for 50 patients, randomly selected from the clinic's records, are as follows:

21	20	31	24	15	21	24	18	33	8
26	17	27	29	24	14	29	41	15	11
13	28	22	16	12	15	11	16	18	17
29	16	24	21	19	7	16	12	45	24
21	12	10	13	20	35	32	22	12	10

a   Determine the average treatment time for the sample of 50 patients. Interpret this statistic.

b   Find the median treatment time. Interpret this statistic.

c   Must the sample mean and sample median be equal? Explain.

d   Choose a suitable graphical display for these data.

e   What insights into the data have you gained from your numerical and graphical summaries?

4.71   Refer to the Crimes Perception database on the data disk (or in Appendix 1). A sample of 200 students at a west coast university participated in a study of young adults' perceptions of acts that may have constituted crimes. Each student was asked the following:

"Which of the following acts do you personally think should be publicly regarded as crimes?" The acts presented were: aggravated assault, armed robbery, arson, atheism, automobile theft, burglary, civil disobedience, drug use, embezzlement, forcible rape, gambling, hijacking, homosexuality, land fraud, masturbation, Nazism, payola (kickbacks), price fixing, prostitution, sexual abuse of children, sexual discrimination, shoplifting, strip mining, treason, and vandalism.

Refer to the variable "crimes"—the number of acts each student regarded as a crime.

a   Identify the type of data collected on crimes.

b   Choose an appropriate graphical method to summarize the "crimes" data. Describe (in plain English) what you see.

c   Compute the mean, the median, and the standard deviation. How do these numerical measures help describe the data? Do they add anything to what you learned from part b?

4

Background For Analyzing Data

5

Probability and Probability Distributions

5.1 Introduction

We stated in Chapter 1 that statistics deals with making sense of data. We gather sample data in a survey or experimental study in order to describe (summarize) and to make inferences about (analyze) the underlying population of measurements, recognizing that only partial information about the populations is contained in the sample. There is necessarily some risk or uncertainy associated with inferences made from the sample data. Most management decisions must be made in the presence of uncertainty. Prices and models for new automobiles must be selected on the basis of shaky forecasts of consumer preferences, national economic trends, and competitive actions. The size and allocation of a hospital staff must be decided with limited information on patient load. The inventory of a product must be set in the face of uncertainty about demand. Probability, the language of uncertainty, can help in making inferences. Consider the following example.

Martha Jones, a candidate for Congress, publicly announces that her forthcoming election is a guaranteed success, and she forecasts victory by a substantial margin in all precincts of her district. Somewhat doubtful about her claims, a local television station randomly selects 20 names from the voter registration list, calls these voters, and asks them for whom they will vote in the upcoming election. Not one of the 20 voters states that he or she will vote for Jones; all favor her opponent. What do you conclude about Jones's claim to victory in the sampled area?

If Jones were correct in predicting victory, at least half the voters in the district would have favored her, and somewhat near this same proportion should have been observed in the sample. As it turned out, none of the voters in the sample favored Jones—a result sharply at variance with her claim. Hence, we infer that the proportion of voters in the population (the district) favoring Jones is less than $\frac{1}{2}$ and that she will lose the district. We conclude that Jones will lose because the sample yielded results highly contradictory to her claim. By "contradictory" we do not mean that it would be impossible to select at random 20 voters, none of whom favor Jones, assuming that Jones's claim of victory was correct. We mean, rather, that such a random draw would be highly *improbable*. Thus, we measure the degree of contradiction to Jones's claim of victory in terms of the probability of the observed sample.

To get a better view of the role that probability plays in making this inference, suppose that the sample instead produced 9 voters in favor of Jones and 11 in favor of her opponent. Would we consider the result highly improbable and reject Jones's claim? How about 7 in favor and 13 against, or 5 in favor and 15 against? Where do we draw the line? At what point do we decide that the result of the observed sample is so improbable, assuming that Jones's claim is correct, that we must reject her claim? To answer this question we need to know how to find the probability of obtaining a particular sample outcome. Knowing this probability, we can determine whether we should view Jones's claim as plausible or implausible. Probability is the tool that enables us to make the necessary inference.

Since probability is the tool for making inferences, we might ask: What is probability? In the preceding discussion, we used the term *probability* in its everyday sense. Let us now examine this idea more closely.

5.2 Interpretations of Probability

Observations of phenomena can result in many different outcomes, some of which are more likely than others. Numerous attempts have been made to give a precise definition for the probability of an outcome. We will cite a few of these.

classical interpretation of probability

The first interpretation of probability, called the **classical interpretation of probability**, arose from games of chance. Typical probability statements of this type are "the probability that a flip of a balanced coin will show 'heads' is 1/2," and "the probability of drawing an ace when a single card is drawn from a well-shuffled standard deck of 52 cards is 4/52." The numerical values for these probabilities arise from the nature of the games. A coin flip has two possible outcomes (a head or tail); the probability of a head should then be 1/2 (1 out of 2). Similarly, there are 4 aces in a standard deck of 52 cards, so the probability of drawing an ace in a single draw is 4/52 or 4 out of 52.

outcome

event

In the classical interpretations of probability, each possible, distinct result is called an **outcome**; an **event** is identified as a collection of outcomes. The probability of an event E under the classical interpretation of probability is computed by taking the ratio of the number N_E of outcomes favorable to event E in the total number N of possible outcomes:

$$P(\text{event } E) = \frac{N_E}{N}$$

The applicability of this interpretation depends on the assumption that all outcomes are equally likely. If this assumption does not hold, the probabilities indicated by the classical interpretation of probability will not be accurate.

relative frequency concept

A second interpretation of probability is called the **relative frequency concept** of probability. If an experiment is repeated a large number of times and event E occurs 30% of the times, then .30 should be a very good approximation to the probability of event E. Symbolically, if an experiment is conducted n different times and if event E occurs on n_E of these trials, then the probability of event E is approximately

$$P(\text{event } E) \approx \frac{n_E}{n}.$$

We say "approximately" because we think of the actual probability P (event E) as being the relative frequency of the occurrence of event E over a very large number of observations or repetitions of the phenomenon. The fact that we can check probabilities that have a relative frequency interpretation (by simulating many repetitions of the experiment) makes this interpretation very appealing and practical.

The third interpretatiaon of probability can be used for problems in which it is difficult to imagine a repetition of the experiment. These are "one-shot" situations. For example, the director of a state welfare agency who estimates the probability that a proposed revision in eligibility rules will be passed by the state legislature is not thinking in terms of a long series of trials. Rather, the director uses a personal or **subjective probability** to make a one-shot statement of belief regarding the likelihood of passage of the proposed legislative revision. The problem with subjective probabilities is that they may vary from person to person and they cannot be checked.

subjective probability

Of the three interpretations presented, the relative frequency concept seems to be the most reasonable one, since it provides a pratical interpretation of the probability for most events of interest. Even though we will never run the necessary number of repetitions of the experiment to determine the exact probability of an event, the fact that we could check the probability of an event gives meaning to the relative frequency concept. Throughout the remainder of this text, we will lean heavily on this interpretation of probability.

Exercises

Basic Techniques

5.1　Indicate which interpretation of the probability statement seems most appropriate.

　a　The National Angus Association has stated that there is a 60/40 chance that wholesale beef prices will rise by the summer; that is, there is a .60 probability of an increase and a .40 probability of a decrease or no change in price.

　b　The quality-control section of a large chemical manufacturing company has undertaken an intensive process-validation study. From this study the quality-control section claims that the probability is .998 that the shelf life of a newly released batch of a particular chemical will exceed the minimum time specified.

　c　A new blend of coffee is being contemplated for release by the marketing division of a large corporation. Preliminary marketing survey results indicate that 550 of a random sample of 1000 potential users rated this new blend as being better than a brand-name competitor. The probability of this happening by chance is approximately .001, assuming that there is really no difference in consumer preference for the two brands.

　d　The probability of receiving a busy signal when attempting to access the company WATS line during the 3:00–5:00 P.M. time frame is .58.

　e　The probability that it will rain tomorrow is .30.

　f　In a certain city, the probability of selecting a household at random within which the head of the household is unemployed is .12.

5.2　Give your own personal subjective probability for each of the following situations. It would be instructive to tabulate these probabilities for the entire class. In which cases did you get large disagreements?

　a　The federal buget will be balanced in the next fiscal year.

　b　You will receive a B or higher in this course.

　c　Two or more individuals in the classroom will have the same birthday.

　d　The Washington Redskins will win the Super Bowl next year.

　e　The total production of Florida oranges next year will exceed this year's production.

5.3　Finding the Probability of an Event

In the preceding section we discussed three different interpretations of probability. We will use the classical interpretation and the relative frequency concept to illustrate how to compute the probability of an outcome or an event. Consider an experiment that consists of tossing two coins—a penny and a dime—and observing their upturned faces. There are four possible outcomes:

　TT: Tails for both coins

　TH: Tails for the penny, and heads for the dime

　HT: Heads for the penny, and tails for the dime

　HH: Heads for both coins

What is the probability of observing the following event: exactly one heads from the two coins?

This probability can be obtained easily if we can assume that all four outcomes are equally likely. In this case, that seems quite reasonable. There are $N = 4$ possible outcomes, and $N_E = 2$ of these (outcomes TH and HT) are favorable for the event of interest—observing exactly one heads. Hence, by the classical interpretation of probability,

$$P(\text{exactly 1 heads}) = \frac{2}{4} = \frac{1}{2} = .5.$$

Since the event of interest also has a relative frequency interpretation, we could obtain the same result empirically, by using the relative frequency concept. Suppose that a penny and dime are tossed 2000 times, with the results shown in Table 5.1. Notice that this approach yields approximate probabilities that are in general agreement with out intuition; that is, intuitively we might expect each of the four outcomes to be equally likely, so that each would occur with a probability equal to 1/4 or .25. This assumption was made for the classical interpretation.

T A B L E 5.1
Results of 2000 Tosses of a Penny and a Dime

Outcome	Frequency	Relative Frequency
TT	474	474/2000 = .237
TH	502	502/2000 = .251
HT	496	496/2000 = .248
HH	528	528/2000 = .264

If we wish to find the probability of observing exactly one heads on a particular toss of two coins, we have, from Table 5.1,

$$P(\text{exactly 1 heads}) \approx \frac{502 + 496}{2000} = .499.$$

This is very close to the theoretical probability, which we have shown to be .5.

The probability of any event—say, event A—will always satisfy the property

$$0 \le P(A) \le 1.$$

In other words, the probability of an event must lie somewhere in the interval from 0 (the occurrence of the event is impossible) to 1 (the occurrence of the event is a "sure thing").

Relations between two events and probabilities associated with these relations will be discussed in the next section.

5.4   Basic Event Relations and Probability Laws

Suppose that A and B represent two experimental events and that you are interested in a new event—the event that A or B occurs. For example, suppose that we toss a pair of dice and define the following events:

A: A total of 7 shows.

B: A total of 11 shows.

Then the event "either A or B occurs" is the event that you toss a total of either 7 or 11 with the pair of dice.

Notice that, for this example, the events A and B are mutually exclusive; that is, if you observe event A (a total of 7), you cannot at the same time observe event B (a total of 11). Thus, if A occurs, B cannot occur (and vice versa).

DEFINITION 5.1

Mutually Exclusive Events Two events A and B are said to be **mutually exclusive** if (when the experiment is performed a single time) the occurrence of one of the events excludes the possibility of the occurrence of the other event. ■

mutually exclusive events

The concept of **mutually exclusive events** is used to specify a second property that the probabilities of events must satisfy. When two or more events are mutually exclusive, the probability that any one of the events will occur is the sum of the event probabilities.

DEFINITION 5.2

Probability of Mutually Exclusive Events If two events, A and B, are mutually exclusive, the **probability** that either event will occur is P(either A or B) $= P(A) + P(B)$. ■

The definition of additivity of probabilities for mutually exclusive events can be extended beyond two events. For example, when we toss a pair of dice, the sum S of the numbers appearing on the dice can assume any one of the values $S = 2, 3, 4, \ldots, 11, 12$. On a single toss of the dice, we can observe only one of these values. Therefore, the values $2, 3, \ldots, 12$ represent mutually exclusive events. It follows that the probability of tossing a sum less than or equal to 4, is

$$P(S \leq 4) = P(2) + P(3) + P(4).$$

For this particular experiment, the dice can fall in 36 equally likely different ways. For example, we can observe a 1 on the first die and a 1 on the second die, denoted by the symbol $(1, 1)$; we can observe a 1 on the first die and a 2 on the second die, denoted by $(1, 2)$; and so on. In other words, for this experiment the possible outcomes are

(1, 1)	(2, 1)	(3, 1)	(4, 1)	(5, 1)	(6, 1)
(1, 2)	(2, 2)	(3, 2)	(4, 2)	(5, 2)	(6, 2)
(1, 3)	(2, 3)	(3, 3)	(4, 3)	(5, 3)	(6, 3)
(1, 4)	(2, 4)	(3, 4)	(4, 4)	(5, 4)	(6, 4)
(1, 5)	(2, 5)	(3, 5)	(4, 5)	(5, 5)	(6, 5)
(1, 6)	(2, 6)	(3, 6)	(4, 6)	(5, 6)	(6, 6)

As you can see, only one of these events, $(1, 1)$, will result in a sum equal to 2. Therefore, we would expect a sum of 2 to occur with a relative frequency of 1/36 in a long series of repetitions of the experiment, so we let $P(2) = 1/36$. The sum

$S = 3$ will occur if we observe either of the outcomes $(1, 2)$ or $(2, 1)$. Therefore, $P(3) = 2/36 = 1/18$. Similarly, we find $P(4) = 3/36 = 1/12$. It follows that

$$P(S \leq 4) = P(2) + P(3) + P(4) = \frac{1}{36} + \frac{1}{18} + \frac{1}{12} = \frac{1}{6}.$$

A third property of event probabilities is stated in terms of an event and its complement.

complement

DEFINITION   5.3

Complement   The **complement** of an event A is the event that A does not occur. The complement of A is denoted by the symbol $\overline{A}$.   ▪

Thus, if we define the complement of an event A as a new event—namely, "A does not occur"—if follows that

$$P(A) + P(\overline{A}) = 1$$

For example, refer again to the two-coin-toss experiment. Suppose that, in many repetitions of the experiment, the proportion of times you observe event A, "two heads show," is 1/4; then it follows that the proportion of times you observe the event $\overline{A}$, "two heads do not show," is 3/4. Thus $P(A)$ and $P(\overline{A})$ always sum to 1.

The three properties that the probabilities of events must satisfy can be summarized as follows.

Properties of Probabilities

If A and B are any two mutually exclusive events associated with an experiment, the $P(A)$ and $P(B)$ must satisfy the following properties:

1   $0 \leq P(A) \leq 1$ and $0 \leq P(B) \leq 1$.
2   $P(\text{either } A \text{ or } B) = P(A) + P(B)$.
3   $P(A) + P(\overline{A}) = 1$ and $P(B) + P(\overline{B}) = 1$.

union
intersection

We can now define two additional event relations: the **union** and the **intersection** of two events.

DEFINITION   5.4

Union   The **union** of two events A and B is the set of all outcomes that are included in either A or B (or both). The union is denoted as $A \cup B$.   ▪

DEFINITION   5.5

Intersection   The **intersection** of two events A and B is the set of all outcomes that are included in both A and B. The intersection is denoted as $A \cap B$.   ▪

These definitions, along with the definition of the complement of an event, formalize some simple concepts. The event $\overline{A}$ occurs when A *does not*; $A \cup B$ occurs when *either A or B* occurs: and $A \cap B$ occurs when *both A and B* occur.

The additivity of probabilities for mutually exclusive events, called the *addition law for two mutually exclusive events*, can be extended to yield the general addition law for probabilities from two events, called the probability of the union of two events.

DEFINITION   5.6

Probability of the Union   Consider two events A and B; the **probability of the union** of A and B is

$$P(A \cup B) = P(A) + P(B) - P(A \cap B). \quad \blacksquare$$

The reason for subtracting $P(A \cap B)$ when computing $P(A \cup B)$ is that, by adding $P(A)$ and $P(B)$, we end up "double-counting" the area of intersection between A and B. Hence, to get rid of the repetition, we must subtract $P(A \cap B)$ from the sum of $P(A)$ and $P(B)$. This can be seen in the figure accompanying Example 5.1.

EXAMPLE   5.1   Events and event probabilities are shown in the following Venn diagram.

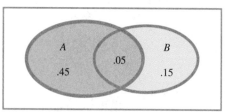

Use this diagram to determine the probabilities listed.

a $P(A), P(\overline{A})$

b $P(B) < P(\overline{B})$

c $P(A \cap B)$

d $P(A \cup B)$

Solution   From the Venn diagram, we are able to determine the following probabilities.

a $P(A) = .5$, therefore, $P(\overline{A}) = 1 - .5 = .5$.

b $P(B) = .2$, therefore, $P(\overline{B}) = 1 - .2 = .8$.

c $P(A \cap B) = .05$.

d $P(A \cup B) = P(A) + P(B) - P(A \cap B) = .5 + .2 - .05 = .65.$ $\quad \blacksquare$

5.5 Conditional Probability and Independence

Consider a situation involving the examination of a large number of insurance claims, categorized according to type of insurance and according to whether the claim was fraudulent. Imagine that the examination produced the results shown in Table 5.2, and suppose that you are responsible for checking insurance claims—in particular, for detecting fraudulent claims. When you examine the next claim that is to be processed, what is the probability of event F—"the claim is fraudulent"? To answer the question, examine Table 5.2 and notice that 10% of all claims are fraudulent. Thus, assuming that the percentages given in the table are reasonable approximations to the true probabilities of receiving specific types of claims, it follows that $P(F) = .10$. Would you say that the risk that the next claim you examine is fraudulent has a probability of precisely .10? We think not, because you have recourse to additional information that may alter your assessment of $P(F)$. For example, you can take into consideration the type of policy involved in the claim you are examining (fire, auto, or other).

TABLE 5.2
Categorization of Insurance
Claims

Category	Type of Policy			Total
	Fire	Auto	Other	
Fraudulent	6%	1%	3%	10%
Nonfraudulent	14%	29%	47%	90%
Total	20%	30%	50%	100%

Suppose that the claim you are examining involves a fire policy. Checking Table 5.2, you can see that 20% (or .20) of all claims are associated with fire policies and that 6% (or .06) of all claims are fraudulent fire policy claims. It follows that the probability that the claim is fraudulent, given that it involves a fire policy, is

$$P(F|\text{Fire policy}) = \frac{\text{Proportion of claims that are fraudulent fire policy claims}}{\text{Proportion of claims involving fire policies}}$$

$$= \frac{.06}{.20} = .30.$$

conditional probability   This probability, $P(F|\text{fire policy})$, is called a **conditional probability** of the event F; that is, it expresses the probability of event F, given the fact that the event "Fire policy" has already occurred. This tells you that 30% of all fire policy claims are fraudulent. The vertical bar in the expression $P(F|\text{fire policy})$ represents the phrase "given that," or simply "given." Thus, the expression is read, "the probability of the event F, given the event Fire policy."

unconditional probability   The probability $P(F) = .10$, called the **unconditional** or **marginal probability** of the event F, gives the proportion of times a claim is fraudulent; that is, it expresses the proportion of times event F occurs in a very large (infinitely large) number of repetitions of the experiment (receiving an insurance claim and determining whether the claim is fraudulent). In contrast, the conditional probability of F, given that the claim is for a fire policy—$P(F|\text{Fire policy})$—gives the proportion of fire policy claims that are fraudulent. Clearly, the various conditional probabilities of F, given the types of policies involved, are of much greater assistance in measuring the risk of fraud than is the unconditional probability of F.

DEFINITION   5.7

Conditional Probability   Consider two events A and B with nonzero probabilities, $P(A)$ and $P(B)$. The **conditional probability** of event A, given event B, is

$$P(A|B) = \frac{P(A \cap B)}{P(B)}.$$

The conditional probability of event B, given event A, is

$$P(B|A) = \frac{P(A \cap B)}{P(A)}.$$ ■

This definition of conditional probabilities gives rise to the multiplication law for probabilities called the probability of the interstion of two events.

DEFINITION   5.8

Probability of the Intersection   The **probability of the intersection** of two events A and B is

$$P(A \cap B) = P(A)P(B|A)$$
$$= P(B)P(A|B).$$ ■

The only difference between Definitions 5.7 and 5.8, both of which involve conditional probabilities, relates to what probabilities are known and what probability needs to be calculated. When the intersection probability $P(A \cap B)$ and the individual probability $P(A)$ are known, we can compute $P(B|A)$. When we know $P(A)$ and $P(B|A)$, we can compute $P(A \cap B)$.

EXAMPLE   5.2   Two supervisors in a company are to selected as saftey representatives within the company. Given that there are six supervisors in research and five in development, and given that each group of two supervisors has the same chance of being selected, find the probability that both supervisors will be chosen from research.

Solution   Let A be the event that the first supervisor selected is from research, and let B be the event that the second supervisor is also from research. Clearly, we want to find $P(A \cap B) = P(A)P(B|A)$.

For this example,

$$P(A) = \frac{6}{11} \quad \text{and} \quad P(B|A) = \frac{5}{10}.$$

Then

$$P(A \cap B) = \left(\frac{6}{11}\right)\left(\frac{5}{10}\right) = \frac{30}{110} = .27. \quad \blacksquare$$

Suppose that the probability of event A is the same regardles of whether event B has occurred; that is, suppose that

$$P(A|B) = P(A).$$

Then we say that the occurrence of event A is not dependent on the occurrence of event B or, more simply, that A and B are **independent events**.

independent events

DEFINITION 5.9   **Independent Events**   Two events A and B are **independent events** if

$$P(A|B) = P(A) \qquad \text{or if} \qquad P(B|A) = P(B).$$

(*Note:* You can show that, if $P(A|B) = P(A)$ then $P(B|A) = P(B)$; and vice versa.) $\blacksquare$

The concept of independence is of particular importance in sampling. Subsequently, we will draw samples from two (or more) populations in order to compare population means, population variances, or some other population parameters. For most of these applications, we will select samples in such a way that the observed values in one sample are independent of the values that appear in another sample. We call these **independent samples**.

independent samples

Exercises

Basic Techniques

5.3   A coin is to be flipped three times. List the possible outcomes in the following form: (result on toss 1, result on toss 2, result on toss 3).

5.4   For the data in Exercise 5.3, assume that each outcome has probability $\frac{1}{8}$ of occurring. Find the following probabilities.

a A: Observe exactly 1 heads.

b B: Observe 1 or more heads.

c C: Observe no heads.

5.5   Refer to Exercise 5.4.

a   Compute the probability of the complement of event A, event B, and event C.

b   Determine whether events A and B are mutually exclusive.

5.6   Determine the following conditional probabilities for the events of Exercise 5.4.

a $P(A|B)$

b $P(A|C)$

c $P(B|C)$

5.7   Refer to Exercise 5.6. Are events A and B independent? Why or why not? What about A and C? What about B and C?

5.8   A die is to be rolled and we are to observe the number that falls face up. Find the probabilities for these events.

a A: Observe a 6.

b B: Observe an even number.

c C: Observe a number greater than 2.

d D: Observe an even number and a number greater than 2.

5.9   Refer to Exercise 5.8. Which of the events (A, B, and C) are independent? Which are mutually exclusive?

5.10   Consider the following outcomes for an experiment:

Outcome	1	2	3	4	5
Probability	.20	.25	.15	.10	.30

Let event A consist of outcomes 1, 3, and 5, and let event B consist of outcomes 4 and 5.

a   Find $P(A)$ and $P(B)$.

b   Find P(both A and B occur).

c   Find P(either A or B occurs).

5.11   Refer to Exercise 5.10. Does P(either A or B occurs) $= P(A) + P(B)$? Why or why not?

Applications

5.12   A student has to take an accounting course and an economics course next term. Assuming that there are no schedule conflicts, describe the possible outcomes for selecting one section of the accounting course and one of the economics course if there are four possible accounting sections and three possible economics sections.

5.13   The emergency room of a hospital has two backup generators, either of which can supply enough electricity for basic hospital operations. We define events A and B as follows:

Event A: Generator 1 works properly.

Event B: Generator 2 works properly.

Describe the following events in words:

a   Complement of A

b $B|A$

c   Either A or B

5.14   A survey of a number of large corporations gave the following probability table for events related to the offering of a promotion involving a transfer.

Promotion Transfer	Married		Unmarried	Total
	Two-career Marriage	One-career Marriage		
Rejected	.184	.0555	.0170	.2565
Accepted	.276	.3145	.1530	.7435
Total	.46	.37	.17	

Use the probabilities to answer the following questions.

a What is the probability that a professional (selected at random) will accept the promotion? Reject it?

b What is the probability that a professional (selected at random) is part of a two-career marriage? A one-career marriage?

5.15 An institutional investor is considering a large investment in two of five companies. Suppose that, unknown to the investor, two of the five firms are on shaky ground with regard to the development of new products.

a List the possible outcomes for this situation.

b Determine the probability of choosing two of the three firms that are on better ground.

c What is the probability of choosing one of the two firms on shaky ground?

d What is the probability of choosing both of the two shaky firms?

5.16 A survey of workers at two manufacturing sites of a firm included the following question: How effective is manangement in responding to legitimate grievances of workers? The results are shown here.

	Number Surveyed	Number Responding "Poor"
Site 1	192	48
Site 2	248	80

Let A be the event that a worker comes from Site 1, and let B be the event that the response is "poor." Compute $P(A)$, $P(B)$, and $P(A \cap B)$.

5.17 Refer to Exercise 5.16.

a Are events A and B independent?

b Find $P(B|A)$ and $P(B|\overline{A})$. Are they equal?

5.18 A large corporation has spent considerable time developing employee performance rating scales to evaluate employees' job performance on a regular basis, so that major adjustments can be made when needed and so that employees who should be considered for "fast-track" advancement can be isolated. Keys to the latter determination are ratings on the employee's ability to perform to his or her capabilities and on his or her formal training for the job.

	Formal Training			
Workload Capacity	None	Little	Some	Extensive
Low	.01	.02	.02	.04
Medium	.05	.06	.07	.10
High	.10	.15	.16	.22

The probabilities for being placed on a fast track are as indicated for the 12 categories of workload capacity and formal training. The following three events (A, B, and C) are defined:

A: An employee works at the high-capacity level.

B: An employee falls into the highest (extensive) formal training category.

C: An employee has little or no formal training and works below high capacity.

a Find $P(A)$, $P(B)$, and $P(C)$.

b Find $P(A|B)$, $P(A|\overline{B})$, and $P(\overline{B}|C)$.

c Find $P(A \cup B)$, $P(A \cap C)$, and $P(B \cap C)$.

5.19 The utility company in a large metropolitan area finds that 70% of its customers pay a given monthly bill in full.

a Suppose that two customers are chosen at random from the list of all customers. What is the probability that both customers will pay their latest monthly bill in full?

b   What is the probability that at least one of them will pay in full?

5.20   Refer to Exercise 5.19. A more detailed examination of the company records indicates that 95% of the customers who pay one monthly bill in full will pay the next monthly bill in full, too; only 10% of those who pay less than the full amount one month will pay in full the next month.

a   Find the probability that a customer selected at random will pay two consecutive months in full.

b   Find the probability that a customer selected at random will pay neither of two consecutive months in full.

c   Find the probability that a customer chosen at random will pay exactly one month in full.

5.6 Random Variables

The basic language of probability developed in this chapter deals with many different kinds of events. We are interested in calculating the probabilities associated with both quantitative and qualitative events. For example, we developed techniques that could be used to calculate the probability that a person selected at random for a Nielsen survey of television viewing habits would favor the ABC nightly news program (as opposed to that of CBS or NBC). These same techniques are also applicable to finding the probability that a person selected for the Nielsen survey watches television more than 30 hours per week.

These qualitative and quantitative events can be classifed as events (or outcomes) associated with qualitative and quantitative variables. For example, in the Nielsen survey, responses to the question "Which evening television news program do you prefer: ABC, CBS, or NBC?" are observations on a qualitative variable, since the possible responses vary in kind but not in any numerical degree. Because we cannot predict with certainty what a particular person's response will be, the variable is classified as a **qualitative random variable**. Other qualitative random variables that are commonly measured include political party affiliation, socioeconomic status, geographic location, and sex/race classification.

qualitative random variable

A finite (and typically quite small) number of possible outcomes are associated with any qualitative variable. Using the methods of this chapter, it is possible to calculate the probabilities associated with these events.

Many times the events of interest in an experiment are quantitative outcomes associated with a **quantitative random variable**, since the possible responses vary in numerical magnitude. For example, in a Nielsen survey, responses to the question "How many hours a week do you watch television?" are observations on a quantitative random variable. Events of interest, such as viewing television more than 30 hours per week, are measured by this quantitative random variable. Other quantitative random variables include the change in earnings per share of a stock over the next year, the increase in total sales over the next year, and the number of persons voting for the incumbent in an upcoming election. Again, the methods of this chapter can be applied to calculate the probability associated with any particular event.

quantitative random variable

There are major advantages to dealing with quantitative random variables. The numerical yardstick underlying a quantitative variable makes the mean and the standard deviation (for instance) sensible. With qualitative random variables, there isn't much more to be said than has already been said. The methods of this chapter can be used to calculate the probabilities of various events, and that's about all. With quantitative random variables we can do much more: we can average the resulting quantities, find standard deviations, and assess probable errors, among other things. Hereafter, we use the phrase *random variable* to mean a quantitative random variable; virtually all texts on probability theory use this phrase the same way we do.

Most events of interest result in numerical observations or measurements. If a quantitative variable measured (or observed) in an experiment is denoted by the symbol y, we are interested in the values that x can assume. These values are called numerical outcomes. The number of students in a class of 50 who earn an A in their biology course is a numerical outcome. The percentage of registered voters who cast ballots in a given election is also a numerical outcome. The quantitative variable y is called a random variable because the value that y assumes in a given experiment is a chance or random outcome.

Random variables are classified as one of two types.

DEFINITION   5.10

> **Discrete Random Variable**   When observations on a quantitative random variable can assume only a countable number of values, the variable is called a **discrete random variable**. ▪

These are examples of discrete random variables:

- The number of bushels of apples per acre for a given orchard this year
- The number of accidents per year at an intersection
- The number of voters in a sample favoring candidate Jones

Notice that it is possible to count the number of values that each of these random variables can assume.

DEFINITION   5.11

> **Continuous Random Variable**   When observations on a quantitative random variable can assume any of the countless number of values in a line interval, the variable is called a **continuous random variable**. ▪

discrete random variable
continuous random
variable

For example, the daily maximum temperature in Rochester, New York, can assume any of the infinitely many values on a line interval. It could be 89.6, 89.799, or 89.7611114. Typical continuous random variables are temperature, pressure, height, weight, and distance.

The distinction between **discrete** and **continuous random variables** is pertinent when we are seeking the probabilities associated with specific values of a random variable. The need for the distinction will become apparent when we discuss probability distributions in later sections of this chapter.

5.7 Probability Distributions for Discrete Random Variables

probability distribution

As was previously stated, we need to know the probability of observing a particular sample outcome in order to make an inference about the population from which the sample was drawn. To do this, we need to know the probability associated with each value of the variable y. Viewed as relative frequencies, these probabilities generate a distribution of theoretical relative frequencies called the **probability distribution** of y. Probability distributions differ for discrete and continuous variables, but the interpretation is essentially the same for both.

The *probability distribution for a discrete random variable* displays the probability $P(y)$ associated with each value of y. This display can be presented as a table, a graph, or a formula. To illustrate, consider the tossing of two coins in Section 5.2, and let y be the number of heads observed. Then y can take the values 0, 1, or 2. From the data of Table 5.1, we can determine the approximate probability for each value of y, as given in Table 5.3. We should point out that the relative frequencies in the table are very close to the theoretical relative frequencies (probabilities), which can be shown to be .25, .50, and .25 by using the classical interpretation of probability. If we had employed 2,000,000 tosses of the coins instead of 2000, the relative frequencies for $y = 0$, 1, and 2 would be indistinguishable from the theoretical probabilities.

TABLE 5.3
Empirical Sampling Results for
y: the Number of Heads in
2000 Tosses of Two Coins

y	Frequency	Relative Frequency
0	474	.237
1	998	.499
2	528	.264

The probability distribution for y, the number of heads in the toss of two coins, is shown in Table 5.4. It is presented graphically as a *probability histogram* in Figure 5.1.

TABLE 5.4
Probability Distribution for the
Number of Heads When Two
Coins Are Tossed

y	$P(y)$
0	.25
1	.50
2	.25

The probability distribution for this simple discrete random variable illustrates three important properties of discrete random variables.

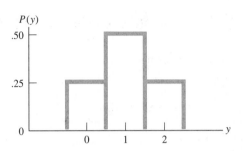

Properties of Discrete Random Variables

1   The probability associated with every value of y lies between 0 and 1.

2   The sum of the probabilites for all values of y is equal to 1.

3   The probabilities for a discrete random varable are additive. Hence, the probability that $y = 1$ or 2 is equal to $P(1) + P(2)$.

5.8 A Useful Discrete Random Variable: The Binomial

Many populations of interest to businesspersons and scientists can be represented as large sets of 0s and 1s. For example, consider the set of responses of all adults in the United States to the question, "Do you favor the development of nuclear energy?" If we disallow "no opinion," the responses will constitute a set of "yes" responses and "no" responses. If we assign a 1 to each yes and a 0 to each no, the population will consist of a set of 0s and 1s, and the sum of the 1s will equal the total number of persons favoring the development. The sum of the 1s divided by the total number of adults in the United States will equal the proportion of people who favor the development.

Gallup and Harris polls are examples of the sampling of 0, 1 populations. People are surveyed, and their opinions are recorded. Based on the sample responses, Gallup and Harris estimate the proportions of people in the population who favor some particular issue or possess some particular characteristic.

Similar surveys are conducted in the biological sciences, in engineering, and in business, but they may be called experiments rather than polls. For example, experiments are conducted to determine the effect of new drugs on small animals, such as rats or mice, before progressing to larger animals and, eventually, to human subjects. Many of these experiments bear a marked resemblance to a poll, in that the experimenter records only whether the drug was effective. Thus, is 300 rats are injected with a drug and 230 show a favorable response, the experimenter has conducted a "poll"—a poll of rat reaction to the drug, with 230 "in favor" and 70 "opposed."

Similar "polls" are conducted by most manufacturers to determine the fraction of a product that is of good quality. Samples of industrial products are collected before shipment, and each item in the sample is judged "defective" or "acceptable" according to criteria established by the company's quality control department. Based on the number of defectives identified in the sample, the company can decide whether the product is suitable for shipment. Notice that this example, as well as those preceding, has the practical objective of making an inference about a population based on information contained in a sample.

binomial experiment

The public opinion poll, the consumer preference poll, the drug-testing experiment, and the industrial sampling for defectives are all examples of a common, frequently conducted sampling situation known as a **binomial experiment**. The binomial experiment is conducted in all areas of science and business; it only differs from one situation to another in the nature of the objects being sampled (people, rats, electric light bulbs, oranges). Thus, it is useful for us to define its characteristics. We can then apply our knowledge of this one kind of experiment to a variety of sampling experiments.

For all practical purposes, the binomial experiment is identical to the cointossing example of previous sections. Here, n different coins are tossed (or a single coin is tossed n times), and we are interested in the number of heads observed. We assume that the probability of tossing heads on a single trial is p (p may equal .50, as it would for a balanced coin, but in many practical situations $\frac{p}{n}$ will take some other value between 0 and 1). We also assume that the outcome for any one toss is unaffected by the results of any preceding tosses. These characteristics can be summarized as shown.

DEFINITION   5.12

> **Binomial Experiment**   A **binomial experiment** has the following properties:
>
> **1** The experiment consists of n identical trials.
>
> **2** Each trial results in one of two outcomes. We will label on outcome a success and the other a failure.
>
> **3** The probability of success on a single trial is equal to p, and p remains the same from trial to trial.[†]
>
> **4** The trials are independent; that is, the outcome of one trial does not influence the outcome of any other trial.
>
> **5** The random variable y is the number of successes observed during the n trials. ▪

EXAMPLE   5.3   A survey of 500 farmers is conducted to determine the proportion who favor additional price supports for dairy products. Does this survey satisfy the properties of a binomial experiment?

Solution   To answer this question, we check each of the five characteristics of the binomial experiment, to determine if they are satisfied.

[†]Some textbooks and computer programs use the letter π rather than p.

1   Are there n identical trials? Yes. There are $n = 500$ interviews, all the same.

2   Does each trial result in one of two outcomes? Yes. Each farmer interviewed either favors or does not favor additional price supports.

3   Is the probability of success the same from trial to trial? Yes. If we let "success" denote a farmer who favors additional supports, then, assuming that the list of farmers from which the sample was drawn is large, the probability of success will (for all practical purposes) remain constant from trial to trial.

4   Are the trials independent? Yes. The outcome of one interview is unaffected by the results of the other interviews.

5   Is the random variable of interest to the experimenter the number of successes y in the sample? Yes. We are interested in the number of farmers in the sample of 500 who favor additional price supports for dairy products.

Since all five characteristics are satisfied, the survey represents a binomial experiment. ▪

E X A M P L E **5.4**   An economist interviews 75 students in a class of 100, to estimate the proportion of students who expect to obtain a C or better in the course. Is this a binomial experiment?

Solution   Check this experiment against the five characteristics of a bionomial.

1   Are there identical trials? Yes. Each of 75 students is interviewed.

2   Does each trial result in one of two outcomes? Yes. Each student either does or does not expect to obtain a grade of C or higher.

3   Is the probability of success the same from trial to trial? No. If we let success denote a student who expects to obtain a C or higher, then the probability of success can change considerably from trial to trial. For example, suppose that, unknown to the professor, 75 of the 100 students expect to obtain a grade of C or higher. Then p, the probability of success for the first student interviewed is $75/100 = .75$. If that student is a failure (does not expect a C or higher), the probability of success for the next student is $75/99 = .76$. Suppose that, after 70 students have been interviewed, 60 were successes and 10 were failures. Then the probability of success for the next (71st) student is $15/30 = .50$.

This example shows how the probability of success can change substantially from trial to trial in situations where the sample size is a relatively large portion of the total population size. This experiment does not satisfy the properties of a binomial experiment. ▪

It should be noted that very few real-life situations perfectly satisfy the requirements stated in Definition 5.12, but for many the lack of agreement is so small that the binomial experiment still provides a very good model for reality.

Having defined the binomial experiment and suggested several practical applications, we now examine the probability distribution for the binomial random variable y—the number of successes observed in n trials. Although it would be possible to approximate $P(y)$, the probability associated with a value of y in a binomial experiment, by using a relative frequency approach, it is easier to make use of a general formula for binomial probabilities.

Formula for Computing $P(y)$ in a Binomial Experiment

The probability of observing y successes in n trials of a binomial experiment is

$$P(y) = \frac{n!}{y!(n-y)!} p^y q^{n-y}$$

where

n = number of trials

p = probability of success on a single trial

$q = 1 - p$ = probability of failure on a single trial

y = number of successes in n trials

$n! = n(n-1)(n-2) \cdots (3)(2)(1).$

As indicated above, the notation $n!$ (referred to as n factorial) is used for the product

$$n! = n(n-1)(n-2) \cdots (3)(2)(1).$$

For $n = 3$,

$$n! = 3! = (3)(3-1)(3-2) = (3)(2)(1) = 6.$$

Similarly, for $n = 4$,

$$4! = (4)(3)(2)(1) = 24.$$

We also note that $0!$ is defined to be equal to 1.

To see how the formula for binomial probabilities can be used to calculate the probability for a specific value of y, consider the following examples.

EXAMPLE 5.5   An experiment consists of tossing a coin two times. If the probability of heads is .5, compute the probability distribution for y, the total number of heads, using the binomial formula $P(y)$. Compare your results to those given in Table 5.4.

Solution   Using the formula

$$P(y) = \frac{n!}{y!(n-y)!} p^y q^{n-y}$$

and substituting for $n = 2$, $p = .5$, $q = .5$, and $y = 0, 1, 2$, we obtain

$$P(y = 0) = \frac{2!}{0!2!}(.5)^0(.5)^2 = .25$$

$$P(y = 1) = \frac{2!}{1!1!}(.5)(.5) = .50$$

$$P(y = 2) = \frac{2!}{2!0!}(.5)^2(.5)^0 = .25.$$

Notice that these results are identical to those presented in Table 5.4.   ▪

EXAMPLE   5.6   A survey is conducted to determine the proportion of adults in a certain locale who favor raising the legal drinking age from 19 to 21. A random sample of 300 adults is selected from the list of registered voters. They are interviewed, and the number of those favoring the change is recorded. Is this a binomial experiment?

Solution   To answer this question, we will check each of the five characteristics of a binomial experiment, to determine if they are satisfied.

1   Are there n identical trials? Yes; $n = 300$ interviews are conducted in an indentical manner.

2   Does each trial result in one of two outcomes? Yes; each adult interviewed either favors or does not favor the change.

3   Is the probability of success the same from trial to trial? Yes; if we let success denote a person favoring the change, then, assuming that the list of registered voters is large, the probability of selecting an adult who favors the change will remain (for all practical purposes) constant from trial to trial.

4   Are the trials independent? Yes; the outcome of one interview is unaffected by the results of the other interviews.

5   For this experiment, the random variable of interest to the experimenter is the number of successes in the sample.

Since all five characteristics are present, the survey represents a binomial experiment.   ▪

EXAMPLE   5.7   Suppose that a sample of households is randomly selected from all the households in a city, in order to estimate the percentage in which the head of the household is unemployed. To illustrate how to compute a binomial probability, suppose that the unknown percentage is actually 10% and that a sample of $n = 5$ (we are selecting a small sample to make the calculation magnageable) is selected from the population. What is the probability that all 5 heads of the sample households are employed?

Solution   We must carefully define which outcome we wish to call a success. For this example, we will define a success as being employed. Then the probability of success when one person is selected from the population is $p = .9$ (because the proportion unemployed is .1). We wish to find the probability that $y = 5$ (all 5 are employed) in 5 trials.

$$P(y = 5) = \frac{5!}{5!(5 - 5)!}(.9)^5(.1)^0$$

$$= \frac{5!}{5!0!}(.9)^5(.1)^0$$

$$= (.9)^5 = .590$$

The binomial probability distribution for $n = 5$ and $p = .9$ is shown in Figure 5.2. The probability of observing employed heads of household in a sample of 5 is shaded in the figure.   ∎

FIGURE 5.2
The Binomial Probability
Distribution for $n = 5$ and
$p = .9$

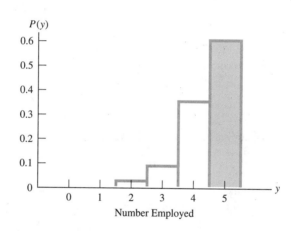

E X A M P L E   5.8   Refer to Example 5.7, and calculate the probability that exactly one person in the sample of 5 heads of household is unemployed. What is the probability of one or fewer being unemployed?

Solution   Since y is the number of employed heads of household in the sample of 5, one unemployed person would correspond to 4 employed ($y = 4$). Then

$$P(4) = \frac{5!}{4!(5 - 4)!}(.9)^4(.1)^1$$

$$= \frac{(5)(4)(3)(2)(1)}{(4)(3)(2)(1)(1)}(.9)^4(.1)$$

$$= 5(.9)^4(.1)$$

$$= .328$$

Thus, the probability of selecting 4 employed heads of household in a sample of 5 is .328, or roughly one chance in three.

The outcome "one or fewer unemployed" is the same as the outcome "4 or 5 employed." Since y represents the number employed, we seek the probability that $y = 4$ or 5. Because the values associated with a random variable represent mutually exclusive events, the probabilities for discrete random variables are additive. Thus we have

$$P(y = 4 \text{ or } 5) = P(4) + P(5)$$
$$= .328 + .590$$
$$= .918$$

That is, the probability that a random sample of 5 households will yield either 4 or 5 employed heads of household is .918. This high probability is consistent with our intuition: we can expect the number of employed heads of household in the sample to be large if 90% of all heads of household in the city are employed.   ▪

Like any relative frequency histogram, a binomial probability distribution possesses a mean μ and a standard deviation σ. Although we omit the derivations, we give the formulas for these parameters.

Mean and Standard Deviation of the Binomial Probability Distribution

$$\mu = np \qquad \text{and} \qquad \sigma = \sqrt{npq}$$

where p is the probability of success in a given trial, n is the number of trials in the binomial experiment, and $q = 1 - p$.

Knowing p and the sample size n, we can calculate μ and σ to locate the center and describe the variability for a particular binomial probability distribution. Thus we can quickly determine the values of y that are probable and the ones that are improbable.

E X A M P L E   5.9   Calculate the mean and the standard deviation for a binomial probability distribution with $p = .5$ and $n = 20$. The probability distribution for the number of successes is shown in Figure 5.3.

Solution   Substituting into the formulas, we obtain

$$\mu = np = 20(.5) = 10$$
$$\sigma = \sqrt{npq} = \sqrt{(20)(.5)(.5)} = \sqrt{5} = 2.24$$

Notice that $y = 0$ is more than 4σ away from the mean $\mu = 10$. If we apply the Empirical Rule to this mound-shaped distribution, we see that it is highly improbable

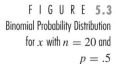

F I G U R E   **5.3**
Binomial Probability Distribution
for x with $n = 20$ and
$p = .5$

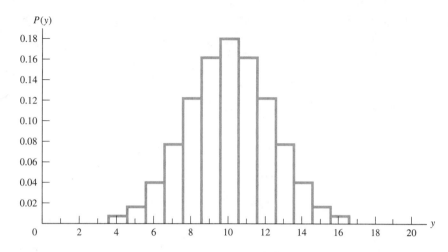

that, in 20 trials, we would observe such a small value of y if p really were equal to .5.   ▪

E X A M P L E   **5.10**   A poll shows that 516 of 1218 voters favor the reelection of a particular political candidate. Do you think that the candidate will win?

Solution   To win the election, the candidate needs at least 50% of the votes cast in a two-person race. Let us see whether $y = 516$ is too small a value of y to imply a value of p (the proportion of voters favoring the candidate) equal to .5 or larger. If $p = .5$, then

$$\mu = np = (1218)(.5) = 609$$
$$\sigma = \sqrt{npq} = \sqrt{(1218)(.5)(.5)}$$
$$= \sqrt{304.5} = 17.45$$

and $3\sigma = 52.35$.

You can see from Figure 5.4 that $y = 516$ is more than 3σ, or 52.35, away from $\mu = 609$. In fact, $y = 516$ is more than 5σ away from $\mu = 609$, the value of μ if p were really equal to .5. Thus is appears that the number of voters in the sample who favor the candidate is much too small for a candidate who does, in fact, enjoy the support of a majority favoring reelection. Consequently, we conclude that he or she will lose. (Notice that this conclusion is based on the assumption that the set of voters from which the sample was drawn is the same as the set of voters who will vote. We also must assume that the opinions of the voters will not change between the time of sampling and the date of the election.)   ▪

FIGURE 5.4
Location of the Observed Value
of $y(y = 516)$ Relative to
μ

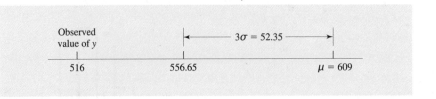

This section has been devoted to presenting the binomial probability distribution so that you can see how binomial probabilities are calculated and so that you can calculate them for small values of n, if you so desire. In practice, n is usually large (in national surveys, sample sizes as large as 1500 are common), and the computation of the binomial probabilities is very tedious.

Fortunately, these computations can be avoided. Table 1 of Appendix 3 gives individual binomial probabilities $P(y)$ for various values of n, y, and p. For example, from Table 1, $P(y = 2)$ for $n = 10$ and $p = .4$ is found by reading down the x column on the left for $p = .4$. The correct probability is .1209. For $p > .50$, you must refer to the y column on the right. So for $n = 10$, $p = .6$, and $P(y = 2)$, we locate the $p = .6$ column and read up the y column on the right. The correct value is .0106. For most purposes, exact values of the probabilities are unnecessary. We will present a simple procedure in Chapter 6 for obtaining approximate values to the probabilities we need in making inferences. We can also use some very rough procedures for evaluating probabilities by using the mean and the standard deviation of the binomial random variable y, along with the Empirical Rule.

Exercises

Basic Techniques

5.21 Consider the following class experiment: Toss three coins and observe the number of heads, y. Let each student repeat the experiment 10 times; then combine the class results, and construct a relative frequency table for y. Notice that these frequencies give approximations to the actual probabilities that $y = 0, 1, 2,$ or 3. (*Note:* Calculate the actual probabilities by using the binomial formula $P(y)$ to compare the approximate results with the actual probabilities.)

5.22 Let y be a binomial random variable. Compute $P(y)$ for each of the following situations.

a $n = 10$, $p = .2$, $y = 3$

b $n = 4$, $p = .4$, $y = 2$

c $n = 16$, $p = .7$, $y = 12$

5.23 Let y be a binomial random variable with $n = 8$ and $p = .4$. Find the following values.

a $P(y \leq 4)$

b $P(y > 4)$

c $P(y \leq 7)$

d $P(y > 6)$

Applications

5.24   An appliance store has the following probabilities for y, the number of major appliances sold on a given day.

y	$P(y)$
0	.100
1	.150
2	.250
3	.140
4	.090
5	.080
6	.060
7	.050
8	.040
9	.025
10	.015

a   Construct a graph of $P(y)$.

b   Find $P(y \leq 2)$.

c   Find $P(y \geq 7)$.

d   Find $P(1 \leq y \leq 5)$.

5.25   The weekly demand for copies of a popular word-processing program at a computer store has the probability distribution shown here.

y	$P(y)$
0	.06
1	.14
2	.16
3	.14
4	.12
5	.10
6	.08
7	.07
8	.06
9	.04
10	.03

a   What is the probability that three or more copies will be demanded in a particular week?

b   What is the probability that the demand will be for at least two but no more than six copies?

c   If the store has eight copies of the program available at the beginning of each week, what is the probability that the demand will exceed the supply in a given week?

5.26   A biologist randomly selects 10 portions of water, each equal to $.1 \text{ cm}^3$ in volume, from the local reservoir and counts the number of bacteria present in each portion. The biologist then totals the number of bacteria for the ten portions to obtain an estimate of the number of bacteria per cubic centimeter present in the reservoir water. Is this a binomial experiment?

5.27   Examine the accompanying newspaper clipping. Does the sampling described appear to satisfy the characteristics of a binomial experiment?

> *Poll Finds Opposition to Phone Taps* New York—People surveyed in a recent poll indicated they are 81% to 13% against having their phones tapped without a court order. The people in the survey, by 68% to 27%, were opposed to letting the

government use a wiretap on citizens suspected of crimes, except with a court order. The survey was conducted for 1,495 households and also found the following results:

—The people surveyed are 80% to 12% against the use of any kind of electronic spying device without a court order.

—Citizens are 77% to 14% against allowing the government to open their mail without court orders.

—They oppose, by 80% to 12%, letting the telephone company disclose records of long-distance phone calls, except by court order.

For each of the questions, a few of those in the survey had no responses.

5.28 A survey is conducted to estimate the percentage of pine trees in a forest that are infected by the pine shoot moth. A grid is placed over a map of the forest, dividing the area into 25-foot by 25-foot square sections. Then 100 of the squares are randomly selected, and the number of infected trees is recorded for each square. Is this a binomial experiment?

5.29 A survey was conducted to investigate the attitudes of nurses working in Veterans Administration (VA) hospitals. A sample of 1000 nurses received a mailed questionnaire, and the number favoring or opposing a particular issue was recorded. If we confine our attention to the nurses' responses to a single question, does this sampling represent a binomial experiment? As with most mail surveys, some of the nurses did not respond. What effect might nonresponses in the sample have on the estimate of the percentage of all VA nurses who favor the particular proposition?

5.30 A random sample of 10 members was obtained to ascertain opinions concerning a new wage package proposal to a local union by union leaders. If we assume that $p = .6$ of all the members who disagree with the wage package, compute the following probabilities.

a All disagree.

b Exactly six disagree.

c Six or more disagree.

d All agree.

5.31 Refer to Exercise 5.30.

a Compute the probabilities for parts a through d if $p = .3$.

b Indicate how you would compute $P(y \le 100)$ for $n = 1000$ and $p = .3$.

5.32 An experiment is conducted to test the effect of an anticoagulant drug on rats. A random sample of four rats is employed in the experiment. If the drug manufacturer claims that 80% of the rats will be favorably affected by the drug, what is the probability that none of the four experimental rats will be favorably affected? One of the four? One or fewer?

5.33 A criminologist claims that the probability of "reform" for a first-offense embezzler is .9. Suppose that we define "reform" as meaning that the person commits no criminal offenses within a 5-year period. Three paroled embezzlers were randomly selected from the prison records, and their behavioral histories were examined for the 5-year period following their prison release. If the criminologist's claim is correct, what is the probability that all three were reformed? At least two?

5.34 Consider the following experiment: Toss three coins and observe the number of heads, y. Repeat the experiment 100 times, and construct a relative frequency table for y. Notice that these frequencies give approximations to the exact probabilities that $y = 0, 1, 2$, and 3. (*Note:* These probabilities can be shown to be 1/8, 3/8, 3/8, and 1/8 respectively.)

5.35 Refer to Exercise 5.34. Use the formula for the binomial probability distribution to show that $P(0) = 1/8$, $P(1) = 3/8$, $P(2) = 3/8$, and $P(3) = 1/8$.

5.36 Suppose that you match coins with another person a total of 1000 times. What is the mean number of matches? The standard deviation? Calculate the interval $\mu \pm 3\sigma$. (*Hint:* The probability of a match in the toss of a single pair of coins is $p = .5$.)

```
BINOMIAL PROBABILITIES FOR n=10 AND P(SUCCESS) = 0.6

ROW      K          PDF          CDF

  1      0        0.000105      0.00010
  2      1        0.001573      0.00168
  3      2        0.010617      0.01229
  4      3        0.042467      0.05476
  5      4        0.111477      0.16624
  6      5        0.200658      0.36690
  7      6        0.250823      0.61772
  8      7        0.214991      0.83271
  9      8        0.120932      0.95364
 10      9        0.040311      0.99395
 11     10        0.006047      1.00000
```

5.37   Refer to Exercise 5.30. Indicate how you could compute $P(y \leq 100)$ if $n = 1000$ for $p = .6$.

5.38   Over a long period of time in a large multinational corporation, 10% of all sales trainees are rated as outstanding, 75% are rated as excellent/good, 10% are rated as satisfactory, and 5% are considered unsatisfactory. Find the following probabilities for a sample of ten trainees selected at random.

a   Two are rated as outstanding.

b   Two or more are rated as outstanding.

c   Eight of the ten are rated as either outstanding or excellent/good.

d   None of the trainees is rated as unsatisfactory.

5.39   A new technique, balloon angioplasty, is being widely used to open clogged heart valves and vessels. The balloon is inserted via a catheter and is inflated, opening the vessel; thus, no surgery is required. Left untreated, 50% of the people with heart-valve disease die within about 2 years. If experience with this technique suggests that approximately 70% of patients treated with it live for more than 2 years, would the next five patients treated with balloon angioplasty at a hospital constitute a binomial experiment with $n = 5$ and $p = .70$? Why or why not?

5.40   A prescription drug firm claims that only 12% of all new drugs shown to be effective in animal tests ever make it through a clinical testing program and onto the market. If a firm has 15 new compounds that have shown effectiveness in animal tests, find the following probabilities.

a   None reach the market.

b   One or more reach the market.

c   Two or more reach the market.

5.41   Does Exercise 5.40 satisfy the properties of a binomial experiment? Why or why not?

5.9   Probability Distributions for Continuous Random Variables

Discrete random variables (such as the binomial) have possible values that are distinct and separate, such as 0 or 1 or 2 or 3. Other random variables are most usefully considered to be *continuous*: their possible values form a whole interval (or range or continuum). For instance, the 1-year return per dollar invested in a common stock could range from 0 to some quite large value. In practice, virtually all random variables assume a discrete set of values; thus, the return per dollar of a million-dollar common-stock investment could be $1.06219423 or $1.06219424 or $1.06219425 or. . . . But when there are a great many possible values for a random variable, it is sometimes mathematically useful to treat the random variable as continuous.

Theoretically, then, a continuous random variable is one that can assume values associated with infinitely many points in a line interval. We state, without elaboration, that it is impossible to assign a small amount of probability to each value of y (as was done for a discrete random variable) and retain the property that the probabilities sum to 1.

We overcome this difficulty by resorting to the concept of the relative frequency histogram of Chapter 3, where we talked about the probability that will fall into a given interval. Recall that the relative frequency histogram for a population containing a large number of measurements almost forms a smooth curve, because the number of class intervals can be made large and the width of the intervals can be decreased. Thus we envision a smooth curve that provides a model for the population relative frequency distribution generated by repeated observation of a continuous random variable. This will resemble the curve shown in Figure 5.5.

F I G U R E 5.5   Probability Distribution for a Continuous Random Variable

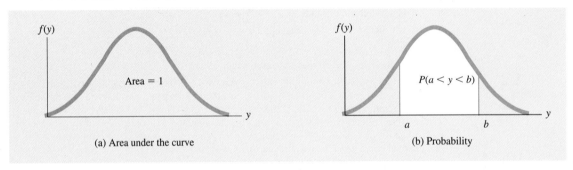

(a) Area under the curve   (b) Probability

Recall that the histogram relative frequencies are proportional to areas over the class intervals and that these areas possess a probabilistic interpretation. If a measurement is randomly selected from the set, the probability that it will fall in an interval is proportional to the histogram area above the interval. Since a population is the whole (100%, or 1), we want the total area under the probability curve to equal 1. If we let the total area under the curve equal 1, then the areas over intervals are exactly equal to the corresponding probabilities.

The graph for the probability distribution for a continuous random variable is shown in Figure 5.5. The ordinate (height of the curve) for a given value of y is denoted by the symbol $f(y)$. Many people are tempted to say that $f(y)$, like $P(y)$ for the binomial random variable, designates the probability associated with the continuous random variable y. But as mentioned before, it is impossible to assign a positive probability to each of the infinitely many possible values of a continuous random variable. Thus, all we will say is that $f(y)$ represents the height of the probability distribution for a given value of y.

The probability that a continuous random variable falls in an interval—say, between two points a and b—follows directly from the probabilistic interpretation given to the area over an interval for the relative frequency histogram (Section 3.7) and is equal to the area under the curve over the interval a to b, as shown in Figure 5.5. This probability is written $P(a < y < b)$.

Curves of many shapes can be used to represent the population relative frequency distribution for measurements associated with a continuous random variable. Fortunately, the areas under some of these curves have been tabulated and are ready for use. Thus, if we know that student examination scores possess a particular probability distribution, as in Figure 5.6, and if areas under the curve have been tabulated, we can find the probability that a particular student will score more than 80%, by looking up the tabulated (shaded) area.

FIGURE 5.6
Hypothetical Probability
Distribution for Student
Examination Scores

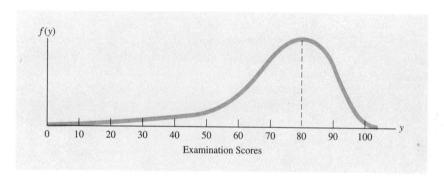

We will find that data collected on many continuous variables in nature possess mound-shaped frequency distributions and that many of these are nearly bell-shaped. A continuous variable (the normal) and its probability distribution (the bell-shaped normal curve) provide a good model for these types of data. The normally distributed variable also plays a very important role in statistical inference. We will study its bell-shaped probability distribution in detail in the next section.

5.10   A Useful Continuous Random Variable: The Normal

normal curve

Many textbooks on introductory statistics note that the distribution of heights for males (or females) can be approximated by a smooth bell-shaped curve that is known as a **normal curve** or a normal distribution. Searching throught the literature, we found that Newman & White (1951) studied the physical characteristics of army personnel. Figure 5.7 shows a histogram based on this study, displaying the relative frequency distribution for the heights (in inches) of a sample of 24,404 United States Army males at the time of their release from the service. A normal curve is

F I G U R E 5.7
Relative Frequency Histogram
for the Heights of 24,404
Servicemen, with Normal Curve
Superimposed

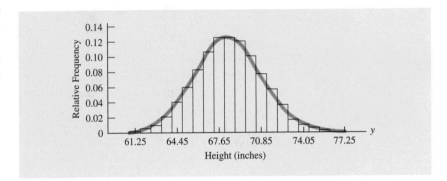

superimposed over the histogram to show the close approximation to it obtained from the sample of servicemen.

Many other commonly occurring and important continuous random variables possess a normal probability distribution. The **normal**, or Gaussian, **distribution** (named for the famous mathematician Karl Friedrich Gauss, 1777–1855) is a continuous bell-shaped curve, as shown in Figure 5.8. The total area under the normal curve is equal to 1, and the probability that a normal random variable y assumes a value in a particular interval, say $a < y < b$, is the area under the curve that lies over the interval (see the shaded area).

normal distribution

F I G U R E 5.8
A Normal Probability Distribution

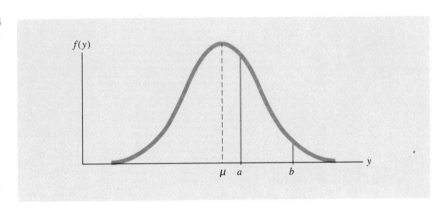

There are infinitely many normal curves, one corresponding to each pair of values that you might assign to μ and σ. But all are symmetrical about the mean and are bell-shaped, and for all such curves the areas within a specified number of standard deviations of the mean are identical. For example, the area within one standard deviation of the mean will always (to two decimal places) equal .68, as represented in Figure 5.9(a). Likewise, the area within two standard deviations of the mean will always (to two decimal places) equal .95, as depicted in Figure 5.9(b).

The probability that a normal random variable y assumes values in an interval, say $a < y < b$, can be obtained by using a **table of areas** under the normal curve. Table 2 in Appendix 3 gives areas under the normal curve between the mean and a

table of areas

FIGURE 5.9 Characteristics of a Normal Probability Distribution

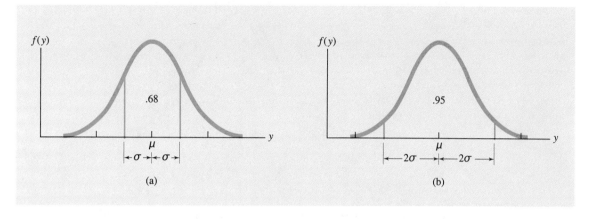

(a)

(b)

point any number—say, z—of standard deviations $(z\sigma)$ to the right of μ. A partial reproduction of this table is shown in Table 5.5. The tabulated area is shown in Figure 5.10.

TABLE 5.5 Format of the Table of Normal Curve Areas, Table 2 in Appendix 3

z	.00	.01	.02	.03	.04	.05	.06	.07	.08	.09
0.0	.0000	.0040	.0080	.0120	.0160	.0199	.0239	.0279	.0319	.0359
0.1	.0398	.0438	.0478	.0517	.0557	.0596	.0636	.0675	.0714	.0753
0.2	.0793	.0832	.0871	.0910	.0948	.0987	.1026	.1064	.1103	.1141
0.3	.1179	.1217	.1255	.1293	.1331	.1368	.1406	.1443	.1480	.1517
0.4	.1554	.1591	.1628	.1664	.1700	.1736	.1772	.1808	.1844	.1879
⋮	⋮	⋮	⋮	⋮	⋮	⋮	⋮	⋮	⋮	
1.0	.3413	.3438	.3461	.3485	.3508	.3531	.3554	.3577	.3599	.3621
1.1	.3643	.3665	.3686	.3708	.3729	.3749	.3770	.3790	.3810	.3830
1.2	.3849	.3869	.3888	.3907	.3925	.3944	.3962	.3980	.3997	.4015
⋮	⋮	⋮	⋮	⋮	⋮	⋮	⋮	⋮	⋮	
1.6	.4452	.4463	.4474	.4484	.4495	.4505	.4515	.4525	.4535	.4545
⋮	⋮	⋮	⋮	⋮	⋮	⋮	⋮	⋮	⋮	
2.0	.4772	.4778	.4783	.4788	.4793	.4798	.4803	.4808	.4812	.4817

In the table, areas to the left of the mean need not be tabulated, because the normal curve is symmetric about the mean. Thus the area between the mean and a point 2σ to the right of the mean is the same as the area between the mean and a similar point 2σ to the left.

The number z of standard deviations is given to the nearest tenth in the left-hand column of Table 2 in Appendix 3. Adjustments to take z to the nearest hundredth are given in the top row of the table. Entries in the table represent the proportional areas corresponding to particular values of z. For example, the area between the mean and

FIGURE 5.10
Tabulated Area Under the
Normal Curve

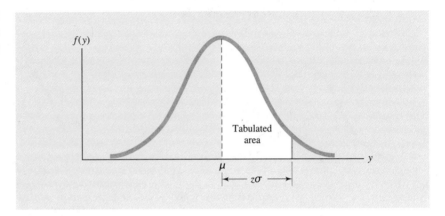

a point $z = 2$ standard deviations to the right of the mean is shown in the second column of the table opposite $z = 2.0$. This area, shaded in Figure 5.11(a), is .4772. Likewise, the area between the mean and a point two standard deviations to the left of the mean, shown in Figure 5.11(b), is also .4772. Therefore, the area within two standard deviations of the mean is $2(.4772) = .9544$. This explains the origin of the figure "approximately 95%" in the Empirical Rule.

FIGURE 5.11   Tabulated Area Corresponding to $z = 2$

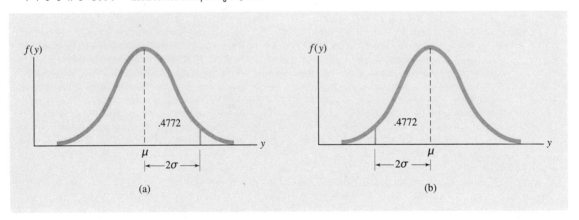

(a)   (b)

Similarly, the area between the mean and a point one standard deviation to the right of the mean (that is, $z = 1$) is .3413. Consequently, the area within one standard deviation of the mean is .6828, or approximately 68%, as stated in the Empirical Rule. This area is shown in Figure 5.12.

Suppose that we wish to find the area corresponding to $z = 1.64$. Proceed down the left column of the table to the row $z = 1.6$ and across the top of the table to the

FIGURE 5.12
Area Within One Standard
Deviation of the Mean

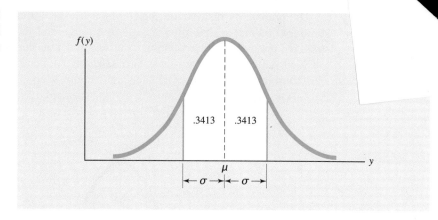

FIGURE 5.13
Area Corresponding to
$z = 1.64$

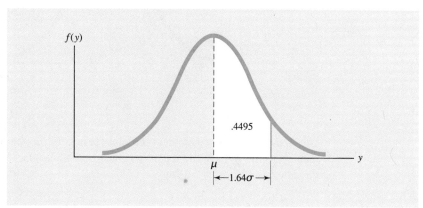

.04 column. The intersection of the $z = 1.6$ row and the .04 column gives the desired area, .4495. This area is shown in Figure 5.13.

To determine how many standard deviations a measurement y lies from the mean μ, we must first determine the distance between y and μ. Recall that this distance can be represented as

$$\text{Distance} = y - \mu.$$

Then the distance between y and μ can be converted into a number of standard deviations by dividing by σ, the standard deviation of y. This standardized distance is often called a **z-score**.

z-score

$$z = \frac{\text{Distance}}{\text{Standard deviation}} = \frac{y - \mu}{\sigma}.$$

The probability distribution for z, which has $\mu = 0$ and $\sigma = 1$, is called the **standard normal distribution** (see Figure 5.14), and the variable z is called a **standard normal random variable**. The area under the curve between $z = 0$ and a specified value of z—say, z_0— has been tabulated in Table 2 of Appendix 3 and is shown in Figure 5.14. These tabulated areas can be used to find the area under any normal curve if you know the mean μ and the standard deviation σ.

FIGURE 5.14
Standard Normal Distribution

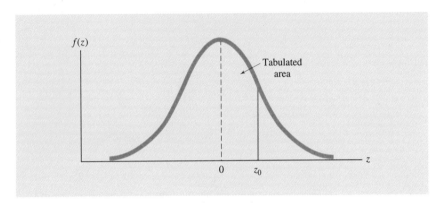

To calculate the area under a normal curve between the mean μ and a specified value y to the right of the mean, we must first determine the number z of standard deviations y lies from μ using

$$z = \frac{y - \mu}{\sigma}.$$

Then we refer to Table 2 in Appendix 3 and obtain the entry corresponding to the calculated value of z. This entry is the desired area (probability) under the curve between μ and the specified value of y.

We illustrate the use of the table of normal curve areas with a simple example, and then we proceed to more practical applications.

EXAMPLE 5.11   Suppose that y is a normally distributed random variable with mean $\mu = 8$ and standard deviation $\sigma = 2$. Find the probability that y lies in the interval from 8 to 11; that is, find the fraction of the total area under the curve that lies between 8 and 11 (see the shaded portion of Figure 5.15).

Solution   To determine the desired area, we compute the number of standard deviations that separate $y = 11$ from the mean $\mu = 8$:

$$z = \frac{y - \mu}{\sigma} = \frac{11 - 8}{2} = 1.5.$$

The corresponding area can then be determined from the entry in Table 2 of Appendix 3 opposite $z = 1.5$. The desired area is .4332. Therefore, if a single value of y is selected at random from its population, the probability that y lies between 8 and 11 is equal to .4332. ∎

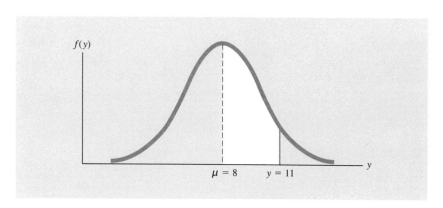

FIGURE **5.15**
Area Under the Curve Over the
Interval from $\mu = 8$ to
$y = 11$

EXAMPLE **5.12** The quantitative portion of a nationally administered achievement test is scaled so that the mean score is 500 and the standard deviation is 100.

a If we assume that the distribution of scores is normal (bell-shaped), what percentage of the students throughout the country should score between 500 and 682?

b What percentage should score between 340 and 682?

Solution Consider Figure 5.16.

a To determine the percentage of students who should score between 500 and 682, we must compute the area A_1 between $\mu = 500$ and $y = 682$:

$$z = \frac{y - \mu}{\sigma} = \frac{682 - 500}{100} = 1.82.$$

The tabulated area for this value of z (Table 2, Appendix 3) is $A_1 = .4656$. Thus, we expect 45.56% of the students to score between 500 and 682.

FIGURE 5.16
Area Between $y = 340$ and
$y = 682$

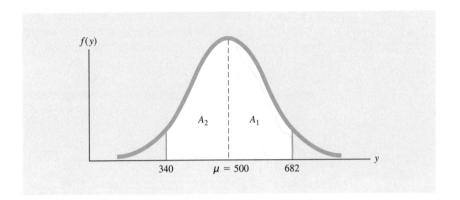

b To determine the percentage of students who should score between 340 and 682, notice that the area between 340 and 682 is equal to the sum of A_1 and A_2 in Figure 5.16. To find A_2, we compute the number of standard deviations that separate $y = 340$ from $\mu = 500$. Hence,

$$z = \frac{y - \mu}{\sigma} = \frac{340 - 500}{100} = -1.6.$$

Negative values of z indicate a point ot the left of the mean. The appropriate area can be found, using Table 2 of Appendix 3, by ignoring the negative sign for z. This area is .4452. Thus, we would expect $A_1 + A_2 = .4656 + .4452 = .9108$ or 91.08% of the students to score between 340 and 682 on the examination. ▪

E X A M P L E   5.13   Records maintained by the budget office of a particular state indicate that the amount of time elasped between submission of travel vouchers and reimbursement of funds has an approximately normal distribution with a mean equal to 45 days and a standard deviation equal to 5 days.

a What is the probability that the elapsed time between submission and reimbursement will exceed 58 days for a travel expense report selected at random?

b If you submittted a travel expense report and did not receive reimbursement within 58 days, what would you conclude?

Solution   A sketch of the desired area A_2 is shown in Figure 5.17.

a To determine the area A_2, we must first compute a z score:

$$z = \frac{y - \mu}{\sigma} = \frac{58 - 45}{5} = 2.6$$

Thus, the value 58 is 2.6 standard deviations above the mean $\mu = 45$. The area under a normal curve from the mean to a point 2.6 standard deviations above the mean is, from Table 2 of Appendix 3, .4953. This area is indicated by A_1 in

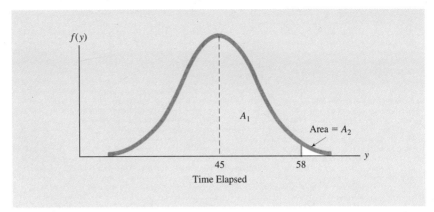

Figure 5.17. Since the total area under a normal curve to the right of the mean is .5, we can compute A_2 by subtracting $A_1 = .4953$ from .5. Thus,

$$A_2 = .5 - .4953 = .0047$$

b   Since the probability of having to wait more than 58 days is so small, .0047, we would conclude that something happened to the travel expense report—that it was lost in the mail or misplaced.   ∎

In this section we have computed areas under a normal curve. To do this, we converted the distance between a measurement y and the mean μ into a number of standard deviations and then referred to Table 2 in Appendix 3. The resulting areas are equal to the probabilities that randomly selected measurements will fall in particular intervals.

Exercises

Basic Techniques

5.42   Use Table 2 of Appendix 3 to find the area under the normal curve between each pair of values.
a $z = 0$ and $z = 1.3$
b $z = 0$ and $z = -1.9$

5.43   Repeat Exercise 5.42 for the following values.
a $z = 0$ and $z = .7$
b $z = 0$ and $z = -1.2$

5.44   Repeat Exercise 5.42 for the following values.

   a $z = 0$ and $z = 1.29$

   b $z = 0$ and $z = -.77$

5.45   Repeat Exercise 5.42 for the following values.

   a $z = -.21$ and $z = 1.35$

   b $z = .37$ and $z = 1.20$

5.46   Repeat Exercise 5.46 for the following values.

   a $z = 1.43$ and $z = 2.01$

   b $z = -1.74$ and $z = -.75$

5.47   Find the probability that z is greater than 1.75.

5.48   Find the probability that z is less than 1.14.

5.49   Find a value for z—say, z_0—such that $P(z > z_0) = .5$.

5.50   Find a value for z—say, z_0—such that $P(z > z_0) = .025$.

5.51   Find a value for z—say, z_0—such that $P(z > z_0) = .0089$.

5.52   Find a value for z—say, z_0—such that $P(z > z_0) = .05$.

5.53   Find a value for z—say, z_0—such that $P(-z_0 < z < z_0) = .95$.

5.54   Let y be a normal random variable with mean equal to 100 and standard deviation equal to 8. Find these probabilities.

   a $P(y > 100)$

   b $P(y > 110)$

   c $P(y < 115)$

   d $P(88 < y < 12)$

   e $P(100 < y < 108)$

Applications

5.55   The College Boards, which are administered each year to many thousands of high school students, are scored in such a way as to yield a mean of 500 and a standard deviation of 100. These scores are close to being normally distributed. What percentage of the scores can be expected to be

   a greater than 600?

   b greater than 700?

   c less than 450?

   d between 450 and 600?

5.56   Let y be a normal random variable with $\mu = 20$ and $\sigma = 4$. Find the following probabilities.

   a $P(y < 15)$

   b $P(y > 25)$

   c $P(y > 30)$

   d $P(15 < y < 25)$

5.57   Sales figures (on a monthly basis) for a particular food industry tend to be normally distributed with mean of 150($1000s) and a standard deviation of 35($1000s). Compute the following probabilities.

   a $P(y > 200)$

   b $P(y > 220)$

   c $P(y < 120)$

   d $P(100 < y < 200)$

5.58 Refer to Exercise 5.55. An exclusive club wishes to invite those who score in the top 10% on the College Board to join.

a What minimum score is required to be invited to join the club?

b What score separates the top 60% of the population from the bottom 40%? What do we call this value?

5.59 The mean for a normal distribution is 50, and the standard deviation is 10.

a At what percentile does the value 38 lie? Choose the appropriate answer.

$$88.49 \qquad 38.49 \qquad 49.99 \qquad 0.01 \qquad 11.51$$

b Which of the following is the z-score corresponding to the 67th percentile?

$$1.00 \qquad 0.95 \qquad 0.44 \qquad 2.25 \qquad \text{none of these}$$

5.60 The distribution of weights of a large group of high school boys is normal with $\mu = 120$ lb and $\sigma = 10$ lb. Which of the following is true?

a About 16% of the boys will be over 130 lb.

b Probably fewer than 2.5% of the boys will be below 100 lb.

c Half of the boys can be expected to weigh less than 120 lb.

d All of the above are true.

Summary

In this chapter we introduced some of the basic concepts of probability and the concept of a random variable. The probability distributions for two particularly important random variables, the binomial and the normal, were presented along with examples showing how these can be applied.

Probability, which measures the strength of our belief in the occurrence of a particular outcome of an experiment, is viewed conceptually as being the relative frequency of occurrence of the outcome (event) when the experiment is repeated over and over again. Probability is important to statistics because the observation of a sample selected from a population is an experiment. Based on the probability of an observed sample, statisticians (and you) make inferences about the population from which the sample was selected. For example, if we theorize that only 5% of all telephone subscribers are delinquent in paying their telephone bills, and we then sample 10 bills and find that all are overdue, we would have to reject our theory and infer that the proportion of overdue bills in the population is larger than 5%. We reach this conclusion not because it is impossible to draw 10 unpaid bills in a sample of 10, assuming that the 5% theory is correct, but because it is *highly improbable*.

Three probabilistic concepts play important roles in sampling and statistical inferences. These are the additive property of probabilities for mutually exclusive events, conditional probability, and independent events. As noted in this chapter, most experiments result in numerical events obtained by the observation of random variables. These events, the values that a random variable may assume, are mutually exclusive. Therefore, to find the probability that a random variable, when observed, will assume one of a set of specific values, we sum the probabilities corresponding to these values.

The concept of independence plays an important role in statistics, especially as it relates to random sampling (to be discussed in Chapter 6), because the manner in which a measurement is selected (sampled) affects the probability of the occurrence of events of interest.

Random variables observed in real-life experiments or surveys can be discrete (usually data representing counts of whole objects, such as the number of bacteria per cubic centimeter of water) or continuous (such as the length of time required to obtain service at a medical clinic). Two of the most important are the discrete binomial random variable and the continuous normal random variable. In addition to noting the characteristics of these two random variables and the situations to which they might apply, we gave their probability distributions. The probability distribution for the binomial random variable gives the probability associated with each of the values 0, 1, 2, . . . , n that the random variable can assume. To find the probability that a discrete random variable y will assume one of a set of these mutually exclusive numerical events (one event corresponding to each value), we simply sum the probabilities corresponding to the values. Thus, for example,

$$P(3 \leq y \leq 5) = P(3) + P(4) + P(5)$$

The probability distribution for the normal random variable is a smooth bell-shaped curve that is essentially equal to the theoretical relative frequency distribution for a normally distributed population of measurements. For a continuous random variable y, the probability that y will assume a value in the interval $a \leq y \leq b$ is identical to the proportional area under the probability distribution curve over the interval $a \leq y \leq b$. And since no area lies above a single point— say, $y = a$— it follows that $P(y = a) = 0$. However, in practical problems, we are interested in probabilities of the form $y \leq a$, $y \geq a$, or $a \leq y \leq b$, rather than in the probability that y assumes some specific value—say, $y = a$.

In later chapters we will employ sample statistics to make inferences about population parameters. The probability distribution of a sample statistic, called its *sampling distribution*, plays a key role in selecting good statistics and in evaluating their reliability. Sampling distributions are the topic of Chapter 6.

Key Terms

outcome	unconditional probability
event	independent events
classical interpretation of probability	qualitative random variable
conditional probability	quantitative random variable

discrete random variable	complement
continuous random variable	binomial experiment
probability distribution	normal curve
relative frequency concept	normal distribution
independent samples	table of areas
subjective probability	z-score
mutually exclusive events	standard normal distribution
union	standard normal random variable
intersection	

Key Formulas

1 For two independent events A and B:

$$P(A|B) = P(A) \quad \text{and} \quad P(B|A) = P(B)$$

2 Binomial probability distribution:

$$P(y) = \frac{n!}{y!(n-y)!} p^y q^{n-y}$$

3 Mean and standard deviation of the binomial distribution:

$$\mu = np \qquad \sigma = \sqrt{npq}$$

4 z-score:

$$z = \frac{y - \mu}{\sigma}$$

Supplementary Exercises

Class Exercise

5.61 Each student should perform the following experiment 20 times: Toss 5 coins and observe the number of heads.

The possible values of y, the number of heads in 5 coin tosses, are 0, 1, 2, 3, 4, and 5. Each student should keep track of the number of times each outcome is observed and then should combine his or her results with those of the rest of the class to construct the following table.

y	Frequency	Relative Frequency (approximate probability)	$P(y)$
0			.031
1			.156
2			.313
3			.313
4			.156
5			.031

Notice that the exact value of the probability, $P(y)$, associated with each value of y is listed in the last column of the table. Your relative frequencies computed on the basis of data from this experiment should be approximately the same as the exact probabilities. Remember that the degree of accuracy increases as the number of repetitions of the experiment increases.

5.62 Refer to Exercise 5.61 and the observed relative frequencies. Approximate the probability of observing y equal to 0 or 1 heads. What is the actual probability of observing 0 or 1 heads in a toss of 5 coins?

5.63 Identify the following variables as being either discrete or continuous.

a The number of patient arrivals per hour at a medical clinic.

b The number of accidents at a given intersection each year.

c The average amount of electricity (measured in kilowatt-hour units) consumed per household per month in New York City.

d The number of deaths per year attributed to lung cancer.

e The age of freshman United States senators when they take the oath of office.

f The gross national product for the United States per year.

5.64 Some parts of California are particularly earthquake-prone. Suppose that, in one such area, 40% of all homeowners are insured against earthquake damage. Twelve homeowners are to be selected at random.

a What is the probability that exactly 3 of the 12 homeowners will have earthquake insurance? (Use the formula for $P(y)$.)

b What is the probability that at least 7 of the 12 homeowners will have earthquake insurance? (Use Table 2 of Appendix 3.)

c Determine the mean and the standard deviation of the random variable y, which represents the number among the 12 who have earthquake insurance.

5.65 The mean contribution per person for a college alumni fund drive was $101, with a standard deviation of $32. Assuming that contributions follow a mound-shaped distribution, what percentage of the contributions

a were between $37 and $165?

b exceeded $165?

c were between $37 and $133?

d were less than $69?

5.66 The lifetime of a color TV picture tube is normally distributed, with a mean of 7.2 years and a standard deviation of 2.3 years.

a What is the probability that a randomly chosen picture tube will last more than 10 years?

b If the manufacturer guarantees the picture tubes for 3 years, what is the chance that a randomly chosen picture tube will wear out before the guarantee has expired?

c If the manufacturer is willing to replace only 2% of all picture tubes due to early failure, how should it change the guarantee?

5.67   A brand of water-softener salt comes in packages marked "net weight 40 lb." The company claims that the bags contain an average of 40 lb of salt, with a standard deviation in the weights of 1.5 lb. Furthermore, it is known that the weights are normally distributed. What is the probability that the weight of a randomly selected bag will be 39 lb or less, if the company's claim is true?

5.68   A data-processing company requires applicants for computer programming positions to take a test. The company policy is to reject all applicants whose score is below 50. The mean score for applicants has been 60 with a standard deviation of 10. The distribution of the applicants' scores is mound-shaped. Approximately what percentage of the applicants have been rejected by this test?

5.69   The following table relates to the members of the local chapter of the Building Trades Union. Use the data to answer the following questions. (As usual, carry out all your calculations and your answers to at least four places.)

Employment Status	Carpenter	Plumber	Bricklayer	Total
Employed	81	48	34	163
Unemployed	56	42	49	147
Total	137	90	83	310

a   Find each of the following two probabilities.

P(Employed or Bricklayer).

P(Carpenter or Employed).

b   Are the two events Unemployed and Plumber independent? Carefully explain you answer.

5.70   An accounting office has six incoming telephone lines. The probability distribution of busy lines, y, at any given time during business hours is

y	$P(y)$
0	0.052
1	0.154
2	0.232
3	0.240
4	0.174
5	0.105
6	0.043

a   Is y a discrete or a continuous random variable?

b   What is the probability that at least 4 lines will be busy?

c   What is the probability that between 2 and 4, inclusive, will be busy?

d   At least one?

5.71   The accompanying table records the votes cast in a recent election for an increase in the tax levy for the school district comprising the towns of Bazetta, Delightful, and Shihola.

Vote	Bazetta	Delightful	Shihola	Total
For the levy	192	84	381	657
Against the levy	448	940	145	1533
Total	640	1024	526	2190

Consider the following events.

B: The voter lives in Bazetta.

D: The voter lives in Delightful.

S: The voter lives in Shihola.

F: The voter voted for the levy.

A: The voter voted against the levy.

a   Find each of the following probabilites:

$$P(S) \qquad P(D|A) \qquad P(D \text{ or } A)$$

b   Are the events *F* and *B* *independent*? Carefully explain your answer.

5.72   Define each of the underlined terms.

The events *A* and *B* are independent.

The events *A* and *B* are mutually exclusive.

5.73   In the mining town of Wounded Brook, Colorado, many of the mines have been closed because of new environmental regulations. Not surprisingly, 80% of the people in the town are against the new regulations. What is the probability of randomly selecting five people from Wounded Brook and discovering that three of the five are *in favor* of the regulations? Show your formulas and computations.

5.74   Suppose that a person responds to a set of five true/false test questions on a random basis.

a   What is the person's probability of getting exactly two questions correct? (Choose the correct answer.)

.0312   .3125   .0500   .625

b   What is the probability of the person's getting three or more questions correct? (Choose the correct answer.)

.0312   .1875   .5000   .6875

5.75   Suppose that a telephone survey is to be done in the Atlanta area. Previous telephone surveys indicate that 60% of the time the phone will be answered by a woman. Suppose that we make three calls initially. What is the probability that:

a   a woman will answer for each call?

b   a woman will answer for exactly one call?

c   a woman will not answer for any of the three calls?

5.76   A man's blood test shows a cholesterol count of 310 milligrams per deciliter. The man's doctor says that the average cholesterol count for men of a similar age is 200 mg/dl and that only about $2\frac{1}{2}$% of the male population will have a value higher than 310.

a   Determine the standard deviation, assuming that the counts are normally distributed.

b   What chance is there of finding a male with a cholesterol count of less than 100?

5.77   Pay rates for hourly employees in a particular industry are assumed to be normally distributed, with a mean of $7.00 and a standard deviation of $1.50.

a   What percentage of employees have hourly pay rates of between $7 and $10?

b   What is the probability of finding an employee with a rate of greater than $8.50?

5.78   The average weight of newborn, full-term babies at a local hospital is 7.5 lb, with a standard deviation of 1.6 lb. Assuming that the distribution of weights is approximately normal, find the probability that a newborn weighs

a   less than 6 lb.

b   more than 8 lb.

c   between 6 and 8 lb.

5.79 Weather forecasters have been the object of many jokes due to errors in forecasting. Of course, these errors are not completely avoidable. After obtaining information (measurements) on many different variables such as wind direction, wind velocity, and barometric pressure from local sources and satellite communications, the forecaster must interpret these data and supply an inference (weather forecast). How effectively weather forecasters have employed statistics in preparing their forecasts is open to question, but it is clear that probability has become an integral part of such forecasts. We've all heard weather forecasts that state, "There is a 50 percent chance of rain this morning, decreasing to a 30 percent chance this afternoon and evening." Give you interpretation of this statement.

Area Forecast

Partly cloudly to occasionally cloudy through tomorrow with a chance of thundershowers mainly during the afternoons and evenings. Low today and tomorrow will be near 80. High will be near 90. Winds will be southwest to west 5 to 15 mph, stronger and gusty near showers. Rain probability is 50 percent today and 30 percent tonight.

5.80 Read the following news clipping. Explain why this might or might not be a binomial experiment. What information, missing in the article, is needed to enable you to conclude firmly that the survey is a binomial experiment?

Study of Divided Families Shows Positive Attitudes

Chicago—A study of divorced mothers and their children has revealed some positive attitudes among members of divided families. Perhaps a broken home is not the psychological disaster for family members that society has suspected.

The study, involving 20 mothers with one or more children between the ages of 6 and 18, was conducted to determine the basic concerns of divorced mothers and their children. There were 20 mothers and 35 children involved in the study.

All the women were working full time. Most of them had made plans toward bettering their earning power. The women had been divorced from 3 months to 15 years. The educational level of women in the study was high, compared to the national average: 12 years to 18 years of education.

A key aim of the study was to determine the feelings of the women and their children about their acceptance in society.

Eighty-six percent of the children felt that at school they were treated the same as children whose parents were married. Children aged 10 through 12 especially preferred that teachers and friends be told about the home situation. They wanted news of the divorce not to come as a surprise to others or to be a source of embarrassment for them.

In general, the children were doing well in school and even excelled in some areas.

Although the trend among most of the women was to socialize mainly with single persons, 80% of them felt accepted in their neighborhoods. Half of them said they felt accepted at church.

Among the children, 91% indicated they were treated no differently at Sunday school. Ninety percent of the sample were active church members.

Most of the women, 85%, said that after their divorces their attitudes toward divorce had shifted from negative to positive. The same proportion saw advantages for their children, in terms of understanding life and people, as a result of the divorce.

5.81 Suppose that you are the personnel manager for a manufacturing concern and are responsible for safety procedures in your plant. Records are maintained on the number of accidents on a daily basis, and these are totaled by the month. Explain why these data are or are not measurements on a binomial random variable.

5.82 A recent survey suggests that Americans anticipate a reduction in living standards, and that a steadily increasing level of consumption may no longer be deemed as important as it was in the past. Suppose that a poll of 2000 people found 1373 in favor of forcing a reduction in the size of American automobiles by legislative means. Would you expect to observe as many as 1373 in favor of this proposition if, in fact, the general public was split 50—50 on the issue? Why?

5.83 An experiment was conducted to test for the presence or absence of fungus on tabacco plants. Four hundred plants were observed to determine whether they had been infected by the fungus.

a Does this appear to be a binomial experiment? Explain why it might or might not satisfy the characteristics of a binomial experiment.

b Suppose that the characteristics of a binomial experiment are satisfied. What interpretation can you give to p?

c Previous experience suggests that the fungus affects 50% of a planting of tobacco seedlings. What is the mean values of y, the number of plants infected by the fungus? The standard deviation of y? If p really equals .5, is it probable that the observed number of infected plants could be as large as (or larger than) $y = 242$? Explain.

5.84 Answer the following questions for the survey discussed in the accompanying news article.

a Does this appear to be a binomial experiment?

b Explain why it might or might not satisfy the five characteristics of a binomial experiment.

> ### Alcoholism Reported Up in Army
>
> Large numbers of young American soldiers are becoming alcoholics. This parallels the increase of alcohol use among young civililans, researchers report.
>
> In a study of 1873 Army men randomly selected from bases of the United States, nearly 2 out of every 5 soldiers were found to be either actual alcoholics, borderline alcoholics, or potential alcoholics.
>
> The study showed that the largest percentage of problem drinkers were under age 20 and had ranks below sergeant.

5.85 Experience has shown that a lie detector will show a positive reading (indicate a lie) 10% of the time when a person is telling the truth and 95% of the time when a person is lying. Suppose that a sample of five suspects is subjected to a lie detector test regarding a recent one-person crime. What is the probability of observing no positive reading if all five suspects plead innocent and are telling the truth?

5.86 A large stadium utilizes floodlights to illuminate the field. The supplier of these lights claims that the time to failure is approximately normal, with mean 40 hours and standard deviation 4 hours.

a What is the probability that a randomly selected floodlight will burn for at least 30 hours?

b If the stadium buys 1500 floodlights, how many would you expect to last at least 30 hours?

c What might you conclude if only 1400 lights lasted at least 30 hours?

5.87 Find the value of z such that 5% of the area under the curve lies to its right.

5.88 Find the value of z such that 2.5% of the area under the curve lies to its right.

5.89 A normally distributed variable y possesses a mean and a standard deviation equal to 7 and 2, respectively. Find the z-value corresponding to $y = 6$.

5.90 Refer to Exercise 5.89. Find the value of z corresponding to $y = 8.5$.

5.91 Refer to Exercise 5.89. Find the probability that y lies in the interval 6 to 8.5.

5.92   The dollar sales per salesperson for a large company averaged $60,000 per year, with a standard deviation of $7,000. What fraction of the salespeople might be expected to total less than $50,000 in sales per year?

5.93   The length of time required to complete a standard achievement test has a mean of 58 minutes, with a standard deviation of 9.5 minutes. If the professor wants to time the exam so that it will be completed by 90% of the students, how long an examination period must be scheduled? (*Hint:* Using Table 2 of Appendix 3, we can find the z-value corresponding to an area of .4). Then, from the formula

$$z = \frac{y - \mu}{\sigma},$$

we can solve for the required value of y (see Figure 5.18).

FIGURE **5.18**
Probability Distribution of the Length of Time to Complete an Achievement Test (Exercise 5.93)

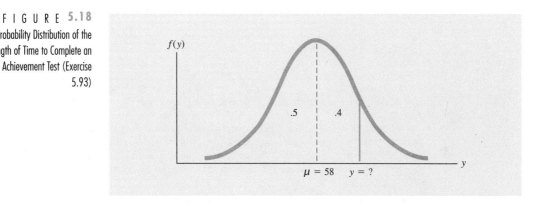

5.94   Refer to Exercise 5.93. Find the 80th percentile for the length of time required to complete the test. (*Note:* The 80th percentile here means that 80% of the students would have completion times less than or equal to this value.) Find the lower quartile.

5.95   Give the mean and the standard deviation of a binomial random variable with $n = 30$ and $p = .2$.

5.96   There is one disadvantage of z-scores: they are difficult to explain to a person who is not well versed in statistics. A college professor, in a stroke of genius that turned out to be highly misguided, once decided to report the results of an exam as z-scores. He was quickly besieged by anxious students who did not understand what a z-score of 0 meant. The professor knows that you are taking a course in statistics, so he asks you to explain what a z-score of 0 actually represents in terms of this exam. Give a *clear* but *short* explanation.

5.97   Weights of college students on a large campus are assumed to be normally distributed, with $\mu = 150$. In this experiemnt, 97.5% of the students weighed less than 210 pounds.

a   What is the standard deviation of the normal distribution?

b   If four students are randomly selected, what is the probability that all four of them will weigh less than 180 pounds each?

c   What percentage of the students weigh less than 105 pounds?

5.98   The probability of surviving a rare disease of the nervous system is one in twenty. A new drug has been developed to combat the disease and is tried on three patients. One out of the three survives. To some experimenters, this survival rate (33%) may appear to be a tremendous improvement over the old established rate of 5%. But remember, the 33% survival rate occurred only in a sample of three, not in a large number of cases.

If the survival rate, using the new drug, is still only 5% (that is, if the drug offers no improvement on existing treatments), what is the probability of observing one or more survivals in a sample of three? Examining this probability, what would you be inclined to say (as a matter of intuition) about the effectiveness of the new drug? Explain your reasoning.

5.99   An immunologist claims that a flu shot is 80% effective against the flu. This means that, of those having the shot, 80% will be immune to the flu while the remaining 20% will be vulnerable to contracting the flu. If 8000 people receive the flu shot and all are exposed to the disease, what is the expected number y who will not get the flu? What is the standard deviation of x? Suppose that 6200 survive the winter without contracting the flu. Is this value of $y(y = 6200)$ improbable, assuming that the flu shot is 80% effective in immunizing people against the flu?

Class Exercise

5.100   Suppose that 50% of all apartment dwellings in a city contain one or more obvious violations of the fire code. If 10 apartments are selected at random, what is the probability that 7 or more will be found to be in violation of the fire code?

Although we can solve this problem by using the formula of Section 5.8, we can easily obtain an approximate answer (and thereby verify our formula) by means of a coin-tossing experiment. Tossing a single coin once is analogous to selecting a single apartment in the population described above. An outcome of heads could correspond to an apartment in violation of the fire code; and tails to one that is in conformance with the code. Flipping the coin 10 times (or tossing 10 coins once) would be equivalent to sampling 10 apartments from the population.

Select 10 coins, mix them thoroughly, toss them, and record the number y of heads (apartments in violation of the code) in the sample. Record whether y is 7 or larger. Repeat this process $n = 100$ times, and count the number of times y is 7 or larger. This will give you the number of times n_A your event of interest was observed. Then calculate an approximate value for the probability of observing 7 or more apartments in violation of the code in a sample of 10, using

$$P(A) = P(7 \text{ or more}) \approx \frac{n_A}{n} = \frac{n_A}{100}.$$

This approximation may be poor because $n = 100$ repetitions of the experiment is not large enough to acquire an accurate approximation to $P(A)$.

To obtain a more accurate approximation to $P(A)$, combine the data collected by all the members of your class, and use these data to approximate $P(A)$. This value y should be close to the exact value of $P(A)$, which is .172. (You can compute the exact value by using the binomial formula $P(y)$ for $y = 7, 8, 9,$ and 10).

To see how the experimentally obtained approximations for $P(A)$ vary, collect the approximations from each member of your class and describe them by using a relative frequency histogram. Notice how they cluster about the exact value $P(A) = .172$. Although we only have one approximation based on the larger value of n for the combined data, you can see that this value will tend to fall closer to .172 than the approximations based on $n = 100$ repetitions of the experiment.

5.101   The relative frequencies shown in the accompanying table were generated from a survey of the employees of a large insurance company in the midwest.

Hair Color by Eye Color of
Employees

Hair Color	Black	Blue	Brown	Green	Hazel	Total
	Eye Color					
Black	.0106	.0053	.0582	.0053	.0053	.0847
Blonde	.0000	.1429	.0053	.0476	.0265	.2223
Brown	.0000	.1746	.2751	.1005	.1164	.6666
Red	.0000	.0052	.0159	.0053	.0000	.0264
Total	.0106	.3280	.3545	.1587	.1482	1.0000

Assuming that the relative frequencies in the above table are population probabilities, determine the following probabilities.

a An employee is either green-eyed or hazel-eyed.

b An employee is either blue-eyed or black-haired or both.

c Given that an employee is brown-haired, the employee has brown eyes.

d Illustrate whether the events "green-eyed" and "brown-haired" are independent or not independent.

5.102 A survey has shown that .30 of the customers of a Rockford, Illinois, bank consider the bank's accuracy to be "good." If a sample of ten customers of the bank is randomly selected, what is the probability that two or fewer will consider the accuracy of the bank to be "good"?

5.103 If the mean and the standard deviation of the distribution of prices for new textbooks in a student bookstore are $26.80 and $4.14, respectively, and if the mean and the standard deviation for used textbooks are $17.80 and $2.80, respectively, which would be the more unusual price—$16.00 for a new book or $12.50 for a used book? Assume that the bookstore prices are approximately normally distributed. Determine what percentage of the prices of new texts fall in the range form $20.00 to $30.00.

5.104 Getting from here to there could be a problem for some University of Kentucky students, according to a news release. Out of 2735 students entering introductory level geography courses, 793 could not locate Lexington, Kentucky, the home of the University of Kentucky, on a map of Kentucky. What is the probability that a random sample of 5 students entering an introductory geography course at the university would show 0 who couldn't locate Lexington on a map of Kentucky?

Experiences with Real Data

5.105 Refer to the HAM-D total scores for the four treatment groups in the Clinical Trials database on the data disk (or in Appendix 1). If we assume that the mean for the combined scores is 14.2 and the standard deviation is 6.3, determine z-scores for total scores of 5, 16, and 24.

5.106 Refer to Exercise 5.105. What total scores correspond to z-scores of -1, 0, and 2, respectively? (*Hint*: Use the formula for z and solve for y.)

5.107 Refer to the Clinical Trials database and Exercise 5.105. If the combined HAM-D total scores are assumed to be normally distributed, what is the probability of having a total score higher that 24? Less than 10?

6

Sampling Distributions

6.1   Introduction

A statistic is a numerical descriptive measure of a sample, whereas a parameter is a numerical descriptive measure of a population. We will use statistics to make inferences about population parameters—particularly to *estimate* the value of a population parameter or to make a *decision* about its value. As you will learn, we have to

sampling distribution   know a statistic's probability distribution, called its **sampling distribution**, in order to evaluate the reliability of the inferences we draw from it. For example, we must determine the probability that a statistic will give an estimate that falls close to (say, within some specified distance of) the actual value of the population parameter. Or if we plan to base a decision about a population parameter on the observed value of a sample statistic, we must establish the probability that the sample statistic will produce an incorrect decision.

This chapter is about sampling and sampling distributions. We will define the simplest and most common method of sampling and will give the sampling distributions for some important statistics computed from this type of sample. As we proceed through this chapter, you will see how probability and probability distributions (Chapter 5) play key roles in statistical inference.

6.2 Random Sampling

In Chapter 2, we discussed random samples and introduced various sampling schemes for gathering data. What is the importance of random sampling? We must know how a sample was selected so that we can determine probabilities associated with various sample outcomes. The probabilities of samples selected *in a random manner* can be determined, and we can use these probabilities to make inferences about the population from which the sample was drawn.

Sample data selected in a nonrandom fashion are frequently distorted by a *selection bias*. A selection bias exists whenever there is a systematic tendency to overrepresent or underrepresent some part of the population. For example, a survey of households conducted during the week entirely between the hours of 9 A.M. and 5 P.M. would be severely biased toward households in which at least one household member regularly stayed at home. Hence, any inferences made from the sample data would be biased toward the attributes or opinions of families in which at least one member stayed at home, and they may not be truly representative of the population of households in the region.

random sample   Now we turn to a definition of a **random sample** of n measurements selected from a population containing N measurements ($N > n$). (*Note:* This is a simple random sample, as discussed in Chapter 2. Since most of the random samples discussed in this text will be simple random samples, we'll drop the adjective *simple* unless it is needed for clarification.)

DEFINITION 6.1

Random Sample   A sample of n measurements selected from a population is a **random sample** if every different sample of size n from the population has an equal probability of being selected. ▪

EXAMPLE 6.1   Suppose that a population consists of the six measurements 1, 2, 3, 4, 5, 7. List all possible different samples of two measurements that could be selected from the population. Give the probability associated with each sample in a random sample of $n = 2$ measurements selected from the population.

Solution   All possible samples are listed in the accompanying table.

Sample	Measurements
1	1,2
2	1,3
3	1,4
4	1,5
5	1,7
6	2,3
7	2,4
8	2,5
9	2,7
10	3,4
11	3,5
12	3,7
13	4,5
14	4,7
15	5,7

Now let us suppose that we draw a single sample of $n = 2$ measurements from the 15 possible samples of two measurements. The sample selected is called a random sample if every sample had an equal probability $\left(\frac{1}{15}\right)$ of being selected. ∎

It is rather unlikely that we will ever achieve a truly random sample, because the probabilities of selection are not always exactly equal. But we do the best we can. One of the simplest and most reliable ways to select a random sample of n measurements from a population is to use a table of random numbers (see Table 11 in Appendix 3). **random number tables** **Random number tables** are constructed in such a way that, no matter where you start in the table and no matter what direction you move in, the digits occur randomly and with equal probability. Thus, if we wished to choose a random sample of $n = 10$ measurements from a population containing 100 measurements, we could label the measurements in the population from 0 to 99 (or 1 to 100). Then, by referring to Table 11 in Appendix 3 and choosing a random starting point, we could take the next 10 two-digit numbers going across the page as the labels of the particular measurements to be included in the random sample. Similarly, by moving up or down the page, we could also obtain a random sample.

EXAMPLE 6.2　A small community consists of 850 families. We wish to obtain a random sample of 20 families to ascertain public reaction to a wage and price freeze. Refer to Table 11 in Appendix 3 to determine which families should be sampled.

Solution　Assuming that a list of all families in the community is available (such as in a telephone directory), we could label the families from 0 to 849 (or, equivalently, from 1 to 850). Then, referring to Table 11 in Appendix 3, we choose a starting point.

Suppose that we decide to start at line 1, column 3. Going down the page, we choose the first 20 three-digit numbers between 000 and 849. From Table 11, we have

256	067	214	669
192	081	196	607
270	650	009	629
429	625	521	012
357	478	496	754

These 20 numbers identify the 20 families that are to be included in our sample.   ■

A telephone directory is not always the best source for names, especially in surveys related to economics or politics. In the 1936 presidential campaign, Franklin Roosevelt was running as the Democratic candidate against the Republican candidate, Governor Alfred Landon of Kansas. This was a difficult time for the nation; the country had not yet recovered from the Great Depression of the 1930s, and 9 million people were unemployed.

The *Literary Digest* set out to sample the voting public and predict the winner of the election. Using names and addresses taken from telephone books and club memberships, the *Literary Digest* sent out 10 million questionnaires and got 2.4 million back. Based on the responses to the questionnaire, the *Digest* predicted a Landon victory by 57% to 43%.

At this time, George Gallup was starting his survey business. He conducted two surveys. The first one, based on 3,000 people, predicted what the results of the *Digest* survey would be long before the *Digest* results were published; the second survey, based on 50,000 people, was used to forecast *correctly* the Roosevelt victory.

Where did the *Literary Digest* go wrong? The first problem was a severe selection bias. By taking the names and addresses from telephone directories and club memberships, it systematically excluded the poor from its survey. And unfortunately for the *Digest*, the vote was split along economic lines: the poor gave Roosevelt a large majority, whereas the more affluent tended to vote for Landon. A second source of error might have been *nonresponse bias*. Because only 20% of the 10 million people returned their surveys, and because approximately half of those who responded favored Landon, one might suspect that the nonrespondents had different preferences from the respondents. This was, in fact, true.

How, then, does one obtain a random sample? Careful planning and a certain amount of ingenuity are required to have even a decent chance to approximate random sampling. This is especially true when the universe of interest involves people. People can be difficult to work with; they have a tendency to discard mail questionnaires and to refuse to participate in personal interviews. Unless we are very careful, the data we obtain may be full of biases having unknown effects on the inferences we are attempting to make.

We do not have sufficient time to explore the topic of random sampling further in this text; entire courses at the undergraduate and graduate levels can be devoted to sample survey research methodology. The important point to remember is that data from a random sample provide the foundation for making statistical inferences.

Random samples are not easy to obtain, but with care we can avoid many potential biases that could affect the inferences we make.

Exercises

Basic Techniques

6.1 Define what is meant by a random sample. Is it possible to draw a truly random sample? Comment.

6.2 Suppose that we want to select a random sample of $n = 10$ persons from a population of 800. Use Table 11 in Appendix 3 to identify the persons to appear in the sample.

6.3 Refer to Exercise 6.2. Identify the elements of a population of $N = 1000$ to be included in a random sample of $n = 15$.

Applications

6.4 City officials want to sample the opinions of homeowners in a community regarding the desirability of increasing local taxes to improve the quality of the public schools. If a random number table is used to identify the homes to be sampled, and if a home is discarded if the homeowner is not home when visited by the interviewer, is it likely that this process will approximate random sampling? Explain.

6.5 A local TV network wants to run an informal survey of individuals who exit from a local voting station to ascertain early results on a proposal to raise funds to move the city-owned historical museum to a new location. How might the network sample voters to approximate random sampling?

6.6 A psychologist was interested in studying women who are in the process of obtaining a divorce, to determine whether there are significant attitudinal changes after the divorce has been finalized. Existing records from the geographic area in question show that 798 couples have recently filed for divorce. Assume that a sample of 25 women is needed for the study, and use Table 11 in Appendix 3 to determine which women should be asked to participate in the study. (*Hint:* Begin in column 2, row 1, and proceed down.)

6.7 Refer to Exercise 6.6. As is the case in most surveys, not all persons chosen for a study will agree to participate. Suppose that 5 of the 25 women selected refuse to participate. Determine 5 more women to be included in the study.

6.8 Suppose that you have been asked to run a public opinion poll related to an upcoming election. There are 1000 registered voters in a specific precinct, and you wish to obtain a random sample of 50 persons. Use a computer program to indicate which individuals are to be included in the sample. A Minitab program is shown here for purposes of illustration. (*Note:* We assume that a list exists of the 1000 voters, with the numbers 1 to 1000 corresponding to people on the list.)

```
MTB > RANDOM 50 C1;
SUBC> INTEGER 1 TO 1000.
MTB > PRINT C1

C1
    491     286     222     133     303      68     330     685     851     678     757
    535     941     920     823     401     901     484     457     152     783     706
    775     109     174     417     180     902     332     214     456     997     863
     14     845     471     669     207     537     149     375     282     584     667
    952     248     679     589     772     704

MTB > STOP
```

6.3 The Sampling Distribution for ȳ and the Central Limit Theorem

The numerical value that a sample statistic will have cannot be predicted exactly in advance. Even if we knew that a population mean μ was 216.37 and that the population standard deviation σ was 32.90 (that is, even if we knew the complete population distribution), we could not say that the sample mean $\bar{y}$ would be exactly equal to 216.37. A sample statistic is a random variable; it is subject to random variation because it is based on a random sample of measurements selected from the population of interest. And, like any other random variable, a sample statistic has a probability distribution. We call the probability distribution of a sample statistic the *sampling distribution* of that statistic. Stated differently, the sampling distribution of a statistic is the population of all values for that statistic.

The actual mathematic derivation of sampling distributions is one of the basic problems of mathematical statistics. Let's look at how the **sampling distribution of ȳ** can be obtained for a simplified population.

sampling distribution of ȳ

EXAMPLE   6.3   The sample mean $\bar{y}$ is to be calculated from a random sample of size 2 taken from a population consisting of the five values ($2, $3, $4, $5, $6). Find the sampling distribution of $\bar{y}$, based on a sample of size 2.

Solution   One way to find the sampling distribution is by counting. There are ten possible samples of two items from the population of five items:

Possible Samples of Size 2	Value of ȳ
2, 3	2.5
2, 4	3
2, 5	3.5
2, 6	4
3, 4	3.5
3, 5	4
3, 6	4.5
4, 5	4.5
4, 6	5
5, 6	5.5

Assuming that each sample of size 2 is equally likely, it follows that the sampling distribution for $\bar{y}$ based on $n = 2$ observations selected from this population is as follows:

$\bar{y}$	$P(\bar{y})$
2.5	1/10
3	1/10
3.5	2/10
4	2/10
4.5	2/10
5	1/10
5.5	1/10

This sampling distribution is shown as a graph in Figure 6.1. The graph illustrates that the distribution of measurements is symmetrical, with a mean of 4.0 and a standard deviation of about 1.0 (the range divided by 4). ▪

F I G U R E **6.1**
Sampling Distribution for $\bar{y}$ in Example 6.3

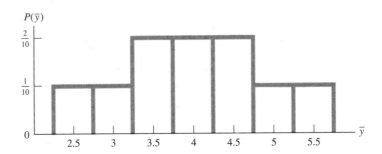

Interpretations for Sampling Distributions

Sampling distributions can be interpreted in several different ways. One way makes use of the classical interpretation of probability. Imagine listing all possible samples that can be drawn from a population; the probability that a sample statistic has a particular value (say $\bar{y} = 3.5$) is equal to the proportion of all possible samples that yield that value. In Example 6.3, $P(\bar{y} = 3.5) = 2/10$ corresponds to the fact that 2 of the 10 possible samples have a sample mean of 3.5.

A second interpretation uses the long-run relative frequency approach. Imagine taking repeated samples of a fixed size from a population and calculating the value of the sample statistic for each sample. In the long run, the relative frequencies for the possible values of the sample statistic will approach the corresponding sampling distribution probabilities. For example, if one took a large number of samples from the population of Example 6.3 and computed the sample mean for each sample, approximately 20% would have $\bar{y} = 3.5$.

In practice, though, a sample is taken only once, and only one value of the sample statistic is calculated. Consequently, a sampling distribution is not something you see in practice; rather it is a theoretical concept. However, if we know what would have happened had we repeated the process many times, we can make inferences based on the results of a single sample.

Explanation for Why Some Sampling Distributions Are Normal

Quite a few sample statistics will have a normal (bell-shaped) sampling distribution. A very plausible explanation for this is offered by the Central Limit Theorem. We will illustrate its applicability and then state it formally.

Central Limit Theorem

The **Central Limit Theorem**, one of the most important theorems in statistics, provides information on the sampling distribution of $\bar{y}$. It states that the sampling distribution of sums, or means, based on repeated random samples of measurements from a population will be approximately bell-shaped. This idea can best be illustrated with an example. Recall that a die is a cube whose faces show from one to six dots. Thus there are six possible values that can appear face up when a die is rolled (tossed). We can simulate a population of outcomes by throwing a die a large number of times. The resulting relative frequency histogram for $\bar{y}$ (the number of dots appearing face up) would be as indicated in Figure 6.2.

FIGURE 6.2
Relative Frequency Distribution for a Population of Die Tosses

Suppose we now draw samples of five measurements ($n = 5$) from this distribution by tossing the die five times and recording the value of y observed each time (see Table 6.1). Note that the values of y observed for the first sample of five measurements are 3, 5, 1, 3, and 2, respectively. We repeat this process 100 times to obtain 100 samples of five measurements. The sum Σy and the sample mean $\bar{y}$ are shown for each sample in columns three and seven, and four and eight, respectively.

We can construct a relative frequency histogram for $\bar{y}$ (the mean of the five sample measurements) by first preparing a frequency table from the data in Table 6.1. The tabulation of relative frequencies is shown in Table 6.2, and the resulting relative frequency histogram is shown in Figure 6.3(b). According to the Central Limit Theorem, this relative frequency histogram of $\bar{y}$ should be approximately normal.

Notice that, although the relative frequencies for the individual values of y are equal and hence the relative frequency distribution is flat (see Figure 6.3(a)), the distribution of the sample means is mound-shaped and even somewhat bell-shaped (see Figure 6.3(b)). You can visually compare these two relative frequency distributions by observing Figure 6.3. The irregularities in the curve of Figure 6.3(b) are due to the small number of samples used to illustrate this concept. These irregularities would be less obvious if the sampling were conducted a large number of times. (Such extensive sampling is a time-consuming task to perform manually, but it can be done easily with the aid of a computer). The result would verify the Central Limit Theorem, which we now state as it applies to means.

TABLE 6.1
Sums and Means for 100
Samples of Five Die Tosses

Sample Number	Sample Measurements	$\sum y$	$\bar{y}$	Sample Number	Sample Measurements	$\sum y$	$\bar{y}$
1	3, 5, 1, 3, 2	14	2.8	51	2, 3, 5, 3, 2	15	3.0
2	3, 1, 1, 4, 6	15	3.0	52	1, 1, 1, 2, 4	9	1.8
3	1, 3, 1, 6, 1	12	2.4	53	2, 6, 3, 4, 5	20	4.0
4	4, 5, 3, 3, 2	17	3.4	54	1, 2, 2, 1, 1	7	1.4
5	3, 1, 3, 5, 2	14	2.8	55	2, 4, 4, 6, 2	18	3.6
6	2, 4, 4, 2, 4	16	3.2	56	3, 2, 5, 4, 5	19	3.8
7	4, 2, 5, 5, 3	19	3.8	57	2, 4, 2, 4, 5	17	3.4
8	3, 5, 5, 5, 5	23	4.6	58	5, 5, 4, 3, 2	19	3.8
9	6, 5, 5, 1, 6	23	4.6	59	5, 4, 4, 6, 3	22	4.4
10	5, 1, 6, 1, 6	19	3.8	60	3, 2, 5, 3, 1	14	2.8
11	1, 1, 1, 5, 3	11	2.2	61	2, 1, 4, 1, 3	11	2.2
12	3, 4, 2, 4, 4	17	3.4	62	4, 1, 1, 5, 2	13	2.6
13	2, 6, 1, 5, 4	18	3.6	63	2, 3, 1, 2, 3	11	2.2
14	6, 3, 4, 2, 5	20	4.0	64	2, 3, 3, 2, 6	16	3.2
15	2, 6, 2, 1, 5	16	3.2	65	4, 3, 5, 2, 6	20	4.0
16	1, 5, 1, 2, 5	14	2.8	66	3, 1, 3, 3, 4	14	2.8
17	3, 5, 1, 1, 2	12	2.4	67	4, 6, 1, 3, 6	20	4.0
18	3, 2, 4, 3, 5	17	3.4	68	2, 4, 6, 6, 3	21	4.2
19	5, 1, 6, 3, 1	16	3.2	69	4, 1, 6, 5, 5	21	4.2
20	1, 6, 4, 4, 1	16	3.2	70	6, 6, 6, 4, 5	27	5.4
21	6, 4, 2, 3, 5	20	4.0	71	2, 2, 5, 6, 3	18	3.6
22	1, 3, 5, 4, 1	14	2.8	72	6, 6, 6, 1, 6	25	5.0
23	2, 6, 5, 2, 6	21	4.2	73	4, 4, 4, 3, 1	16	3.2
24	3, 5, 1, 3, 5	17	3.4	74	4, 4, 5, 4, 2	19	3.8
25	5, 2, 4, 4, 3	18	3.6	75	4, 5, 4, 1, 4	18	3.6
26	6, 1, 1, 1, 6	15	3.0	76	5, 3, 2, 3, 4	17	3.4
27	1, 4, 1, 2, 6	14	2.8	77	1, 3, 3, 1, 5	13	2.6
28	3, 1, 2, 1, 5	12	2.4	78	4, 1, 5, 5, 3	18	3.6
29	1, 5, 5, 4, 5	20	4.0	79	4, 5, 6, 5, 4	24	4.8
30	4, 5, 3, 5, 2	19	3.8	80	1, 5, 3, 4, 2	15	3.0
31	4, 1, 6, 1, 1	13	2.6	81	4, 3, 4, 6, 3	20	4.0
32	3, 6, 4, 1, 2	16	3.2	82	5, 4, 2, 1, 6	18	3.6
33	3, 5, 5, 2, 2	17	3.4	83	1, 3, 2, 2, 5	13	2.6
34	1, 1, 5, 6, 3	16	3.2	84	5, 4, 1, 4, 6	20	4.0
35	2, 6, 1, 6, 2	17	3.4	85	2, 4, 2, 5, 5	18	3.6
36	2, 4, 3, 1, 3	13	2.6	86	1, 6, 3, 1, 6	17	3.4
37	1, 5, 1, 5, 2	14	2.8	87	2, 2, 4, 3, 2	13	2.6
38	6, 6, 5, 3, 3	23	4.6	88	4, 4, 5, 4, 4	21	4.2
39	3, 3, 5, 2, 1	14	2.8	89	2, 5, 4, 3, 4	18	3.6
40	2, 6, 6, 6, 5	25	5.0	90	5, 1, 6, 4, 3	19	3.8
41	5, 5, 2, 3, 4	19	3.8	91	5, 2, 5, 6, 3	21	4.2
42	6, 4, 1, 6, 2	19	3.8	92	6, 4, 1, 2, 1	14	2.8
43	2, 5, 3, 1, 4	15	3.0	93	6, 3, 1, 5, 2	17	3.4
44	4, 2, 3, 2, 1	12	2.4	94	1, 3, 6, 4, 2	16	3.2
45	4, 4, 5, 4, 4	21	4.2	95	6, 1, 4, 2, 2	15	3.0
46	5, 4, 5, 5, 4	23	4.6	96	1, 1, 2, 3, 1	8	1.6
47	6, 6, 6, 2, 1	21	4.2	97	6, 2, 5, 1, 6	20	4.0
48	2, 1, 5, 5, 4	17	3.4	98	3, 1, 1, 4, 1	10	2.0
49	6, 4, 3, 1, 5	19	3.8	99	5, 2, 1, 6, 1	15	3.0
50	4, 4, 4, 4, 4	20	4.0	100	2,4, 3, 4, 6	19	3.8

T A B L E **6.2**
Relative Frequency Table for
100 Values of $\bar{y}$

Class	Class Boundaries	Frequency	Relative Frequency
1	1.3–1.5	1	1/100
2	1.5–1.7	1	1/100
3	1.7–1.9	1	1/100
4	1.9–2.1	1	1/100
5	2.1–2.3	3	3/100
6	2.3–2.5	4	4/100
7	2.5–2.7	6	6/100
8	2.7–2.9	10	10/100
9	2.9–3.1	7	7/100
10	3.1–3.3	9	9/100
11	3.3–3.5	11	11/100
12	3.5–3.7	9	9/100
13	3.7–3.9	11	11/100
14	3.9–4.1	10	10/100
15	4.1–4.3	7	7/100
16	4.3–4.5	1	1/100
17	4.5–4.7	4	4/100
18	4.7–4.9	1	1/100
19	4.9–5.1	2	2/100
20	5.1–5.3	0	0
21	5.3–5.5	1	1/100

T H E O R E M   **6.1**

standard error of $\bar{y}$

The Central Limit Theorem, Applied to Means   If random samples containing a fixed number n of measurements are repeatedly drawn from a population with finite mean μ and standard deviation σ, then, if n is large, the sample means will have a distribution that is approximately normal (bell-shaped), with mean $\mu_{\bar{y}} = \mu$ and standard deviation (called the **standard error of $\bar{y}$**) $\sigma_{\bar{y}} = \sigma/\sqrt{n}$. ■

We hope that the die-tossing data of Table 6.1 clarify for you the meaning of the Central Limit Theorem. Although proof is omitted, it can be shown that the mean and the standard deviation for y in the die-tossing distribution (Figure 6.3(a)) are $\mu = 3.50$ and $\sigma = 1.71$. For these values, the Central Limit Theorem states that the sampling distribution for sample means based on $n = 5$ measurements possesses the same mean as the original population: $\mu_{\bar{y}} = \mu = 3.5$. The standard error is

$$\sigma_{\bar{y}} = \frac{\sigma}{\sqrt{n}} = \frac{1.71}{\sqrt{5}} = .76$$

Our experimental die-tossing sample verifies these results. You can see that the mean of the distribution of $\bar{y}$ (Figure 6.3(b)) is approximately 3.5. In addition the range of $\bar{y}$ is $(5.4 - 1.4) = 4.0$, so the standard error is approximately $(4.0)/4$, or 1.0. Hence, the standard error of the distribution of 100 sample means (Figure 6.3(b)), is near the value stated by the Central Limit Theorem, $\sigma/\sqrt{n} = .76$. As noted earlier, the

FIGURE 6.3
Illustration of the Central Limit
Theorem

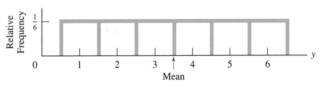

(a) Relative Frequency Distribution for y

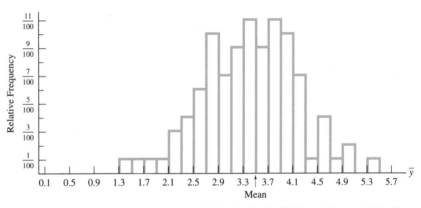

(b) Relative Frequency Distribution for the 100 Sample Means in Table 6.3

relative frequency histogram of Figure 6.3(b) is approximately bell-shaped. Thus the die-tossing sample experiment provides us with practical verification of the Central Limit Theorem.

The Central Limit Theorem also applies to the distribution of sums of sample observations, and the die-tossing data can be used to provide empirical evidence to support this version of the Central Limit Theorem. We leave it to you to construct a histogram of the sample sums $\Sigma \bar{y}$ if you seek graphical evidence of the validity of the Central Limit Theorem as it applies to sums.

THEOREM   6.2   **The Central Limit Theorem, Applied to Sums**   If random samples containing a fixed number n of measurements are repeatedly drawn from a population with finite mean μ and standard deviation σ, then, if n is large, the sums of the sample measurements will have a distribution that is approximately normal (bell-shaped), with mean equal to $n\mu$ and standard error equal to $\sigma\sqrt{n}$.   ▪

Broad Applicability of the Central Limit Theorem

A careful reading of the Central Limit Theorem suggests its broad applicability: it applies to the distribution of sample means drawn from any population with a finite mean μ and standard deviation σ. The resulting distribution of means is approximately normal, with mean and standard error related not only to the mean and standard deviation of the population from which the samples are drawn but also to the sample size n.

Notice that the normal approximation to the distribution of sample means or sums becomes more and more accurate as the sample size n increases. Nonetheless, the approximation is quite good for an n as small as 5 (as in the die-tossing experiment).

The significance of the Central Limit Theorem is twofold:

1 It explains why many measurements have bell-shaped frequency distributions. For example, we might imagine that the test score for an individual on a national aptitude test is influenced by random factors or variables, such as the amount of sleep the individual had the night before, the length of time the individual spent preparing for the exam, the individual's IQ, and so forth. If each of these factors in some way affects the final score, that score is the sum of random variables. The Central Limit Theorem may then help to explain why such scores are approximately normally distributed.

2 It is useful in statistical inference. Many estimators of population parameters used for purposes of statistical inference are sums of averages of sample measurements. To illustrate, we will use the sample mean $\bar{y}$ to estimate a population mean μ. Where sums or averages are involved and the sample size n is large, the many estimates generated in repeated sampling can be expected to possess a bell-shaped normal distribution. Then we can use the properties of the normal distribution to describe the behavior of $\bar{y}$. (This application of the Central Limit Theorem is explained in Chapter 7.)

To summarize, the Central Limit Theorem tells us the nature of the sampling distribution of $\bar{y}$ when the sample size is large. This information can be broken down as follows.

Properties of the Sampling Distribution of Sample Mean $\bar{y}$

1 The sampling distribution of $\bar{y}$ is approximately normal for large sample sizes (n is large).

2 The mean of the sampling distribution is equal to the population mean: $\mu_{\bar{y}} = \mu$.

3 The standard error of the sampling distribution is $\sigma_{\bar{y}} = \sigma/\sqrt{n}$, where σ is the standard deviation of the sampled population.

An obvious question is: How large should the sample size be in order for the Central Limit Theorem to hold? Numerous studies on this question have been conducted over the years, and the results of these studies show that, in general, the Central Limit Theorem will hold for $n > 30$. However, one should not apply the rule blindly. If the population is heavily skewed, the sampling distribution for $\bar{y}$ will remain skewed even for $n > 30$. On the other hand, if the population is symmetric, the Central Limit Theorem will hold for $n < 30$. For example, the Central Limit Theorem worked very well for a sample size of $n = 5$ in the die-tossing experiment. So take a look at the sample data. If the sample histogram is heavily skewed, then probably the population will also be skewed. Consequently a value of n that is much larger than 30 may be required before the distribution of $\bar{y}$ becomes nearly normal.

EXAMPLE 6.4   Reclaimed phosphate land in Polk County, Florida, has been found to emit a higher mean radiation level than other nonmining land in the county. Suppose that the radiation level for the reclaimed land has a distribution with a mean $\mu = 5.0$ working levels (WL) and a standard deviation of .5 WL. Suppose that 20 houses built on reclaimed land are randomly selected and the radiation level is measured in each. What is the probability (approximately) that the sample mean for the 20 houses will exceed 5.2 WL?

Solution   According to the Central Limit Theorem, the sample mean of $n = 20$ randomly selected radiation level measurements will be approximately normally distributed, with

$$\mu_{\bar{y}} = \mu = 5.0 \text{ WL}$$

$$\sigma_{\bar{y}} = \frac{\sigma}{\sqrt{n}} = \frac{.5}{\sqrt{20}} = .11.$$

The distribution will be as shown in Figure 6.4.

The probability that $\bar{y}$ exceeds 5.2 WL is the shaded area shown in Figure 6.4. To find this area, we must determine how many standard errors ($\sigma_{\bar{y}} = .11$) the point 5.2 lies to the right of $\mu = 5.0$. This number is

$$z = \frac{\bar{y} - \mu}{\sigma_{\bar{y}}} = \frac{5.2 - 5.0}{.11} = 1.82$$

Turning to Table 2 of Appendix 3, we find that the area A corresponding to $z = 1.82$ is

$$A = .4656.$$

Since the total area to the right of the mean is equal to .5, the probability that $\bar{y}$ will exceed 5.2 WL is P, where

$$P = .5 - A = .5 - .4656 = .0344. \quad \blacksquare$$

FIGURE 6.4
Sampling Distribution of $\bar{y}$ for Samples of $n = 20$ Randomly Selected from a Population with $\mu = 5.0$ and $\sigma = .5$

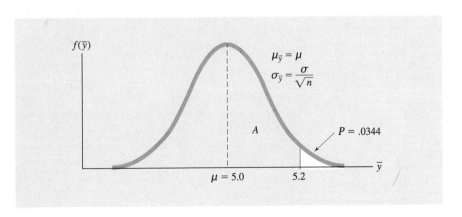

Exercises

Basic Techniques

6.9   A population consists of the five measurements 2, 6, 8, 0, and 1. Use the Central Limit Theorem to answer the following questions.

a   How many different samples of size $n = 2$ can be drawn from the population? (*Hint:* List them.)

b   What is the mean of the sampling distribution of $\bar{y}$? (*Hint:* Compute the mean for the population.)

c   What is the standard error of the sampling distribution of $\bar{y}$ if $\sigma = 4.43$?

6.10   Complete the following sentence. As the sample size increases, the standard error of $\bar{y}$ _____.

6.11   A random sample of $n = 100$ measurements is obtained from a population with $\mu = 55$ and $\sigma = 20$. Describe the sampling distribution for $\bar{y}$, giving $\mu_{\bar{y}}$ and $\sigma_{\bar{y}}$.

6.12   A random sample of $n = 60$ measurements is obtained from a population with $\mu = 192$ and $\sum = 43$. Describe the sampling distribution for $\bar{y}$.

6.13   Describe the sampling distribution for $\sum y$ in Exercise 6.12.

6.14   Refer to Table 6.1. Using the sample data from the 100 samples of $n = 5$ die tosses, construct a frequency table similar to Table 6.2 for $\sum y$, the sum of the sample measurements. Use an interval width of 1, with a starting point of 6.5.

6.15   Using the frequency table of Exercise 6.14, construct a relative frequency histogram for $\sum y$. The mean of the distribution of $\bar{y}$, the number of dots appearing on the upper face, can be shown to be $\mu = 3.5$, and the size of each sample is $n = 5$. Thus, by the Central Limit Theorem, the relative frequency histogram should be mound-shaped, with a mean approximately equal to $n(\mu) = 5(3.5) = 17.5$. Because the relative frequency histogram of $\sum y$ is based on only 100 samples, the approximation could be improved by using more repetitions of the experiment.

6.16   A random sample of $n = 25$ measurements is selected from a population with mean equal to 80 and standard deviation equal to 7.

a   What is the probability that the sample mean will exceed 81?

b   What is the probability that the sample mean will fall in the interval $79 \leq \bar{y} \leq 81$?

c   What is the probability that the sample mean will be less than 78?

6.17   A random sample of 16 measurements is drawn from a population with a mean of 60 and a standard deviation of 5. Describe the sampling distribution of $\bar{y}$, the sample mean. Within what interval would you expect $\bar{y}$ to lie approximately 95% of the time?

6.18   Refer to Exercise 6.17. Describe the sampling distribution for the sample sum $\sum y_i$. Is it unlikely (improbable) that $\sum y_i$ will be more than 70 units away from 960? Explain.

Applications

6.19   Psychomotor retardation scores for a large group of manic-depressive patients were found to be approximately normal with a mean of 930 and a standard deviation of 130.

a   What fraction of the patients scored between 800 and 1100?

b   Less than 800?

c   Greater than 1200?

6.20   Refer to Exercise 6.19. Find the 90th percentile for the distribution of manic-depressive scores. [*Hint:* Solve for y in the expressions $z = (y - \mu)/\sigma$, where z is the number of standard deviations the 90th percentile lies above the mean μ.]

6.21   The oxygen content in water must exceed some minimum value in order to support aquatic life. Suppose that this value is approximately 6.0 parts per million (ppm). In one experiment, $n = 5$

jars of water are randomly selected from a stream. Records from previous months suggest that the mean oxygen content is $\mu = 6.0$ ppm and the standard deviation is $\sigma = .7$ ppm.

a   What is the probability that the sample mean exceeds 6.5 ppm?

b   Suppose that the sample mean equals 7.0 ppm. Intuitively, what would you conclude about the mean oxygen content (μ) of the stream? Has the situation changed from previous months? (*Note:* In Chapter 7 we will answer this question by using a statistical decision procedure.)

6.22   Federal resources have been tentatively approved to fund the construction of an outpatient clinic. In order for the designers to present plans for a facility that will handle patient load requirements while staying within a limited budget, a study of patient demand was made. By studying a similar facility in the area, researchers found that the distribution of the number of patients requiring hospitalization during a week could be approximated by a normal distribution with a mean of 125 and a standard deviation of 32.

a   Use the Empirical Rule to describe the distribution of y, the number of patients requesting service in a week.

b   If the facility is built with a 160-patient capacity, in what fraction of all weeks might the clinic be unable to handle the demand?

6.23   Refer to Exercise 6.22. What size facility should be built so that the probability that the patient load will exceed the clinic's capacity is .05? .01?

6.4   Normal Approximation to the Binomial

The Central Limit Theorem just discussed allows us to calculate probabilities for a binomial random variable by approximating the binomial distribution with a normal curve and then using normal curve areas as approximations to the desired probabilities. We have seen that probabilities associated with values of y can be computed for a binomial experiment for any values of n or p, but the task becomes more difficult when n gets large. For example, suppose that a sample of 1000 voters is polled to determine sentiment toward the consolidation of a city and county government. What would be the probability of observing 460 or fewer favoring consolidation, if we assume that 50% of the entire population favor the change? Here we have a binomial experiment with $n = 1000$ and p, the probability of selecting a person favoring consolidation, equal to .5. To determine the probability of observing 460 or fewer favoring consolidation in the random sample of 1000 voters, we could compute $P(y)$, using the binomial formula for $y = 460, 459, \ldots, 0$. The desired probability would then be

$$P(y = 460) + P(y = 459) + \cdots + P(y = 0)$$

There would be 461 probabilities to calculate, and each one would be somewhat difficult due to the factorials. For example, the probability of observing 460 favoring consolidation is

$$P(y = 460) = \frac{1000!}{460! \, 540!}(.5)^{460}(.5)^{540}$$

A similar calculation would be needed for all other values of y.

The normal distribution can be used in many situations to approximate the binomial probability distribution, and areas under the normal curve can be used to approximate the actual binomial probabilities. This process of approximating a binomial distribution with a normal distribution is called a **normal approximation to the binomial**. The normal distribution that provides the best approximation to the binomial probability distribution has a mean and a standard deviation given by the following formulas.

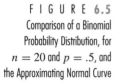
normal approximation to
the binomial

Normal Approximation to the Binomial Probability Distribution

$$\mu = np \qquad \sigma = \sqrt{npq} \quad \text{where} \quad q = 1 - p$$

This approximation can be used if

$$n \geq \frac{5}{\min\,(p, q)},$$

that is, if n is greater than or equal to 5 divided by the minimum of p and q. Equivalently, the normal approximation can be used if $np \geq 5$ and $nq \geq 5$.

To illustrate how well the distribution of y can be approximated by a normal distribution, we show the binomial probability distribution for $n = 20$ and $p = .5$ in Figure 6.5, with the approximating normal curve superimposed. For this situation, the mean and standard deviation are

$$\mu = np = (20)(.5) = 10$$
$$\sigma = \sqrt{npq} = \sqrt{(20)(.5)(.5)} = 2.24.$$

FIGURE 6.5
Comparison of a Binomial
Probability Distribution, for
$n = 20$ and $p = .5$, and
the Approximating Normal Curve

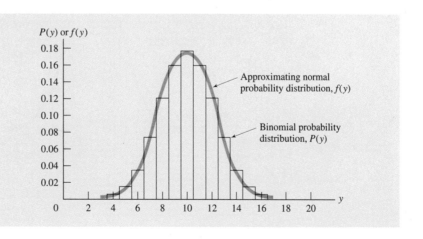

As you can see, this approximation is surprisingly good, considering the small value of n. However, when p is close to 0 or 1, we would need a much larger sample in order for the normal distribution to provide a good approximation to the binomial probability distribution.

E X A M P L E   6.5   Use the normal approximation to the binomial probability distribution (Figure 6.5) to find the probability that $y \geq 13$.

Solution   The probability that we wish to approximate is the sum of the probability rectangles (Figure 6.5) corresponding to $y = 13, 14, \ldots, 20$. To approximate the area corresponding to these rectangles, we need to find the area under the normal curve to the right of $y = 12.5$ (see Figure 6.6). Notice that using only the area to the right of $y = 13$ would be incorrect, because then we would be omitting the area corresponding to the left half of the rectangle for $y = 13$.

To find the probability P that $y \geq 13$, we must first determine how many standard deviations the point $y = 12.5$ lies to the right of $\mu = np = 10$. This is

$$z = \frac{y - \mu}{\sigma} = \frac{12.5 - 10}{2.24} = 1.12.$$

Then the area A between the mean of the standard normal distribution and the point $z = 1.12$ is .3686 (see Figure 6.7). Therefore, the probability is

$$P = .5 - A = .5 - .3686 = .1314. \quad \blacksquare$$

F I G U R E   6.6
Required Area for Example 6.5

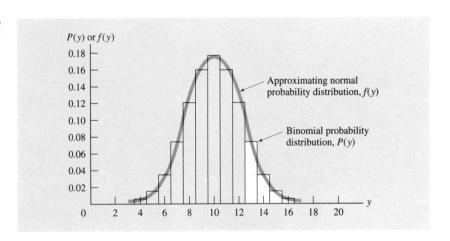

FIGURE 6.7
Normal Curve for Example 6.5

FIGURE 6.7
Normal Curve for Example 6.5

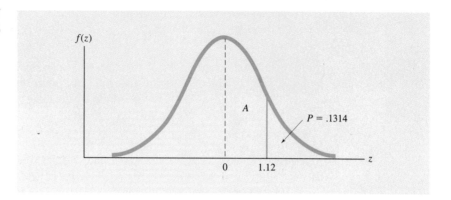

EXAMPLE   6.6   Refer to the example discussed at the beginning of this section. Use the normal approximation to the binomial to compute the probability of observing 460 or fewer in a sample of 1000 favoring consolidation, if we assume that 50% of the entire population favor the change.

Solution   The normal distribution used to approximate the binomial distribution will have

$$\mu = np = 1000(.5) = 500$$
$$\sigma = \sqrt{npq} = \sqrt{1000(.5)(.5)} = 15.8.$$

Notice that

$$n \geq \frac{5}{.5} = 10.$$

Hence we may use the normal approximation to the binomial. The desired probability is represented by the shaded area shown in Figure 6.8.

FIGURE 6.8
Approximating Normal
Distribution for the Binomial
Distribution of Example 6.6 with
$\mu = 500$ and $\sigma = 15.8$

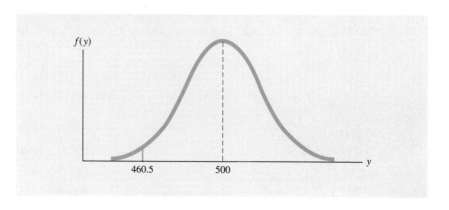

As noted earlier, we could compute

$$P(y \le 460) = P(y = 460) + P(y = 459) + \cdots + P(y = 0),$$

using the binomial probability distribution with $n = 1000$ and $p = .5$. This same probability can be approximated by using the area under the normal curve to the left of $y = 460.5$ (see Figure 6.8). The z-score corresponding to $y = 460.5$ is

$$z = \frac{y - \mu}{\sigma} = \frac{460.5 - 500}{15.8} = -2.50.$$

Referring to Table 2 of Appendix 3, we find that the area under the normal curve between 460.5 and 500 (that is, for $z = -2.50$), is .4938. Thus, the probability of observing 460 or fewer favoring consolidation is approximately $.5 - .4938 = .0062.$ ▪

EXAMPLE 6.7   Refer to Example 6.6. Suppose that 460 of a sample of 1000 potential voters favor the consolidation of the city and county governments. Would you expect the consolidation issue to pass?

Solution   In Example 6.6 we computed $P(y \le 460)$ to be approximately .0062 when $p = .5$ and $n = 1000$. Since this probability is so small, an observed value of y equal to 460 contradicts the assumption that $p = .5$ (or more) of the voters favor consolidation. Because of this contradiction, we conclude that the consolidation issue will not pass.   ▪

6.5 Sampling Distribution for $\hat{p}$

Binomial populations are sampled in order to make inferences about p, the proportion of successes in the population. The most obvious statistic to select to make these inferences is the proportion of successes in the sample, denoted by the symbol $\hat{p}$ (read "p hat"; the "hat" over the symbol p is used to denote "estimator of p"). What can we say about the sampling distribution of the sample proportion $\hat{p}$?

If we assign the value 0 to each trial that fails and the value 1 to each successful trial in a binomial experiment (as was suggested in the previous chapter), then the number of successes y in the sample of n trials is equal to the sum of the n sample measurements. Therefore, it follows that the sample proportion

$$\hat{p} = \frac{y}{n} = \frac{\text{Sum of the sample measurements}}{n}$$

sampling distribution of $\hat{p}$ is the sample mean and that the **sampling distribution of $\hat{p}$** will be approximately normally distributed when n is large (because of the Central Limit Theorem). It can be shown (proof omitted) that the mean and the standard deviation of this approximating normal distribution are

$$\mu_{\hat{p}} = p$$

$$\sigma_{\hat{p}} = \sqrt{\frac{pq}{n}}.$$

The sampling distribution for the sample proportion $\hat{p}$ is as shown in Figure 6.9.

FIGURE 6.9
Sampling Distribution for $\hat{p}$

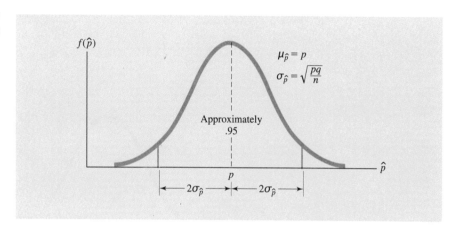

Properties of the Sampling Distribution of a Binomial Sample Proportion $\hat{p}$

1 The sampling distribution for $\hat{p}$ is approximately normal for large sample sizes (n is large).

2 The mean $\mu_{\hat{p}}$ of the sampling distribution is equal to p, the population proportion of successes.

3 The standard error of the sampling distribution for $\hat{p}$ is equal to

$$\sigma_{\hat{p}} = \sqrt{\frac{pq}{n}}.$$

In the next example, we show how this information can be used in a practical application.

EXAMPLE 6.8   Two thousand new automobile steering mechanisms were tested in order to estimate the proportion p of all the steering mechanisms that might be faulty. If the true proportion is $p = .03$, what is the probability that the sample proportion will be within .01 of p?

Solution   The 2000 steering mechanisms can be viewed as constituting a random sample. Thus the sampling distribution for $\hat{p}$ will appear as shown in Figure 6.10, with mean

$$p = .03$$

and standard error

$$\sigma_{\hat{p}} = \sqrt{\frac{pq}{n}} = \sqrt{\frac{(.03)(.97)}{2000}} = .0038.$$

The probability that the sample proportion $\hat{p}$ falls within .01 of the population proportion $p = .03$ is the shaded area shown under the sampling distribution of $\hat{p}$ in Figure 6.10. If A is the area over the interval .03 to .04, then the probability that p falls in the interval .02 to .04 is $2A$.

FIGURE 6.10
Sampling Distribution for $\hat{p}$, with $n = 2000$ and $p = .03$

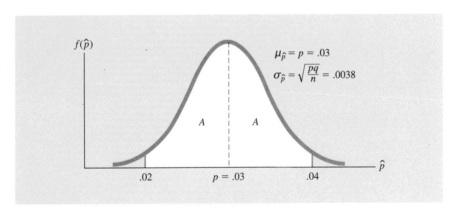

To find the number of standard deviations between $p = .03$ and a value of $\hat{p}$ equal to .04, we calculate the z-score:

$$z = \frac{\hat{p} - p}{\sigma_{\hat{p}}} = \frac{.04 - .03}{.0038} = 2.63.$$

Since the area A to the right of the mean, corresponding to $z = 2.63$, is .4957, it follows that

$$P(.02 \leq \hat{p} \leq .04) = 2A = 2(.4957) = .9914.$$

In other words, if the population proportion of steering mechanisms that are defective is $p = .03$, the proportion of steering mechanisms that malfunction in a sample of $n = 2000$ should be within .01 of p with a high probability—namely, .9914. Therefore, in this case the sample proportion should provide an accurate estimate of p.   ▪

Exercises

Basic Techniques

6.24 A random sample of $n = 1000$ measurements is obtained from a binomial population with $p = .7$. Describe the sampling distribution for the sample proportion $\hat{p}$.

6.25 Refer to Exercise 6.24. Use the sampling distribution of $\hat{p}$ to find the probability that $\hat{p}$ lies in the interval from .6 to .8.

6.26 Suppose that you select a sample of $n = 10$ from a binomial population with $p = .5$.

   a Use the binomial probability distribution to calculate the probability that y falls in the interval $3 \leq y \leq 5$.

   b Use the normal approximation to the binomial probability distribution to find the approximate probability that $3 \leq y \leq 5$. Compare this result with your answer in part a.

6.27 An insurance company states that 10% of all fire insurance claims are fraudulent. Suppose that the company is correct and that it receives 100 claims.

   a What is the probability that more than 12 will be fraudulent?

   b What is the probability that 15 or more will be fraudulent?

6.28 Refer to Exercise 6.27.

   a What assumptions must be true in order for your answers to be valid?

   b Suppose that only one claim in the sample was fraudulent. Would you have any doubts about the company's statement? Explain.

6.29 Suppose that you wish to conduct a poll to estimate the proportion of adult Americans who favor less governmental control of private business. Further, suppose that this unknown proportion is really $p = .5$. If you draw a random sample of $n = 1000$ adults from the United States population, what is the probability that the sample proportion will differ from the population proportion ($p = .5$) by more than .02?

6.30 Suppose that 2% of all babies in the United States are born with one or more congenital malformations, and that 3 million babies are born in a given year.

   a Determine the mean and the standard deviation of the number of malformed infants born in that year.

   b Determine the probability of observing 50,000 or fewer babies with malformations in that year.

6.31 Refer to Exercise 6.30.

   a Determine the sampling distribution for $\hat{p}$, the proportion of malformed infants observed in the sample.

   b What is the probability of observing 3% or more sampled infants with malformations? 4% or more?

6.6 Using Computers to Help Make Sense of Data

The MINITAB software system was used to simulate the sampling distribution for $\bar{y}$ based on 50 samples of $n = 5$ measurements drawn from a normal population with $\mu = 20$ and $\sigma = 4$. As you can see, we had to use the statements RANDOM, NORMAL, and RMEAN (row mean) to generate each sample of $n = 5$ measurements and to compute the sample mean. The rows of the output correspond to the samples. Columns C1 to C5 give the five sample measurements; column C6 gives the sample mean for the five measurements in that sample (row). For example, the five sample measurements for sample 1 are 16.8741, 25.3995, 16.9347, 21.4610, and 18.1604. The sample mean for these five measurements is 19.7659 (in column 6). Numerical descriptive measures of the sampling distribution and a histogram are shown in the output. The mean and the standard deviation of the 50 sample means can be computed by using column C6. Notice that the mean of the 50 sample means (20.013) is very close to the theoretical value, $\mu = 20$; similarly, the standard deviation of the $\bar{y}$ values, called the standard error of the mean, is 1.635, which is close to the theoretical value, $\sigma/\sqrt{n} = 1.79$. These two values would get closer and closer to the theoretical values with more and more samples of size $n = 5$.

The MINITAB software system can be used for such simulations as the one illustrated here. Another example is shown in Exercise 6.67. Further details about simulations of other distributions are discussed in *MINITAB Handbook* (1985).

The MINITAB program (together with resulting output) for generating a random sample of $n = 50$ observations from a normal distribution with $\mu = 20$ and $\sigma = 4$ is as follows.

MINITAB Output

```
MTB > DESCRIBE C6

                    N      MEAN    MEDIAN    TRMEAN    STDEV    SEMEAN
C6                 50    20.013    19.888    20.047    1.635     0.231

                  MIN       MAX        Q1        Q3
C6             15.816    23.491    18.987    21.091

MTB > HISTOGRAM C6

Histogram of C6    N = 50

Midpoint    Count
      16        2    **
      17        2    **
      18        4    ****
      19       10    **********
      20       11    ***********
      21       11    ***********
      22        7    *******
      23        3    ***

MTB > STOP
```

MINITAB Output (Continued)

```
MTB > RANDOM 50 C1-C5;
SUBC> NORMAL MU=20 SIGMA=4.
MTB > RMEAN C1-C5 INTO C6
MTB > PRINT C1-C6
```

ROW	C1	C2	C3	C4	C5	C6
1	16.8741	25.3995	16.9347	21.4610	18.1604	19.7659
2	19.4324	15.8662	16.3463	22.0762	21.2142	18.9871
3	21.2565	20.4629	25.5044	19.9371	20.8196	21.5961
4	18.9032	23.2406	18.2825	16.9949	22.0357	19.8914
5	25.3764	19.2741	18.5989	16.7060	19.4684	19.8848
6	28.3111	21.8848	21.0318	22.1336	20.8780	22.8479
7	19.4931	21.5756	14.8059	22.4375	8.5457	17.3715
8	18.8756	27.7662	14.9073	19.8881	26.7723	21.6419
9	18.7657	18.6138	27.7414	18.5340	22.5266	21.2363
10	16.1338	16.6816	20.6520	25.5658	14.5431	18.7153
11	21.8564	19.1099	24.3206	13.4724	20.7221	19.8963
12	28.5122	18.1532	23.6170	20.1738	18.4358	21.7784
13	22.1693	17.2697	18.9918	8.3434	25.1159	18.3780
14	18.0749	22.9861	14.2990	29.0757	20.9143	21.0700
15	27.2761	16.6008	17.4737	25.8921	15.9889	20.6463
16	15.7957	22.5431	16.7043	18.4629	24.7385	19.6489
17	17.8824	20.5549	21.0722	19.6192	23.6858	20.5629
18	15.1843	20.3594	19.3868	16.2928	20.7562	18.3959
19	18.4177	15.7609	21.6672	14.4446	26.9868	19.4554
20	16.6753	23.0347	17.8983	13.0907	24.2285	18.9855
21	25.7549	21.6419	25.3913	19.8639	21.6751	22.8654
22	27.4655	18.1044	26.1283	19.1541	21.0821	22.3869
23	20.1849	24.4747	24.1054	16.7468	20.2561	21.1536
24	21.1502	19.1478	22.2134	21.2329	15.4588	19.8406
25	13.7164	22.2291	19.1040	24.9974	23.0158	20.6125
26	18.3187	22.6083	25.2441	24.2121	18.1341	21.7034
27	17.4527	15.9950	14.5016	26.0450	22.3829	19.2754
28	14.8372	12.8505	16.6534	26.9380	22.6204	18.7799
29	23.4141	15.5976	19.9732	19.2032	24.5812	20.5539
30	14.8103	18.7435	21.8222	21.6948	19.7429	19.3627
31	22.1664	26.7094	18.8926	14.4512	20.3430	20.5125
32	18.5790	12.8160	17.3507	14.8919	18.1710	16.3617
33	18.8711	19.2198	29.7562	20.7049	23.1339	22.3372
34	19.0054	20.9936	16.0830	21.4640	23.9231	20.2938
35	19.3211	17.7350	18.1622	18.2605	14.9564	17.6871
36	17.3581	20.5371	27.1294	20.3896	13.1959	19.7220
37	17.2472	21.1953	19.4368	17.2755	18.9192	18.8148
38	21.1997	20.1846	10.8977	19.9111	24.1360	19.2658
39	19.2758	3.8455	23.8584	20.1019	11.9972	15.8158
40	14.7697	21.4131	18.9148	19.0714	17.0307	18.2399
41	19.9498	19.1230	16.4114	24.4762	16.1187	19.2158
42	16.6260	22.9204	20.8912	29.7361	13.8969	20.8141
43	18.7749	12.7132	24.6322	14.6196	15.4005	17.2281
44	24.0786	19.8415	19.1932	16.4600	25.3107	20.9768
45	24.0907	26.4468	25.9627	19.5786	21.3782	23.4914
46	17.9675	22.7537	17.4523	20.6101	23.3425	20.4252
47	24.8894	21.5625	23.1996	17.5499	18.1108	21.0624
48	23.5743	23.8954	17.9681	16.4051	17.0662	19.7818
49	21.6814	21.9238	17.2023	29.3187	18.3077	21.6868
50	20.5072	18.1959	12.6280	26.4176	20.4563	19.6410

Summary

In this chapter we introduced random sampling and sampling distributions—two important background topics for developing statistical inference.

To develop the methods for analyzing data presented in Chapters 8 through 17, we assume that the sample data we discuss have been drawn "at random" from the population of interest. It follows that the data-gathering step in making sense of data must involve random sampling. We discussed ways to accomplish random sampling through the use of a table of random numbers.

The second major topic, the sampling distribution of a statistic, was discussed in some detail. Problems in analyzing data involve selecting a random sample of measurements and computing sample statistics (such as $\bar{y}$ and $\hat{p}$) from the sample data. In a real-life situation, we will *not* repeat this sampling/computation process over and over again; but if we know the sampling distribution of the sample statistic, we know what would have happened if we had repeated the sampling/computation process many more times.

At the heart of making sense of data at the data analysis step is our ability to make statistical inferences about the population from which the sample data were drawn. Knowledge of the sampling distribution of the sample statistic is essential for making these inferences. And since many of the simple statistics used in statistical inferences are sums or averages of sample measurements (obtained in a random sample), the Central Limit Theorem helps explain why their sampling distributions (for sufficiently large sample sizes) are normally distributed.

Chapters 5 and 6 have provided the background material for analyzing data and drawing statistical inferences. We will now discuss methods for analyzing data and making inferences about specific population parameters.

Key Terms

sampling distribution

random sample

random number table

sampling distribution of $\bar{y}$

Central Limit Theorem

standard error of $\bar{y}$

normal approximation to the binomial

sampling distribution of $\hat{p}$

Key Formulas

1 Sampling distribution for $\bar{y}$:

Mean: μ

Standard error: $\sigma_{\bar{y}} = \dfrac{\sigma}{\sqrt{n}}$

2 Sampling distribution for $\sum y$:

    Mean: $n\mu$

    Standard error: $\sqrt{n}\sigma$

3 Normal approximation to the binomial:

    Mean: np

    Standard deviation: $\sqrt{npq}$

    provided $n \geq \dfrac{5}{\min(p, q)}$

4 Sampling distribution for $\hat{p}$:

    Mean: p

    Standard error: $\sigma_{\hat{p}} = \dfrac{\sqrt{pq}}{n}$

Supplementary Exercises

6.32 A random sample of $n = 25$ measurements is selected from a population with mean $\mu = 3$ and standard deviation $\sigma = 1$.

    a Find the approximate probability that $\bar{y} \geq 3.1$.

    b Find the approximate probability that $2.8 \leq \bar{y} \leq 3.2$.

    c Find the approximate probability that $\sum y \geq 80$.

6.33 A marketing research firm believes that approximately 25% of all persons mailed a "sweep-stakes" offer will respond. Suppose that a preliminary mailing of 5000 is conducted in a region.

    a What is the probability that 1000 or fewer will respond?

    b What is the probability that 3000 or more will respond?

6.34 Let y be the IQ of any college student. It is believed that y has a normal distribution, with a mean of 107 and a standard deviation of 15. If a sample of 25 college students is selected at random, find the probability that the *sample mean* will be:

    a Greater than 110.

    b Between 104 and 113.

    c Below 102.

6.35 **a** Simulate 100 times the experiment of rolling a pair of dice and finding the sum of the two numbers. Obtain a histogram of the sum.

    b Simulate 1000 times the experiment of rolling a pair of dice and finding the sum of the two numbers. Obtain a histogram of the sum.

    c Describe and explain the differences between the histograms in parts a and b.

6.36 Suppose that, in the past year, 40% of all cars sold by a dealership were small cars. Assume that the current population of car buyers has not changed in its preference for small cars.

    a If 8 people enter the dealership to look at cars, what is the probability that exactly 5 of them will prefer small cars?

    b If 800 people enter the dealership in a month, what are the mean and the standard deviation of the number who will prefer small cars?

6.37 At Central Library, the mean number of books checked out in a day is 320, and the standard deviation is 75. If we choose 30 days at random from the year, what is the approximate probability that the average number of books checked out in these 30 days will be between 335 and 350?

6.38 Medical studies indicate that the lead level in a child's body is dangerous if it is over .30 mg/mℓ in a blood sample. Suppose that, in New York City, the proportion of children with lead level above the danger point is .13. Take a random sample of 700 children in New York City.

a What is the chance that the proportion of children in the sample with lead levels above the danger point is higher than .16?

b What is the chance that the proportion of children in the sample with lead levels above the danger point is within .02 of the true population proportion?

6.39 At a large bank, the amounts of money in personal savings accounts have a mean of $289.56 and a standard deviation of $124.

a Assuming that these amounts are approximately normally distributed, find the percentage of accounts that have a balance in excess of $250.

b If 120 of these accounts are selected randomly, what is the probability that the sample mean will be in excess of $250?

6.40 The town of Centerville has 48,500 adult residents. Of those residents, 31,040 are registered to vote. If we were to take a random sample of 200 adults (with replacement), what is the approximate probability that more than 135 of them are registered to vote?

6.41 Circle exactly one number in each part.

a As the sample size n increases, the standard deviation of the sampling distribution:

1 increases.

2 stays the same.

3 decreases.

4 Not enough information to say for sure.

b As the sample size n increases, the mean of the sampling distribution:

1 increases.

2 stays the same.

3 decreases.

4 Not enough information to say for sure.

c As the sample size n increases, the sampling distribution:

1 looks more and more like the distribution from which the samples were drawn.

2 looks more and more like a normal distribution.

3 becomes more and more tightly clustered about its mean.

4 Both (2) and (3) above.

5 None of the above.

6.42 Let us define an English word as being "short" if it contains five or fewer letters; otherwise, the word is "long." Suppose that 40% of all English words are short. Find the probability that 200 randomly selected words contain at most 85 short words.

6.43 The values 1, 3, 7, and 9 constitute a population with $\mu = 5$ and $\sigma = 3.65$.

a Construct a frequency distribution for the population.

b Determine the sampling distribution of $\bar{y}$, based on $n = 2$.

c Graph the sampling distribution.

d Determine the mean and the standard error of the sampling distribution based on part b. Does this agree with what the Central Limit Theorem states?

6.44   Last year a company initiated a program to compensate its employees for unused sick days, paying each employee a bonus of one-half the usual wage earned for each unused sick day. The question that naturally arises is "Did this policy motivate employees to use fewer allotted sick days?" Before last year, employees averaged 7 sick days per year, with a standard deviation of 2. Assuming that these parameters did not change last year, find the approximate probability that the sample mean number of sick days used by 100 employees chosen at random would be less than or equal to 6.4 last year.

6.45   A student taking a sampling course is required to conduct a survey of 200 fellow students to "estimate" the proportion of students from the entire student body who favor a semester system as compared to a school year based on quarters.

    a   How might the student select a random sample of 200 students from the student body? (Use your university as an example.)

    b   Why does the student need a random sample? Why can't he/she take a group of friends and use their opinions to ascertain the proportion of the entire student body favoring a semester system?

6.46   The daily shrinkage due to theft in the inventory of a men's department store possesses a probability distribution whose mean equals $320 and whose standard deviation equals $80. The store reports the shrinkage as a total T for a 4-week period. Describe the probability distribution for the shrinkage for the 4-week period and justify your conclusions.

6.47   Refer to Exercise 6.46. Suppose that the mean daily shrinkage μ is unknown to you and that you randomly sample $n = 50$ days to estimate its value. What is the probability that the sample mean $\bar{y}$ will deviate from μ by more than $20?

6.48   Due to temperature variations, the expansion or contraction of a gas pipeline per 1000 feet is normally distributed, with mean $\mu = 1$ in. and standard deviation $\sigma = .5$ in. If the pipeline is 50,000 ft long, what is the probability that the pipe might expand by more than 55 in.?

6.49   From returns in previous years, it has been found that approximately 70% of all tax returns in a given income category are incorrectly filed. A spot check of 5000 returns drawn at random shows that only $y = 2600$ have been filed incorrectly. Assuming that 70% of all returns will be incorrect this year too, find the mean and the standard deviation of the random variable y, and use this information to describe its variability in repeated sampling.

6.50   Judging from the results you obtained in Exercise 6.49, would you anticipate that approximately 70% of the returns this year will be incorrectly filed? Explain.

6.51   The distribution of milkfat percentages for milk from Holstein cattle in a particular state during the 1960s was approximately normal, with a mean of 3.7% and a standard deviation of .3.

    a   What percentage of the Holsteins gave milk that had a milkfat percentage of less than 3%?

    b   Greater than 4.5%?

6.52   Refer to Exercise 6.51.

    a   Find the limits within which 90% of the milkfat percentages fell.

    b   Compute the 95th percentile for the distribution of milkfat percentages.

6.53   Refer to Exercise 6.51. Suppose a random sample of $n = 25$ Holsteins is selected from the population of Holstein cattle in the state.

    a   Describe the distribution of $\bar{y}$, the mean milkfat percentage for the milk yielded by the sample of 25 cattle.

    b   Compare the distribution of $\bar{y}$ in part a to a distribution of $\bar{y}$ from a sample of 100 Holsteins.

    c   What is the probability that the sample mean milkfat percentage will exceed 4 in part a?

6.54   A manufacturer claims that 95% of the components that it supplies for a new jet transport meet a specified rigid standard of performance. Suppose that 400 of these components are tested. Find the probability of observing that 30 or more do not meet the standard of performance, assuming that the manufacturer's claim is correct.

6.55 Refer to Exercise 6.54. Suppose that the test of 400 components is performed and that 30 of the components fail to meet the standard of performance. What might you conclude about the manufacturer's claim?

 6.56 An airline has found over the past several years that 10% of the persons who book reservations on a particular flight will not show up at flight time. On a given day, records indicate that the flight is fully booked at 300, with 35 or more people waiting on standby. What is the probability that all 35 people on standby will have a seat available at flight time?

 6.57 Data collected over a long period of time indicate that a particular birth defect occurs in one of every 1000 live births. Data collected from a medical center in a particular section of the country found 10 children with the birth defect from the total of 20,000 birth records examined. If we assume that the 20,000 records examined represent a random sample of birth records, what is the probability of observing 10 or fewer children with birth defects in the sample?

 6.58 Research on sales displays in grocery stores has shown that 70% of all people who pick up a particular item will purchase it. As part of a class project in a small community, a random sample of 300 shoppers is observed to see how many people handle and then purchase the sale item on display at the front of the grocery store. Assuming that 70% of all shoppers who handle the item will eventually purchase it, what is the probability of observing fewer than 200 shoppers who handle and then buy the item in the sample of 300 shoppers?

6.59 If you randomly sample $n = 400$ observations from a binomial population with parameter $p = .3$, what is the probability that the sample proportion will differ from $p = .3$ by more than .02?

 6.60 A supplier of a particular semiconductor claims that no more than .5% are defective. If you randomly sample and test 10,000 of these semiconductors and find that .9% are defective, what would you conclude about the manufacturer's claim? Explain your reasoning. (*Note:* In later chapters we will give a statistical procedure for making the decision.)

6.61 **Class Exercise** Each three-digit number shown in a random number table appears with a probability of near .001; that is, the numbers possess a probability distribution as shown in Figure 6.11.

FIGURE 6.11
Probability Distribution for
Three-digit Random Numbers
Selected from a Random
Number Table

Simulate random sampling from a population possessing the probability distribution of Figure 6.11 by selecting $n =$ three-digit random numbers from Table 11 of Appendix 3, and calculate the sample range:

$$\text{Range} = \text{Difference between the largest and the smallest measure-} \\ \text{ments in the sample}$$

Repeat this process 200 times to obtain 200 sample ranges, and construct the relative frequency histogram. This histogram will be similar to the sampling distribution for the sample range based on samples of $n = 3$ measurements selected from a population possessing a relative frequency histogram similar to the one shown in Figure 6.11. You could obtain a better approximation to this sampling distribution by increasing the number of samples used to construct your histogram—say, 1000 samples of $n = 3$ measurements each, rather than 200 samples.

As was noted in Chapter 4, the range is a good measure of data variation for small samples. In fact, for small samples it is easier to compute and is almost as good an estimator of the population standard deviation σ as is the sample standard deviation s. For this reason the range is often used to measure data variation in industrial quality control, where a manufacturing process is monitored by taking frequent small samples of one or more measures of product quality over time (for example, one sample every hour). The sampling distribution that you have derived would be useful in evaluating the properties of the sample range for quality-control measurements on a random variable that possess the probability distribution shown in Figure 6.11.

6.62   If 65% of all married women who work do so primarily for the money, what is the probability that a random sample of 100 married women who work will show 68 or more who work primarily for the money? What proportion of the time will a sample of 100 show exactly 65 who work primarily for the money?

6.63   Today's new fathers are much more willing to help with the family chores than were their counterparts in the past, according to sociologists. As an example, studies have shown that 80% of new fathers will change diapers for their children. If this is correct, what is the probability that, in a sample of 1000, there are 700 or fewer who will change diapers for their children?

6.64   The average daily temperature in January in Seattle, Washington, is 44.4°F, with a standard deviation of 1.8 Fahrenheit degrees. If a sample of 30 January days is drawn, what is the probability that the sample mean will be greater than 44.4°F? Greater than 45°F?

6.65   The playing time of cuts on phonograph records of popular recording artists averages 180 sec, with a standard deviation of 48 sec. A random sample of 40 selections from various albums showed a mean time of 206 sec. Is this a likely outcome if the true mean is 180 sec? Take a random sample of 30 selections from records you or your friends own, and see how they compare with the overall average of 180 sec.

6.66   A survey taken by the *New York Times* in 1993 showed that 90% of those surveyed thought that there would be better government if there were more females in Congress. If .90 is the true population proportion, what is the probability that a sample of 400 would show a proportion of .80 or less?

6.67   Use a computer program to simulate the sampling distribution for $\bar{y}$, based on 40 samples of size $n = 16$ drawn from a normal population with $\mu = 60$ and $\sigma = 5$. A portion of a MINITAB program is shown here.

```
MTB > RANDOM 40 C1-C16;
SUBC> NORMAL MU=60 SIGMA=5.
MTB > RMEAN C1-C16 INTO C17
MTB > PRINT C17

C17
   59.2758     61.3386     59.6048     60.6651     59.0729     61.4702     60.2351
   60.7723     58.7352     59.6414     58.4658     61.4506     57.9570     60.9287
   60.4606     58.5009     61.5293     61.9023     58.0660     59.0477     62.1565
   58.5997     58.8638     58.9735     58.9055     60.5827     59.6956     60.1475
   60.2772     59.5553     61.1230     60.9288     59.7171     59.9432     62.7418
   57.9715     59.9170     58.6996     61.4800     60.0822

MTB > MEAN C17

   MEAN    =       59.987

MTB > STOP
```

Experiences with Real Data

6.68 Refer to the Murder Rate database and the Murder Rate Sample Means database on the data disk (or in Appendix 1).

a Construct a stem-and-leaf plot for each database.

b Compare the two plots. Does the Central Limit Theorem appear to apply? Explain.

6.69 Refer to the two databases of Exercise 6.68.

a Compute the mean and the standard deviation for the Murder Rate database. (Assuming that this is the population of measurements, these values represent μ and σ, respectively.)

b Compute the mean and the standard deviation for the Murder Rate Sample Means database. (These values should approximate $\mu_{\bar{y}}$ and $\sigma_{\bar{y}}$, the mean and standard error for the sampling distribution of $\bar{y}$.)

c Based on the results in part a, compute $\mu_{\bar{y}}$ and $\sigma_{\bar{y}}$. Compare these values to those you obtained in part b. What results have we confirmed with this exercise?

6.70 A manager of a supermarket has monitored the distribution of checkout times for nonexpress checkout lanes at a particular branch. A random sample of 25 checkout times yielded the following data (times are in minutes and are organized from lowest to highest): .4, .4, .5, .5, .5, .5, .6, .6, .7, .8, .9, 1.1, 1.2, 1.4, 1.5, 1.8, 2.0, 2.3, 2.6, 2.9, 3.4, 4.2, 5.0, 6.6, 9.2, and 16.3.

a Comment on the data without summarizing them.

b Summarize the data graphically and numerically.

c Does the hypothetical distribution (population) of checkout times from which the sample of 25 measurements was obtained appear to be approximately normal? Explain.

Step Three:
Analyzing Data: Means,
Variances, and Proportions

7

Inferences About μ

7.1 Introduction

Previous chapters have set the stage for considering methods for the analysis of data—Step 3 in making sense of data. We've considered methods for gathering data from a survey or experimental study. In most cases, the data-gathering stage involves obtaining a sample of measurements from an underlying population. The graphical and numerical methods of Chapters 3 and 4 equip us with useful ways to summarize the sample data; now we turn to data analysis methods. In the next four chapters, we will deal with data analysis methods that allow us to make inferences from the sample data about specific parameters from the underlying population.

Inferences—specifically, decisions and predictions—are centuries old and play a very important role in our lives. Each of us is faced daily with personal decisions and situations that require us to make predictions about the future. The government is concerned with predicting the flow of gold to Europe. A stockbroker wants to know how the stock market will behave. A metallurgist would like to use the results of an experiment to determine whether a new type of steel is more resistant than another type to temperature changes. A veterinarian investigates the effectiveness of a new product for treating worms in cattle. The inferences that these individuals make should be based on relevant facts, which we call observations or data.

In many practical situations, the relevant facts are abundant, seemingly inconsistent, and (in many respects) overwhelming. As a result, a careful decision or prediction is often little better than an outright guess. You need only refer to the "Market Views" section of the *Wall Street Journal* to observe the diversity of expert opinion concerning future stock market behavior. Similarly, a visual analysis of data by scientists and engineers often yields conflicting opinions regarding conclusions to be drawn from an experiment.

Although many individuals feel that they can make good decisions "on their own," experience suggests that most people are incapable of assimilating sample data, mentally weighing each bit of relevant information, and arriving at a good inference without following some structured approach to decision-making or predictions. (You may test your own inference-making ability in connection with exercises in this chapter and in Chapters 8 through 10. Scan the data and make an inference before you use the appropriate statistical procedure; then compare the results.) A statistician, rather than relying on intuition, uses structured methods for data analysis to aid in making inferences. Although we have touched upon some of the notions involved in statistical inference in preceding chapters, we will now collect our ideas in a presentation of some of the basic ideas involved in statistical inference.

At the heart of making sense of data, we are usually interested in making an inference (decision, prediction, estimate) about a population based on information contained in the sample. Populations are characterized by numerical descriptive measures called *parameters*. Typical population parameters are the mean μ, the standard deviation σ, and the population proportions p. Most practical inferential problems you will encounter can be phrased to imply an inference about one or more parameters of a population. For example, in an experiment in which we wish to predict the average amount of money paid to welfare recipients in a given year, the population of interest is the set of all yearly welfare payments, and we are interested in estimating the value of the population mean μ.

estimation
hypothesis testing

Methods for making inferences about parameters fall into one of two categories. Either we will **estimate** (predict) the value of the population parameter of interest, or we will **test a hypothesis** about the value of the parameter. These two methods of statistical inference—estimation and hypothesis testing—employ different procedures, and (more important) they answer two different questions about the parameter. In estimating a population parameter, we are answering the question, "What is the value of the population parameter?" In testing a hypothesis, we are answering the question, "Is the parameter value equal to this specific value?"

Consider a study in which an investigator is interested in examining the effectiveness of a drug in reducing anxiety levels of anxious patients. A screening procedure is employed to identify a group of anxious patients. After the patients are admitted into the study, each one's anxiety level is measured on a rating scale immediately before he or she receives the first dose of the drug and then at the end of one week of drug therapy. These sample data can be used to make inferences about the population from which the sample was drawn, either by estimation or by a statistical test:

Estimation: Information from the sample can be used to estimate (or predict) the mean decrease in anxiety ratings for the set of all anxious patients who may conceivably be treated with the drug.

Statistical test: Information from the sample can be used to determine whether the population mean decrease in anxiety ratings is greater than zero.

Notice that the inference related to estimation is aimed at answering the question, "What is the mean decrease in anxiety ratings for the population?" In contrast, the statistical test attempts to answer the question, "Is the mean drop in anxiety ratings greater than zero?"

In this chapter, we will consider estimation of a population mean μ and a statistical test about μ.

Exercises

Basic Techniques

7.1 A researcher is interested in estimating the percentage of registered voters in her state who have voted in at least one election over the past 2 years.

  a Identify the population of interest to the researcher.

  b How might you select a sample of voters to gather this information?

7.2 Refer to Exercise 7.1. Is the researcher faced with a problem related to estimation or to testing a hypothesis? What is the parameter of interest?

7.3 A manufacturer claims that the average lifetime of a particular fuse is 1500 hours. Information from a sample of 35 fuses shows that the average lifetime is 1380 hours. What can be said about the manufacturer's claim?

  a Identify the population of interest to us.

  b Would answering the question posed involve estimation or testing a hypothesis?

7.4 Refer to Exercise 7.3. How might you select a sample of fuses from the manufacturer to test the claim?

7.2 Estimating μ

The simplest statistical inference problem is point estimation, where we compute a single value (statistic) from the sample data to estimate a population parameter. Suppose that we are interested in estimating a population mean and that we are willing to assume that the underlying population is normal. One natural statistic we could use to estimate the population mean is the sample mean; but alternatively we could use the median and the trimmed mean. Which sample statistic should we use?

A whole branch of mathematical statistics deals with problems related to developing point estimators (the formulas for calculating specific point estimates from sample data) of parameters from various underlying populations and determining whether a particular point estimator has certain desirable properties. Fortunately, we will not have to derive these point estimators—they'll be given to us for each parameter. Then, knowing which point estimator (formula) to use for a given parameter, we can focus on developing confidence intervals (interval estimates) for these same parameters.

In this section, we deal with point and interval estimations of a population mean μ. Tests of hypotheses about μ are covered in Section 7.5.

For most problems in this text, the sample mean $\bar{y}$ will be used as a point estimate of μ; it is also used to form an interval estimate for the population mean μ. From the Central Limit Theorem for the sample mean (given in Chapter 5) we know that, for large n (crudely, $n > 30$), $\bar{y}$ will be approximately normally distributed, with a mean μ and a standard error $\sigma_{\bar{y}}$. Then from our knowledge of the Empirical Rule and areas under a normal curve, we know that the interval $\mu \pm 2\sigma_{\bar{y}}$, or more precisely, the interval $\mu \pm 1.96\sigma_{\bar{y}}$, includes 95% of the $\bar{y}$'s in repeated sampling, as shown in Figure 7.1.

FIGURE 7.1
Sampling Distribution for $\bar{y}$

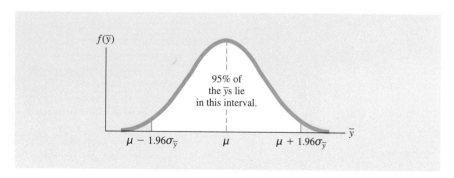

FIGURE 7.1
Sampling Distribution for $\bar{y}$

$f(\bar{y})$

95% of
the $\bar{y}$s lie
in this interval.

$\mu - 1.96\sigma_{\bar{y}}$ μ $\mu + 1.96\sigma_{\bar{y}}$ $\bar{y}$

Consider the interval $\bar{y} \pm 1.96\sigma_{\bar{y}}$. Any time $\bar{y}$ lies in the interval $\mu \pm 1.96\sigma_{\bar{y}}$, the interval $\bar{y} \pm 1.96\sigma_{\bar{y}}$ will contain the parameter μ (see Figure 7.2) and this will occur with probability .95. The interval $\bar{y} \pm 1.96\sigma_{\bar{y}}$ represents an interval estimate of μ.

We evaluate the goodness of an interval estimation procedure by examining the fraction of times in repeated sampling that interval estimates would encompass the parameter to be estimated. This fraction, called the **confidence coefficient**, is .95 when the formula used is $\bar{y} \pm 1.96\sigma_{\bar{y}}$; that is, 95% of the time in repeated sampling, intervals calculated using the formula $\bar{y} \pm 1.96\sigma_{\bar{y}}$ will contain the mean μ.

confidence coefficient

F I G U R E 7.2
When the Observed Value of $\bar{y}$
Lies in the Interval
$\mu \pm 1.96\sigma_{\bar{y}}$, the Interval
$\bar{y} \pm 1.96\sigma_{\bar{y}}$ Contains the
Parameter μ

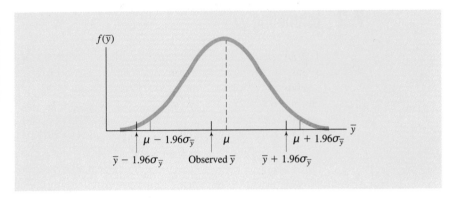

This idea is illustrated in Figure 7.3. Twenty different samples are drawn from a population with mean μ and variance σ^2. For each sample, an interval estimate is computed, using the formula $\bar{y} \pm 1.96\sigma_{\bar{y}}$. Notice that, although the intervals bob about, most of them capture the parameter μ. In fact, if we repeated the process of drawing samples and computing confidence intervals, 95% of the intervals so formed would contain μ.

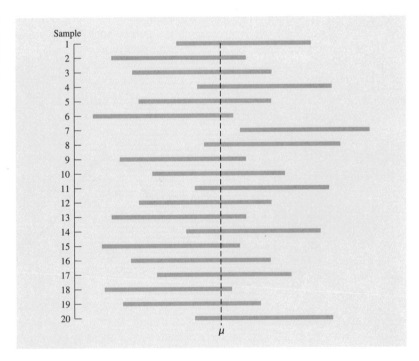

In a given experimental situation, we calculate only one such interval. This interval, called a **95% confidence interval**, represents an interval estimate of μ.

95% confidence interval

EXAMPLE 7.1   In a random sample of $n = 36$ parochial schools throughout the South, the average number of pupils per school is 379.2, with a standard deviation of 124. Use the sample to construct a 95% confidence interval for μ, the mean number of pupils per school for all parochial schools in the South.

Solution   The sample data indicate that $\bar{y} = 379.2$ and $s = 124$. The appropriate 95% confidence interval is then computed by using the formula

$$\bar{y} \pm 1.96\sigma_{\bar{y}},$$

where $\sigma_{\bar{y}} = \sigma/\sqrt{n}$. In Section 7.8, we present a procedure for obtaining a confidence interval for μ when σ is unknown. However, for all practical purposes, if the sample size is 30 or more, we can estimate the population standard deviation σ with s in the confidence interval formula. With s replacing σ, our interval is

$$379.2 \pm 1.96\frac{124}{\sqrt{36}} \quad \text{or} \quad 379.2 \pm 40.51.$$

The interval from 338.69 to 419.71 forms a 95% confidence interval for μ. In other words, we are 95% sure that the average number of pupils per school for parochial schools throughout the South lies between 338.69 and 419.71.   ■

There are many different confidence intervals for μ, depending on the confidence coefficient we choose. For example, the interval $\mu \pm 2.58\sigma_{\bar{y}}$ includes 99% of the values of $\bar{y}$ in repeated sampling (see Figure 7.4), and the interval $\bar{y} \pm 2.58\sigma_{\bar{y}}$ forms a **99% confidence interval** for μ.

99% confidence interval

FIGURE 7.4
Sampling Distribution of $\bar{y}$

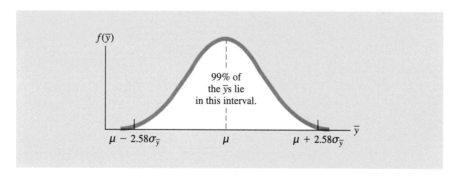

We can state a general formula for a confidence interval for μ with a **confidence coefficient of** $(1 - \alpha)$, where α (Greek letter alpha) is between 0 and 1. For a specified value of $(1 - \alpha)$, a $100(1 - \alpha)\%$ confidence interval for μ is given by the following formula. Here we assume that σ is known or that the sample size is large enough to permit us to replace σ with s.

$(1 - \alpha)$ confidence coefficient

Confidence Interval for μ, with σ Known

$$\bar{y} \pm z_{\alpha/2}\sigma_{\bar{y}}, \quad \text{where} \quad \sigma_{\bar{y}} = \frac{\sigma}{\sqrt{n}}$$

$z_{\alpha/2}$

The quantity $z_{\alpha/2}$ is a value of z having a tail area of $\alpha/2$ to its right. In other words, at a distance of $z_{\alpha/2}$ standard deviations to the right of μ, an area of $\alpha/2$ lies under the normal curve. Values of $z_{\alpha/2}$ can be obtained from Table 2 in Appendix 3 by looking up the z-value corresponding to an area of $(1 - \alpha)/2$ (see Figure 7.5). Common values of the confidence coefficient $(1 - \alpha)$ and $z_{\alpha/2}$ are given in Table 7.1.

FIGURE 7.5
Interpretation of $z_{\alpha/2}$ in the
Confidence Interval Formula

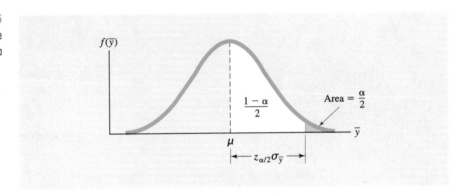

TABLE 7.1
Common Values of the
Confidence Coefficient
$(1 - \alpha)$ and the
Corresponding z-Value $(z_{\alpha/2})$

Confidence Coefficient $(1 - \alpha)$	Area in Table 2 $(1 - \alpha)/2$	Value of $\alpha/2$	Corresponding z-Value $(z_{\alpha/2})$
.90	.45	.05	1.645
.95	.475	.025	1.96
.98	.49	.01	2.33
.99	.495	.005	2.58

EXAMPLE   7.2   A forester is interested in estimating the average number of "count trees" per acre (trees larger than a specified size) on a 2000-acre plantation. She can then use this information to determine the total timber volume for trees in the plantation. A random sample of $n = 50$ 1-acre plots is selected and examined. The average (mean) number of count trees per acre is found to be 27.3, with a standard deviation of 12.1. Use this information to construct a 99% confidence interval for μ, the mean number of count trees per acre for the entire plantation.

Solution   We use the general confidence interval with confidence coefficient equal to .99 and a $z_{\alpha/2}$-value equal to 2.58 (see Table 7.1). Substituting the appropriate values into the formula $\bar{y} \pm 2.58\sigma_{\bar{y}}$ and replacing σ with s in $\sigma_{\bar{y}} = \sigma/\sqrt{n}$, we have

$$27.3 \pm 2.58\frac{12.1}{\sqrt{50}}.$$

This corresponds to the confidence interval 27.3 ± 4.42—that is, the interval from 22.88 to 31.72. Thus, we can be 99% sure that the average number of count trees per acre is between 22.88 and 31.72.   ■

The discussion in this section has included one rather unrealistic assumption: namely, that the population standard deviation is known. In practice, it's difficult to find situations in which the population mean is unknown, but the standard deviation is known. Usually both the mean and the standard deviation must be estimated from the sample. Since σ is estimated by the sample standard deviation s, the actual standard error of the mean, $\sigma/\sqrt{n}$, is naturally estimated by $s/\sqrt{n}$. This estimation introduces another source of random error (s varies randomly, from sample to sample, around σ) and, strictly speaking, invalidates our confidence interval formula. Fortunately, the formula is still a very good approximation for large sample sizes. As a very rough rule, we can use this formula when n is larger than 30; a better way to handle this issue is described in Section 7.8.

Statistical inference-making procedures differ from ordinary procedures in that we not only make an inference, but also provide a measure of how good that inference is. For interval estimation, both the width of the confidence interval and the confidence coefficient measure the goodness of the inference. Obviously, for a given confidence coefficient, the smaller the width of the interval, the better the inference. The confidence coefficient, on the other hand, is set by the experimenter to express how much assurance he or she places in the prospect that the interval estimate encompasses the parameter of interest.

Exercises

Basic Techniques

7.5   The sample mean and the sample standard deviation based on a sample of 50 measurements are $\bar{y} = 105$ and $s = 11$.

   a   Calculate a 95% confidence interval for μ.

   b   Calculate a 99% confidence interval for μ.

7.6   Give a careful verbal interpretation of the confidence interval in part a of Exercise 7.5.

7.7   Refer to Exercise 7.5.

a   Discuss the impact of doubling the sample size from $n = 50$ to $n = 100$ on the 95% confidence interval. Assume for discussion purposes that $\bar{y}$ and s are still 105 and 11, respectively.

b   What impact would quadrupling the sample size have? (*Note:* Answer this question without doing the calculations.)

Applications

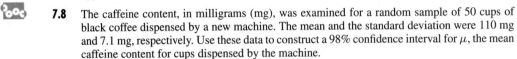

7.8   The caffeine content, in milligrams (mg), was examined for a random sample of 50 cups of black coffee dispensed by a new machine. The mean and the standard deviation were 110 mg and 7.1 mg, respectively. Use these data to construct a 98% confidence interval for μ, the mean caffeine content for cups dispensed by the machine.

7.9   A random sample of year-end statements from 22 small businesses (under $500,000 in sales) in a city shows that the mean gross profit margin was 5.2% (of sales) with a standard deviation of 3.3%. Use these data to place a 90% confidence interval for μ. Assume $\sigma \approx 3.3$.

7.10   Recent data from a national survey of 1350 women indicated that the average woman goes to a hair salon once every 5 weeks and spends on the average $26.40 per visit. With a standard deviation of $12.00, use these data to construct a 99% confidence interval for μ.

7.11   A social worker is interested in estimating the average length of time spent outside prison by first offenders who later commit a second crime and are sent to prison again. A random sample of $n = 150$ prison records in the county courthouse indicates that the average length of prison-free life between first and second offenses is 3.2 years, with a standard deviation of 1.1 years. Use the sample information to estimate μ, the mean prison-free life between first and second offenses for all prisoners on record in the county courthouse. Construct a 95% confidence interval for μ. Assume that σ can be replaced by s.

7.12   Refer to Exercise 7.9. What impact would a doubling of the sample size have on the confidence interval?

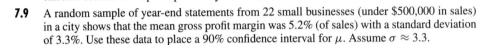

7.13   The rust mite, a major pest of citrus fruits in Florida, punctures the cells of the leaves and fruit. Damage by rust mites is readily recognizable because injured fruit display a brownish (rust) color and are somewhat smaller than normal, depending on the severity of the attack. If the rust mites are not controlled, affected groves suffer a substantial reduction in both fruit yield and fruit quality. In either case, the citrus grower suffers financially, since the produce is of a lower grade and sell for less on the fresh fruit market. This year, more and more citrus growers have gone to a preventive program of maintenance spraying for rust mites. In evaluating the effectiveness of the program, a random sample of 60 10-acre plots, one plot from each of 60 groves, is selected. These show an average yield of 850 boxes, with a standard deviation of 100 boxes. Give a 95% confidence interval for μ, the average (10-acre) yield for all groves utilizing such a maintenance spraying program. Assume that σ can be replaced by s.

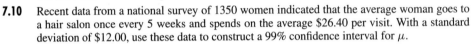

7.14   An experiment is conducted to examine the susceptibility of root stocks of a variety of lemon trees to a specific larva. Forty of the plants are subjected to the larvae and examined after a fixed period of time. The response of interest is the logarithm of the number of larvae per gram counted on each root stock. For these 40 plants, the sample mean is 9.02 and the standard deviation is 1.12. Use these data to construct a 90% confidence interval for μ, the mean susceptibility of the population of lemon tree root stocks from which the sample was drawn. Assume that σ can be replaced by s.

7.15   A mobility study is conducted among a random sample of 900 high-school graduates of a particular state over the past 10 years. For each of the persons sampled, the distance between the high school attended and the present permanent address is recorded. For these data, $\bar{y} = 430$ miles and $s = 262$ miles. Using a 95% confidence interval, estimate the average number of miles between a person's high school and present permanent address for high school graduates of the state over the past 10 years. Assume that σ can be replaced by s.

7.16   A problem of interest to the United States, other governments, and world councils concerned with the critical shortage of food throughout the world involves finding a method to estimate the total amount of grain crops that will be produced throughout the world in a particular year.

One method of predicting total crop yields is based on satellite photographs of the earth's surface. Because a scanning device reads the total acreage of a particular type of grain with error, the device must read many equal-sized plots of a particular planting in order to calibrate the reading on the scanner with the actual acreage. Satellite photographs of 100 50-acre plots of wheat were read by the scanner and gave a sample average and sample standard deviation of

$$\bar{y} = 3.27 \quad \text{and} \quad s = .23,$$

respectively. Find a 95% confidence interval for the mean scanner reading for the population of all 50-acre plots of wheat. Explain the meaning of this interval.

7.17   Another agricultural problem involves the production of protein, an important component of human and animal diets. Although it is common knowledge that grains and legumes contain high amounts of protein, it is not as well known that certain grasses also provide a good source of protein. For example, Bermuda grass contains approximately 20% protein by weight. In a study to verify these results, 100 1-pound samples were analyzed for protein content. The mean and the standard deviation of the sample were

$$\bar{y} = 18 \text{ pounds} \quad \text{and} \quad s = .08 \text{ pounds},$$

respectively. Estimate the mean protein content per pound for the Bermuda grass from which this sample was selected. Use a 95% confidence interval. Explain the meaning of this interval.

7.3   Choosing the Sample Size for Estimating μ

How can we determine the number of observations to include in the sample? The implications of such a question are clear. Data collection costs money. If the sample is too large, time and talent are wasted. Conversely, it is wasteful if the sample is too small, because inadequate information is purchased for the time and effort expended. Moreover, it may be impossible to increase the sample size at a later time. Hence, the number of observations to include in the sample depends on the amount of information the experimenter wants to buy.

Suppose that we want to estimate the average amount for accident claims filed against an insurance company. To decide how many claims must be examined, we would have to determine how accurate the company wants the estimate to be. For example, the company might indicate that the tolerable error is 10 units (± 5 units) or less. Then we would want the confidence interval to be of the form $\bar{y} \pm 5$.

There are two considerations in choosing the appropriate sample size for estimating μ using a confidence interval: tolerable error and confidence level. The tolerable error establishes the desired width of the confidence interval. A wide confidence interval is not very informative, but the cost of obtaining a narrow confidence interval can be quite large. Similarly, having too low a confidence level (say, 50%) means that the stated confidence interval is likely to be in error, but obtaining a higher level of confidence is more expensive.

What constitutes reasonable certainty? In most situations, the confidence level is set at 95% or 90%, partly because of tradition and partly because these levels represent (to some people) a reasonable level of certainty. The 95% (or 90%) level translates into a long-run chance of 1 in 20 (or 1 in 10) that the reported results will not cover the population parameter. This seems reasonable and is comprehensible, whereas 1 chance in 1,000 or 1 in 10,000 is just too small.

The tolerable error depends heavily on the context of the problem, and only someone who is familiar with the situation can make a reasonable judgment about its magnitude.

For a confidence interval for a population mean μ, the plus-or-minus term of the confidence interval is

$$z_{\alpha/2}\sigma_{\bar{y}} \quad \text{where} \quad \sigma_{\bar{y}} = \sigma/\sqrt{n}.$$

Three quantities determine the value of the plus-or-minus term: the desired confidence level (which determines the z-value used), the standard deviation (σ), and the sample size (which together with σ determines the standard error $\sigma_{\bar{y}}$). Usually, a guess must be made about the size of the population standard deviation. (Sometimes an initial sample is taken to estimate the standard deviation; this estimate provides a basis for determining the additional sample size needed.) For a given tolerable error, once the confidence level is specified and an estimate of σ has been supplied, the required sample size can be calculated using the preceding formula.

If a 95% confidence interval is to be of the form $\bar{y} \pm E$, then we solve the expression

$$1.96\sigma_{\bar{y}} = E$$

for n. The width of the interval is $2E$.

In general, if we want to estimate μ using a $100(1 - \alpha)\%$ confidence interval of the form $\bar{y} \pm E$, where E is specified, then we solve the equation

$$z_{\alpha/2}\sigma_{\bar{y}} = E$$

for n.

Sample Size Required for a $100(1 - \alpha)\%$ Confidence Interval for μ of the Form $\bar{y} \pm E$

$$n = \frac{(z_{\alpha/2})^2 \sigma^2}{E^2}$$

Notice that determining a sample size to estimate μ requires knowledge of the population variance σ^2 (or standard deviation σ). We can obtain an approximate sample size by estimating σ^2, using one of two methods:

1   Employ information from a prior experiment to calculate a sample variance s^2. This value is used to approximate σ^2.

2   Use information on the range of the observations to obtain an estimate of σ.

We can then substitute the estimated value of σ^2 in the sample-size equation to determine an approximate sample size n.

We illustrate the procedure for choosing a sample size with two examples.

EXAMPLE 7.3   Union officials are concerned about reports of inferior wages being paid to a company's employees under their jurisdiction. They decide to take a random sample of n wage sheets from the company to estimate the average hourly wage. If wages in the company are known to have a range of $10 per hour, determine the sample size required to estimate the average hourly wage μ using a 95% confidence interval with width equal to $1.20.

Solution   We want a 95% confidence interval with width $1.20, so $E = \$.60$. The value we use to substitute for σ is Range/4 = 2.50. Substituting into the formula for n, we have

$$n = \frac{(1.96)^2(2.5)^2}{(.60)^2} = 66.69.$$

To be on the safe side, we round this number up to the next integer. A sample size of 67 should give a 95% confidence interval with the desired width of $1.20.   ▪

EXAMPLE 7.4   A federal agency has decided to investigate the advertised weight printed on cartons of a certain brand of cereal. The company in question periodically samples cartons of cereal coming off the production line, to check their weight. A summary of 1500 of the weights made available to the agency indicates a mean weight of 11.80 ounces per carton and a standard deviation of .75 ounce. Use this information to determine the number of cereal cartons the federal agency must examine to estimate the average weight of cartons being produced now, using a 99% confidence interval of width .50.

Solution   The federal agency has specified that the width of the confidence interval is to be .50, so $E = .25$. Assuming that the weights made available to the agency by the company are accurate, we can take $\sigma = .75$. The required sample size with $z_{\alpha/2} = 2.58$ is

$$n = \frac{(2.58)^2(.75)^2}{(.25)^2} = 59.91.$$

That is, the federal agency must obtain a random sample of 60 cereal cartons to estimate the mean weight to within $\pm.25$.   ▪

Exercises

Basic Techniques

7.18   Refer to Example 7.3.

a   How large a sample is needed to obtain a 90% confidence interval with width $.60? $.30? $.15?

b   In general, for a given confidence level, by how much must you increase the sample size to cut the interval width in half?

Applications

7.19 The giant size of a new "tough cleaning" laundry detergent has a listed net weight of 42 ounces. If the variability in weight has a standard deviation of 2 ounces, how many boxes must be sampled to estimate the average fill weight to within $\pm.25$ ounce, using a 95% confidence interval?

7.20 Refer to Exercise 7.19. Determine the effects of imposing first a 90% and then a 99% confidence level on the required sample size.

7.21 A biologist would like to estimate the effect of an antibiotic on the growth of a particular bacterium by examining the mean amount of bacteria present per plate of culture when a fixed amount of the antibiotic is applied. Previous experimentation with the antibiotic on this type of bacterium indicates that the standard deviation of the amount of bacteria present is approximately 13 cm^2. Use this information to determine the number of observations (cultures that must be developed and then tested) to estimate the mean amount of bacteria present, using a 99% confidence interval with a half-width of 3 cm^2.

7.22 Investigators would like to estimate the average annual taxable income of apartment dwellers in a city to within $500, using a 95% confidence interval. If we assume that the annual incomes range from $0 to $40,000, determine the number of observations that should be included in the sample.

7.23 Refer to Exercise 7.22. Determine the required sample size if the desired error in a 95% confidence interval is $E = 250$. Do the same for $E = 1000$. Compare your results to those of Exercise 7.22.

7.24 As part of a much larger study of trends in long-distance telephone usage, a study is to be conducted this month of residential homes occupied by married couples who are between 25 and 40 years of age. How large a sample should be taken if the mean number of long-distance calls for the month is to be estimated to within one call, using a 90% confidence interval? Assume $\sigma \approx 4.0$.

7.4 Quality Control: $\bar{y}$-Charts

control chart

We can extend the notion of a confidence interval for μ to obtain a **control chart**. As consumers, we are vitally interested in product quality. We expect product quality for a particular item to be uniform from one time period to another, and we expect it to live up to the product description advertised by the manufacturer. For example, in buying paint from a paint store, we expect different gallons of the same color to be uniform in color, and we expect the color to be identical to that advertised in the paint-sample brochure. Similarly, the Food and Drug Administration (FDA) not only expects but demands that medicinal drugs have uniform potency and meet the standards advertised by the pharmaceutical firm.

Consumers are not the only people interested in product quality. Reputable manufacturers are also concerned that their products meet the standards they have claimed. If the quality of a product falls below the standards advertised by the company, there is a risk that consumers will reject the product and buy from a competitor. Similarly, if product quality drifts above the standards established by the company, it is in the company's interest to upgrade their advertising to reflect the increase in quality.

quality control

Quality-control techniques have been developed to monitor the ongoing quality of a manufacturing process in order to maintain uniform quality or at least to detect any shift in product quality. We can monitor the product quality of a production process by using a graph called a *control chart*. Thus, we could graph the sample mean or sample range for samples collected over a period of time to monitor product quality. We discuss the $\bar{y}$-chart in this section; it is used to examine whether the mean output of a process is "in control." In Chapter 9 we discuss r-charts and s-charts for process variability.

Typically a control chart consists of three lines: a center line, an upper control line, and a lower control line. In a control chart for the mean, successive sample means would be plotted much as they appear in Figure 7.6. The sample means are shown by the dots in this figure. If one of the sample means falls outside either the upper or the lower control line, the process is judged to be out of control; that is, it appears that product quality has shifted. At this point, company officials and production personnel would try to establish the cause of the shift and would initiate corrective changes in the production process.

FIGURE 7.6
$\bar{y}$-Chart for Sample Means

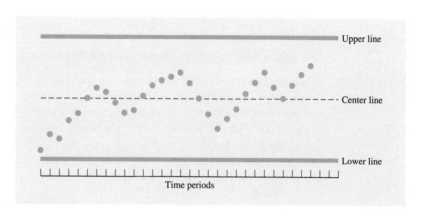

Time periods

center line

Establishing the three control lines is quite simple. The **center line** (denoted by $\bar{y}_c$) represents the average of k sample means, each based on n observations. We generally recommend taking $k \geq 25$ and $n > 3$. These samples should be taken at some time when the process is judged to be under control (stable and predictable). Then, if we let y_{ij} denote the jth observation in sample i and we let

$$\bar{y}_i = \sum_j y_{ij}/n$$

denote the mean for sample i, the average of the k sample means is

$$\bar{y}_c = \sum_i \frac{\bar{y}_i}{k} = \sum_{ij} \frac{y_{ij}}{nk}$$

UCL and LCL for mean quality

The **upper control limit (UCL)** and the **lower control limit (LCL)** are computed as follows:

$$\text{UCL} = \bar{y}_c + 3\frac{\sigma}{\sqrt{n}} \qquad \text{and} \qquad \text{LCL} = \bar{y}_c - 3\frac{\sigma}{\sqrt{n}}.$$

In accordance with the Empirical Rule, the interval $\bar{y}_c \pm (3\sigma/\sqrt{n})$ should contain nearly all the sample means $\bar{y}_i$ in repeated sampling. If a sample mean falls outside this interval, either we have observed an extremely unlikely event or the process quality has changed and $\bar{y}_c$ is no longer an accurate measure of the actual mean product quality. The latter conclusion is more realistic and is used to signal a manufacturing process out of control.

The standard deviation σ in the formulas for the upper and lower control lines can be estimated either by using a "pooled" sample variance from the k samples or, more quickly, by using the k sample ranges. We will employ the latter procedure. Letting r_i denote the range for the n sample measurements in sample i and letting $\bar{r}$ denote the average of the k sample ranges, we can estimate σ by

$$\hat{\sigma} = \frac{\bar{r}}{d_n},$$

where d_n is obtained from Table 13 of Appendix 3. For example, suppose that we have $k = 20$ different samples of $n = 7$ observations per sample and $\bar{r} = 5$. Then $d_7 = 2.704$ and $\hat{\sigma} = 5/2.704 = 1.849$.

EXAMPLE 7.5 A company that dyes rugs is interested in monitoring the color uniformity of its product over time. Although maintaining uniform color is somewhat important for patterned or multicolored rugs, it is much more important for solid-colored rugs, where minor changes in solid colors are readily recognizable. Rug-color quality can be monitored by taking readings on a colorimeter. Twenty-five samples of five measurements each from a rug being dyed red yielded the data listed in Table 7.2. These data were obtained during a period when the manager believed that the process was in control.

Use the data of Table 7.2 to construct a control chart for the mean colorimeter reading.

Solution From Table 7.2 we have

$$\bar{y}_c = \sum_{ij} \frac{y_{ij}}{nk} = \frac{258.2}{5(25)} = 2.07$$

$$\bar{y} = \frac{\sum_i r_i}{k} = \frac{56.8}{25} = 2.27.$$

From Table 13 in Appendix 3, we have $d_5 = 2.326$ and, hence,

$$\hat{\sigma} = \frac{\bar{r}}{d_n} = \frac{2.27}{2.326} = .98.$$

The center line is therefore 2.07, with upper and lower control lines given by

$$\text{UCL} = \bar{y}_c + 3\frac{\hat{\sigma}}{\sqrt{n}} = 2.07 + \frac{3(.98)}{\sqrt{5}} = 3.38$$

$$\text{LCL} = \bar{y}_c - 3\frac{\hat{\sigma}}{\sqrt{n}} = 2.07 - \frac{3(.98)}{\sqrt{5}} = .76.$$

TABLE **7.2**
Colorimeter Readings for the 25
Samples of Example 7.5

Sample	Observation	Sample Sum	Sample Range
1	2.4, 1.8, 0.7, 1.0, 2.5	8.4	1.8
2	2.3, 3.0, 2.5, 1.2, 3.1	12.1	1.9
3	1.3, 1.2, 0.9, 1.2, 3.0	7.6	2.1
4	0.5, 2.2, 2.4, 1.5, 3.0	9.6	2.5
5	2.8, 1.9, 2.6, 1.3, 2.9	11.5	1.6
6	2.4, 3.1, 1.7, 3.3, 2.6	13.1	1.6
7	2.5, 2.9, 1.4, 4.0, 2.1	12.9	2.6
8	1.1, 2.9, 3.0, 1.4, 2.8	11.2	1.9
9	3.3, 2.2, 2.7, 2.8, 2.1	13.1	1.2
10	0.8, 4.2, 2.3, 1.4, 2.1	10.8	3.4
11	0.2, 2.6, 2.3, 0.7, 4.2	10.0	4.0
12	1.8, 1.6, 2.3, 2.1, 1.7	9.5	.7
13	0.1, 3.9, 2.3, 1.4, 1.0	8.7	3.8
14	1.1, 3.1, 1.8, 0.9, 1.8	8.7	2.2
15	0.5, 0.9, 4.0, 2.2, 2.8	10.4	3.5
16	2.9, 3.3, 1.9, 3.1, 2.3	13.5	1.4
17	3.5, 2.0, 2.5, 2.0, 0.3	10.3	3.2
18	2.5, 2.1, 2.7, 1.7, 1.5	10.5	1.2
19	1.1, 3.9, 2.7, 1.2, 1.3	10.2	2.8
20	1.4, 2.0, 2.5, 4.2, 2.4	12.5	2.8
21	2.2, 1.9, 0.7, 1.3, 1.4	7.5	1.5
22	1.5, 1.5, 1.1, 2.3, 2.4	8.8	1.3
23	2.2, 1.3, 2.5, 1.9, 0.7	8.6	1.8
24	1.7, 0.1, 1.8, 0.7, 2.1	6.4	2.0
25	3.7, 1.5, 1.9, 0.6, 4.6	12.3	4.0
Totals		258.2	56.8

Notice that this is a conceptually different situation from the classical confidence interval for μ, where we are trying to "discover" (estimate) a single value of μ. Here a process can change over time, and we are trying to "track" the value of μ.   ∎

An observation that falls outside one of the control lines is a signal that something has changed. If σ is known and the control lines are computed on the basis of the known value of σ, a value outside a control line suggests that the process mean has shifted. Unfortunately, when σ is unknown and must be estimated, a value outside one of the control lines could suggest a shift in the mean quality, an increase in σ, or both.

To protect ourselves, we should also keep a control chart on product quality variability. Additional information about these control charts for process variability (r-charts) is presented in Chapter 9.

Exercises

Basic Techniques

7.25 Refer to Example 7.5. Graph the upper and lower control limits and the center line. Plot the sequence of sample means listed below to determine whether and when the process is out of control. (*Note:* Each mean is based on five measurements.)

2.0   1.9   1.6   1.5   1.7   1.8   2.2   2.1   2.0   2.3   2.4   2.7   2.8   2.9

7.26 Fifty random samples of size $n = 8$ are selected from a process. The means of the sample means and ranges $\bar{y}_c$ and $\bar{r}$ are found to be .640 and .015, respectively. Construct a y-chart for μ for this process.

7.27 Refer to Exercise 7.26. A sample of size $n = 8$ is selected from the process after the control chart has been constructed. The sample mean is found to be .647. Does the process seem to be in control? Why or why not?

Applications

7.28 The labeled amount of ingredient A in a marketed cough drop is 1 in 1500 parts of the total labeled weight (2.2 g). The assay for the ingredient is based on ten cough drops that have been dissolved. A sequence of 48 sample means is shown here, expressed as a percentage of the total labeled weight. If $\bar{r} = 4.1(\%)$, determine the center line and the upper and lower control limits for μ.

108	110	110
108	107	112
109	107	111
107	108	111
109	108	110
108	109	110
109	108	112
109	110	110
107	109	110
107	108	112
109	107	110
111	110	111
109	111	113
107	111	111
106	110	110
107	111	110

7.5 A Statistical Test for μ

The second type of inference-making procedure is statistical testing (or hypothesis testing). As with estimation procedures, we make an inference about a population parameter, but here the inference is of a different sort. In this section, we present a statistical test that leads to an answer to the question, "Is the population mean equal to a specified value μ_0?" For example, in studying the antipsychotic properties of an experimental compound, we might ask whether the average shock-avoidance response for rats treated with a specific dose is 60, the same value that has been observed after extensive testing using a suitable standard drug.

A statistical test of hypothesis can be likened to a criminal trial. We begin with a research hypothesis, something that we wish to support. For example, in the trial the research hypothesis of the prosecution is that the defendant is guilty. The prosecuting attorney attempts to support this hypothesis by showing that its antithesis—that the defendant is not guilty—is false. The latter hypothesis, called a null hypothesis, is the crux of a statistical test. If we can collect evidence to show that the null hypothesis is false, we can conclude that the research hypothesis (or alternative hypothesis) is true.

How can we (or the court) decide which hypothesis is true—the null hypothesis or the alternative hypothesis? In both cases evidence is collected; the evidence in a statistical test is the information contained in a sample selected from the population. This evidence is then weighed (or considered), so that a decision can be made. In the criminal trial a jury functions as the decision maker, weighing the evidence to reach a decision. In a statistical test of a hypothesis we utilize a test statistic (some quantity computed from the sample measurements) to assist us in reaching a decision. We use this quantity in the following way: if the test statistic takes a value that is contradictory to the null hypothesis, we reject the null hypothesis and conclude that the research (or alternative hypothesis) is true. Similarly, in a court trial, if the evidence presented to a jury is highly contradictory to the hypothesis of innocence (null hypothesis), the jury rejects the null hypothesis and concludes that the defendant is guilty (that is, that the research hypothesis is true).

statistical test

A **statistical test** is composed of the five parts listed in the following box.

Five Parts of a Statistical Test

1   Null hypothesis, denoted by H_0

2   Research hypothesis (also called the alternative hypothesis), denoted by H_a

3   Test statistic, denoted by T.S.

4   Rejection region, denoted by R.R.

5   Conclusion

research hypothesis (H_a)

For example, in setting up a statistical test of the mean yield per acre (in bushels) for a particular variety of soybeans, we may be interested in the **research hypothesis** that the mean yield per acre μ is greater than 520 bushels, the average observed for farms throughout a particular state in the past several years. To verify the research hypothesis, we try to contradict another hypothesis, called the **null hypothesis**, that $\mu = 520$ (that is, that the correct figure equals the highest average yield that still differs from the research hypothesis).

null hypothesis (H_0)

Having stated the null and research hypotheses, we obtain a random sample of 1-acre yields from farms throughout the state and compute $\bar{y}$ and s, the sample mean and sample standard deviation, respectively. The decision to accept the null hypothesis or to reject it in favor of the research hypothesis is based on a **test statistic** or decision maker computed from the sample data. If the population can be assumed

test statistic (T.S.)

to be more or less mound-shaped, a logical choice as a decision maker for μ would be $\bar{y}$ or some function of the sample mean.

If we choose $\bar{y}$ as the test statistic, we know that the sampling distribution of $\bar{y}$ (assuming that the null hypothesis is true) is approximately normal, with mean $\mu = 520$. Values of $\bar{y}$ that contradict the null hypothesis and favor the research hypothesis will lie in the upper tail of the distribution of $\bar{y}$ (see Figure 7.7). These contradictory values form a **rejection region** for our statistical test. If the observed value of $\bar{y}$ falls in the region of Figure 7.7, we reject the null hypothesis that the mean yield per acre is $\mu = 520$ in favor of the research hypothesis that $\mu > 520$. Notice that we are supporting the research hypothesis by contradicting the null hypothesis. If the observed value of $\bar{y}$ falls in the acceptance region rather than in the rejection region, we do not reject the null hypothesis. However, this does not mean that we automatically *accept as true* the null hypothesis that H_0: $\mu = 520$ (exactly). We return to the notion of acceptance of H_0 after we discuss the two types or errors that can be made.

rejection region (R.R.)

F I G U R E **7.7**
Assuming That H_0 is True, Contradictory Values of $\bar{y}$ Are in the Upper Tail

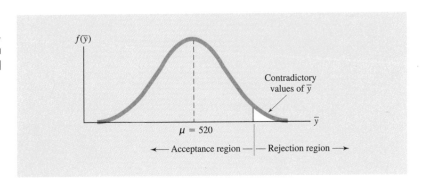

As with any two-way decision process, we can make an error by falsely rejecting the null hypothesis or by falsely accepting the null hypothesis. We give these errors the special names **Type I error** and **Type II error**, respectively.

Type I error
Type II error

D E F I N I T I O N **7.1** **Type I Error**   A **Type I error** is committed if we reject the null hypothesis when it is true. The probability of a Type I error is denoted by the symbol α. ▪

D E F I N I T I O N **7.2** **Type II Error**   A **Type II error** is committed if we accept the null hypothesis when it is false and the research hypothesis is true. The probability of a Type II error is denoted by the symbol β (Greek letter beta). ▪

The two-way decision process is shown in Table 7.3, with corresponding probabilities associated with each situation.

Although it would be desirable to determine the acceptance and rejection regions so as to minimize both α and β simultaneously, this is not possible. The probabilities

TABLE 7.3
Two-Way Decision Process

Decision	Null Hypothesis	
	True	False
Reject H_0	Type I error (α)	Correct $(1 - \beta)$
Accept H_0	Correct $(1 - \alpha)$	Type II error (β)

associated with Type I and Type II errors are inversely related. For a fixed sample size n, as we change the rejection region to increase α, then β decreases; and vice versa (see Figure 7.8).

FIGURE 7.8
Relationship Between α and β

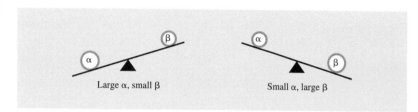

associated Large α, small β Small α, large β

To alleviate what appears to be an impossible bind, the experimenter specifies a tolerable probability for a Type I error of the statistical test. Thus, the experimenter may choose α to be .01, .05, .10, and so on. Specification of a value for α then locates the rejection region. Determination of the associated probability of a Type II error is more complicated and will be delayed until later in this chapter.

Let us now see how the choice of α locates the rejection region. Returning to our soybean example, we reject the null hypothesis for large values of the sample mean $\bar{y}$. Suppose that we have decided to take a sample of $n = 36$ 1-acre plots, and from these data we compute $\bar{y} = 573$ and $s = 124$. Can we conclude that the mean yield for all farms is above 520?

specifying α   Before answering this question, we must **specify** α. If we are willing to take the risk that 1 time in 40 we will incorrectly reject the null hypothesis, then $\alpha = 1/40 = .025$. An appropriate rejection region can be specified for this value of α by referring to the sampling distribution of $\bar{y}$. Assuming that the null hypothesis is true and that σ can be replaced by s, then $\bar{y}$ is normally distributed, with $\mu = 520$ and $\sigma_{\bar{y}} = 124/\sqrt{36} = 20.67$. Since the shaded area of Figure 7.9 corresponds to α, locating a rejection region with an area of .025 in the right tail of the distribution of $\bar{y}$ is equivalent to determining the value of z that has an area of .025 to its right. Referring to Table 2 in Appendix 3, we find that this value of z is 1.96. Thus, the rejection region for our example is located 1.96 standard errors ($1.96\sigma_{\bar{y}}$) above the mean $\mu = 520$. If the observed value of $\bar{y}$ is greater than 1.96 standard errors above $\mu = 520$, we reject the null hypothesis, as shown in Figure 7.9.

FIGURE 7.9
Rejection Region for the
Soybean Example when
$\alpha = .025$

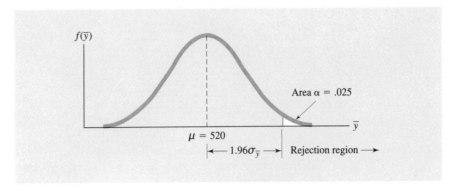

EXAMPLE   7.6   Set up all the parts of a statistical test for the soybean example, and use the sample data to reach a decision on whether to accept or reject the null hypothesis. Set $\alpha = .025$. Assume that σ can be estimated by s.

Solution   The first four parts of the test are as follows:

H_0: $\mu = 520$

H_a: $\mu > 520$

T.S.: $\bar{y}$

R.R.: For $\alpha = .025$, reject the null hypothesis if $\bar{y}$ lies more than 1.96 standard errors above $\mu = 520$.

The computed value of $\bar{y}$ was 573. To determine the number of standard errors that $\bar{y}$ lies above $\mu = 520$, we compute a z score for $\bar{y}$, using the formula

$$z = \frac{\bar{y} - \mu_0}{\sigma_{\bar{y}}},$$

where $\sigma_{\bar{y}} = \sigma/\sqrt{n}$. Substituting into the formula,

$$z = \frac{\bar{y} - \mu_0}{\sigma_{\bar{y}}} = \frac{573 - 520}{124/\sqrt{36}} = 2.56.$$

Conclusion: Since the observed value of $\bar{y}$ lies more than 1.96 standard errors above the hypothesized mean $\mu = 520$, we reject the null hypothesis in favor of the research hypothesis and conclude that the average soybean yield per acre is greater than 520.   ▪

one-tailed test   The statistical test conducted in Example 7.6 is called a **one-tailed test**, because the rejection region is located in only one tail of the distribution of $\bar{y}$. If our research hypothesis were H_a: $\mu < 520$, suitably small values of $\bar{y}$ would support rejection of the null hypothesis. This test would also be one-tailed, but the rejection region would be located in the lower tail of the distribution of $\bar{y}$. Figure 7.10 displays the rejection region for the alternative hypothesis $H_a: \mu < 520$ when $\alpha = .025$.

FIGURE 7.10
Rejection Region for H_a:
$\mu < 520$ When $\alpha = .025$
for the Soybean Example

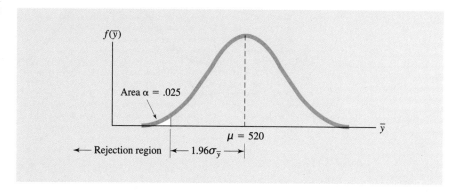

We can formulate a **two-tailed test** for the research hypothesis $H_a: \mu \neq 520$, where we are interested in detecting whether the mean yield per acre of soybeans is either substantially greater or substantially less than 520. Clearly, both large and small values of $\bar{y}$ contradict the null hypothesis, and we locate the rejection region in both tails of the distribution of $\bar{y}$. A two-tailed rejection region for $H_a: \mu \neq 520$ and $\alpha = .05$ is shown in Figure 7.11.

two-tailed test

FIGURE 7.11
Two-Tailed Rejection Region for
$H_a: \mu \neq 520$ When
$\alpha = .025$ for the Soybean
Example

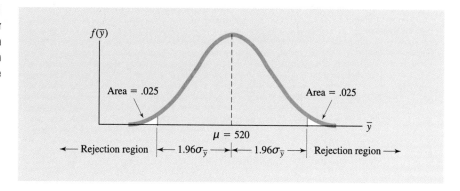

EXAMPLE 7.7   A corporation maintains a large fleet of company cars for its salespeople. To check the average number of miles driven per month per car, a random sample of $n = 40$ cars is examined. The mean and the standard deviation for the sample are 2572 miles and 350 miles, respectively. Records for previous years indicate that the average number of miles driven per car per month was 2600. Use the sample data to test the research hypothesis that the current mean μ differs from 2600. Set $\alpha = .05$ and assume that σ can be estimated by s.

FIGURE 7.12
Rejection Region for H_a:
$\mu \neq 2600$ When
$\alpha = .05$, Example 7.7

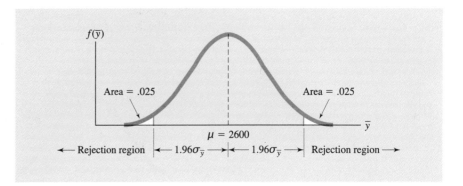

Solution   The research hypothesis for this statistical test is H_a: $\mu \neq 2600$, and the null hypothesis is H_0: $\mu = 2600$. Using $\alpha = .05$, the two-tailed rejection region for this test is located as shown in Figure 7.12.

To determine how many standard errors our test statistic $\bar{y}$ lies away from $\mu = 2600$, we compute

$$z = \frac{\bar{y} - \mu_0}{\sigma/\sqrt{n}} = \frac{2752 - 2600}{350/\sqrt{40}} = 2.75.$$

The observed value for $\bar{y}$ lies more than 1.96 standard errors above the mean, so we reject the null hypothesis in favor of the alternative H_a: $\mu \neq 2600$. Since the computed value of $\bar{y}$ is greater than the hypothesized mean $\mu = 2600$, we conclude that the mean number of miles driven is greater than 2600.   ∎

The mechanics of the statistical test for a population mean can be greatly simplified if we use z rather than $\bar{y}$ as a test statistic. If we used

H_0: $\mu = \mu_0$ (where μ_0 is some specified value)

H_a: $\mu > \mu_0$

and the test statistic

$$z = \frac{\bar{y} - \mu_0}{\sigma/\sqrt{n}},$$

then for $\alpha = .025$ we would reject the null hypothesis if $z > 1.96$—that is, if $\bar{y}$ lies more than 1.96 standard errors above the mean. Similarly, for the same null hypothesis, $\alpha = .05$, and H_a: $\mu \neq \mu_0$, we would reject the null hypothesis if the computed value of z is greater than 1.96 or less than -1.96, or equivalently, if $|z| > 1.96$.

test for population mean   The statistical **test for a population mean** is summarized next. For H_0: $\mu = \mu_0$, three different alternatives are given, with their corresponding rejection regions. In a given situation, you will choose only one of the three alternatives, with its associated rejection region.

Summary of a Statistical Test for μ with σ Known

H_0: $\mu = \mu_0$ (μ_0 is specified)

H_a: 1 $\mu > \mu_0$
       2    $\mu < \mu_0$
       3    $\mu \neq \mu_0$

T.S.: $z = \dfrac{\bar{y} - \mu_0}{\sigma/\sqrt{n}}$

R.R.:   For a probability α of a Type I error,
       1   reject H_0 if $z > z_\alpha$.
       2   reject H_0 if $z < -z_\alpha$.
       3   reject H_0 if $|z| > z_{\alpha/2}$.

Note: For the time being, if σ is unknown but $n \geq 30$, you may replace σ by s in the standard error $\sigma_{\bar{y}} = \sigma/\sqrt{n}$ and proceed with the test. A more detailed discussion of inferences about μ when σ is unknown is presented later in this chapter.

EXAMPLE   7.8   The average (mean) live weights of a farmer's steers prior to slaughter was 380 pounds in past years. This year his 50 steers were fed on a new diet. Suppose that we consider these 50 steers fed on the new diet as a random sample taken from a population of all possible steers that may be fed the diet now or in the future. Use the sample data given here and $\alpha = .01$ to test the research hypothesis that the mean live weight for steers on the new diet is greater than 380. The sample data are $n = 50$; $\bar{y} = 390$; $s = 35.2$.

Solution   Using the sample data with $\alpha = .01$, we can determine that the five parts of a statistical test are as follows:

H_0: $\mu = 380$

H_a: $\mu > 380$

T.S.: $z = \dfrac{\bar{y} - \mu_0}{\sigma/\sqrt{n}} = \dfrac{390 - 380}{35.2/\sqrt{50}} = \dfrac{10}{35.2/7.07} = 2.01$

R.R.:   For $\alpha = .01$, and a one-tailed test, we reject H_0 if $z > z_{.01}$, where $z_{.01} = 2.33$.

Conclusion: Since the observed value of z, 2.01, does not exceed 2.33, we might be tempted to accept the null hypothesis that $\mu = 380$. The only problem with this conclusion is that we do not know β, the probability of incorrectly accepting the null hypothesis. To hedge somewhat in situations where z does not fall in the rejection region and β has not been calculated, we recommend stating that there is insufficient evidence to support rejecting the null hypothesis. To reach a conclusion about whether to accept H_0, the experimenter would have to compute β. If β is small for reasonable alternative values of μ, then H_0 is accepted. Otherwise, the experimenter should conclude that there is insufficient evidence to support rejecting the null hypothesis. ∎

computing β

We can illustrate the **computation of β**, the probability of a Type II error or equivalently the *power* $(1 - \beta)$, using the data in Example 7.8. If the null hypothesis is H_0: $\mu = 380$, the probability of incorrectly accepting H_0 depends on how close the actual mean is to 380. For example, if the actual mean live weight is 400 pounds for steers on the new diet, we would expect β to be much smaller than if the actual mean live weight is 387. The whole process of determining β or the power $(1 - \beta)$ of a test is a "what-if" type of process. We look at β (or $1 - \beta$) for several possible alternative values of μ.

Suppose that the actual live mean weight is 395. What is β? With the null and research hypotheses as before,

$$H_0: \quad \mu = 380$$
$$H_a: \quad \mu > 380$$

and with $\alpha = .01$, we use Figure 7.13(a) to display β. The shaded portion of Figure 7.13(a) represents β, the probability of $\bar{y}$ falling in the acceptable region when the null hypothesis is false and μ is actually 395. Similarly, the power of the test for detecting H_a: $\mu = 395$ is $1 - \beta$, the area in the rejection region.

Now consider two other possible values for μ—namely, 387 and 400. The corresponding values of β are shown as the shaded portions of Figure 7.13(b) and (c), respectively; power is the unshaded portion in the rejection region of Figure 7.13(b) and (c). The three situations illustrated in Figure 7.13 confirm that the probability of a Type II error β decreases (and hence power increases) the farther μ lies away from the hypothesized mean under H_0.

μ_0, μ_a

We can readily calculate β for a test involving μ if we adopt the following notation. Let μ_0 denote the hypothesized mean under H_0 and let μ_a denote the actual mean. The procedure for calculating β is then as summarized in the accompanying box. Although we never really know the actual mean, we can calculate β for any specified value of μ. The decision about whether or not to accept H_0 depends on the magnitude of β for one or more reasonable alternative values. For a one-tailed test of H_0: $\mu = \mu_0$, β is the probability that z is less than

$$z_\alpha - \frac{|\mu_0 - \mu_a|}{\sigma_{\bar{y}}}.$$

This probability is written as

$$P\left[z < z_\alpha - \frac{|\mu_0 - \mu_a|}{\sigma_{\bar{y}}}\right].$$

Formulas for β are given here for one- and two-tailed tests. Examples using these formulas follow.

FIGURE 7.13
The Probability β of a Type II
Error

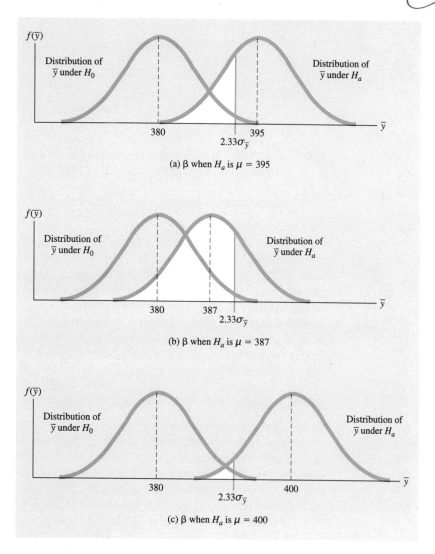

(a) β when H_a is $\mu = 395$

(b) β when H_a is $\mu = 387$

(c) β when H_a is $\mu = 400$

Calculation of β for $H_0: \mu = \mu_0$ When μ_a is the Actual Mean

1   One-tailed test:

$$\beta = P\left(z < z_\alpha - \frac{|\mu_0 - \mu_a|}{\sigma_{\bar{y}}}\right); \qquad \text{power} = 1 - \beta.$$

2   Two-tailed test:

$$\beta \approx P\left(z < z_{\alpha/2} - \frac{|\mu_0 - \mu_a|}{\sigma_{\bar{y}}}\right); \qquad \text{power} = 1 - \beta.$$

EXAMPLE 7.9 Compute β and the power for the test in Example 7.8 if the actual mean live weight of steers is 395.

Solution The research hypothesis for Example 7.8 was $H_a: \mu > 380$. Using $\alpha = .01$ and the computing formula for β with $\mu_0 = 380$ and $\mu_a = 395$, we have

$$\beta = P\left[z < z_{.01} - \frac{|\mu_0 - \mu_a|}{\sigma_{\bar{y}}}\right] = P\left[z < 2.33 - \frac{|380 - 395|}{35.2/\sqrt{50}}\right]$$
$$= P[z < 2.33 - 3.01] = P[z < -.68].$$

Referring to Table 2 in Appendix 3, we can see that the area corresponding to $z = .68$ is .2483. Hence, $\beta = .2483$ and Power $= 1 - .2483 = .7517$. ∎

Previously, when $\bar{y}$ did not fall in the rejection region, we concluded that there was insufficient evidence to support rejecting H_0 because β was unknown. Now when $\bar{y}$ falls in the acceptance region, we can compute β corresponding to one (or more) alternative values for μ that appear reasonable in light of the experimental setting. Then, provided that we are willing to tolerate a specified probability of falsely accepting the null hypothesis equal to the computed value of β for the alternative value(s) of μ considered, our decision is to accept the null hypothesis. Thus, in Example 7.9, if we are willing to risk a β error of about .25 of falsely accepting the null hypothesis, we can accept the null hypothesis $\mu = 380$.

EXAMPLE 7.10 Prospective salespeople for an encyclopedia company are being offered a sales training program. Previous data indicate that the average number of sales per month for those who do not participate in the program is 33. To determine whether the training program is effective, a random sample of 35 new employees is given the sales training and then sent out into the field. One month later, the mean and the standard deviation for the number of sets of encyclopedias sold are 35 and 8.4, respectively. Do the data present sufficient evidence to indicate that the training program enhances sales? Use $\alpha = .05$.

Solution The five parts to our statistical test are as follows:

H_0: $\mu = 33$

H_a: $\mu > 33$

T.S.: $z = \dfrac{\bar{y} - \mu_0}{\sigma_{\bar{y}}} = \dfrac{35 - 33}{8.4/\sqrt{35}} = 1.41$

R.R.: For $\alpha = .05$, we will reject the null hypothesis if $z > z_{.05} = 1.645$.

Conclusion: Since the observed value of z does not fall into the rejection region, we reserve judgment on accepting H_0 until we calculate β; that is, we conclude that there is insufficient evidence to support rejecting the null hypothesis that persons who participated in the sales program have the same mean number of sales per month as those who did not.   ∎

E X A M P L E   **7.11**   Refer to Example 7.10. Suppose that the encyclopedia company thinks that the cost of financing the sales program will be offset by increased sales if those who participated in the program average 38 sales per month. Compute β for $\mu_a = 38$ and, based on the value of β, indicate whether you would accept the null hypothesis.

Solution   Using the computational formula for β with $\mu_0 = 33$, $\mu_a = 38$, and $\alpha = .05$, we have

$$\beta = P\left[z < z_{.05} - \frac{|\mu_0 - \mu_a|}{\sigma_{\bar{y}}}\right] = P\left[z < 1.645 - \frac{|33 - 38|}{8.4/\sqrt{35}}\right]$$
$$= P[z < -1.88].$$

The area corresponding to $z = 1.88$ in Table 2 of Appendix 3 is .0301. Hence,

$$\beta = .0301; \quad \text{Power} = 1 - .0301 = .9699.$$

Because β is relatively small, we accept the null hypothesis and conclude that the training program has not increased the average sales per month above the point at which increased sales offset the cost of the training program.   ∎

In Section 7.2, we discussed how to measure the goodness of interval estimates. The goodness of a statistical test can be measured by the magnitudes of the Type I and Type II errors, α and β. When α is preset at a tolerable level by the experimenter, β is a function of the sample size for a fixed value of μ_a. The larger the sample size, the more information we have concerning μ and hence, the smaller the value of β. We will consider now the problem of designing an experiment for testing $H_0: \mu = \mu_0$ when α is specified and β is preset for a fixed actual value of μ_a. This problem reduces to determining the sample size needed for the fixed values of α and β.

7.6   Choosing the Sample Size for Testing μ

The quantity of information available for a statistical test about μ is measured by the magnitudes of the Type I and Type II error probabilities, α and β. Suppose that we are interested in testing

$$H_0: \quad \mu = \mu_0$$

against a one-sided alternative

$$H_a: \quad \mu > \mu_0$$

In addition, suppose that we want the probability of a Type I error to be α and the probability of a Type II error to be β or less when the actual value of μ lies a distance of Δ (delta) or more above μ_0. The sample size necessary to meet these requirements is shown in the following box.

Sample Size for a One-Sided Test of μ

$$n = \sigma^2 \frac{(z_\alpha + z_\beta)^2}{\Delta^2}$$

Note: If σ^2 is unknown, substitute an estimated value to get an approximate sample size.

The same formula applies to the one-sided alternative $H_a: \mu < \mu_0$, except that here we want the probability of a Type II error to be of magnitude β or less when the actual value of μ lies a distance of Δ or more below μ_0.

E X A M P L E   7.12   A cereal packager is concerned that one of his machines has a mean fill per package of more than 16 ounces, the labeled net weight. While this is not bad from a public relations standpoint, it could cost the packager a great deal of money. Previous experience suggests that the standard deviation of the package fill weights is approximately .225. For

$$H_0: \mu = 16$$
$$H_a: \mu > 16$$

with $\alpha = .05$, determine the sample size required to make $\beta = .01$ or less if the actual mean is 16.1 ounces or more. By putting this restriction on β, the packager is saying that it wants a very small probability of falsely accepting $H_0: \mu = 16$, when in fact the actual mean is 16.1 ounces or more.

Solution   From previous data, the fill weights have a standard deviation approximately equal to .225. The appropriate z-values, $z_{.05}$ and $z_{.01}$, for $\alpha = .05$ and $\beta = .01$ are 1.645 and 2.33, respectively. Using $\Delta = 16.1 - 16 = .1$, the required sample size is

$$n = \frac{(.225)^2(1.645 + 2.33)^2}{(.1)^2} = 79.99 \approx 80.$$

Therefore, the packager must obtain a random sample of $n = 80$ cartons to conduct this test under the specified conditions.

Suppose that, after obtaining the sample, we find that the computed value of

$$z = \frac{\bar{y} - 16}{\sigma_{\bar{y}}}$$

does not fall in the rejection region. What is our conclusion? In similar situations in previous sections, our conclusion has been that there was insufficient evidence to support rejecting H_0. Now, however, knowing that $\beta \leq .01$ when $\mu \geq 16.1$, we can feel safe in our conclusion to accept H_0: $\mu = 16$. No further testing is required.   ▪

With a slight modification of the sample size formula for the one-tailed tests, we can test

$$H_0: \mu = \mu_0$$
$$H_a: \mu \neq \mu_0$$

for a specified α and β with $\Delta = |\mu - \mu_0|$. A formula for an approximate sample size to use when testing μ is presented here.

Approximate Sample Size for a Two-sided Test of H_0: $\mu = \mu_0$

$$n = \frac{\sigma^2}{\Delta^2}(z_{\alpha/2} + z_\beta)^2$$

Note: If σ^2 is unknown, substitute an estimated value to get an approximate sample size.

Exercises

Basic Techniques

7.29   Consider the data of Example 7.12, and compute the sample size required for testing H_0: $\mu = 16$ against H_a: $\mu \neq 16$ for $\alpha = .05$ and $\beta \leq .01$ when the actual value of μ lies more than .1 unit away from $\mu_0 = 16$.

7.30   A random sample of 50 measurements from a population yielded $\bar{y} = 40.1$ and $s = 5.6$. Use these data to test the null hypothesis H_0: $\mu = 38$ against the alternative hypothesis H_a: $\mu > 38$. Use $\alpha = .05$ and draw a conclusion. Could you have made a Type II error in this situation? Explain.

7.31   For the data of Exercise 7.30, determine the power of rejecting H_0: $\mu = 38$, given that the alternative hypothesis is true and $\mu_a = 40$. Do the same for $\mu_a = 42$ and 44, in order to sketch the power of this test for the various alternatives.

7.32   The mean and the standard deviation of a random sample of $n = 50$ measurements are $\bar{y} = 63.7$ and $s = 14.2$. Conduct a statistical test of H_0: $\mu = 68$ against the alternative hypothesis H_a: $\mu < 68$, using $\alpha = .05$.

7.33   Refer to Exercise 7.32. Will your conclusion be different if you select $\alpha = .01$? Explain.

Applications

7.34   The administrator of a nursing home would like to do a time-and-motion study of staff time spent per day performing nonemergency-type chores. In particular, she would like to test the null hypothesis H_0: $\mu = 16$ (person-hours per day) against H_a: $\mu < 16$. The value of 16 arose from a previous study prior to the introduction of some efficiency measures. How many days must be sampled to test the proposed hypothesis if $\alpha = .05$ and $\beta \leq .10$ when the actual value of μ is 12 hours (a 25% decrease from previous results) or less? Assume $\sigma^2 = 7.64$.

7.35 Refer to Exercise 7.34. Determine the sample size for testing H_0: $\mu = 16$ and $\alpha = .05$ if the power of detecting a mean of 13 or less is .80 or more.

7.36 The increase in exercise capacity (in minutes) was recorded for each of 90 adult male patients following treatment for congestive heart failure. Given that the sample results yield $\bar{y} = 2.2$ and $s = 1.05$, use these data to test the null hypothesis H_0: $\mu = 2.0$ versus H_a: $\mu > 2.0$. Use $\alpha = .05$ to draw a conclusion.

7.37 Refer to Exercise 7.36. Sketch a power curve (power versus μ_a) for this test based on $\mu_a = 2.1, 2.2, 2.3,$ and 2.5.

7.38 To evaluate the success of a one-year experimental program designed to increase the mathematical achievement of underprivileged high-school seniors, the mathematics scores for a sample of $n = 100$ underprivileged seniors were obtained for comparison with the previous year's statewide average of 525 for underprivileged seniors. You wish to examine whether the mean achievement level has increased over last year's statewide average. Discuss whether you would use a one-tailed or a two-tailed test. Set up all parts of the statistical test for μ, using $\alpha = .05$.

7.39 Refer to Exercise 7.38. Suppose that you wish to examine whether the mean achievement has changed (up or down) over the past year. Would you use a two-tailed test? Explain. Set up all parts of the statistical test for μ, using $\alpha = .01$.

7.40 To study the effectiveness of a weight-reducing agent, a clinical trial was conducted in which 35 overweight males were placed on a fixed diet. After a two-week period, each male was weighed and then given a supply of the weight-reducing agent. The diet was to be maintained; in addition, a single dose of the weight-reducing agent was to be taken each day. At the end of the next two-week period, weights were again obtained. Set up all parts of the statistical test for the alternative hypothesis that μ, the average weight loss, is greater than 0. Why is a one-tailed test appropriate? Use $\alpha = .05$.

7.41 Refer to Exercise 7.40.

a   The average weight loss for the second two-week period was $\bar{y} = 10.3$ pounds, and the standard deviation was $s = 4.6$. Perform a statistical test and draw conclusions. Use $\alpha = .05$.

b   Based on the results for part a, can you conclude that the weight-reducing agent is effective? Explain.

7.42 Transportation—getting people to their destination and home again—is a national problem. One aspect of this problem currently being studied by the Federal Highway Administration is how to successfully merge automobiles entering at high speed with congested interstate traffic. To study this problem, an automobile merging system was installed at the entrance to I-75 in Tampa, Florida. Through the use of a series of display lights, a driver is told whether he or she is traveling at an appropriate speed to merge successfully into the existing traffic on the highway. Prior to installation of the system, investigators measured the stress levels of many drivers merging onto the highway during the 4 to 6 P.M. rush hour period. Similar testing on a random sample of 50 drivers was conducted after the merging system was installed.

For purposes of illustration, suppose that the average stress level prior to the installation of the system was 8.2 (measured on a 10-point scale). Set up appropriate null and alternative hypotheses to test the research hypothesis that the average stress level for drivers under the merging system is less than that observed prior to the installation of the system. Is this a one- or two-tailed test?

7.43 Refer to Exercise 7.42. Suppose that the sample mean and standard deviation for the 50 drivers tested using the merging system were, respectively, 7.6 and 1.8. Use these data to test the alternative hypothesis of Exercise 7.42. Use $\alpha = .05$.

7.44 Tooth decay generally develops first on teeth that have irregular shapes (typically molars). The most susceptible surfaces on these teeth are the chewing surfaces. Usually the enamel on these surfaces contains tiny pockets that tend to hold food particles. Bacteria begin to eat the food particles, creating an environment in which the tooth surface decays.

Of particular importance in the decay rate of teeth, in addition to the natural hardness of the teeth, is the form of food eaten by the individual. Some forms of carbohydrates are particularly detrimental to dental health. Many studies have been conducted to verify these findings, and we can imagine how the study might have been run. A random sample of 60 adults was obtained from a given locale. Each person was examined and then maintained a diet supplement with a sugar solution at all meals. At the end of a one-year period, the average number of newly decayed teeth for the group was .70, and the standard deviation was .4. Do these data present sufficient evidence to indicate that the mean number of newly decayed teeth for people whose diet includes a sugar solution is greater than .30, a rate that had been shown to apply to persons whose diet did not contain the sugar solution supplement? Why would a two-tailed test be inappropriate? Use $\alpha = .05$.

7.7 The Level of Significance of a Statistical Test

In Section 7.6, we introduced hypothesis testing along rather traditional lines: we defined the parts of a statistical test along with the two types of errors and their associated probabilities α and β. In recent years, many statisticians and other users of statistics have objected to the decision-based approach to hypothesis testing. Rather than running a statistical test with a preset value of α, they argue, we should specify the null and alternative hypotheses, collect the sample data, and determine the weight of the evidence for rejecting the null hypothesis. This weight, given in terms of a probability, is called the **level of significance** (or p-value) of the statistical test.

level of significance

We illustrate the calculation of a level of significance with an example.

E X A M P L E   7.13   Refer to Example 7.8.

    a Rather than specifying a preset value for α, determine the level of significance for the statistical test.

    b How would the level of significance change if $\bar{y}$ had been 397 rather than 390?

Solution **a** The null and alternative hypotheses are

$$H_0: \mu = 380$$
$$H_a: \mu > 380.$$

From the sample data, the computed value of the test statistic is

$$z = \frac{\bar{y} - 380}{s/\sqrt{n}} = \frac{390 - 380}{35.2/\sqrt{50}} = 2.01.$$

The level of significance for this test (that is, the weight of evidence for rejecting H_0) is the probability of observing a value of $\bar{y}$ greater than 390, assuming that the null hypothesis is true. This value can be computed by using the z-value of the test statistic, 2.01, and referring to Table 2 in Appendix 3 to determine the probability of observing a z-value greater than 2.01. This probability, which is sometimes designated by the letter p, is seen to be .0222. This value is represented by the shaded area in Figure 7.14. Thus, we would say that the level of significance for this test is .0222.

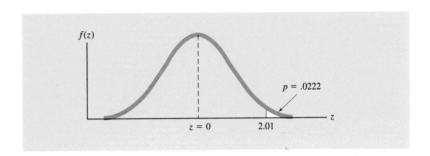

b For $\bar{y} = 397$, the corresponding value of the z-statistic is 3.42. Since the largest value of z in Table 2 in Appendix 3 is 3.09, the p-value is less than .001. We show this as $p < .001$. ∎

As Example 7.13 shows, the level of significance represents the probability of observing a sample outcome more contradictory to H_0 than the observed sample result. *The smaller the value of this probability, the heavier the weight of the sample evidence against H_0.* For example, a statistical test with a level of significance of $p = .01$ shows considerably more evidence for the rejection of H_0 than does another statistical test with $p = 20$.

Suppose that the null and alternative hypotheses in Example 7.13 were

$$H_0: \mu = 380$$
$$H_a: \mu < 380$$

and the computed value of z had been $z = -2.01$. The level of significance would still be $p = .0222$ (see Figure 7.15).

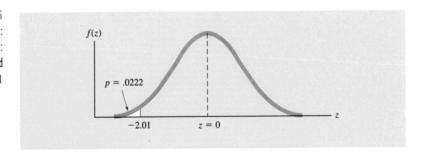

p-value for one-tailed
test

To summarize, the **level of significance for a one-tailed test** can be computed as follows:

For H_a: $\mu > \mu_0$,

$$p = P[z > \text{Computed } z].$$

For H_a: $\mu < \mu_0$,

$$p = P[z < \text{Computed } z].$$

p-value for two-tailed test

For two-tailed tests (as determined by the form of H_a), we still compute the probability of obtaining a sample outcome more contradictory to H_0 than the observed result, but the level of significance is commonly taken to be twice this probability. For a two-tailed test, the level of significance can be written as

$$p = 2P[z > |\text{Computed } z|].$$

E X A M P L E   7.14   Determine the level of significance for the data of Example 7.8 if the null hypothesis and alternative hypothesis are

$$H_0: \mu = 380$$
$$H_a: \mu \neq 380.$$

Solution   The computed value of z is 2.01. Since the probability of observing a value of z greater than 2.01 is .0222, the level of significance for the two-tailed statistical test is $p = 2(.0222) = .0444$. ∎

There is much to be said in favor of this approach to hypothesis testing. Rather than reaching a decision directly, the statistician (or person performing the statistical test) presents the experimenter with the weight of evidence for rejecting the null hypothesis. The experimenter can then draw his or her own conclusion. Some experimenters will reject a null hypothesis if $p = .10$, whereas others will require $p < .05$ or $p < .01$ for rejecting the null hypothesis. The experimenter is left to make the decision based on what he or she believes is enough evidence to support rejection of the null hypothesis.

Many professional journals have followed this approach by reporting the results of a statistical test in terms of its level of significance. Thus, we might read that a particular test was significant at the $p = .05$ level or perhaps the $p < .01$ level. By reporting results this way, the reader is left to draw his or her own conclusion.

One word of warning is needed here. The p-value of .05 has become a magic level, and many people seem to feel that a particular null hypothesis should not be rejected unless the test achieves the .05 level or lower. This has resulted in part from the decision-based approach, with α preset at .05. Try not to fall into this trap when reading journal articles or reporting the results of your statistical tests. After all, statistical significance at a particular level does not dictate importance or practical significance. Rather, it means that a null hypothesis can be rejected with a specified low risk of error. For example, suppose that a company is interested in determining whether the average number of miles driven per car per month for the sales force has risen above 2600. Sample data from 400 cars show that $\bar{y} = 2640$ and $s = 35$. For these data, the z-statistic for H_0: $\mu = 2600$ is $z = 22.86$ based on $\sigma = 35$; the level of significance is $p < .0000000001$. Thus, even though there has only been a 1.5% increase in the average monthly miles driven for each car, the result is (highly) statistically significant. Is this increase of any practical significance? Probably not. What we have done is proved *conclusively* that the mean μ has increased slightly.

Throughout this text we will conduct statistical tests from both the decision-based approach and from the level-of-significance approach, to familiarize you with both avenues of thought. For either approach, remember to consider the practical significance of your findings after drawing conclusions based on the statistical test.

Exercises

Basic Techniques

7.45   Sample data for a statistical test of H_0: $\mu = 40$ yielded a z-score of 1.86.

    a   Determine the level of significance for a test of H_a: $\mu > 40$.

    b   Determine the level of significance for a test of H_a: $\mu \neq 40$.

Applications

7.46   A random sample of 36 cigarettes of a certain brand was tested for nicotine content. The sample mean and sample standard deviation (in milligrams) are, respectively, 15.1 and 3.8. Give the level of significance of the statistical test of H_0: $\mu = 14$ (the claimed nicotine content) against the alternative hypothesis H_a: $\mu > 14$.

7.47   A psychological experiment was conducted to investigate the length of time (time delay) between administration of a stimulus and observation of a specified reaction. Each of 36 persons in a random sample was subjected to the stimulus and observed for the time delay. The sample mean and standard deviation were 2.2 and .57 seconds, respectively. Test the null hypothesis that the mean time delay for the hypothetical set of all persons who may be subjected to the stimulus is $\mu = 1.6$ against the alternative hypothesis that the mean time delay differs from 1.6. Use $\alpha = .05$.

7.8 Inferences About μ, with σ Unknown

Th estimation and test procedures about μ presented earlier in this chapter were based on the assumption that the population variance was known or that we had enough observations to allow s to be a reasonable estimate of σ. In this section, we will present a test that can be applied when σ is unknown, no matter what the sample size. For example, in determining the average concentration of a drug in the bloodstream one hour after patients suffering from a rare disease are treated with the drug, it might be impossible to obtain a random sample of 30 or more observations at a given time. What test procedure could be used to make inferences about μ?

W. S. Gosset faced a similar problem around the turn of the century. As a chemist for Guinness Breweries, he was asked to make judgments about the mean quality of various brews, but he was not supplied with large sample sizes to help him reach his conclusions.

Gosset thought that when he used the test statistic

$$z = \frac{\bar{y} - \mu_0}{\sigma/\sqrt{n}},$$

with σ replaced by s for small sample sizes, he was falsely rejecting the null hypothesis $H_0: \mu = \mu_0$ at a slightly higher rate than that specified by α. This problem intrigued him, and he set out to derive the distribution and percentage points of the test statistic

$$\frac{\bar{y} - \mu_0}{s/\sqrt{n}}$$

for $n < 30$.

For example, suppose that an experimenter sets α at a nominal level—say, .05. Then he or she expects to falsely reject the null hypothesis approximately 1 time in 20. However, Gosset proved that the actual probability of a Type I error for this test was somewhat higher than the nominal level designated by α. He published the results of his study under the pen name Student, because it was against company policy for him to publish his results in his own name at that time. The quantity

$$\frac{\bar{y} - \mu_0}{s/\sqrt{n}}$$

Student's t

is called the t statistic, and its distribution is called the *Student's t distribution* or, simply, **Student's** t (see Figure 7.16).

FIGURE 7.16
A Student's t Distribution with a Normal Distribution Superimposed

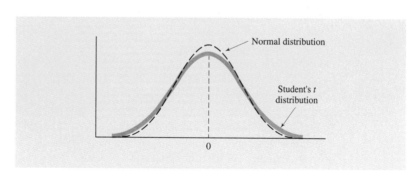

Although the quantity

$$\frac{\bar{y} - \mu_0}{s/\sqrt{n}}$$

possesses a t distribution only when the sample is selected from a normal population, the t distribution provides a reasonable approximation to the distribution of

$$\frac{\bar{y} - \mu_0}{s/\sqrt{n}}$$

when the sample is selected from a population with a mound-shaped distribution. We summarize the properties of t next.

Properties of Student's t Distribution

1 The t distribution, like that of z, is symmetrical about 0.

2 The t distribution is more variable than the z distribution (see Figure 7.15).

3 There are many different t distributions. We specify a particular one by means of a parameter called the **degrees of freedom** (df). Thus, we specify

$$t = \frac{\bar{y} - \mu_0}{s/\sqrt{n}}, \qquad df = n - 1.$$

4 As n (or equivalently, df) increases, the distribution of t approaches the distribution of z.

The phrase "degrees of freedom" sounds awfully mysterious, but the idea will eventually become second nature to you. The technical definition requires advanced mathematics, which we will avoid; on a less technical level, the basic idea is that degrees of freedom are pieces of information for estimating σ using s. The standard deviation s for a sample of n measurements is based on the deviations $y_i - \bar{y}$. Because $\sum(y_i - \bar{y}) = 0$ always, it follows that, if $n - 1$ of the deviations are known, the last (nth) is fixed mathematically to make the sum equal 0. It is therefore noninformative. So, in a sample of measurements, there are $n - 1$ pieces of information (degrees of freedom) about σ.

Because of the symmetry of t, only upper-tail percentage points (probabilities or areas) of the distribution of t have been tabulated; these appear in Table 3 in Appendix 3. The degrees of freedom (df) are listed along the left column of the page. An entry in the table specifies a value of t—say, t_a—such that an area a lies to its right (see Figure 7.17). Various values of a appear across the top of Table 3 in Appendix 3. Thus, for example, with df $= 7$, the value of t with an area .05 to its right is 1.895 (found in the $a = .05$ column and df $= 7$ row).

FIGURE 7.17
Area Tabulated in Table 3 in
Appendix 3 for the t Distribution

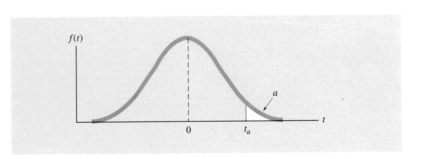

We can use the t distribution to make inferences about a population mean μ. The sample test concerning μ is summarized next. The only difference between the z test discussed earlier in this chapter and the test given here is that t replaces z. The t test (rather than the z test) should be used any time σ is unknown and the distribution of y-values is mound-shaped.

Statistical Test About μ, σ Unknown

H_0: $\mu = \mu_0$

H_a: **1** $\mu > \mu_0$
          2 $\mu < \mu_0$
          3 $\mu \neq \mu_0$

T.S.: $t = \frac{\bar{y} - \mu_0}{s/\sqrt{n}}$

R.R.:   For a probability α of a Type I error and df $= n - 1$,
        1   reject H_0 if $t > t_\alpha$.
        2   reject H_0 if $t < -t_\alpha$.
        3   reject H_0 if $|t| > t_{\alpha/2}$.

Recall that a denotes the area in the tail of the t distribution. For a one-tailed test in which the probability of a Type I error is equal to α, we locate the rejection region by using the value from Table 3 in Appendix 3 for $a = \alpha$ and df $= n - 1$. But for a two-tailed test, we use the t-value from Table 3 corresponding to $a = \alpha/2$ and df $= n - 1$.

Thus, for a one-tailed test we reject the null hypothesis if the computed value of t is greater than the t-value from Table 3 in Appendix 3, and if $a = \alpha$ and df $= n - 1$. Similarly, for a two-tailed test, we reject the null hypothesis if $|t|$ is greater than the t-value from Table 3 for $a = \alpha/2$ and df $= n - 1$.

E X A M P L E   7.15   A tire company guarantees that a particular brand of tire has a mean useful lifetime of 42,000 miles or more. A consumer test agency, wishing to verify this claim, observed $n = 10$ tires on a test wheel that simulated normal road conditions. The lifetimes (in thousands of miles) were as follows:

$$42 \quad 36 \quad 46 \quad 43 \quad 41 \quad 35 \quad 43 \quad 45 \quad 40 \quad 39$$

Use these data to determine whether there is sufficient evidence to contradict the manufacturer's claim. Set $\alpha = .05$.

Solution   The null and research hypotheses for this example are

$$H_0: \quad \mu = 42$$

and

$$H_a: \quad \mu < 42.$$

Notice that we are giving the manufacturer the benefit of the doubt by setting $\mu = 42$ for H_0.

Before setting up the test statistic and rejection region, we must compute the sample mean and sample standard deviation. You can verify that

$$\sum_i y_i = 410 \quad \text{and} \quad \sum_i y_i^2 = 16,926.$$

Then

$$y_i = \frac{\sum_i y_i}{10} = \frac{410}{10} = 41.$$

Similarly, substituting into the shortcut formula for s^2, we find

$$s^2 = \frac{1}{9}\left[\sum_i y_i^2 - \frac{(\sum_i y_i)^2}{10}\right] = 12.89$$

$$s = \sqrt{12.89} = 3.59.$$

The test statistic, then, is

$$t = \frac{\bar{y} - \mu_0}{s/\sqrt{n}} = \frac{41 - 42}{3.59/\sqrt{10}} = -.88$$

and the rejection region is

$$\text{R.R.: Reject } H_0 \text{ if } t < -t_{.05}.$$

From Table 3 in Appendix 3, we find that the critical t-value with $a = .05$ and df $= 9$ is 1.833, so $-t_{.05}$ is -1.833. Since the observed value of t is not less than -1.833, we have insufficient evidence to support the conclusion that the mean lifetime of this brand of tires is less than 42,000 miles.

At this point, someone might suggest calculating β, the probability of a Type II error, to see whether we can accept the manufacturer's claim. Unfortunately, this is a much more difficult task for a small-sample test than it is for a large-sample test, and it is beyond the scope of this text. (If you are interested in pursuing the topic, consult *Biometrika Tables for Statisticians*, Volume I.) Our conclusion is that there is insufficient evidence to justify rejecting the company's claim, and we should continue sampling. ∎

E X A M P L E   **7.16**   Refer to Example 7.15. Rather than performing the statistical test with a preset α level, give the level of significance for the test.

Solution   For the one-tailed lower-tail test, the computed t-value is $t = -.88$. If we had an entire table of t areas for each df, this would be no problem. Because of space limitations, we show only a few areas (a) for each df. The best we can do for $t = -.88$ and df $= 9$ is to say that $p > .10$. Based on this probability, the experimenter would probably conclude that there was insufficient evidence to justify rejecting the null hypothesis. If you think that the level of significance should be given more precisely, you can refer to more detailed tables of the t distribution in the *Biometrika Tables for Statisticians* or any of several statistical software packages that compute p-values for various test procedures.   ■

In addition to being able to run a statistical test for μ when σ is unknown, we can construct a confidence interval using t. The confidence interval for μ with σ unknown is identical to the corresponding confidence interval for μ when σ is known, with z replaced by t and σ replaced by s.

100(1 − α)% Confidence Interval for μ, with σ Unknown

$$\bar{y} \pm t_{\alpha/2} \frac{s}{\sqrt{n}}$$

Note: df $= n - 1$, and the confidence coefficient is $(1 - \alpha)$.

EXAMPLE 7.17   In a psychological depth-perception test, each person in a random sample of $n = 14$ airline pilots was asked to judge the distance between two markers at the other end of a laboratory. The sample data (recorded in feet) are listed below.

2.7   2.4   1.9   2.6   2.4   1.9   2.3
2.2   2.5   2.3   1.8   2.5   2.0   2.2

Use the sample data to place a 95% confidence interval on μ, the average recorded distance for this psychological test.

Solution   Before setting up a 95% confidence interval on μ, we must compute $\bar{y}$ and s. You can verify that

$$\sum_i y_i = 31.70 \quad \text{and} \quad \sum_i y_i^2 = 72.79.$$

The sample mean, variance, and standard deviation are then

$$\bar{y} = \frac{\sum_i y_i}{14} = \frac{31.70}{14} = 2.26$$

$$s^2 = \frac{1}{13}\left[72.79 - \frac{(31.7)^2}{14}\right] = .078$$

$$s = \sqrt{.078} = .28.$$

Referring to Table 3 in Appendix 3, we find that the t-value corresponding to $a = .025$ and df $= 13$ is 2.160. Hence, the 95% confidence interval is

$$\bar{y} \pm t_{\alpha/2} \frac{s}{\sqrt{n}} \quad \text{or} \quad 2.26 \pm (2.160) \frac{.28}{\sqrt{14}},$$

which is the interval $2.26 \pm .16$, or 2.10 to 2.42. Thus, we are 95% confident that the interval from 2.10 to 2.42 will encompass the mean μ. ∎

Exercises

Basic Techniques

7.48 Why is the z-test of Section 7.5 inappropriate for testing H_0: $\mu = \mu_0$ when $n < 30$?

7.49 Set up the rejection region based on t for H_0: $\mu = \mu_0$ when $\alpha = .05$ and the following conditions hold:

   a H_a: $\mu < \mu_0$, $n = 15$

   b H_a: $\mu \neq \mu_0$, $n = 23$

   c H_a: $\mu > \mu_0$, $n = 6$

7.50 Repeat Exercise 7.49 with $\alpha = .01$.

7.51 The sample data for a t-test of H_0: $\mu = 15$ and H_a: $\mu > 15$ are $\bar{y} = 16.2$, $s = 3.1$, and $n = 18$. Use $\alpha = .05$ to draw your conclusions.

Applications

7.52 Each person in a random sample of 10 students in a fourth-grade reading class was thoroughly tested to determine reading speed and reading comprehension. Based on a fixed-length standardized test reading passage, the following speeds (in minutes) and comprehension scores (based on a 100-point scale) were obtained.

Student	1	2	3	4	5	6	7	8	9	10
Reading Speed	5	7	15	12	8	7	10	11	13	9
Reading Comprehension	60	76	96	100	81	75	85	88	98	83

   a Use the reading speed data to place a 95% confidence interval on μ, the average speed for all fourth-grade students in the large school from which the sample was drawn.

   b Interpret the interval estimate in part a.

   c How would your inference change if you used a 98% confidence interval?

7.53 Refer to Exercise 7.52. Using the reading comprehension data, test the research hypothesis that the mean for all fourth graders on the standardized examination is greater than 80, the statewide average for comparable students the previous year. Give the level of significance for your test. Interpret your findings.

7.54 Refer to Exercise 7.53.

   a Set up all parts for a statistical test of the research hypothesis that the mean score for all fourth-graders is different from 80, the statewide average the previous year.

   b Give the level of significance for this test.

7.55   The amount of sewage and industrial pollutants dumped into a body of water affects the health of the water by reducing the amount of dissolved oxygen available for aquatic life. Suppose that weekly readings are taken from the same location in a river over a two-month period. Use the summary data given here to conduct a statistical test of the research hypothesis that the mean dissolved oxygen content is less than 5.0 parts per million, a level some scientists think is marginal for supplying enough dissolved oxygen for fish to survive.

5.100000000
4.900000000
5.600000000
4.200000000
4.800000000
4.500000000
5.300000000
5.200000000

8.000000000	sample size
4.950000000	$\bar{y}$
.2028571428	s^2
.4503966505	s

7.56   A dealer in recycled paper places empty trailers at various sites; these are gradually filled by individuals who bring in old newspapers and the like. The trailers are picked up (and replaced by empties) on several schedules. One such schedule involves pickup every second week. This schedule is desirable if the average amount of recycled paper is more than 1600 cubic feet per two-week period. The dealer's records for 18 two-week periods show the following volumes (in cubic feet) at a particular site:

1660	1820	1590	1440	1730	1680	1750	1720	1900
1570	1700	1900	1800	1770	2010	1580	1620	1690

$(\bar{y} = 1718.3, s = 137.8)$

Assume that these figures represent the results of a random sample. Do they support the research hypothesis that $\mu > 1600$, using $\alpha = .10$? Write out all parts of the hypothesis-testing procedure.

7.57   Place an upper bound on the p-value of Exercise 7.56. Would you say that $\mu > 1600$ is strongly supported?

7.58   A federal regulatory agency is investigating an advertised claim that a certain device can increase the gasoline mileage of cars. Seven such devices are purchased and installed in seven cars belonging to the agency. Gasoline mileage for each of the cars under standard conditions is recorded both before and after installations.

				Car			
	1	2	3	4	5	6	7
Mi/gal before	19.1	19.9	17.6	20.2	23.5	26.8	21.7
Mi/gal after	20.0	23.7	18.7	22.3	23.8	19.2	24.6
Change	.9	3.8	1.1	2.1	.3	−7.6	2.9

The mean change is .50 miles per gallon, and the standard deviation is 3.77.

a   Formulate appropriate null and research hypotheses.

b   Is the advertised claim supported at $\alpha = .05$? Carry out the steps of a hypothesis test, and draw a conclusion.

7.9 Assumptions Underlying the Analysis Methods of This Chapter

Now that we have looked at the first methods for analyzing data (dealing in particular with methods and inferences concerning a population mean μ), what assumptions must we make in order to use these methods and draw these inferences?

Any statistical method for analyzing data involves assumptions. Some assumptions are general and apply to many different methods for analyzing data; others are specific to a particular method. What assumptions apply to the methods of this chapter?

First, the methods of this chapter apply only to data obtained from a random sample. For example, the plus-or-minus factor that is part of a confidence interval is an allowance for random error, not an allowance for biases inherent in the method of data collection. This underscores how important the first step—collecting the data—is to the validity of the inferences we derive from the sample data. If the data have been collected in a haphazard way, the confidence interval (or test) computed from these data is likely to be in error, simply because of biases in the data collection. There are no known ways to compensate for the biases inherent in badly chosen samples.

If the data collection stage has been handled well and you have indeed obtained a random sample, two assumptions must be considered for the analysis methods of this chapter: *independence* of the measurements within a sample, and *normality* of the underlying population from which the sample measurements were drawn.

All the methods described in this chapter assume that the measurements in a sample are independent of each other. Not all random-sampling methods yield independent observations. For example, suppose that a real-estate assessor chooses 22 city blocks of homes to evaluate from the tax lists of a city, and then assesses the market value of all homes in each block. Assuming that the assessor does, in fact, choose the blocks randomly, there is no systematic bias in favor of low-value homes or high-value homes. But there is a dependence problem. Given the well-established tendency of high-value homes to cluster together (and also of low-value homes to occur in bunches), if one home in the sample has higher-than-average value, so do adjacent homes. The assessment may involve 300 homes, but the method does not give 300 separate, independent measurements of home values. In fact, the data arising from the assessor's evaluations would be more appropriately evaluated by the cluster-sampling methods described in Chapter 2.

The most common problem with the assumption of independence occurs in time-series data—data collected in a well-defined chronological order. Suppose, for example, that we measure the dollar volume of back orders for a particular manufacturer on twenty consecutive Friday afternoons. It is reasonable to suppose that a high back-order volume on one Friday is likely to be followed by high back-order volumes on succeeding Fridays; and the same goes for low volumes. The standard error formulas that we use in confidence intervals depend heavily on the assumption of independence of observations. When there is dependence, the standard error formulas may underestimate the actual uncertainty in an estimate. Indeed, even for most dependencies, the degree of underestimation may be serious.

Beyond the assumption of independence, methods for analyzing data concerning a population mean μ involve an assumption that the underlying population is normally distributed. In practice no population is exactly normal.

How restrictive is this assumption? In practice, provided that the population of measurements is symmetric, the results based on a t distribution will hold. This property of the t distribution and the common occurrence of mound-shaped distributions in practice make the Student's t an invaluable tool for use in statistical inference.

How does one know what to do in practice? The answer is quite simple—look at the data. If a plot of the data (such as a histogram) suggests any gross skewness, there is a good chance that the population is skewed, and hence the t method should not be used. In situations such as this, you should consult a professional statistician for an alternative method.

7.10 Using Computers to Help Make Sense of Data

The MINITAB, EXECUSTAT, and SAS software systems have been used here to show how computers can help make sense of data by performing the calculations required for making inferences. The data from Exercise 7.52 were used with these software systems to illustrate the following confidence intervals and statistical tests:

Reading Speed: 90% and 95% confidence interval for μ.

Reading Comprehension: 90% and 95% confidence interval for μ.

Reading Comprehension: Statistical test of H_0: $\mu = 80$ versus H_a: $\mu > 80$, and a statistical test of H_0: $\mu = 80$ versus H_a: $\mu \neq 80$.

In the MINITAB output that follows, the 90% and 95% confidence interval for μ for reading speed and reading comprehension are listed and clearly marked under the appropriate "T INTERVAL." Thus the 95% confidence interval for μ on reading comprehension is 75.46 to 92.94. The statistical test results for reading comprehension are also clearly marked with t-values and p-values.

The EXECUSTAT output follows the MINITAB output; and it, too, clearly identifies the 90% and 95% confidence limits for u based on reading speed and reading comprehension. You can ignore the intervals shown in the output for the variance and for the standard deviation; these will be discussed in Chapter 9. The EXECUSTAT output for the two statistical tests based on reading comprehension are shown with the null hypothesis, alternative hypothesis, computed t-value, and corresponding p-value. Notice that the MINITAB and EXECUSTAT outputs give identical results for the confidence intervals and tests, except that MINITAB usually carries fewer places to the right of the decimal point than EXECUSTAT does.

The SAS output follows the output for EXECUSTAT. At first glance it may seem more complicated. Actually, though, the output shows the formulas for the 90% and 95% confidence intervals for reading speed and reading comprehension, as well as the computed confidence limits. The only statistical test shown for SAS is H_0: $\mu = 80$ versus H_a: $\mu \neq 80$ for reading comprehension. Except for minor differences in the fourth and fifth decimal places, the results of the EXECUSTAT and SAS programs are identical; and rounded to two decimal places, all three systems give exactly the same limits, t-values, and p-values.

The important thing to remember is that these systems do the calculations, but you must make the inferences. We don't yet have expert systems that can do everything for us. Consequently, you must make sense of the data; these computer software systems can help by doing the (sometimes long and tedious) calculations.

MINITAB Output

```
MTB > NAME C1 'SPEED' C2 'COMP'
MTB > READ C1 C2
DATA>   5    60
DATA>   7    76
DATA>  15    96
DATA>  12   100
DATA>   8    81
DATA>   7    75
DATA>  10    85
DATA>  11    88
DATA>  13    98
DATA>   9    83
DATA> END
     10 ROWS READ

MTB > TINTERVAL 90 PERCENT C1

               N      MEAN    STDEV   SE MEAN    90.0 PERCENT C.I.
SPEED         10     9.700    3.093     0.978   (   7.907,  11.493)

MTB > TINTERVAL 95 PERCENT C1

               N      MEAN    STDEV   SE MEAN    95.0 PERCENT C.I.
SPEED         10     9.700    3.093     0.978   (   7.487,  11.913)

MTB > TINTERVAL 90 PERCENT C2

               N      MEAN    STDEV   SE MEAN    90.0 PERCENT C.I.
COMP          10     84.20    12.22     3.86   (  77.12,   91.28)

MTB > TINTERVAL 95 PERCENT C2

               N      MEAN    STDEV   SE MEAN    95.0 PERCENT C.I.
COMP          10     84.20    12.22     3.86   (  75.46,   92.94)

MTB > TTEST MU=80 C2;
SUBC> ALTERNATIVE +1.

TEST OF MU = 80.000 VS MU G.T. 80.000

               N      MEAN    STDEV   SE MEAN        T   P VALUE
COMP          10    84.200   12.218     3.864     1.09      0.15

MTB > TTEST MU=80 C2

TEST OF MU = 80.000 VS MU N.E. 80.000

               N      MEAN    STDEV   SE MEAN        T   P VALUE
COMP          10    84.200   12.218     3.864     1.09      0.31

MTB > STOP
```

EXECUSTAT Output

```
One Sample Analysis for SPEED
Confidence Intervals

90% confidence intervals
    Mean: (7.90704,11.493)
    Variance: (5.09395,25.9708)
    Std. deviation: (2.25698,5.09616)

95% confidence intervals
    Mean: (7.48739,11.9126)
    Variance: (4.52677,32.1914)
    Std. deviation: (2.12762,5.67375)


One Sample Analysis for COMP
Confidence Intervals

90% confidence intervals
    Mean: (77.1172,91.2828)
    Variance: (79.4917,405.278)
    Std. deviation: (8.91581,20.1315)

95% confidence intervals
    Mean: (75.4595,92.9405)
    Variance: (70.6407,502.35)
    Std. deviation: (8.4048,22.4132)

Hypothesis Test - Mean COMP

Null hypothesis: mean = 80
Alternative: greater than

Computed t statistic = 1.08702
            P value = 0.1526


Hypothesis Test - Mean COMP

Null hypothesis: mean = 80
Alternative: not equal

Computed t statistic = 1.08702
            P value = 0.3053
```

SAS Output

```
OPTIONS NODATE NONUMBER PS=60 LS=78;
TITLE1 'FOURTH-GRADERS READING SPEED AND COMPREHENSION';

DATA READING;
INPUT SPEED COMP;
CARDS;
 5    60
 7    76
15    96
12   100
 8    81
 7    75
10    85
11    88
13    98
 9    83
;
PROC PRINT N;
TITLE2 'LISTING OF THE DATA';

PROC MEANS / NOPRINT;
 VAR SPEED COMP;
 OUTPUT OUT=DATA1 MEAN=MSPEED MCOMP STDERR=SSPEED SCOMP;

DATA DATA2;
  SET DATA1;
  L90SPEED=MSPEED - (1.833)*SSPEED;
  H90SPEED=MSPEED + (1.833)*SSPEED;
  L95SPEED=MSPEED - (2.262)*SSPEED;
  H95SPEED=MSPEED + (2.262)*SSPEED;
  L90COMP=MCOMP - (1.833)*SCOMP;
  H90COMP=MCOMP + (1.833)*SCOMP;
  L95COMP=MCOMP - (2.262)*SCOMP;
  H95COMP=MCOMP + (2.262)*SCOMP;

PROC PRINT;
TITLE2 'CONFIDENCE INTERVALS';
TITLE3 '90% AND 95% FOR SPEED AND COMP';
  VAR L90SPEED H90SPEED L95SPEED H95SPEED
      L90COMP H90COMP L95COMP H95COMP;

DATA DATA3;
  SET READING;
  NEWCOMP = COMP-80;

PROC MEANS T PRT;
TITLE1 'COMPREHENSION T-TEST';
TITLE2 'NULL HYPOTHESIS EQUAL 80';
TITLE3 'ALTERNATIVE HYPOTHESIS NOT EQUAL 80';
  VAR NEWCOMP;

RUN;
```

```
FOURTH-GRADERS READING SPEED AND COMPREHENSION
             LISTING OF THE DATA

              OBS      SPEED      COMP

               1         5         60
               2         7         76
               3        15         96
               4        12        100
               5         8         81
               6         7         75
               7        10         85
               8        11         88
               9        13         98
              10         9         83

                      N = 10
```

```
     FOURTH-GRADERS READING SPEED AND COMPREHENSION
                  CONFIDENCE INTERVALS
            90% AND 95% FOR SPEED AND COMP

OBS L90SPEED H90SPEED L95SPEED H95SPEED L90COMP H90COMP L95COMP H95COMP

 1   7.90715  11.4928  7.48755  11.9124  77.1177 91.2823 75.4601 92.9399
```

```
                 COMPREHENSION T-TEST
               NULL HYPOTHESIS EQUAL 80
          ALTERNATIVE HYPOTHESIS NOT EQUAL 80

             Analysis Variable : NEWCOMP


             N Obs            T   Prob>|T|
             -----------------------------
              10      1.0870150     0.3053
             -----------------------------
```

Summary

Chapter 7 deals with the third step in making sense of data—analyzing data. We introduced you to important concepts about making inferences (in the form of estimates or decisions) based on sample data. First we considered how to construct a confidence interval for μ when σ is known. The corresponding statistical test for μ when σ is known is composed of five parts: null hypothesis, alternative (or research) hypothesis, test statistic, rejection region, and conclusion. It employs the technique of proof by contradiction.

To support the research hypothesis, we gather information to contradict the null hypothesis $H_0: \mu = \mu_0$. As with any two-decision problem, two types of errors can be committed: the rejection of H_0 when H_0 is true (a Type I error) and the acceptance of H_0 when H_0 is false and some alternative is true (a Type II error). The probabilities for these errors, designated by α and β, respectively, measure the goodness of the test procedure.

We also considered an alternative to the traditional decision-based approach for a statistical test for a hypothesis. Rather than relying on a preset level of α, we compute the weight of evidence for rejecting the null hypothesis. This weight, expressed in terms of a probability, is called the level of significance for the test. Most professional journals summarize the results of a statistical test using the level of significance.

The final topic we discussed involved inferences about μ when σ is unknown (which is almost always the case). Through use of the t distribution, we can construct both confidence intervals and a statistical test for μ. Since the t-values of the t distribution approach the z-values of a normal distribution as the sample size n increases, and since σ is almost never known, it is convenient to use t results for all inferences about μ (large or small sample).

Key Terms

estimation	statistical test
hypothesis testing	specifying α
confidence coefficient	Type I error
95% confidence interval	Type II error
99% confidence interval	two-tailed test
$(1 - \alpha)$ confidence coefficient	one-tailed test
$z_{\alpha/2}$	test for population mean
control chart	computing β
quality control	μ_0
center line	μ_a
lower confidence limit (LCL)	level of significance
upper confidence limit (UCL)	p
research hypothesis (H_a)	p-value for one-tailed test
null hypothesis (H_0)	p-value for two-tailed test
test statistic (T.S.)	Student's t
rejection region (R.R.)	degrees of freedom

Key Formulas

1 $100(1 - \alpha)\%$ confidence interval for μ (σ known):

$$\bar{y} \pm z_{\alpha/2}\sigma_{\bar{y}}, \quad \text{where} \quad \sigma_{\bar{y}} = \sigma/\sqrt{n}.$$

2 $100(1 - \alpha)\%$ confidence interval for μ (σ unknown):

$$\bar{y} \pm t_{\alpha/2}s/\sqrt{n}, \quad \text{df} = n - 1$$

3 Sample size for estimating μ with $100(1 - \alpha)\%$ confidence interval, $\bar{y} \pm E$:

$$n = \frac{(z_{\alpha/2})^2 \sigma^2}{E^2}.$$

4 Statistical test for μ (σ known):

$$H_0: \quad \mu = \mu_0$$
$$\text{T.S.:} \quad z = \frac{\bar{y} - \mu_0}{\sigma/\sqrt{n}}.$$

5 Statistical test for μ (σ unknown):

$$H_0: \quad \mu = \mu_0$$
$$\text{T.S.:} \quad t = \frac{\bar{y} - \mu_0}{s/\sqrt{n}}, \qquad \text{df} = n - 1.$$

6 Calculation of β (and equivalently power) for a test on μ:

   a One-tailed test:

$$\beta = P\left(z < z_\alpha - \frac{|\mu_0 - \mu_a|}{\sigma_{\bar{y}}}\right), \quad \text{where} \quad \sigma_{\bar{y}} = \sigma/\sqrt{n}.$$

   b Two-tailed test:

$$\beta \approx P\left(z < z_{\alpha/2} - \frac{|\mu_0 - \mu_a|}{\sigma_{\bar{y}}}\right)$$

7 Sample size for a statistical test on μ:

   a One-tailed test:

$$n = \frac{\sigma^2 (z_\alpha + z_\beta)^2}{\Delta^2}$$

   b Two-tailed test:

$$n = \sigma^2 \frac{(z_{\alpha/2} + z_\beta)^2}{\Delta^2}$$

Supplementary Exercises

7.59 To test the effectiveness of a new spray for controlling rust mites, we would like to compare the average yield for treated citrus groves with the average yield for untreated groves in previous years. A random sample of 30 1-acre groves is chosen and sprayed according to a recommended schedule. The average yield for the 30-grove sample turns out to be 830 boxes, with a standard deviation of 91. Yields from groves in the same area without rust mite maintenance spraying have averaged 760 boxes over previous years. Do these data present sufficient evidence to indicate that the mean yield for groves sprayed with the new preparation is higher than 760 boxes, the average over previous years without spraying? Is this a one-tailed or two-tailed test? Use $\alpha = .05$.

7.60   A wine manufacturer sells a cabernet wine whose label asserts that it has an alcohol content of 11%. Fifteen bottles of this wine are selected at random and analyzed for alcohol content, with a resulting mean of 10.2% and a standard deviation of 1.2%.

    a   Find a 95% confidence interval for the mean alcohol content.

    b   Based on your answer to part a, do you think the label is correct? Explain.

7.61   A paint manufacturer wishes to validate its claim that 1 gallon of its paint covers 400 square feet, so it sets up a test based on a random sample of 50 1-gallon cans of paint. The hypothesis to be tested is $H_0: \mu = 400$ versus $H_a: \mu > 400$; the significance level is $\alpha = .05$.

    a   In words, what is the parameter of interest (μ)?

    b   Give the rejection region (including critical value) for the test.

    c   If the sample of 50 cans showed an average coverage of 412 square feet and a standard deviation of 38 square feet, what would you conclude?

    d   Find the p-value of the test.

7.62   A study of the operation of a parking garage showed that, in the past, average parking duration was 220 minutes. Recently, the garage was remodeled and charges increased. The management wants to know if the changes have had any effect on mean parking duration. Thus, we wish to test $H_0: \mu = 220$ versus $H_a: \mu \neq 220$, and we will use $\alpha = .05$. A random sample of 50 cars had an average parking time of 208 minutes and a standard deviation of 40 minutes.

    a   Give the rejection region, including test-statistic formula and critical value.

    b   Give the observed value of the test statistic.

    c   Give your conclusion about changes in parking time.

    d   Give the significance level (p-value) of the test.

7.63   A random sample of 35 city buses showed the mean number of passengers (per day, per bus) to be 225, with a standard deviation of 60 passengers.

    a   Find a 95% confidence interval for the average number of passengers.

    b   In words, describe the parameter of interest in this problem, and give its value (if it is known).

    c   In words, describe a sample statistic in this problem, and give its value (if it is known).

7.64   An office manager wishes to estimate the mean time required to handle a customer complaint. A sample of 38 complaints shows a mean handling time of 28.7 minutes and a standard deviation of 12 minutes.

    a   Give a point estimate for the true mean time required to handle customer complaints.

    b   Construct a 90% confidence interval estimate for the true mean time required to handle customer complaints.

7.65   The concentration of mercury in a lake has been measured many times. This population of measurements has an average of 1.20 mg/m^3 (milligrams per cubic meter) with a standard deviation of .30 mg/m^3. Following an accident at a smelter on the shore of the lake, nine more measurements were taken. These have an average mercury concentration of 1.45 mg/m^3. Report the level of significance of the evidence from this sample that the mean mercury concentration in the lake has increased.

7.66   Answer "true" or "false" for each question.

    a   Given any particular random sample, if we form the 95% confidence interval for the sample mean, there is a 95% chance that the population mean lies in this confidence interval.

    b   If a large number of random samples are selected, and we form the 95% confidence interval for each sample mean, the population mean will lie in about 95% of these confidence intervals.

    c   If a sample size is larger than 30, there is a 95% chance that the sample mean equals the population mean.

d   If a very large number of random samples are selected, there is a 95% chance that one of the sample means is equal to the population mean.

e   The 95% confidence interval around a given sample mean is wider than the 90% confidence interval around that mean.

f   In order to prove that $\mu = \mu_0$ with a Type I error of .05, we must select a sample and fail to reject the null hypothesis H_0: $\mu = \mu_0$, using $\alpha = .05$.

g   To find the critical value for a *two-tailed* test with a Type I error of .04, we can look in Table 2 of Appendix 3 for the z-score corresponding to the area .4800.

h   To find the critical value for a *one-tailed* test with a Type I error of .02, we can look in Table 2 of Appendix 3 for the z-score corresponding to the area .4800.

i   If we rejected the null hypothesis at the $\alpha = .05$ level, we would also have rejected it at the $\alpha = .01$ level.

7.67   Answer "true" or "false" for each question. If your answer is "false," change the statement to make it true. Change only the *underlined* words.

a   A Type I error is committed when we fail to reject the null hypothesis H_0 when H_0 is actually false.

b   If we make a Type II error, we have missed detecting an event or effect when there actually was one.

c   The probability of making a Type I error is equal to β.

d   If we increase the probability of making a Type II error, we will increase the probability of making a Type I error.

7.68   Over the years, projected due dates for expectant mothers have been notoriously bad. In a recent survey of 100 mothers, the average number of days to birth beyond the projected due date was 9.2, with a standard deviation of 12.4. Use these data to find a 95% confidence interval for the mean number of days to birth beyond the due date.

7.69   Refer to Exercise 7.68. Use these data to find a 90% confidence interval for the mean number of days to birth beyond the due date.

7.70   A corporation maintains a large fleet of company cars for its salespeople. In order to determine the average number of miles driven per month by all salespeople, a random sample of 70 records was obtained. The mean and the standard deviation for the number of miles were 3250 and 420, respectively. Estimate μ, the average number of miles driven per month for all the salespeople within the corporation, using a 99% confidence interval.

7.71   The length of time required to assemble an electronic fuse was measured for 50 assemblers. The mean and the standard deviation were 3.2 minutes and .3 minutes, respectively. Give a 90% confidence interval for the mean length of time required to assemble a fuse.

7.72   The diameter of extruded plastic pipe varies about a mean value that is controlled by a machine setting. A random sample of the diameter of 50 pieces of plastic pipe gave a mean and a standard deviation of 4.05 in. and .12 in., respectively. Do the data present sufficient evidence to indicate that the mean diameter is different from 4 in.? Use $\alpha = .05$.

7.73   The manufacturer of an automatic control device claims that the device will maintain a mean room humidity of 80%. The humidity in a controlled room was recorded for a period of 30 days, and the mean and the standard deviation were found to be 78.3% and 2.9%, respectively. Do the data present sufficient evidence to contradict the manufacturer's claim? Use $\alpha = .05$.

7.74   A buyer wishes to determine whether the mean sugar content per orange shipped from a particular grove is less than .027 lb. A random sample of 50 oranges produced a mean sugar content of .025 lb and a standard deviation of .003 lb. Do the data present sufficient evidence to indicate that the mean sugar content is less than .027 lb? Use $\alpha = .05$.

7.75   One method of dealing with the electrical power shortage involves using floating nuclear power plants located a few miles offshore in the ocean. Because there is great concern about the possibility of a ship colliding with the floating (but anchored) power plant, navigation experts have stated that it would be desirable if the average number of ships per day passing within 10 miles of the proposed power site location were less than 7. To verify this hypothesis for the proposed site, a random sample of 60 days was used throughout the peak shipping months. For each day, the number of ships passing within the 10-mile limit was recorded. The sample mean and a standard deviation were 6.3 ships and 2 ships, respectively. Use these data to test the navigation experts' alternative hypothesis. Use $\alpha = .05$.

7.76   Administrative officials for a university are concerned that first-year students who take advantage of off-campus housing facilities have significantly lower grade point averages (GPA) than all first-year students at the school. After the fall quarter, the all-first-year-student average GPA was 2.1 (on a 4-point system). Since it was not possible to isolate grades for all students living in off-campus housing by university records, a random sample of 81 off-campus first-year students was obtained by tracing students through their permanent home addresses. The sample mean and standard deviation were found to be 1.92 and .2, respectively. Do these data present sufficient evidence to indicate that the average GPA for all off-campus first-year students is lower than the all-first-year-student average? Use $\alpha = .05$.

7.77   In a standard dissolution test for tablets of a particular drug product, the manufacturer must obtain the dissolution rate for a batch of tablets prior to release of the batch. Suppose that the dissolution test consists of assays for 36 individual 25-mg tablets. For each test, the tablet is suspended in an acid bath and then assayed after 30 minutes. The sample mean and standard deviation after 30 minutes are 19.8 and .42 mg, respectively. Use these data to test $H_0: \mu = 20$ (80% of the labeled amount in the tablets) against the alternative hypothesis $H_a: \mu < 20$. Use $\alpha = .05$.

7.78   Refer to Exercise 7.77. Give the level of significance for the test when the alternative hypothesis is $H_a: \mu \neq 20$.

7.79   Statistics has become a valuable tool for auditors, especially where large inventories are involved. It would be costly and time-consuming for an auditor to inventory each item in a large operation. Thus, the auditor frequently resorts to obtaining a random sample of items and using the sample results to check the validity of a company's financial statement. For example; a hospital financial statement claims an inventory that averages $300 per item. An auditor's random sample of 20 items yielded a mean and standard deviation of $160 and $90, respectively. Do the data contradict the hospital's claimed mean value per inventoried item and indicate that the average is less than $300? Use $\alpha = .05$.

7.80   Over the past 5 years, the mean time required for a warehouse to fill a buyer's order has been 25 minutes. Officials of the company believe that the length of time has increased recently, either due to a change in the workforce or due to a change in customer purchasing policies. The processing time (in minutes) was recorded for a random sample of 15 orders processed over the past month:

28	25	27	31	10
26	30	15	55	12
24	32	28	42	38

Do the data present sufficient evidence to indicate that the mean time required to fill an order has increased? Use the accompanying computer output to reach a conclusion based on $\alpha = .01$.

```
MTB > NAME C1 'TIME'
MTB > SET C1
MTB > END
MTB > PRINT 'TIME'

TIME
    28     25     27     31     10     26     30     15     55     12     24     32     28
    42     38

MTB > TTEST 25 'TIME';
SUBC> ALTERNATIVE 1.

TEST OF MU = 25.000 VS MU G.T. 25.000

               N      MEAN     STDEV    SE MEAN        T     P VALUE
TIME          15    28.200    11.441      2.954     1.08        0.15

MTB > STOP
```

7.81 Give the level of significance for the statistical test in Exercise 7.80.

7.82 If a new process for mining copper is to be put into full-time operation, it must produce an average of more than 50 tons of ore per day. A 5-day trial period gave the results shown in the accompanying table. Do these figures warrant putting the new process into full-time operation? Test by using $\alpha = .05$.

Day	1	2	3	4	5
Yield (in Tons)	50	47	53	51	52

7.83 A test was conducted to determine the length of time required for a student to read a specified amount of material. All students were instructed to read at the maximum speed at which they could still comprehend the material. Sixteen students took the test, with the following results (in minutes):

25   18   27   29   20   19   25   24
32   21   24   19   23   28   31   22

Estimate the mean length of time required for all students to read the material, using a 95% confidence interval.

7.84 A random sample of eight students participated in a psychological test of depth perception. Two markers, one labeled A and the other B, were arranged a fixed distance apart at the far end of the laboratory. One by one, the students were ushered into the room and asked to judge the distance between the two markers at the other end of the room. The sample data (in feet) were as follows:

2.1   2.2   2.6   2.3
1.8   2.3   2.4   2.5

Construct a 90% confidence interval for μ, the mean judged distance for all students for which the sample is representative.

7.85 The lifetimes (in years) of 10 automobile batteries of a certain brand are

2.4   1.9   2.0   2.1   1.8   2.3   2.1   2.3   1.7   2.0

Estimate the mean lifetime, using a 95% confidence interval.

7.86 An antibiotic drug manufacturer randomly sampled 12 different locations in the fermentation vat to determine average potency for the batch of antibiotics being prepared. Readings were as follows:

8.9   9.0   9.1   8.9
9.1   9.0   9.0   9.0
8.9   8.8   9.1   8.8

Use the accompanying computer output to estimate the mean potency for the batch, based on a 95% confidence interval. Interpret the interval.

```
MTB > SET C1
MTB > END
MTB > PRINT C1

C1
    8.9     9.0     9.1     8.9     9.1     9.0     9.0     9.0     8.9     8.8     9.1
    8.8

MTB > TINTERVAL 95% CONFIDENCE C1

                N       MEAN    STDEV   SE MEAN     95.0 PERCENT C.I.
C1             12      8.9667   0.1073   0.0310   ( 8.8985,  9.0349)

MTB > STOP
```

7.87 In a statistical test about μ, the null hypothesis was rejected. Based on this conclusion, which of the following statements are true?

a A Type I error was committed.

b A Type II error was committed.

c A Type I error could have been committed.

d A Type II error could have been committed.

e It is impossible to have committed both Type I and Type II errors.

f It is impossible that neither a Type I nor a Type II error was committed.

g Whether any error was committed is not known, but if an error was made, it was Type I.

h Whether any error was committed is not known, but if an error was made, it was Type II.

7.88 Answer "true" or "false" for each statement.

a In a test of a hypothesis, a test statistic is computed from the sample data.

b A statistical test of a hypothesis employs the technique of proof by contradiction; that is, we try to show that the alternative hypothesis is true by showing that the null hypothesis is false.

c The sample size n plays an important role in testing hypotheses because it measures the amount of data (and hence information) on which we base a decision. If the data are quite variable and n is too small, it is unlikely that we will reject the null hypothesis even when the null hypothesis is false.

7.89 Complete the following statements (more than one word may be needed).

a If we take all possible samples (of a given sample size) from a population, the distribution of sample means tends to be _____, and the mean of these sample means is equal _____.

b The larger the sample size, other things remaining equal, the _____ the confidence interval.

c The larger the confidence coefficient, other things remaining equal, the _____ the confidence interval.

d The statement "If random samples of a fixed size are drawn from any population (regardless of the form of the population distribution), as n becomes larger, the distribution of sample means approaches normality," is known as the _____.

e By failing to reject a null hypothesis that is false, one makes a _____ error.

7.90 Suppose that the tar content of cigarettes is normally distributed, with a mean of 10 mg and a standard deviation of 2.4 mg. A new manufacturing process is developed for decreasing the tar content. A sample of 16 cigarettes produced by the new process yielded a mean of 8.8 mg. Use $\alpha = .05$.

a Do a test of hypothesis to determine if the new process has significantly *decreased* the tar content. Use the following outline.

> Null hypothesis
>
> Alternative hypothesis
>
> Assumptions
>
> Rejection region(s)
>
> Test statistic and computations
>
> Conclusion
>
> > In statistical terms
> >
> > In plain English

b Based on your conclusion, could you have made

i   a Type I error?

ii a Type II error?

iii neither error?

iv both Type I and Type II errors?

7.91 The board of health of a particular state was called to investigate claims that raw pollutants were being released into the river flowing past a small residential community. By applying financial pressure, the state was able to get the violating company to make major concessions toward installing a new water purification system. In the interim, different production systems were to be initiated to help reduce the pollution level of water entering the river. To monitor the effect of the interim system, a random sample of 50 water specimens was taken throughout the month at a location downstream from the plant. If $\bar{y} = 5.0$ and $s = .70$, use the sample data to determine whether the mean dissolved oxygen count of the water (in ppm) is less than 5.2, the average reading at this location over the past year.

a List the five parts of the statistical test, using $\alpha = .05$.

b Conduct the statistical test, and state your conclusion.

7.92 Refer to Figure 7.13. Compute β for parts (b) and (c), using H_0: $\mu = 380$ and H_a: $\mu > 380$. Recall that $n = 50$ and $s = 35.2$.

7.93 As described in Exercise 7.42, an automatic merging system has been installed at the entrance ramp to a major highway. Prior to installation of the system, investigators found the average stress level of drivers to be 8.2 on a 10-point scale. After installation, a sample of 50 drivers showed $\bar{y} = 7.6$ and $s = 1.8$. Conduct a statistical test of the research hypothesis that the average stress at peak hours for drivers under the new system is less than 8.2 (the average stress level prior to the installation of the automatic merge system). Determine the level of significance of the statistical test. Interpret your findings.

7.94 The search for alternatives to oil as a major source of fuel and energy inevitably poses many environmental challenges. These challenges require solutions to problems in such areas as strip mining. Let us focus on one. If coal is to continue being used as a major source of fuel and energy, we must consider ways to keep large amounts of sulfur dioxide (SO_2) and particulates from getting into the air. This is especially important at large government and industrial operations. Here are several possibilities:

1 Build smokestacks extremely high.

2 Remove the SO_2 and particulates from the coal prior to combustion.

3 Remove the SO_2 from the gases after the coal is burned but before the gases are released into the atmosphere. This is accomplished by using a scrubber.

Several scrubbers have been developed in recent years. Suppose that a new one has been constructed and is set for testing at a given power plant. Fifty samples are obtained at various times from gases emitted from the stack. The mean SO_2 emission is .13 lb per million Btu, with a standard deviation of .05 lb. Use the sample data to construct a statistical test of the null hypothesis H_0: $\mu = .145$, the average emission level for one of the more efficient scrubbers that has been developed. Choose an appropriate alternative hypothesis, with $\alpha = .05$.

7.95 Refer to Exercise 7.15. Construct a 99% confidence interval for μ, the average number of miles between a person's high school and present permanent address for the state sample.

7.96 Refer to Exercise 7.94. Rather than being interested in testing the research hypothesis that $\mu < .145$, the average emission level for one of the more efficient scrubbers, we may wish to estimate the mean emission level for the new scrubber. Use the sample data to construct a 99% confidence interval for μ. Interpret your results.

7.97 As part of an overall evaluation of training methods, an experiment was conducted to determine the average exercise capacity of healthy male army inductees. Each male in a random sample of 35 healthy army inductees exercised on a bicycle ergometer (a device for measuring work done by the muscles) under a fixed workload until he tired. Blood pressure, pulse rate, and other indicators were carefully monitored to ensure that no one's health was in danger. The measured exercise capacities (mean time, in minutes) for the 35 inductees are as follows:

23	19	36	12	41	43	19
28	14	44	15	46	36	25
35	25	29	17	51	33	47
42	45	23	29	18	14	48
21	49	27	39	44	18	13

a Use these data to construct a 95% confidence interval for μ, the average exercise capacity for healthy male inductees. Interpret your findings.

b How would your interval change if you used a 99% confidence interval?

7.98 Using the data of Exercise 7.97, determine the number of sample observations that would be required to estimate μ to within 1 minute, using a 95% confidence interval. (*Hint:* Substitute $s = 12.36$ for σ in your calculations.)

7.99 A study was conducted to examine the effect of a preparation of mosaic virus on tobacco leaves. In a random sample of $n = 32$ leaves, the mean number of lesions was 22, with a standard deviation of 3. Use these data and a 95% confidence interval to estimate the average number of lesions for leaves affected by such a preparation of mosaic virus.

7.100 Refer to Exercise 7.99. Use the sample data to form a 99% confidence interval on μ, the average number of lesions for tobacco leaves affected by a preparation of mosaic virus.

7.101 We all remember being told, "Your fever has subsided, and your temperature has returned to normal." What do we mean by the word *normal*? Most people use the benchmark 98.6°F, but this does not apply to all people, only the "average" person. We might define a person's normal temperature as being his or her average temperature when healthy. But even this definition is cloudy because there is variation in a person's temperature throughout the day. To determine a particular subject's normal temperature, we recorded it for a random sample of 30 days. On each day selected for inclusion in the sample, the temperature reading was made at 7 A.M. The sample mean and sample standard deviation for these 30 readings were, respectively, 98.4 and .15. Assuming that the subject was healthy on all of the days examined, use these data to estimate the person's 7 A.M. "normal" temperature, using a 99% confidence interval.

7.102 Refer to the data of Exercise 7.97. Suppose that the random sample of 35 inductees was selected from a large group of new army personnel being subjected to a new (and hopefully improved) physical fitness program. Assume that previous testing with several thousand personnel over the past several years has shown an average exercise capacity of 29 minutes. Run a statistical test for the research hypothesis that the average exercise capacity is improved for the new fitness program. Give the level of significance for the test. Interpret your findings.

7.103 Refer to Exercise 7.97.

a How would the research hypothesis change if we were interested in determining whether the new program is better or worse than the physical fitness program for inductees?

b What is the level of significance for your test?

 7.104 In a random sample of 40 hospitals from a list of hospitals with over 100 semiprivate beds, a researcher collected information on the proportion of persons whose bills are covered by a group policy under a major medical insurance carrier. The sample proportions are given in the following chart:

.67	.74	.68	.63	.91	.81	.79	.73
.82	.93	.92	.59	.90	.75	.76	.88
.85	.90	.77	.51	.67	.67	.92	.72
.69	.73	.71	.76	.84	.74	.54	.79
.71	.75	.70	.82	.93	.83	.58	.84

Use the sample data to construct a 90% confidence interval on μ, the average proportion of patients per hospital with group medical insurance coverage.

7.105 Refer to Exercise 7.104. Use the same data to construct a 99% confidence interval.

7.106 Faculty members in a state university system who resign within 10 years of initial employment are entitled to receive the money paid into a retirement system, plus 4% per annum. Unfortunately, experience has shown that the state is extremely slow in returning this money. Concerned about this, a local teachers' organization decides to investigate. For a random sample of 50 employees who resigned from the state university system over the past five years, the average time between the termination date and reimbursement was 75 days, with a standard deviation of 15 days. Use the data to estimate the mean waiting time for reimbursement, using a 95% confidence interval.

7.107 Refer to Exercise 7.106. After a confrontation with the teachers' union, the state promised to make reimbursements within 60 days. Monitoring of the next 40 resignations yields an average of 58 days, with a standard deviation of 10 days. If we assume that these 40 resignations represent a random sample of the state's future performance, estimate the mean reimbursement time, using a 99% confidence interval.

7.108 Refer to Example 7.11. Compute β for $\mu = 40$. What would be your conclusion based on the magnitude of β?

7.109 Refer to Exercise 7.108. Using the values of β computed for $\mu_a = 38$ and $\mu_a = 40$, calculate the probability of a Type II error for several other values of μ_a, in order to construct a graph plotting β against μ_a.

7.110 A MINITAB program is shown here for the data of Exercise 7.97. Identify the 90% confidence interval for μ.

```
MTB > SET C6
MTB > END
MTB > PRINT C6

C6
    23      19      36      12      41      43      19      28      14      44      15      46      36
    25      35      25      29      17      51      33      47      42      45      23      29      18
    14      48      21      49      27      39      44      18      13

MTB > ZINTERVAL 90% CONFIDENCE SIGMA=12.36 C6

THE ASSUMED SIGMA =12.4

                N       MEAN      STDEV    SE MEAN    90.0 PERCENT C.I.
C6             35      30.51      12.36      2.09    (   27.07,    33.95)

MTB > STOP
```

7.111 Use the data of Exercise 7.104 and the following MINITAB output to answer questions about the confidence interval for μ.

```
MTB > SET C4
MTB > END
MTB > PRINT C4

C4
   0.67    0.73    0.59    0.93    0.92    0.82    0.75    0.51    0.81    0.54    0.85
   0.68    0.76    0.75    0.58    0.69    0.92    0.82    0.67    0.73    0.71    0.77
   0.91    0.74    0.88    0.74    0.71    0.90    0.83    0.72    0.93    0.70    0.67
   0.79    0.79    0.90    0.63    0.84    0.76    0.84

MTB > STANDARD DEVIATION C4
   ST.DEV. =       0.10861
MTB > ZINTERVAL 98% CONFIDENCE , SIGMA = 0.10861 C4

THE ASSUMED SIGMA =0.109

            N      MEAN     STDEV   SE MEAN     98.0 PERCENT C.I.
C4         40    0.7620    0.1086    0.0172   (  0.7220,   0.8020)

MTB > STOP
```

a What is the sample mean for the data shown in the computer output?

b What is the confidence coefficient for the confidence interval shown?

c What are the confidence limits, and how would you interpret the interval estimate?

7.112 A random sample of birth rates from 40 inner-city areas shows an average of 35 per thousand, with a standard deviation of 6.3. Estimate the mean inner-city birth rate. Use a 95% confidence interval.

7.113 A random sample of 30 standard metropolitan statistical areas (SMSAs) was selected, and the ratio (per 1000) of registered voters to the total number of persons 18 years and over was recorded in each area. Use the following data to test the research hypothesis that μ, the average ratio (per 1000), is different from 675, last year's average ratio. Give the level of significance for your test.

802	497	653	600	729	812
751	730	635	605	760	681
807	747	728	561	696	710
641	848	672	740	818	725
694	854	674	683	695	803

7.114 Improperly filled orders are a costly problem for mail-order houses. To estimate the mean loss per incorrectly filled order, a large firm plans to sample n incorrectly filled orders and to determine the added cost associated with each one. It is estimated that the added cost is between $40 and $400. How many incorrectly filled orders must be sampled to estimate the mean additional cost, using a 95% confidence interval of width $20?

7.115 Records from a particular hospital were examined to determine the average length of stay for patients being treated for lung cancer. Data from a sample of 100 records showed $\bar{y} = 2.1$ months and $s = 2.6$ months.

a Would a confidence interval for μ based on t be appropriate? Why or why not?

b Indicate an alternative procedure for estimating the center of the distribution.

7.116 Refer to Exercise 7.28. Graph the center line and control limits for μ, and plot the sequence of sample means shown here to determine if the process is out of control:

109	113	108	108	110	108	107	108	103	107
109	109								

7.117 Investigators would like to estimate the average annual taxable income of apartment dwellers in a city to within $500, using a 95% confidence interval. If we assume that the annual incomes for apartment dwellers have a range of $40,000, determine the number of observations that should be included in the sample.

7.118 The stated weight on the new giant-sized laundry detergent package is 42 ounces. Also displayed on the box is the following statement: "Individual boxes of this product may weigh slightly more or less than the marked weight due to normal variations incurred with high-speed packaging machines, but each day's production of detergent averages slightly above the marked weight." Discuss how you might attempt to test this claim. Or would it be simpler to modify this claim slightly for testing purposes? State all parts of your test. Would there be any way to determine in advance the sample size required to pick up a specified alternative with power equal to .90, using $\alpha = .05$?

7.119 After a decade of steadily increasing popularity, sales of automatic teller machines (ATMs) have been on the decline. In a recent month, a spot check of a random sample of 40 suppliers indicated that shipments averaged 20% lower than those for the corresponding period one year ago. Assume that the standard deviation is 6.2% and that the percentage data appear mound-shaped. Use these data to construct a 99% confidence interval on the mean percentage decrease in shipments of ATMs.

7.120 Doctors have recommended that we try to limit our caffeine intake to 200 mg or less per day. A sample of 35 office workers were asked to record their caffeine intake for a 7-day period, using the following chart.

coffee (6 oz)	100–150 mg
tea (6 oz)	40–110 mg
cola (12 oz)	30 mg
chocolate cake	20–30 mg
cocoa (6 oz)	5–20 mg
milk chocolate (1 oz)	5–10 mg

After the 7-day period, the average daily intake was obtained for each worker. The sample mean and the sample standard deviation of the daily averages were 560 mg and 160 mg, respectively. Use these data to estimate μ, the average daily intake, using a 90% confidence interval.

7.121 Refer to Exercise 7.120. How many additional observations would be needed to estimate μ to within ± 10 mg with 90% confidence?

7.122 Investigators from the Ohio Department of Agriculture recently selected a junior high school in the area and took samples of the half-pint (8-ounce) milk cartons used for student lunches. Based on 25 containers, the investigators found that the cartons were, on average, .067 ounces short of a half pint, with a standard deviation of .02.

a Use these data to test the hypothesis that the average shortfall is zero against a one-sided alternative. Give the p-value for your test.

b Although .067 ounces is only a few drops, predict the annual savings (in pints) for the dairy if it sells 3 million 8-ounce cartons of milk each year with this shortweight.

Experiences with Real Data

7.123 Refer to the Clinical Trials database on the data disk (or in Appendix 1) to construct a 95% confidence interval for the mean HAM-D total score of treatment group C. How would this interval change for a 99% confidence interval?

7.124   Using the Clinical Trials database, give a 90% confidence interval for the Hopkins Obrist cluster score for all four treatment groups.

7.125   Refer to the Insurance Claims database on your data disk. Assume that these data represent a random sample for all damage claims over the past year.

   a   Construct a 90% confidence interval for the mean damage claim.

   b   Describe, in plain English, what this interval means.

7.126   Refer to Exercise 7.125.

   a   How would your confidence interval change for a 95% confidence interval?

   b   Would your interpretation change, too? Explain.

7.127   Refer to the Patient Treatment Time database on your data disk. Assume that similar data for the entire previous year yielded a mean treatment time of 25 minutes. Use the present data to test the null hypothesis H_0: $\mu = 25$, using a two-tailed test.

   a   Give the p-value for your test.

   b   Interpret your results in English.

   c   What error might you have committed? Explain.

Inferences About $\mu_1 - \mu_2$

8.1 Introduction

The inferences we have made so far have dealt with a parameter from a single population. Quite often, however, we are faced with an inference involving a comparison of parameters from different populations. For example, we might wish to compare the mean corn crop yield for two varieties of corn, the mean nitrogen content of two different lakes, or the mean length of time between administration and eventual relief for two different antivertigo drugs.

In many sampling situations, we select independent random samples from two populations in order to compare the population means or proportions. In many cases, the statistic used for making these inferences represent the difference between the corresponding sample statistics. For example, suppose that we select independent random samples of n_1 observations from one population and n_2 observations from a second population. We will use the difference between the sample means, $(\bar{y}_1 - \bar{y}_2)$, to make an inference about the difference between the population means $(\mu_1 - \mu_2)$.

The following theorem can be used to help find the sampling distribution for the difference between sample statistics computed from independent random samples.

THEOREM 8.1

If two independent random variables y_1 and y_2 are normally distributed, with means and variances (μ_1, σ_1^2) and (μ_2, σ_2^2), respectively, the difference between the random variables will be normally distributed, with mean equal to $(\mu_1 - \mu_2)$ and variance equal to $(\sigma_1^2 + \sigma_2^2)$.

Note: The sum $(y_1 + y_2)$ of the random variables will also be normally distributed, with mean $(\mu_1 + \mu_2)$ and variance $(\sigma_1^2 + \sigma_2^2)$. ▪

Theorem 8.1 can be applied directly to find the sampling distribution of the difference between two independent sample means or two independent sample proportions. The Central Limit Theorem (discussed in Chapter 6) implies that, if independent samples of sizes n_1 and n_2 are selected from two populations 1 and 2, then, where n_1 and n_2 are large, the sampling distributions of $\bar{y}_1$ and $\bar{y}_2$ will be approximately normal, with means and variances $(\mu_1, \sigma_1^2/n_1)$ and $(\mu_2, \sigma_2^2/n_2)$, respectively. Consequently, since $\bar{y}_1$ and $\bar{y}_2$ are independent normally distributed random variables, it follows from Theorem 8.1 that the sampling distribution for the difference in the sample means, $(\bar{y}_1 - \bar{y}_2)$, will be approximately normal, with a mean

$$\mu_{\bar{y}_1 - \bar{y}_2} = \mu_1 - \mu_2,$$

a variance

$$\sigma_{\bar{y}_1 - \bar{y}_2}^2 = \sigma_{\bar{y}_1}^2 + \sigma_{\bar{y}_2}^2 = \frac{\sigma_1^2}{n_1} + \frac{\sigma_2^2}{n_2},$$

and a standard error

$$\sigma_{\bar{y}_1 - \bar{y}_2} = \sqrt{\frac{\sigma_1^2}{n_1} + \frac{\sigma_2^2}{n_2}}.$$

The sampling distribution of the difference between two independent normally distributed sample means is shown in Figure 8.1.

Properties of the Sampling Distribution for the Difference Between Two Sample Means, $(\bar{y}_1 - \bar{y}_2)$

1   The sampling distribution of $(\bar{y}_1 - \bar{y}_2)$ is approximately normal for large samples.

2   The mean of the sampling distribution, $\mu_{\bar{y}_1 - \bar{y}_2}$, is equal to the difference between the population means, $(\mu_1 - \mu_2)$.

3   The standard error of the sampling distribution is

$$\sigma_{\bar{y}_1 - \bar{y}_2} = \sqrt{\frac{\sigma_1^2}{n_1} + \frac{\sigma_2^2}{n_2}}.$$

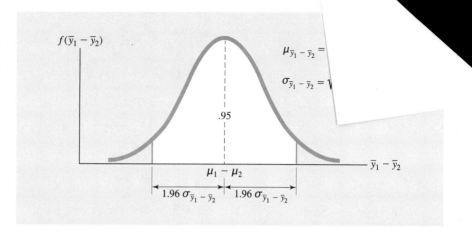

FIGURE 8.1

Sampling Distribution for the Difference Between Two Sample Means

The sampling distribution for the difference between two sample means, $(\bar{y}_1 - \bar{y}_2)$, can be used to answer the types of questions that were asked about the sampling distribution for $\bar{y}$ in Chapter 6. Since sample statistics are used to make inferences about corresponding population parameters, we can use the sampling distribution of a statistic to calculate the probability that the statistic will fall within a specified distance of the population parameter. For example, we could use the sampling distribution of the difference in sample means to calculate the probability that $(\bar{y}_1 - \bar{y}_2)$ will lie within a specified distance of the unknown difference in population means $(\mu_1 - \mu_2)$. Inferences (estimations or tests) about $(\mu_1 - \mu_2)$ will be discussed in succeeding sections of this chapter.

8.2 Inferences About $\mu_1 - \mu_2$: Independent Samples

In situations where we make inferences about $\mu_1 - \mu_2$ based on independent samples, we assume that we are sampling from two normal populations (1 and 2) that have differenct means μ_1 and μ_2 but identical variances σ^2. We than draw independent random samples of sizes n_1 and n_2. The sample means are $\bar{y}_1$ and $\bar{y}_2$; the corresponding sample variances are s_1^2 and s_2^2, respectively. Using the data from the two samples, we would like to make a comparison between the population means μ_1 and μ_2. In particular, we want to estimate and test a hypothesis concerning the difference $\mu_1 - \mu_2$.

A logical point estimate for the difference in population means is the sample difference $\bar{y}_1 - \bar{y}_2$. The standard error for the difference between sample means is more complicated to calculate than the standard error for a single sample mean, but the confidence interval has the same form; point estimate $\pm t$ (standard error). A general confidence interval for $\mu_1 - \mu_2$, with a confidence coefficient of $(1 - \alpha)$, is given next.

Confidence Interval for $\mu_1 - \mu_2$, Given Independent Samples

$$(\bar{y}_1 - \bar{y}_2) \pm t_{\alpha/2} s_p \sqrt{\frac{1}{n_1} + \frac{1}{n_2}},$$

where

$$s_p = \sqrt{\frac{(n_1 - 1)s_1^2 + (n_2 - 1)s_2^2}{n_1 + n_2 - 2}}$$

and

$$df = n_1 + n_2 - 2.$$

weighted average (s_p^2)

The quantity s_p in the confidence interval is an estiamte of the standard deviation σ for the two populations and is formed by combining (pooling) information from the two samples. In fact, s_p^2 is a **weighted average** of the sample variances s_1^2 and s_2^2. For the special case where the sample sizes are the same ($n_1 = n_2$), the formula for s_p^2 reduces to $s_p^2 = (s_1^2 + s_2^2)/2$, the mean of the two sample variances. The degrees of freedom for the confidence interval are a combination of the degrees of freedom for the two samples; that is, $df = (n_1 - 1) + (n_2 - 1) = n_1 + n_2 - 2$.

Recall that we are assuming that the two populations from which we draw the samples have normal distributions, with a common variance σ^2. Of course, if the confidence interval presented were valid only when these assumptions were met exactly, the estimation procedure would be of limited use. But fortunately, the confidence coefficient remains relatively stable if both distributions are mound-shaped and the sample sizes are approximately equal. These assumptions are discussed further at the end of this section.

EXAMPLE 8.1   Company officials wanted to know how long a particular drug retained its potency. A random sample (sample 1) of $n_1 = 10$ bottles of the product was drawn from the production line and analyzed for potency. A second sample (sample 2) of $n_2 = 10$ bottles was obtained and stored in a regulated environment for a period of one year. The readings obtained from each sample are given in Table 8.1.

TABLE 8.1
Potency Reading for Two Samples

Sample 1		Sample 2	
10.2	10.6	9.8	9.7
10.5	10.7	9.6	9.5
10.3	10.2	10.2	9.6
10.8	10.0	10.2	9.8
9.8	10.6	10.1	9.9

Suppose that we let μ_1 denote the mean potency for all bottles that might be sampled coming off the production line and we let μ_2 denote the mean potency for all bottles that may be retained for a period of one year. Estimate $\mu_1 - \mu_2$, using a 95% confidence interval.

Solution   The necessary calculations from the data of Table 8.1 are presented next.

Sample 1	Sample 2
$\sum_j y_{1j} = 103.7$	$\sum_j y_{2j} = 98.3$
$\sum_j y_{1j}^2 = 1076.31$	$\sum_j y_{2j}^2 = 966.81$

Then

$$\bar{y}_1 = \frac{103.7}{10} = 10.37 \qquad\qquad \bar{y}_2 = \frac{98.3}{10} = 9.83$$

$$s_1^2 = \frac{1}{9}\left[1076.31 - \frac{(103.7)^2}{10}\right] = .105 \qquad s_2^2 = \frac{1}{9}\left[966.81 - \frac{(98.3)^2}{10}\right] = .058.$$

The estimate of the common standard deviation σ is

$$s_p = \sqrt{\frac{(n_1 - 1)s_1^2 + (n_2 - 1)s_2^2}{n_1 + n_2 - 2}} = \sqrt{\frac{9(.105) + 9(.058)}{18}},$$

which, for $n_1 = n_2 = 9$, reduces to

$$s_p = \sqrt{\frac{.105 + .058}{2}} = .285.$$

The t-value based on df $= n_1 + n_2 - 2 = 18$ and $a = .025$ is 2.101. A 95% confidence interval for the difference in mean potencies is

$$(10.37 - 9.83) \pm 2.101(.285)\sqrt{1/10 + 1/10}, \quad \text{or} \quad .54 \pm .268.$$

We estimate that the difference in mean potency between the bottles from the production line and those stored for one year, $\mu_1 - \mu_2$, lies in the interval from .272 to .808. ∎

E X A M P L E   8.2   A study was conducted to determine whether persons in suburban district 1 have a different mean income from those in district 2. A random sample of 20 homeowners was taken in district 1. Although 20 homeowners were to be interviewed in district 2 also, 1 person refused to provide the information requested, even though the researcher promised to keep the interview confidential. So only 19 observations were obtained from district 2. The data, recorded in thousands of dollars, produced the sample means and variances shown in Table 8.2. Use these data to construct a 95% confidence interval for $\mu_1 - \mu_2$.

TABLE 8.2
Income Data for Example 8.2

Statistic	District 1	District 2
Sample size	20	19
Sample mean	18.27	16.78
Sample variance	8.74	6.58

Solution Histograms plotted for the two samples suggest that the two populations are mound-shaped (nearly normal). The sample variances are also very similar. The difference between the sample means is

$$\bar{y}_1 - \bar{y}_2 = 18.27 - 16.78 = 1.49.$$

The estimate of the common standard deviation σ is

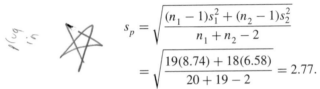

$$s_p = \sqrt{\frac{(n_1 - 1)s_1^2 + (n_2 - 1)s_2^2}{n_1 + n_2 - 2}}$$

$$= \sqrt{\frac{19(8.74) + 18(6.58)}{20 + 19 - 2}} = 2.77.$$

The t-value for $a = \alpha/2 = .025$ and df $= 20 + 19 - 2 = 37$ is not listed in Table 3 of Appendix 3 but taking the labeled value for the nearest df (df $= 40$), we have $t = 2.021$. A 95% confidence interval for the difference between the mean incomes in the two districts is of the form

$$\bar{y}_1 - \bar{y}_2 \pm t_{\alpha/2} s_p \sqrt{\frac{1}{n_1} + \frac{1}{n_2}}.$$

Substituting into the formula, we obtain

$$1.49 \pm 2.021(2.77)\sqrt{\frac{1}{20} + \frac{1}{19}}$$

or

$$1.49 \pm 1.79.$$

Thus, we estimate the difference between the mean incomes to lie somewhere in the interval from $-.30$ to 3.28. If we multiply these limits by $1000, the confidence interval for the difference between the mean incomes is $-$300 to $3280. Since this interval includes both positive and negative values for $\mu_1 - \mu_2$, we are unable to determine whether the mean income for district 1 is larger or smaller than the mean income for district 2. ▪

We can also test a hypothesis about the difference between two population means. As with any test procedure, we begin by specifying a research hypothesis for the difference between the population means. Thus, or example, we might specify that the difference $\mu_1 - \mu_2$ is greater than some value D_0. (*Note: D_0 is often 0.*) The entire test procedure is summarized here.

Statistical Test for $\mu_1 - \mu_2$, Given Independent Samples

H_0: $\mu_1 - \mu_2 = D_0$ (D_0 is specified)

H_a: ① $\mu_1 - \mu_2 > D_0$

     2   $\mu_1 - \mu_2 < D_0$

     3   $\mu_1 - \mu_2 \neq D_0$

T.S.: $t = \dfrac{\bar{y}_1 - \bar{y}_2 - D_0}{s_p\sqrt{1/n_1 + 1/n_2}}$

R.R.:   For a Type I error α and df $= n_1 + n_2 - 2$

    1   reject H_0 if $t > t_\alpha$.

    2   reject H_0 if $t < -t_\alpha$.

    3   reject H_0 if $|t| > t_{\alpha/2}$.

EXAMPLE 8.3   An experiment was conducted to compare the mean number of tapeworms in the stomachs of sheep that had been treated for worms against the mean number in those that were untreated. A sample of 14 worm-infected lambs was randomly divided into two groups. Seven were injected with the drug, and the remainder were left untreated. After a 6-month period, the lambs were slaughtered and the following worm counts were recorded:

Drug-treated Sheep	18	43	28	50	16	32	13
Untreated Sheep	40	54	26	63	21	37	39

a   Test a hypothesis that there is no difference in the mean number of worms between treated and untreated lambs. You may assume that the drug cannot increase the number of worms; hence, adopt the alternative hypothesis that the mean for treated lambs is less that the mean for untreated lambs. Use $\alpha = .05$.

b   Indicate the level of significance for this test.

Solution **a**   The calculations for the samples of treated and untreated sheep are summarized next.

5140

sum of scores

ϕ: 108600

S_2^2 6890.91

S_1^2 83 (1.1

$\mathbf{91.2 = S_2}$

S_1^2

$\left[\dfrac{\left[\sum (y_i)^2 - \dfrac{(\sum y_i)^2}{n}\right]}{n-1} \right]$

Drug-treated Sheep	Untreated Sheep
$\sum_j y_{1j} = 200$	$\sum_j y_{2j} = 280$
$\sum_j y_{1j}^2 = 6906$	$\sum_j y_{2j}^2 = 12{,}492$
$\bar{y}_1 = \dfrac{200}{7} = 28.57$	$\bar{y}_2 = \dfrac{280}{7} = 40.0$
$s_1^2 = \dfrac{1}{6}\left[6906 - \dfrac{(200)^2}{7}\right]$	$s_2^2 = \dfrac{1}{6}\left[12{,}492 - \dfrac{(280)^2}{7}\right]$
$= \dfrac{1}{6}[6906 - 5714.29]$	$= \dfrac{1}{6}[12{,}492 - 11{,}200]$
$= 198.62$	$= 215.33$

Under the assumption of equal population variances, the sample variances are combined to form an estimate of the common population standard deviation σ. This assumption appears reasonable, based on the sample variances. Hence, we have

$$s_p = \sqrt{\frac{(n_1 - 1)s_1^2 + (n_2 - 1)s_2^2}{n_1 + n_2 - 2}} = \sqrt{\frac{6(198.62) + 6(215.33)}{12}} = 14.39.$$

The test procedure for the research hypothesis that the treated sheep have a mean infestation level (μ_1) that is less than the mean level (μ_2) for untreated sheep is as follows:

H_0: $\mu_1 - \mu_2 = 0$ (that is, there is no difference in the mean infestation levels)

H_a: $\mu_1 - \mu_2 < 0$

T.S.: $t = \dfrac{\bar{y}_1 - \bar{y}_2}{s_p\sqrt{1/n_1 + 1/n_2}} = \dfrac{28.57 - 40}{14.39\sqrt{1/7 + 1/7}} = -1.49$

R.R.:   For $\alpha = .05$, the critical t-value for a one-tailed test with $df = n_1 + n_2 - 2 = 12$ can be obtained from Table 3 in Appendix 3, using $a = .05$. We will reject H_0 if $t < -1.782$.

Conclusion: Since the observed value of t, -1.49, does not fall in the rejection region, we have insufficient evidence to support rejecting the hypothesis that there is no difference in the mean number of worms in treated and untreated lambs.

b   Using Table 3 of Appendix 3, with $t = -1.49$ and df $= 12$, we see that the level of significance for this test is in the range $.05 < p < .10$.   ▪

The test procedures for comparing two population means presented in this section are based on several assumptions. The first and most critical of these is that the two samples are <u>independent</u>. Practically, we mean that the two samples are drawn from two different populations and that the elements of one sample are unrelated to those of the second sample. If this assumption is not valid, the t methods of this section are likely to be in error, and other methods (such as those presented in section 8.3) may be appropriate.

The second assumption we make is that the samples are drawn from normal populations. Fortunately, this assumption is less critical. For modest-sized samples, the Central Limit Theorem applies, and the sampling distributions for $\bar{y}_1$ and $\bar{y}_2$ are approximately normal. With independent samples and the combined sample size $n_1 + n_2 \geq 30$, the t methods of this section should be reasonably accurate, even in instances where skewness exists in the two populations. A nonparametric alternative to the t test for independent samples is presented in the next section; this alternative does not require normality.

Our third and final assumption is that the two population variances σ_1^2 and σ_2^2 are equal. For now, just examine the sample variances to see that they are approximately equal; later (in Chapter 9), we give a test for this assumption. Many efforts have been made to investigate the effect of deviations from the equal variance assumption on the t methods for independent samples. The general conclusion is that, for equal sample sizes, the population variances can differ by as much as a factor of 3 (for example, $\sigma_1^2 = 3\sigma_2^2$) and the t methods will still apply. This is remarkable and provides a convincing argument for using equal sample sizes. When the sample sizes are different, the more serious case is when the smaller sample size is associated with the larger variance. In this situation and in others where the sample variances (s_1^2 and s_2^2) suggest that $\sigma_1^2 \neq \sigma_2^2$, we can perform an approximate t test by using the test statistic

$$t' = \frac{\bar{y}_1 - \bar{y}_2}{\sqrt{\dfrac{s_1^2}{n_1} + \dfrac{s_2^2}{n_2}}}.$$

Welch (1938) showed that percentage points of a t distribution with modified degrees of freedom can be used to set the rejection region for $H_0: \mu_1 - \mu_2 = D_0$. This t test is summarized next.

Approximate t Test for Independent Samples, Unequal Variance

H_0: $\mu_1 - \mu_2 = D_0$

H_a: **1** $\mu_1 - \mu_2 > D_0$

        2 $\mu_1 - \mu_2 < D_0$

        3 $\mu_1 - \mu_2 \neq D_0$

T.S.: $t' = \dfrac{\bar{y}_1 - \bar{y}_2 - D_0}{\sqrt{\dfrac{s_1^2}{n_1} + \dfrac{s_2^2}{n_2}}}$

R.R.:   For a specified value of α,

1 reject H_0 if $t' > t_\alpha$,
2 reject H_0 if $t' < -t_\alpha$,
3 reject H_0 if $|t'| > t_{\alpha/2}$,

where

$$df = \frac{(n_1 - 1)(n_2 - 1)}{(n_2 - 1)c^2 + (1 - c)^2(n_1 - 1)}, \quad \text{where} \quad c = \frac{s_1^2/n_1}{\dfrac{s_1^2}{n_1} + \dfrac{s_2^2}{n_2}}.$$

Note: If the computed value of df is not an integer, *round down* to the nearest integer.

separate-variance *t* test

The test based on the t' statistic is sometimes referred to as the **separate-variance t test** because we use the separate sample variances s_1^2 and s_2^2, rather than a pooled sample variance, to calculate t'.

E X A M P L E   8.4   Refer to the situation explained in Example 8.3. Suppose that only 13 animals were available for analysis at the end of the treatment period. These data are shown here:

Drug-treated Sheep	5	13	18	6	4	2	15
Untreated Sheep	40	54	26	63	21	37	

Test the research hypothesis $H_a: \mu_1 - \mu_2 < 0$ under the assumption that the two population variances are different. Use $\alpha = .05$.

Solution   It is easy to verify that

$$\bar{y}_1 = 9.00 \qquad \bar{y}_2 = 40.17$$
$$s_1^2 = 38.67 \qquad s_2^2 = 258.17.$$

Then the statistical test is set up as follows:

$$H_0: \quad \mu_1 - \mu_2 = 0$$
$$H_a: \quad \mu_1 - \mu_2 < 0$$

$$\text{T.S.:} \quad t' = \frac{\bar{y}_1 - \bar{y}_2}{\sqrt{\dfrac{s_1^2}{n_1} + \dfrac{s_2^2}{n_2}}} = \frac{9 - 40.17}{\sqrt{\dfrac{38.67}{7} + \dfrac{258.17}{6}}} = -4.47.$$

In order to compute the rejection region, we need

$$c = \frac{s_1^2/n_1}{s_1^2/n_1 + s_2^2/n_2} = \frac{38.67/7}{38.67/7 + 258.17/6} = .114$$
$$c^2 = .013$$

and

$$df = \frac{(n_1 - 1)(n_2 - 1)}{(n_2 - 1)c^2 + (1 - c)^2(n_1 - 1)} = 6.283, \text{ which is rounded to 6.}$$

R.R.: For $\alpha = .05$ and df $= 6$, reject H_0 if $t' < -1.943$.

Conclusion: Since $t' = -4.47$ is less than -1.943, we reject H_0 and conclude that μ_1, the mean worm count for treated sheep, is less than that for untreated sheep. ■

Computer simulations sometimes can help us to understand some of the assumptions underlying our test procedures. One such study was done to compare the pooled t test and separate-variance t test. We'll illustrate this with an example.

EXAMPLE 8.5   For the simulation study, we assumed that we were sampling from the independent normal populations shown here:

Population 1	$\mu_1 = 100$	$\sigma_1 = 15$	$n_1 = 10$
Population 2	$\mu_2 = 100$	$\sigma_2 = 10$	$n_2 = 20$

The study proceeded as follows: A computer program was used to generate a random sample of $n_1 = 10$ observations from population 1 and a random sample of $n_2 = 20$ from population 2. The sample statistics ($\bar{y}_1, \bar{y}_2, s_1^2$, and s_2^2) were computed, as were the test statistics t and t'. This process was repeated 999 more times; and for these 1000 samples, the program kept track of the number of times t and t' rejected the null hypothesis at the upper and lower .05 levels. These results are summarized here for $H_0: \mu_1 - \mu_2 = 0$ and $\alpha = .05$.

Outcome	Pooled t Test	Separate-variance t Test
H_0 rejected for $H_a: \mu_1 - \mu_2 > 0$	75 (7.5%)	46 (4.6%)
H_0 rejected for $H_a: \mu_1 - \mu_2 < 0$	77 (7.7%)	44 (4.4%)

a   Without running the computer simulations study, which test would you have recommended, even without knowing the underlying populations? Explain.

b   What do the computer simulation results tell you about the choice of t or t'? Do they agree with your recommendation in part a?

Solution   a   With samples of sizes $n_1 = 10$ and $n_2 = 20$, we don't have much protection against the possibility of unequal variances. Hence, the separate-variance t test should be somewhat more reliable.

b   Because we know the underlying populations, one of the assumptions underlying the pooled t test—namely, equal population variances—is violated. Because the population means are equal, H_0 is true and we would have expected our tests to reject H_0 approximately 5% of the time in both the upper and lower tails, due to chance. As can be seen, the pooled t test rejected H_0 more frequently (7.5% above and 7.7% below) than we would have expected. The separate-variance t

test, on the other hand, rejected H_0 about as often as we would have expected. These results agree with our conclusion in part a.   ▪

In this section, we developed pooled-variance t methods based on an assumption of equal population variances. In addition, we introduced the t' statistic for an approximate test when the variances are not equal. Confidence intervals and hypothesis tests based on these different procedures (t and t') need not give identical results. Standard computer packages often report the results of both the pooled-variance and separate-variance t tests. Which should you believe?

If the sample sizes are equal, it doesn't matter. The alternative t tests give algebraically identical results when $n_1 = n_2$; and since the t probabilities are robust to nonnormality and unequal population variances, the test results are quite reliable when $n_1 = n_2$. When n_1 and n_2 are nearly equal, the two results are nearly equal. Only when the sample sizes vary greatly (say 1.5 to 1 or worse) do large differences appear in the results. The evidence in such cases indicates that the separate-variance methods contained in computer packages are somewhat more reliable and more conservative.

Exercises

Basic Techniques

8.1 Set up the rejection regions for testing $H_0: \mu_1 - \mu_2 = 0$ for the following conditions:

a $H_a: \mu_1 - \mu_2 \neq 0, n_1 = 12, n_2 = 14$, and $\alpha = .05$.
b $H_a: \mu_1 - \mu_2 > 0, n_1 = n_2 = 8$, and $\alpha = .01$.
c $H_a: \mu_1 - \mu_2 < 0, n_1 = 6, n_2 = 4$, and $\alpha = .05$.
What assumptions must be made prior to applying a two-sample t test?

8.2 Conduct a test of $H_0: \mu_1 - \mu_2 = 0$ against the alternative hypothesis $H_a: \mu_1 - \mu_2 < 0$ for the sample data shown here. Use $\alpha = .05$.

Statistic	Population	
	1	2
Sample size	16	13
Sample mean	71.5	79.8
Sample variance	68.35	70.26

8.3 Refer to the data of Exercise 8.2. Give the level of significance for your test.

8.4 Set up a 95% confidence interval for using the sample data shown here.

Statistic	Sample 1	Sample 2
Mean	63	54
Standard deviation	25	22
n	15	12

8.5 Refer to Exercise 8.4.

a Perform the calculations for a 95% confidence interval.

b Do the same for a 90% and a 99% confidence interval.

c How do the intervals change?

Applications

8.6 In an effort to link cold environments with hypertension in humans, researchers conducted a preliminary experiment to investigate the effect of cold on hypertension in rats. Two random samples of 6 rats each were exposed to different environments. One sample of rats was held in a normal environment at 26°C. The other sample was held in a cold 5°C environment. Blood pressures and heart rates were measured for all rats in both groups. The blood pressures for the 12 rats are shown in the accompanying table. Do the data provide sufficient evidence to indicate that rats exposed to a 5°C environment have a higher mean blood pressure than rats exposed to a 26°C environment? Test by using $\alpha = .05$.

	26°		5°
Rat	**Blood Pressure**	**Rat**	**Blood Pressure**
1	152	7	384
2	157	8	369
3	179	9	354
4	182	10	375
5	176	11	366
6	149	12	423

8.7 A pollution-control inspector suspected that a riverside community was releasing semi-treated sewage into a river and that this, as a consequence, was changing the level of dissolved oxygen in the river. To check this, he drew five randomly selected speciments of river water at a location above the town and another five specimens below. The dissolved oxygen readings, in parts per million, are given in the accompanying table. Do the data provide sufficient evidence to indicate a difference in mean oxygen content between locations above and below the town? Use $\alpha = .05$.

Above Town	4.8	5.2	5.0	4.9	5.1
Below Town	5.0	4.7	4.9	4.8	4.9

8.8 A petroleum corporation was interested in running some preliminary tests to compare the performance of a new gasoline mixture to one currently on the market. Ten identical new automobiles were randomly assigned, five to gasoline 1 and five to gasoline 2. Gasoline 2 contained a mileage additive, and gasoline 1 was regular gasoline. Each automobile was filled with 10 gallons of gasoline and driven over a test course until it stopped. The mileage was recorded for each in the accompanying table.

Gasoline 1	Gasoline 2
282	284
279	285
280	286
278	277
275	283

$$\bar{y}_1 = 278.80 \qquad \bar{y}_2 = 283.00$$

Use these data to construct a 95% confidence interval for the difference in mean mileage for the two gasolines.

8.9 A sociologist gave a current-events test to four blue-collar workers and four white-collar workers. The blue-collar workers made scores of 23, 18, 22, and 21; the white-collar workers made scores of 17, 22, 19, and 18. Estimate the difference in mean scores for blue-collar and white-collar workers, using a 99% confidence interval.

8.10 Two different emission-control devices were being tested to determine the average amount of nitric oxide being emitted by an automobile over a 1-hour period of time. Twenty cars of the same model and year were selected for the study. Ten cars were randomly selected and equipped with Type I emission-control devices, and the remaining cars were equipped with Type II devices. Each of the 20 cars was then monitored for a 1-hour period to determine the amount of nitric oxide emitted.

Use the following data to test the research hypothesis that the mean level of emission for Type I devices (μ_1) is greater than the mean emission level for Type II devices (μ_2). Use $\alpha = .01$.

Type I Device		Type II Device	
1.35	1.28	1.01	0.96
1.16	1.21	0.98	0.99
1.23	1.25	0.95	0.98
1.20	1.17	1.02	1.01
1.32	1.19	1.05	1.02

8.11 It has been estimated that lead poisoning due to an unnatural craving (pica) for substances such as paint may affect as many as a quarter of a million children each year, causing them to suffer from severe, irreversible retardation. Explanations for why children voluntarily consume lead range from "improper parental supervision" to "a child's need to mouth objects." Some researchers, however, have been investigating whether the habit of eating such substances has some nutritional explanation. One such study involved a comparison of a regular diet and a calcium-deficient diet on the ingestion of a lead-acetate solution in rats. Each rat in a group of 20 rats was randomly assigned to either an experimental or a control group. Those in the control group received a normal diet, while the experimental group received a calcium-deficient diet. Each rat occupied a separate cage and was monitored to observe the quantity of .15% lead-acetate solution consumed during the study period. The sample results are summarized here.

Control Group	5.4	6.2	3.1	3.8	6.5	5.8	6.4	4.5	4.9	4.0
Experimental Group	8.8	9.5	10.6	9.6	7.5	6.9	7.4	6.5	10.5	8.3

a Plot the data for the two samples separately. Is there reason to think the assumptions for a *t* test have been violated?

b Run a test of the research hypothesis that the mean quantity of lead acetate consumed in the experimental group is greater than that consumed in the control group. Use $\alpha = .05$.

8.12 Results of a 3-year study examining the effect of various ready-to-eat breakfast cereals on dental caries (tooth decay) in adolescent children were reported by Rowe, Anderson, and Wanninger (1974). A sample of 375 adolescent children of both genders from the Ann Arbor, Michigan, public schools was enrolled (after parental consent) in the study. Each child was provided with toothpaste and boxes of different varieties of ready-to-eat cereals. Although these were brand-name cereals, each type of cereal was packaged in plain white 7-ounce boxes and labeled as wheat flakes, corn cereal, oat cereal, fruit-flavored corn puffs, corn puffs, cocoa-flavored cereal, and sugared oat cereal. Notice that the last four varieties of cereal had been sugar-coated and the others had not.

Each child received a dental examination at the beginning of the study, twice during the study, and once at the end. The response of interest was the incremental DMF surfaces—that is, the difference between the final (poststudy) and initial (prestudy) number of decayed, missing, and filled (DMF) tooth surfaces. Careful records for each participant were maintained throughout the 3 years; and at the end of the study, a person was classified as "noneater" if he or she had eaten fewer than 28 boxes of cereal throughout the study. All others were classified as "eaters." The incremental DMF surface readings for each group are summarized in the accompanying table. Use these data to test the research hypothesis that the mean incremental DMF surface for noneaters is larger than the corresponding mean for eaters. Give the level of significance for your test. Interpret your findings.

Group	Sample Size	Sample Mean	Sample Standard Deviation
Noneaters	73	6.41	5.62
Eaters	302	5.20	4.67

8.13   The study of concentration of atmospheric trace metals in isolated areas of the world has received considerable attention because of concern that humans might somehow alter the climate of the earth by changing the amount and distribution of trace metals in the atmosphere. Consider a study at the South Pole, where at 10 different sampling periods throughout a 2-month period, 10,000 standard cubic meters (scm) of air were obtained and analyzed for metal concentrations. The results associated with magnesium and europium are listed here. (*Note:* Magnesium results are in units of 10^{-9} g/scm; europium results are in units of 10^{-15} g/scm.) Note that $s > \bar{y}$ for the magnesium data. Would you expect the data to be normally distributed? Explain.

Element	Sample Size	Sample Mean	Sample Standard Deviation
Magnesium	10	1.0	2.21
Europium	10	17.0	12.65

8.14   Refer to Exercise 8.13. Could we run a t test comparing the mean metal concentrations for magnesium and europium? Why or why not?

8.15   The costs of major surgery vary substantially from one state to another due to differences in hospital fees, doctors' fees, malpractice insurance costs, and rent. A study of hysterectomy costs was done in California and Montana. Based on a random sample of 20 patient records from each state, the following sample statistics were obtained. Construct a 95% confidence interval for $\mu_1 - \mu_2$ (the California minus Montana difference).

State	Sample Mean	Sample Standard Deviation
Montana	$ 6458	$520
California	$12,690	$305

8.16   A national educational organization monitors reading proficiency for American students on a regular basis, using a scale that ranges from 0 to 500. Sample results based on 500 students per category are shown here. Use these data to make the inferences that follow. Assume that the pooled standard deviation for any comparison is 100.

Age	Gender	Sample Mean[†]	Standard Deviation
9	Male	210	100
	Female	216	90
13	Male	253	93
	Female	262	92
17	Male	283	80
	Female	293	75

[†]What the scale means: 150—Rudimentary reading skills; can follow basic directions. 200—Basic skills; can identify facts from simple paragraphs. 250—Intermediate skills; can organize information in lengthy passages. 300—Adept skills; can understand and explain complicated information. 350—Advanced skills; can understand and explain specialized materials.

a   Construct a meaningful graph that shows age, gender, and mean proficiency scores.

b   Use the sample data to place a 95% confidence interval on the difference in mean proficiencies for females and males age 17 years.

c   Compare the mean scores for females, age 13 and 17 years, using a 90% confidence interval. Does the interval include 0? Why might these means be different?

8.17   The organization alluded to in Exercise 8.16 also examined the effect of television viewing on reading proficiency scores for students of the same age categories. The sample means and sample sizes are shown here.

Hours of TV Viewing per Day		Age (Years)		
		9	13	17
0–2	$\bar{y}$	220	267	295
	n	300	310	305
3–5	$\bar{y}$	220	262	284
	n	280	260	250
6+	$\bar{y}$	202	246	270
	n	210	220	230

a   Plot the sample means on a graph, with age and hours of TV viewing per day. Can you make some general statements about the sample data? What is the effect of TV viewing on reading proficiency within a given age group? What is the influence of age within a given TV viewing category?

b   Construct 99% confidence intervals for the difference between the 0–2 and the 6+ category in each age category. Assume the same pooled standard deviation shown in Exercise 8.16. What is your conclusion?

8.3 A Nonparametric Alternative: The Wilcoxon Rank Sum Test

The two-sample t test of the previous section was based on several assumptions: independent samples, normality, and equal variances. When the assumptions of normality and equal variances are not valid, but the sample sizes are large, the results using a t (or t') test are approximately correct. There is, however, an alternative test procedure that requires less stringent assumptions. This procedure, called the **Wilcoxon rank sum test**, is discussed in this section.

Wilcoxon rank sum test

The assumptions for this test are that we have indpendent random samples taken from two populations. The Wilcoxon rank sum test provides a procedure for testing whether two populations are identical but not necessarily normal. Since the two populations are assumed to be identical under the null hypothesis, independent random samples from the respective populations should be similar. One way to measure the similarity of the samples is to rank jointly (from lowest to highest) the measurements from the combined samples and to examine the sum of the ranks for measurements in sample 1 (or, equivalently, sample 2). Under the null hypothesis of identical populations, the sum of the ranks for a sample will be proportional to the sample size. We let T denote the sum of the ranks for sample 1. Intuitively, if T is extremely small (or large), we would have evidence to reject the null hypothesis that the two populations are identical.

Under the null hypothesis, the statistic T will have a sampling distribution whose mean and variance are given by

$$\mu_T = \frac{n_1(n_1 + n_2 + 1)}{2}$$

$$\sigma_T^2 = \frac{n_1 n_2}{12}(n_1 + n_2 + 1).$$

If, in addition, both sample sizes are more than 10, the sampling distribution of T is approximately normal; this allows us to use a z-statistic in the Wilcoxon rank sum test.

The theory behind the Wilcoxon rank sum test assumes that the population distributions are continuous, so that there is zero probability that any two observations are identical. In practice, there often are ties—two or more observations with the same value. For these situations, each observation in a set of tied values receives a rank score equal to the average of the ranks for the set. For example, if two observations are tied for the ranks 3 and 4, each is given a rank of 3.5; the next higher value receives a rank of 5, and so on. When there are ties, a correction can be applied to the variance formula (see Ott 1993 for more details). From a practical standpoint, however, unless there are many ties, the correction has very little effect on the value of σ_T^2. The Wilcoxon rank sum test is summarized here.

Wilcoxon Rank Sum Test

H_0:   The two populations are identical.

H_a: **1**   Population 1 is shifted to the right of population 2.

       2   Population 1 is shifted to the left of population 2.

       3   Populations 1 and 2 have different location parameters.
         $(n_1, n_2 > 10)$

T.S.: $z = \dfrac{T - \mu_T}{\sigma_T},$

     where T denotes the sum of the ranks in sample 1.

R.R.:   For a specified value of α:

     1   Reject H_0 if $z > z_\alpha$.

     2   Reject H_0 if $z < -z_\alpha$.

     3   Reject H_0 if $|z| > z_{\alpha/2}$.

Note: This test is equivalent to the Mann–Whitney U test (Conover 1980). For smaller sample sizes, a procedure is given in Ott (1993).

EXAMPLE   8.6   Environmental engineers were interested in determining whether a cleanup project on a nearby lake was effective. Prior to initiation of the project, 12 water samples were obtained at random from the lake and analyzed for dissolved oxygen content (in ppm). Due to diurnal fluctuations in the dissolved oxygen, all measurements were obtained at the 2 P.M. peak period. The before and after data are presented in Table 8.3.

TABLE 8.3
Dissolved Oxygen Measurements
(in ppm) for Example 8.6

Before Cleanup		After Cleanup	
11.0	11.6	10.2	10.8
11.2	11.7	10.3	10.8
11.2	11.8	10.4	10.9
11.2	11.9	10.6	11.1
11.4	11.9	10.6	11.1
11.5	12.1	10.7	11.3

Use $\alpha = .05$ to test the following hypotheses:

H_0: The distributions of measurements for before and 6 months after the cleanup project began are identical.

H_a: The distribution of dissolved oxygen measurements for before the cleanup project is shifted to the right of the corresponding distribution of measurements for 6 months after initiation of the cleanup project. (It should be noted that a cleanup project has been effective in one sense if the dissolved oxygen drops over a period of time, but these levels can't get too low or fish and other aquatic life can't survive. In addition, other measures should be considered in assessing the effectiveness of the cleanup project.)

For convenience, the data have been arranged in ascending order in Table 8.3.

Solution   First we must jointly rank the combined sample of 24 observations by assigning the rank of 1 to the smallest observation, the rank of 2 to the next smallest, and so on. When two or more measurements are the same, we assign all of them a rank equal to the average of the ranks they occupy. The sample measurements and associated ranks (shown in parentheses) are listed in Table 8.4.

Since n_1 and n_2 are both greater than 10, we will use the test statistic z. If we anticipate a shift to the left in the distribution after the cleanup, we should expect the sum of the ranks for the observations in sample 1 to be large. Thus, we should reject H_0 for large values of $z = (T - \mu_T)/\sigma_T$.

Substituting into the formulas for μ_T and σ_T^2, we have

$$\mu_T = \frac{n_1(n_1 + n_2 + 1)}{2} = \frac{12(12 + 12 + 1)}{2} = 150$$

and

$$\sigma_T^2 = \frac{n_1 n_2}{12}(n_1 + n_2 + 1) = \frac{12(12)(25)}{12} = 300$$

Hence, $\sigma_T = \sqrt{300} = 17.32$. The computed value of z is

$$z = \frac{T - \mu_T}{\sigma_T} = \frac{216 - 150}{17.32} = 3.81.$$

TABLE **8.4**
Dissolved Oxygen Measurements
and Ranks for Example 8.6

Before Cleanup		After Cleanup	
11.0	(10)	10.2	(1)
11.2	(14)	10.3	(2)
11.2	(14)	10.4	(3)
11.2	(14)	10.6	(4.5)
11.4	(17)	10.6	(4.5)
11.5	(18)	10.7	(6)
11.6	(19)	10.8	(7.5)
11.7	(20)	10.8	(7.5)
11.8	(21)	10.9	(9)
11.9	(22.5)	11.1	(11.5)
11.9	(22.5)	11.1	(11.5)
12.1	(24)	11.3	(16)
	$T = 216$		

Since this value exceeds 1.645, we reject H_0 and conclude that the distribution of before-cleanup measurements is shifted to the right of the corresponding distribution of after-cleanup measurements; that is, the after-cleanup measurements of dissolved oxygen tend to be smaller than the corresponding before-cleanup measurements. ▪

As an alternative to the two-sample t test, the Wilcoxon rank sum test requires fewer assumptions. In particular, the Wilcoxon test does not require that the two populations be normally distributed—only that they be identical under H_0. When the assumptions underlying a t test hold, the t test is more likely to declare an existing difference. This only seems logical, since the t test uses the magnitudes of observations rather than just their relative magnitudes (ranks). But when the assumptions for a t test are violated, the Wilcoxon rank sum test is the more informative test and is more likely to delare a difference when one exists. This is particularly true when nonnormality of the populations is present in the form of severe skewness or extreme outliers.

Exercises

Applications

8.18   A plumbing contractor was interested in making her operation more efficient by cutting down on the average distance between service calls while maintaining at least the same level of business activity. One plumber (plumber 1) was assigned a dispatcher who monitored all his incoming requests for service and outlined a service strategy for that day. Plumber 2 was to continue as she had in the past, providing service in roughly sequential order for stacks of service calls received. The total daily mileages for these two plumbers are recorded here for a total of 18 days (3 work weeks).

Plumber 1	88.2	94.7	101.8	102.6	89.3	95.7
	78.2	80.1	83.9	86.1	89.4	71.4
	92.4	85.3	87.5	94.6	92.7	84.6

Plumber 2	105.8	117.6	119.5	126.8	108.2	114.7
	90.2	95.6	110.1	115.3	109.6	112.4
	104.6	107.2	109.7	102.9	99.1	111.5

a   Plot the sample data for each plumber, and compute $\bar{y}$ and s.

b   Based on your findings in part a, which procedure appears to be more appropriate for comparing the distributions?

8.19   An experiment was conducted to compare the weights of the combs of roosters fed two different vitamin-supplemented diets. Twenty-eight healthy roosters were randomly divided into two groups, with one group receiving diet I and the other receiving diet II. After the study period, the comb weight (in milligrams) was recorded for each rooster. These data are given here.

Diet I	73	130	115	144	127	126	112	76	68	101	126	49	110	123

Diet II	80	72	73	60	55	74	67	89	75	66	93	75	68	76

a   Use the Wilcoxon rank sum test to determine whether the distributions of comb weights differ for the two groups. Use $\alpha = .05$.

b   Can you suggest other statistical procedures that might be appropriate for analyzing the same data? Which would you suggest?

8.20   Refer to Exercise 8.19. Suppose that the experimenter was interested in determining whether the comb weights for diet I were selected from a distribution shifted above (to the right of) that for comb weights for diet II. Run an appropriate Wilcoxon rank sum test, and give the p-value. Draw a conclusion.

8.4 Inferences About $\mu_1 - \mu_2$: Paired Data

The methods we presented in the earlier sections of this chapter were appropriate for situations in which independent random samples are obtained from two populations. These methods are not appropriate for studies or experiments in which each measurement in one sample is *matched* or *paired* with a particular measurement in the other sample. In this section, we will deal with methods for analyzing "paired" data. We begin with an example.

EXAMPLE   8.7   Insurance adjusters are concerned about the high estimates they are receiving from garage I for auto repairs compared to those from garage II. To verify their suspicions, each of 15 cars recently involved in an accident was taken to both garages for separate estimates of repair costs. Use a two-sample t test to analyze these data.

Solution   To perform our calculations, we'll use the following computer output for these data.

```
MTB > NAME C1 'GARAGE1'
MTB > NAME C2 'GARAGE2'
MTB > NAME C3 'DIFFER'
MTB > SET C1
MTB > END
MTB > SET C2
MTB > END
MTB > LET C3=C1-C2
MTB > PRINT C1-C3
```

ROW	GARAGE1	GARAGE2	DIFFER
1	7.6	7.3	0.3
2	10.2	9.1	1.1
3	9.5	8.4	1.1
4	1.3	1.5	-0.2
5	3.0	2.7	0.3
6	6.3	5.8	0.5
7	5.3	4.9	0.4
8	6.2	5.3	0.9
9	2.2	2.0	0.2
10	4.8	4.2	0.6
11	11.3	11.0	0.3
12	12.1	11.0	1.1
13	6.9	6.1	0.8
14	7.6	6.7	0.9
15	8.4	7.5	0.9

paired mm-end.

```
MTB > TWOSAMPLE 95% CONFIDENCE FOR 'GARAGE1' AND 'GARAGE2'

TWOSAMPLE T FOR GARAGE1 VS GARAGE2
             N      MEAN     STDEV   SE MEAN
GARAGE1     15      6.85      3.20     0.83
GARAGE2     15      6.23      2.94     0.76

95 PCT CI FOR MU GARAGE1 - MU GARAGE2: (-1.69, 2.92)

TTEST MU GARAGE1 = MU GARAGE2 (VS NE): T= 0.55   P=0.59   DF=  27

MTB > STOP
```

From the output we see that $\bar{y}_1 = 6.85$, $\bar{y}_2 = 6.23$, and $\bar{y}_1 - \bar{y}_2 = .62$. But this difference is rather small considering the variability of the measurements ($s_1 = 3.20$, $s_2 = 2.94$). In fact, the computed t-value (.55) has a p-value of .59, indicating very little evidence of a difference in the average claim estimates for the two garages. ■

A review of the data in Table 8.5 indicates that something about the conclusion in Example 8.7 is inconsistent with our intuition. For all but 1 of the 15 cars, the estimate from garage I was higher than that from garage II. As we know from our experience with the binomial distribution, the probability of observing higher estimates by garage I in $y = 14$ or more of the $n = 15$ trials, assuming no difference ($p = .5$) for garages I and II, is

$$P(y = 14 \text{ or } 15) = P(y = 14) + P(y = 15)$$
$$= \binom{15}{14}(.5)^{14}(.5) + \binom{15}{15}(.5)^{15}.$$

TABLE 8.5
Repair Estimates (in Hundreds of Dollars) for Example 8.7

Car	Garage I	Garage II
1	7.6	7.3
2	10.2	9.1
3	9.5	8.4
4	1.3	1.5
5	3.0	2.7
6	6.3	5.8
7	5.3	4.9
8	6.2	5.3
9	2.2	2.0
10	4.8	4.2
11	11.3	11.0
12	12.1	11.0
13	6.9	6.1
14	7.6	6.7
15	8.4	7.5

Totals $\bar{y}_1 = 6.85$ $\bar{y}_2 = 6.23$

This probability is .000, to three decimal places. Using this binomial probability, we would argue that the observed sample results are highly contradictory to the null hypothesis of equality of estimates for the two garages. Where did we go wrong? Why are there such conflicting results?

The explanation of the difference in the conclusion for a t test and the conclusion based on the binomial distribution is that one of the t test's basic assumptions, independent samples, has been violated by the way the experiment was conducted. The adjusters obtained a measurement from both garages for each car, rather than having a random sample of 15 cars examined by garage I and a second sample of cars examined by garage II.

As you can see from the data in Table 8.5, the repair estimates for a given car are about the same but they vary considerably from car to car. These differences caused large variability among estimates for a given garage and tended to mask the relatively small (but consistent) margin of difference between estimates by the two garages. This fact was recognized when the study was planned. By having both garages give an estimate on each car, we can calculate the difference between the two garages for each car and hence avoid the effects of car-to-car variability in required repairs.

A proper analysis of the paired data in Example 8.7 makes use of the 15 difference measurements to test the null hypothesis that the mean difference, μ_d, is D_0. This hypothesis is equivalent to H_0: $\mu_1 - \mu_2 = D_0$. A of the test procedure, called a **paired t test**, is given next.

paired t test

Paired t Test

H_0: $\mu_d = D_0$

H_a: **1** $\mu_d > D_0$

         2 $\mu_d < D_0$

         3 $\mu_d \neq D_0$

T.S.: $t = \dfrac{\bar{d} - D_0}{s_d/\sqrt{n}}$,   where $\bar{d}$ and s_d are the sample mean and standard deviation of the n differences.

R.R.:   For a specified value of α and df $= n - 1$:

1   Reject H_0 if $t > t_\alpha$.

2   Reject H_0 if $t < -t_\alpha$.

3   Reject H_0 if $|t| > t_{\alpha/2}$.

E X A M P L E   **8.8**   Refer to the data of Example 8.7, and perform a paired t test. Draw a conclusion based on $\alpha = .05$.

Solution   For these data, the first four parts of the statistical test are

H_0: $\mu_d = \mu_1 - \mu_2 = 0$

H_a: $\mu_d > 0$

T.S.: $t = \dfrac{\bar{d}}{s_d/\sqrt{n}}$

R.R.:   For df $= n - 1 = 14$, reject H_0 if $t > t_{.05}$.

Before computing t, we must first calculate s_d, the sample standard deviation of the differences. We can calculate s_d by using our shortcut formula for a sample variance or by using a calculator:

$$s_d^2 = \frac{1}{n-1}\left[\sum_i d_i^2 - \frac{(\sum_i d_i)^2}{n}\right].$$

For the data of Table 8.5,

$$\sum_i d_i = .3 + 1.1 + 1.1 + \cdots + .9 = 9.2$$

$$\bar{d} = \frac{9.2}{15} = .61$$

$$\sum_i d_i^2 = (.3)^2 + (1.1)^2 + (1.1)^2 + \cdots + (.9)^2 = 7.82.$$

Hence, for $n = 15$ differences,

$$s_d^2 = \frac{1}{14}\left[7.82 - \frac{(9.2)^2}{15}\right] = .156.$$

$$s_d = \sqrt{.156} = .394.$$

Substituting into the test statistic t, we have

$$t = \frac{\bar{d} - 0}{s_d/\sqrt{n}} = \frac{.61}{.394/\sqrt{15}} = 6.00.$$

Indeed $t = 6.00$ is far beyond all tabulated t values for df $= 14$, so the p-value is less than .005; presumably p is much less than .005. We conclude that the mean repair estimate for garage I is greater than that for garage II. This conclusion agrees with our intuitive finding based on the binomial distribution.

The purpose of this discussion is not to suggest that we typically have two or more analyses that may give *very* conflicting results for a given situation. Rather, it is that the analysis must fit the experimental situation; and for this experiment, the samples are dependent, demanding that we use an analysis appropriate for dependent (paired) data. We also learned from this situation that one "design" may be better than another for a given problem. Thus, it was only possible to assess the garage-to-garage differences when the large car-to-car differences were eliminated. ▪

The corresponding general $100(1 - \alpha)\%$ confidence interval for μ_d based on paired data is shown next.

$100(1 - \alpha)\%$ Confidence Interval for μ_d Based on Paired Data

$$\bar{d} \pm \frac{t_{\alpha/2}s_d}{\sqrt{n}},$$

where n is the number of pairs of observations (and hence the number of differences) and df $= n - 1$.

The use of these t procedures depends on the assumption that the population of *differences* is normally distributed. For small samples, plot the sample differences; if severe skewness or outliers are present, the binomial test or the signed-rank test of Section 8.5 should be used.

Exercises

Basic Techniques

8.21 Consider the paired data shown here.

Pair	y_1	y_2
1	21	29
2	28	30
3	17	21
4	24	25
5	27	33

a Run a paired t test, and give the p-value for the test.

b What would your conclusion be if you used an argument related to the binomial distribution? Does it agree with the result in part a? When might these two approaches not agree?

Applications

8.22 An agricultural experiment station was interested in comparing the yields for two new varieties of corn. Because the investigators thought that there might be a great deal of variability in yield from one farm to another, each variety was randomly assigned to a different 1-acre plot on each of seven farms. The 1-acre plots were planted; the corn was harvested at maturity. The results of the experiment (in bushels of corn) are listed here.

Farm	1	2	3	4	5	6	7
Variety A	48.2	44.6	49.7	40.5	54.6	47.1	51.4
Variety B	41.5	40.1	44.0	41.2	49.8	41.7	46.8

Use these data to test the null hypothesis that there is no difference in mean yields for the two varieties of corn. Use $\alpha = .05$.

8.23 Thirty sets of identical twins were asked to participate in a one-year study designed to measure certain social attitudes. One twin from each set was randomly assigned to live in the home of a minority family, while the other twin stayed at home. After one year, each person was asked to respond to a long questionnaire designed to detect and measure well-defined attitudes. Let sample 1 denote the combined questionnaire scores for the persons who lived at home, and let sample 2 denote the set of scores for those who lived with a family from a minority class.

a Plot the sample differences. Is there any reason to believe that a t test is inappropriate?

b Test the null hypothesis

$$H_0: \quad \mu_1 - \mu_2 = 0 \text{ (the population mean scores for those not exposed and those exposed to a minority environment are identical)}$$

against the alternative hypothesis

$$H_a: \quad \mu_1 - \mu_2 \neq 0 \text{ (the population mean scores are different for the two environments).}$$

Use $\alpha = .05$.

Set of Twins	Home Environment (y_1)	Minority Environment (y_2)	Difference	Set of Twins	Home Environment (y_1)	Minority Environment (y_2)	Difference
1	78	72	7	16	90	88	2
2	75	70	5	17	89	80	9
3	68	66	2	18	73	65	8
4	92	85	7	19	61	60	1
5	55	60	−5	20	76	74	2
6	74	72	2	21	81	76	5
7	65	57	8	22	89	78	11
8	80	75	5	23	82	78	4
9	98	92	6	24	70	62	8
10	52	56	−4	25	68	73	−5
11	67	63	4	26	74	73	1
12	55	52	3	27	85	75	10
13	49	48	1	28	97	88	9
14	66	67	−1	29	95	94	1
15	75	70	5	30	78	75	3

$$\bar{y}_1 = 75.23 \quad \bar{y}_2 = 71.43 \quad \bar{d} = \bar{y}_1 - \bar{y}_2 = 3.8$$

8.24 Suppose that we wish to estimate the difference between the mean monthly salaries of male and female sales representatives. Since there is a great deal of salary variability from company to company, we might try to filter out the variability due to companies by making male–female comparisons within each company, selecting one male and one female with the required background and work experience from each company. If the range of differences in salaries (between males and females) within a company is approximately $300 per month, determine the number of companies that must be examined to estimate the difference in mean monthly salary for males and females. Use a 95% confidence interval with a half width of $5. (*Hint*: Refer to Section 7.3).

8.25 Refer to Exercise 8.24. If $n = 35$, $\bar{d} = 120$, and $s_d = 250$, construct a 90% confidence interval for μ_d, the mean difference in salaries for male and female sales representatives.

8.5 A Nonparametric Alternative: The Wilcoxon Signed-rank Test

Wilcoxon signed-rank test

The **Wilcoxon signed-rank test**, which makes use of the sign and the magnitude of the rank of the differences between pairs of measurements, provides an alternative to the paired t test. The formal null hypothesis for the the wilcoxon sign-rank test is that the population distribution of differences is symmetrical about D_0; the test is sensitive to shifts in the distribution of differences to the right or the left of D_0. In most cases, D_0 is 0; otherwise, we subtract D_0 from every measurement and proceed as if $D_0 = 0$. The test uses the nonzero differences, ranked in absolute value from lowest to highest. If two or more measurements have the same nonzero difference (ignoring sign), we assign each difference a rank equal to the average of the occupied ranks. The appropriate sign is then attached to the rank of each difference.

Before summarizing the Wilcoxon signed-rank test, we define the following notation:

n = number of pairs of observations with a nonzero difference.

T_+ = sum of the positive ranks; if there are no positive ranks, $T_+ = 0$.

T_- = sum of the negative ranks; if there are no negative ranks, $T_- = 0$.

T = smaller of T_+ and T_-, ignoring their signs.

$$\mu_T = \frac{n(n+1)}{4}.$$

$$\sigma_T = \sqrt{\frac{n(n+1)(2n+1)}{24}}.$$

The Wilcoxon signed-rank test is presented here.

Wilcoxon Signed-rank Test

H_0:   The distribution of differences is symmetrical around D_0. (D_0 is specified; usually D_0 is 0.)

H_a: **1**   The differences tend to be larger than D_0.
   2   The differences tend to be smaller than D_0.
   3   Either 1 or 2 is true (two-sided H_a).

($n \leq 50$)

T.S.:   **1** $T = |T_-|$
   2 $T = T_+$
   3 T = smaller of $|T_-|, T_+$

R.R.:   For a specified value of α (one-tailed .05, .025, .01, or .005; two-tailed .10, .05, .02, .01) and a fixed number of nonzero differences n, reject H_0 if the value of T is less than or equal to the appropriate entry in Table 10 of Appendix 3.

($n > 50$)

T.S.:   Compute the test statistic

$$z = \frac{T - \dfrac{n(n+1)}{4}}{\sqrt{\dfrac{n(n+1)(2n+1)}{24}}}.$$

R.R.:   For cases 1 and 2, reject H_0 if $z < -z_\alpha$; for case 3, reject H_0 if $z < -z_{\alpha/2}$.

E X A M P L E   8.9   Two different brands of fertilizer (A and B) were compared on each of 10 different 2-acre plots. Each plot was subdivided into 1-acre subplots, with brand A randomly assigned to one subplot and brand B to the other. A volume of 60 lb per acre of fertilizer was then applied to each subplot. The data for barley yields, in bushels per acre, are listed in Table 8.6 by fertilizer and plot.

Use the Wilcoxon signed-rank test to test the null hypothesis that the distributions of barley yields for the two brands of fertilizer are identical agianst the alternative hypothesis that they are different. Use $\alpha = .05$.

T A B L E 8.6
Barley Yields (in Bushels), by Plot and by Fertilizer, for Example 8.9

	Barley Yield		
Plot	Fertilizer A (y_1)	Fertilizer B (y_2)	Difference $(y_1 - y_2)$
1	312	346	−34
2	333	372	−39
3	356	392	−36
4	316	351	−35
5	310	330	−20
6	352	364	−12
7	389	375	14
8	313	315	−2
9	316	327	−11
10	346	378	−32

Solution   First we must rank (from lowest to highest) the absolute values of the $n = 10$ differences. These ranks appear in column 2 of Table 8.7. The appropriate sign is then attached to each rank (see column 3 in Table 8.7). The sums of the positive and negative ranks are, respectively,

$$T_+ = 4$$
$$T_- = -7 + (-10) + \cdots + (-6) = -51.$$

Thus, T (the smaller of T_+ and T_-, ignoring the sign) is 4. For a two-tailed test with

T A B L E 8.7
Rankings for the Data of Table 8.6

Plot	Rank of Difference $(y_1 - y_2)$	Rank with Appropriate Sign
1	7	−7
2	10	−10
3	9	−9
4	8	−8
5	5	−5
6	3	−3
7	4	4
8	1	−1
9	2	−2
10	6	−6

$n = 10$ and $\alpha = .05$, we see from Table 10 of Appendix 3 that we should reject H_0 if T is less than or equal to 8. Thus, we reject H_0 and conclude that the distributions of barley yields for the two brands of fertilizers are different. Barley yields for fertilizer A tend to be smaller than (to the left of) corresponding yields for fertilizer B.   ∎

The choice of an appropriate paired-sample test follows the guidelines mentioned for unpaired data in Section 8.2. If the assumptions of the t test are satisfied—in particular, if the distribution of differences is roughly normal—the t test is more powerful. If the distribution of differences is grossly skewed, the nominal t probabilities may be misleading. If the distribution is roughly symmetric but has heavy tails (as indicated by the presence of outliers), the signed-rank test may be more powerful. Often, the tests will yield essentially the same conclusion.

Even with this discussion, you might still be confused as to which statistical test (or confidence interval) you should apply in a given situation when you have a choice of two or more methods. When in doubt, do several different tests; computing costs are usually minimal, especially with the availability of many different statistical software packages such as Minitab, SAS, and Execustat. If the results from the different analyses yield different results, you should identify the peculiarities of the data set to understand why the results differ. If the results agree, and no blatant violations of assumptions occur, you should be very confident in your conclusions.

This particular "hedging" strategy is appropriate not only for paired data, but for many of the situations we have discussed. Since computer software makes it easy to run alternative analyses on the same data, concern over assumptions often can be put to rest when the alternative analyses yield essentially the same results.

Exercises

Basic Techniques

8.26   Refer to Exercise 8.23.

a   Using the data in the table, run a Wilcoxon signed-rank test. Give the p-value, and draw a conclusion.

b   Compare your conclusions here to those in Exercise 8.23. Does it make a difference which test (t or signed-rank) is used?

Applications

8.27   The effect of Benzedrine on the heart rate of dogs (in beats per minute) was examined in an experiment on 14 dogs chosen for the study. Each dog was to serve as its own control, with half of the dogs assigned to receive Benzedrine during the first study period and the other half assigned to receive a placebo (saline solution). All dogs were examined to determine their heart rates after 2 hours on the medication. After 2 weeks during which no medication was given, the dogs' regimens were reversed for the second study period. The dogs previously on Benzedrine were given a placebo, and the others received Benzedrine. Again heart rates were measured after 2 hours.

The following sample data are not arranged in the order in which they were taken but have been summarized by regimen. Use these data to test the research hypothesis that the distribution of heart rates for the dogs when receiving Benzedrine is shifted to the right of that for the same animals when on the placebo. Use a one-tailed Wilcoxon signed-rank test, with $\alpha = .05$.

Dog	Placebo	Benzedrine
1	250	258
2	271	285
3	243	245
4	252	250
5	266	268
6	272	278
7	293	280
8	296	305
9	301	319
10	298	308
11	310	320
12	286	293
13	306	305
14	309	313

8.6 Choosing Sample Sizes for Inferences About $\mu_1 - \mu_2$

Sections 7.3 and 7.5 were devoted to sample-size calulations for obtaining a confidence interval about μ with a fixed width and specified degree of confidence or for conducting a statistical test involving μ with predefined levels for α and β. Similar calculations can be made for inferences about $\mu_1 - \mu_2$; either with independent samples or with paired data. Determining the sample size for a $100(1 - \alpha)\%$ confidence interval about $\mu_1 - \mu_2$ of width $2E$ based on independent samples is possible by solving the following expression for n (we will assume that both samples are of the same size):

$$z_{\alpha/2}\sigma\sqrt{\frac{1}{n} + \frac{1}{n}} = E.$$

Notice that, in this formula, σ is the common population standard deviation and we have assumed equal sample sizes.

> **Sample Sizes for a $100(1 - \alpha)\%$ Confidence Interval for $\mu_1 - \mu_2$ of the Form $\bar{y}_1 - \bar{y}_2 \pm E$, Given Independent Samples**
>
> $$n = \frac{2z_{\alpha/2}^2\sigma^2}{E^2}$$
>
> *Note*: If σ is unknown, substitute an estimated value to get an approximate sample size.

The sample sizes obtained by using this formula are usually approximate, because we have to substitute an estimated value of σ (the common population standard deviation). Typically, this estimate must be based on an educated guess supported by information from a previous study or the range of population values.

Corresponding sample sizes for one- and two-sided tests of $H_0: \mu_1 - \mu_2 = D_0$ based on specific values of α and β are shown next.

Sample Sizes for Testing $H_0 : \mu_1 - \mu_2 = D_0$, Given Independent Samples

One-sided test: $n = 2\sigma^2 \dfrac{(z_\alpha + z_\beta)^2}{\Delta^2}$,

Two-sided test: $n = 2\sigma^2 \dfrac{(z_{\alpha/2} + z_\beta)^2}{\Delta^2}$,

where $n_1 = n_2 = n$ and the probability of a Type II error is to be $\leq \beta$ when the true difference $|\mu_1 - \mu_2| \geq \Delta$.

Note: If σ is unknown, substitute an estimated value to obtain an approximate sample size.

E X A M P L E   **8.10**   An experiment was done to determine the effect on dairy cattle of a diet supplemented with liquid whey. While no differences were noted in milk production measurements between cattle given a standard diet (7.5 kg of grain plus hay by choice) with water and those on the standard diet and liquid whey only, a considerable difference between the groups was noted in the amount of hay ingested. Suppose that we now wish to test the null hypothesis of no difference in mean hay consumption for the two diet groups of dairy cattle. For a two-tailed test with $\alpha = .05$, determine the approximate number of dairy cattle that should be included in each group if we want $\beta \leq .10$ for $|\mu_1 - \mu_2| \geq .5$. Previous experimentation has shown σ to be approximately .8.

Solution   From the description of the problem, we have $\alpha = .05$, $\beta \leq .10$ for $\Delta = |\mu_1 - \mu_2| \geq .5$, and $\sigma = .8$. Table 2 of Appendix 3 gives us $z_{.025} = 1.96$ and $z_{.10} = 1.28$. Substituting these values into the formula, we have

$$n \approx \frac{2(.8)^2(1.96 + 1.28)^2}{(.5)^2} = 53.75, \text{ or } 54.$$

We need 54 cattle per group to run the desired test.   ∎

Sample-size calculation can also be done, using the formulas shown, when $n_1 \neq n_2$. In this situation, we let n_2 be some multiple m (e.g., $m = .5$) of n_1; then we substitute $(m + 1)/m$ for 2 in the sample size formulas. After solving for n_1, $n_2 = mn_1$.

Sample sizes for estimating μ_d and for conducting a statistical test for μ_d based on paired data (differences) can be found by using the formulas of Chapter 7 for μ. The only change is that we're working with a single sample of differences rather than with a single sample of y values. For convenience, the appropriate formulas are shown here.

Sample Size Required for a $100(1 - \alpha)\%$ Confidence Interval for μ_d of the Form $\bar{d} \pm E$

$$n = \frac{z_{\alpha/2}^2 \sigma_d^2}{E^2}$$

Note: If σ_d is unknown, substitute an estimated value to obtain an approximate sample size.

Sample Sizes for One- and Two-sided Tests of H_0: $\mu_d = D_0$

One-sided test: $n = \dfrac{\sigma_d^2(z_\alpha + z_\beta)^2}{\Delta^2}$,

Two-sided test: $n = \dfrac{\sigma_d^2(z_{\alpha/2} + z_\beta)^2}{\Delta^2}$,

where the probability of a Type II error is β or less if the true difference $\mu_d \geq \Delta$.

Note: If σ_d is unknown, substitute an estimated value to obtain an approximate sample size.

8.7   Using Computers to Help Make Sense of Data

As in previous chapters, where we've used computers to do the "fuss" (tedious calculations or data manipulation) required in making sense of data, we can use available computer software to perform the data analysis methods required for drawing inferences about the difference in two population means. In this section we illustrate how several different software systems do the calculations for a given method, to familiarize you with different types of output. We also show how the same software system can be used to perform more than one data analysis method on the same data set. This can help us to examine the underlying assumptions for the methods and subsequently to arrive at an appropriate inference (conclusion) based on the sample data.

We now proceed with our example set. Suppose that a firm has a generous but rather complicated policy on end-of-year bonuses for its lower-level managerial personnel. The policy's key factor is a subjective judgment of "contribution to corporate goals." A personnel officer took samples of 24 female and 36 male managers to see if there was any difference in bonuses, expressed as a percentage of yearly salary. The data are listed in Table 8.8.

TABLE 8.8
End-of-Year Company Bonuses, by Gender of Recipient

Gender	Bonus Percentage								
F	9.2	7.7	11.9	6.2	9.0	8.4	6.9	7.6	7.4
	8.0	9.9	6.7	8.4	9.3	9.1	8.7	9.2	9.1
	8.4	9.6	7.7	9.0	9.0	8.4			
M	10.4	8.9	11.7	12.0	8.7	9.4	9.8	9.0	9.2
	9.7	9.1	8.8	7.9	9.9	10.0	10.1	9.0	11.4
	8.7	9.6	9.2	9.7	8.9	9.2	9.4	9.7	8.9
	9.3	10.4	11.9	9.0	12.0	9.6	9.2	9.9	9.0

MINITAB, EXECUSTAT, and SAS output are shown here for a two-sample t test of $H_0: \mu_1 - \mu_2 = 0$.

MINITAB Output

```
MTB > NAME C1 'FEMALE' C2 'MALE'
MTB > SET C1
DATA> END
MTB > SET C2
DATA> END
MTB > PRINT C1 C2

  ROW   FEMALE    MALE

    1      9.2    10.4
    2      7.7     8.9
    3     11.9    11.7
    4      6.2    12.0
    5      9.0     8.7
    6      8.4     9.4
    7      6.9     9.8
    8      7.6     9.0
    9      7.4     9.2
   10      8.0     9.7
   11      9.9     9.1
   12      6.7     8.8
   13      8.4     7.9
   14      9.3     9.9
   15      9.1    10.0
   16      8.7    10.1
   17      9.2     9.0
   18      9.1    11.4
   19      8.4     8.7
   20      9.6     9.6
   21      7.7     9.2
   22      9.0     9.7
   23      9.0     8.9
   24      8.4     9.2
   25              9.4
   26              9.7
   27              8.9
   28              9.3
   29             10.4
   30             11.9
   31              9.0
   32             12.0
   33              9.6
   34              9.2
   35              9.9
   36              9.0
```

```
MTB > TWOSAMPLE T FOR 'FEMALE' VS 'MALE';
SUBC> POOLED.

TWOSAMPLE T FOR FEMALE VS MALE
            N      MEAN    STDEV    SE MEAN
FEMALE     24      8.53     1.19       0.24
MALE       36      9.68     1.00       0.17
```

MINITAB Output (continued)

```
95 PCT CI FOR MU FEMALE - MU MALE: (-1.72, -0.58)

TTEST MU FEMALE = MU MALE (VS NE): T= -4.04  P=0.0002  DF=  58

POOLED STDEV =        1.08

MTB > MANN-WHITNEY FOR 'FEMALE' VS 'MALE'

Mann-Whitney Confidence Interval and Test

FEMALE     N =  24     Median =      8.5500
MALE       N =  36     Median =      9.4000
Point estimate for ETA1-ETA2 is     -1.0000
95.1 pct c.i. for ETA1-ETA2 is (-1.6002,-0.5001)
W = 481.0
Test of ETA1 = ETA2  vs.  ETA1 n.e. ETA2 is significant at 0.0002
The test is significant at 0.0002 (adjusted for ties)

MTB > STOP
```

EXECUSTAT Output

```
Two Sample Analysis for DATA.DATA by DATA.GENDER

                        FEMALE            MALE

Sample size             24                36
Mean                    8.53333           9.68333      diff. = -1.15
Variance                1.41362           1.00771      ratio = 1.4028
Std. deviation          1.18896           1.00385

95% confidence intervals
    mu1 - mu2: (-1.72026,-0.579744) assuming equal variances
    mu1 - mu2: (-1.74424,-0.555757) not assuming equal variances
    variance ratio: (0.676026,3.09135)


Hypothesis Test - Difference of Means

Null hypothesis: difference of means = 0
Alternative: not equal
Equal variances assumed: yes

Computed t statistic = -4.03675
            P value = 0.0002


Hypothesis Test - Difference of Medians

Null hypothesis: difference = 0
Alternative: not equal
Average rank of 24 values in 1 = 20.0417
Average rank of 36 values in 2 = 37.4722

Computed z statistic = 3.80188 (continuity correction applied)
            P value = 0.0001
```

SAS Output

```
OPTIONS NODATE NONUMBER PS=60 LS=78;
TITLE1 'BONUS PERCENTAGE DATA FOR FEMALES AND MALES'

DATA BONUS;
INPUT PERCENT GENDER;
CARDS;
    9.2     1
    7.7     1
   11.9     1
    6.2     1
    9.0     1
    8.4     1
    6.9     1
    7.6     1
    7.4     1
    8.0     1
    9.9     1
    6.7     1
    8.4     1
    9.3     1
    9.1     1
    8.7     1
    9.2     1
    9.1     1
    8.4     1
    9.6     1
    7.7     1
    9.0     1
    9.0     1
    8.4     1
   10.4     2
    8.9     2
   11.7     2
   12.0     2
    8.7     2
    9.4     2
    9.8     2
    9.0     2
    9.2     2
    9.7     2
    9.1     2
    8.8     2
    7.9     2
    9.9     2
   10.0     2
   10.1     2
    9.0     2
   11.4     2
    8.7     2
    9.6     2
    9.2     2
    9.7     2
    8.9     2
    9.2     2
    9.4     2
    9.7     2
    8.9     2
    9.3     2
   10.4     2
   11.9     2
```

SAS Output (continued)

```
                              9.0   2
                             12.0   2
                              9.6   2
                              9.2   2
                              9.9   2
                              9.0   2
                          ;

                          PROC TTEST;
                            CLASS GENDER;
                            VAR PERCENT;
                          RUN;

                          PROC NPAR1WAY  WILCOXON;
                            CLASS GENDER;
                            VAR PERCENT;
                          RUN;
```

BONUS PERCENTAGE DATA FOR FEMALES AND MALES

TTEST PROCEDURE

Variable: PERCENT

GENDER	N	Mean	Std Dev	Std Error
1	24	8.53333333	1.18895887	0.24269521
2	36	9.68333333	1.00384973	0.16730829

Variances	T	DF	Prob>|T|
Unequal	-3.9013	43.6	0.0003
Equal	-4.0367	58.0	0.0002

For H0: Variances are equal, F′ = 1.40   DF = (23,35)   Prob>F′ = 0.3584

BONUS PERCENTAGE DATA FOR FEMALES AND MALES

N P A R 1 W A Y   P R O C E D U R E

Wilcoxon Scores (Rank Sums) for Variable PERCENT
Classified by Variable GENDER

GENDER	N	Sum of Scores	Expected Under H0	Std Dev Under H0	Mean Score
1	24	481.0	732.0	66.1514441	20.0416667
2	36	1349.0	1098.0	66.1514441	37.4722222

Average Scores were used for Ties
Wilcoxon 2-Sample Test (Normal Approximation)
(with Continuity Correction of .5)

S= 481.000   Z= -3.78677   Prob > |Z| =   0.0002

T-Test approx. Significance =   0.0004

Kruskal-Wallis Test (Chi-Square Approximation)
CHISQ= 14.397   DF= 1   Prob > CHISQ=   0.0001

As can be seen from the different outputs of the different software systems, the sample means are 8.533 and 9.683 for females and males, respectively, and the computed value of the test statistic is -4.037, with a two-sided p-value of .0002. Except for minor differences in output format, these same results are obtained from all three software systems. Therefore, based on the two-sample t test, there appears to be a difference in mean bonus levels for female and male personnel: on average, females are receiving a lower bonus percentage than are males. Before we draw our final conclusions, however, let's check these results against those obtained from one or more additional analyses.

The MINITAB output also shows the results of a Mann–Whitney test (which gives identical results to a Wilcoxon rank sum test), and the SAS output shows the results of a Wilcoxon rank sum test comparing the bonus percentage distributions for males and females. Both test results are significant at the $p = .002$ level, suggesting that a difference exists in the distributions of bonuses for females and males. The distribution for females is shifted to the left of that for males. This is consistent with what we observed with the t test comparison of means for females and males.

Other tests shown in the output include the t and t' tests using SAS, corresponding to t tests run under the equal variance assumption (t test) and under the assumption that the variances are unequal (t' test). The similarity of the results suggests that the equal variance assumption is valid.

Other results are shown (a median test in EXECUSTAT and the Kruskal–Wallis test in SAS), but they can be ignored for now.

Since it is very easy to run several different analyses on the same data set once the data have been entered into a software system, these analyses can conveniently be used to confirm the validity of the underlying assumption(s). If, for example, a t test and the nonparametric test yield comparable results, you should be more confident in the conclusions you draw and in the assumptions you made in running the t test.

Summary

In this chapter, we have considered inferences about $\mu_1 - \mu_2$. The first set of methods was based on running a statistical test or constructing a confidence interval for $\mu_1 - \mu_2$ on independent random sample data, using t methods. The Wilcoxon rank sum test, which does not require normality of the underlying populations, was presented as an alternative to the t test.

The second major set of procedures introduced in this chapter can be used to make comparisons between two populations when the sample measurements are paired. In this situation, we no longer have independent random samples, and hence the procedures of Sections 8.2 and 8.3 (t methods and the Wilcoxon rank sum test) are inappropriate. Test and estimation methods for paired data are based on the sample differences for the paired measurements or on the ranks established by the differences. The paired t test and the corresponding confidence interval based on difference measurements were introduced and found to be identical to the single-sample t methods of Chapter 7. The nonparametric alternative to the paired t test is the Wilcoxon signed-rank test.

The material presented in Chapters 7 and 8 lays the foundation for methods of data analysis that use statistical inference (estimation and testing). Many of the data analysis methods presented in Chapters 9–13 build on what we've discussed in these two chapters. It would be a good idea to review the material in this chapter periodically as new topics are introduced, so that you retain the basic elements of statistical inference.

Key Terms

weighted average (s_p^2)
separate-variance t test
Wilcoxon rank sum test
paired t test
Wilcoxon signed-rank test

Key Formulas

1 $100(1 - \alpha)\%$ confidence interval for $\mu_1 - \mu_2$, given independent samples y_1 and y_2 approximately normal, $\sigma_1^2 = \sigma_2^2$:

$$\bar{y}_1 - \bar{y}_2 \pm t_{\alpha/2} s_p \sqrt{\frac{1}{n_1} + \frac{1}{n_2}},$$

where

$$s_p = \sqrt{\frac{(n_1 - 1)s_1^2 + (n_2 - 1)s_2^2}{n_1 + n_2 - 2}} \qquad \text{and} \qquad df = n_1 + n_2 - 2.$$

2 t test for $\mu_1 - \mu_2$, given independent samples, y_1 and y_2 approximately normal, and $\sigma_1^2 = \sigma_2^2$:

$$H_0: \quad \mu_1 - \mu_2 = D_0$$
$$\text{T.S.:} \quad t = \frac{\bar{y}_1 - \bar{y}_2 - D_0}{s_p \sqrt{1/n_1 + 1/n_2}}, \qquad df = n_1 + n_2 - 2.$$

3 t' test for $\mu_1 - \mu_2$, given unequal variance, independent samples, and y_1 and y_2 approximately normal:

$$H_0: \quad \mu_1 - \mu_2 = D_0$$
$$\text{T.S.:} \quad t' = \frac{\bar{y}_1 - \bar{y}_2 - D_0}{\sqrt{\frac{s_1^2}{n_1} + \frac{s_2^2}{n_2}}}, \qquad df = \frac{(n_1 - 1)(n_2 - 1)}{(n_2 - 1)c^2 + (1 - c)^2(n_1 - 1)},$$

where

$$c = \frac{s_1^2/n_1}{\dfrac{s_1^2}{n_1} + \dfrac{s_2^2}{n_2}}.$$

4 Wilcoxon rank sum test for independent samples:

H_0: The two populations are identical,

$(n_1 \leq 10, n_2 \leq 10)$

T.S.: T, the sum of the ranks in sample 1

$(n_1, n_2 > 10)$

T.S.: $z = \dfrac{T - \mu_T}{\sigma_T}$,

where T denotes the sum of the ranks in sample 1,

$$\mu_T = \frac{n_1(n_1 + n_2 + 1)}{2}, \quad \text{and} \quad \sigma_T = \sqrt{\frac{n_1 n_2}{12}(n_1 + n_2 + 1)}.$$

5 Paired t test, given a difference approximately normal:

H_0: $\mu_d = D_0$

T.S.: $t = \dfrac{\bar{d} - D_0}{s_d/\sqrt{n}}$ df $= n - 1$

where n is the number of differences.

6 $100(1 - \alpha)\%$ confidence interval for μ_d, given paired data and differences approximately normal:

$$\bar{d} \pm t_{\alpha/2} s_d/\sqrt{n}.$$

7 Wilcoxon signed-rank test for paired data:

H_0: The distribution of differences is symmetrical about D_0.

T.S.: $n > 50$

$$z = \frac{T - \mu_T}{\sigma_T},$$

where $\mu_T = \dfrac{n(n + 1)}{4}$ and $\sigma_T = \sqrt{\dfrac{n(n + 1)(2n + 1)}{24}}$,

provided that there are no tied ranks.

8 Independent samples: sample sizes for estimating $\mu_1 - \mu_2$ with a $100(1 - \alpha)\%$ confidence interval, $\bar{y}_1 - \bar{y}_2 \pm E$:

$$n = \frac{2z_{\alpha/2}^2 \sigma^2}{E^2}.$$

9 Independent samples: sample sizes for a test of H_0: $\mu_1 - \mu_2 = D_0$:

1   One-sided test:

$$n = \frac{2\sigma^2(z_\alpha + z_\beta)^2}{\Delta^2}$$

2   Two-sided test:

$$n = \frac{2\sigma^2(z_{\alpha/2} + z_\beta)^2}{\Delta^2}.$$

Supplementary Exercises

 8.28   Two alloys, A and B, are used in the manufacture of steel bars. We wish to estimate the difference in load capacity of bars made of each alloy. A sample of 9 bars of alloy A had a mean load capacity of 28.5 tons and a standard deviation of 2.5 tons, whereas a sample of 13 bars of alloy B had an average load capacity of 23.2 tons and a standard deviation of 1.8 tons. Find a 90% confidence interval for the difference.

 8.29   It is thought that exposure to ozone increases lung capacity. To investigate this possibility, a researcher exposed eight rats to ozone in the amount of 2 parts per million for a period of 30 days. The average lung capacity for these rats at the end of the 30 days was 9.4 ml, with a standard deviation of 0.8 ml. A control group of six rats was not exposed to ozone; their lung capacity averaged 8.3 ml, with a standard deviation of 0.7 ml.

    a   Is there sufficient evidence at a 5% significance level to support the original conjecture? Justify your answer with specific numerical values.

    b   Give the *p*-value for the hypothesis test.

 8.30   In a study of factors that may influence the frequency of birds being hit by aircraft (which, ironically, is viewed as a hazard to the aircraft), the noise level of various jets was measured just seconds after their wheels left the ground. The jets were either wide-bodied or narrow-bodied. Twenty-two wide-bodied jets had noise levels averaging 106.4 decibels (dB) and a standard deviation of 3.3 dB, whereas ten narrow-bodied jets had noise levels averaging 114.0 dB and a standard deviation of 2.0 dB. Test whether the average noise levels in the two populations of jets are the same. Report the level of significance of the sample evidence that the two types of jets have different mean noise levels.

 8.31   A farmer wanted to determine which of two soil fumigants, A or B, was more effective in controlling the number of parasites in a particular agricultural crop. To compare the fumigants, four small fields were divided into equal areas: fumigant A was applied to one part, and fumigant B to the other. Crop samples of equal size werre taken from each of the eight plots, and the numbers of parasites per square foot were counted. The data are reported in the following table. Do these data provide sufficient evidence to indicate a difference in the mean level of parasites for the two fumigants?

Field	A	B
1	15	9
2	5	3
3	8	6
4	8	4

 8.32   A psychologist wanted to compare the average length of time it takes individuals to complete two different psychological checklists. From a relatively homogeneous group of 20 individuals, 10 were randomly assigned to list 1 and the other 10 to list 2. The appropriate checklists were then administered, and the amount of time required to complete the task was recorded for each individual. These data are summarized here. Find a 95% confidence interval for $(\mu_1 - \mu_2)$, the difference in mean completion times. What assumptions must you make?

List 1	List 2
$\bar{y}_1 = 54.3$ minutes	$\bar{y}_2 = 48.1$ minutes
$n_1 = 10$	$n_2 = 10$
$s_1^2 = 16.0$	$s_2^2 = 12.2$

8.33 Refer to the data of Exercise 8.32. Construct a 99% confidence interval for $(\mu_1 - \mu_2)$.

8.34 Use the data of Exercise 8.6 to construct a 90% confidence interval for $(\mu_1 - \mu_2)$, the difference in mean blood pressure for rats subjected to the two environments.

8.35 An experiment was conducted to compare the mean lengths of time required for bodily absorption of two drugs, A and B. Ten people were randomly selected and assigned to each drug treatment. Each of the ten persons in the sample received an oral dosage of the assigned drug, and the length of time (in minutes) required for the drug to reach a specified level in the blood was recorded. The means and variances for the two samples are given in the accompanying table. Find a 95% confidence interval for the difference in mean times for absorption.

Statistic	Drug A	Drug B
Sample mean	27.2	33.5
Sample variance	16.36	18.92

8.36 The accompanying SAS computer output reports the drop in blood pressure for three groups of six rats from a strain of hypertensive rats. The six rats in the first group were treated with a low dose of an antihypertensive drug, the second group with a higher dose of the same antihypertensive drug, and the third group with an inert control. Notice that the variability in blood pressure decreased, even for rats in the control group. Also notice that negative values represent increases in blood pressure.

a Draw conclusions in a comparison of the mean drops for the high-dose group and for the control group.

b Is there evidence to indicate a difference between low- and high-dose groups? Explain.

OBS	PRESSURE	DOSE
1	-51.0	LOW
2	15.0	LOW
3	48.0	LOW
4	65.0	LOW
5	-20.0	LOW
6	75.0	LOW
7	69.0	HIGH
8	24.0	HIGH
9	63.0	HIGH
10	87.5	HIGH
11	77.5	HIGH
12	40.0	HIGH
13	9.0	CONTROL
14	12.0	CONTROL
15	36.0	CONTROL
16	77.5	CONTROL
17	-7.5	CONTROL
18	32.5	CONTROL

Analysis Variable : PRESSURE

```
------------------------------- GROUP=LOW -----------------------------------
```

N Obs	N	Mean	Std Dev
6	6	22.0000000	49.9519769

```
------------------------------- GROUP=HIGH ----------------------------------
```

N Obs	N	Mean	Std Dev
6	6	60.1666667	23.8676909

```
------------------------------- GROUP=CONTROL -------------------------------
```

N Obs	N	Mean	Std Dev
6	6	26.5833333	29.6638107

TTEST PROCEDURE

Variable: PRESSURE

GROUP	N	Mean	Std Dev	Std Error	Minimum	Maximum
LOW	6	22.00000000	49.95197694	20.39280919	-51.00000000	75.00000000
HIGH	6	60.16666667	23.86769085	9.74394399	24.00000000	87.50000000

Variances	T	DF	Prob>|T|
Unequal	-1.6887	7.2	0.1342
Equal	-1.6887	10.0	0.1222

For H0: Variances are equal, F' = 4.38   DF = (5,5)   Prob>F' = 0.1308

TTEST PROCEDURE

Variable: PRESSURE

GROUP	N	Mean	Std Dev	Std Error	Minimum	Maximum
HIGH	6	60.16666667	23.86769085	9.74394399	24.00000000	87.50000000
CONTROL	6	26.58333333	29.66381072	12.11020002	-7.50000000	77.50000000

Variances	T	DF	Prob>|T|
Unequal	2.1606	9.6	0.0573
Equal	2.1606	10.0	0.0561

For H0: Variances are equal, F' = 1.54   DF = (5,5)   Prob>F' = 0.6449

8.37   Use the data of Exercise 8.36 to construct a 95% confidence interval for $(\mu_1 - \mu_3)$—the difference in population means for the low-dose group and the control group.

8.38   The elasticity of plastic can vary, depending on the process by which the plastic is prepared. To compute the elasticity of plastic produced by two different processes, six samples from each process were analyzed for elasticity. The resulting data are shown in the accompanying table. Do these data present sufficient evidence to indicate a difference in the mean elasticities for the two processes? Use $\alpha = .05$.

Process 1	Process 2
6.1	9.1
9.2	8.2
8.7	8.6
8.9	6.9
7.6	7.5
7.1	7.9

$$\bar{y}_1 = 7.93 \quad \bar{y}_2 = 8.03$$
$$s_1^2 = 1.46 \quad s_2^2 = .61$$

8.39   The purity of ore can vary greatly from one location to another. Consequently, one determining factor in choosing a site for mining is the metal content of the ore. Three ore samples were obtained from each of two prospective locations and were analyzed to determine their metal content. The results are listed in the accompanying table. Do the data provide sufficient evidence to indicate a difference in the mean metal content for the two locations? Use $\alpha = .01$.

Location 1	50.1	49.6	51.2
Location 2	47.0	46.0	46.4

8.40   Refer to Exercise 8.39. Give the approximate level of significance for your test.

8.41   The amount of work accomplished on a construction job is frequently approximated by a visual estimate of the amount of material used per day. Six experienced men were employed to approximate the number of bricks used on two different jobs. Three men were randomly assigned to job 1, and three to job 2. Each man, independent of the others, approximated the number of bricks used. The approximations (in thousands of bricks) are shown in the accompanying table. Assume that the men have been randomly selected from a very large population of experienced people. Thus, μ_1 is the mean of the large set of approximations produced by people who could have visually estimated the number of bricks in job 1. Similarly, μ_2 is the corresponding mean of a large set of approximations that could have been acquired for job 2. Do these data provide evidence to indicate that the mean number of bricks approximated for job 1 differs from the mean number approximated for job 2? Use $\alpha = .05$.

Job 1	Job 2
107.2	103.2
108.1	105.9
105.7	104.1

$$\bar{y}_1 = 107.00 \quad \bar{y}_2 = 104.40$$
$$s_1^2 = 1.47 \quad s_2^2 = 1.89$$

8.42   Refer to Exercise 8.38. Estimate the difference in the mean elasticities of the two processes, using a 95% confidence interval.

8.43   Refer to Exercise 8.38. Estimate the difference in the mean metal content of the two locations, using a 90% confidence interval.

8.44   Refer to Exercise 8.41. Construct a 95% confidence interval for the difference in the mean estimates for the two jobs.

8.45   An experiment was conducted to investigate the effect of the drug Propranolol in reducing hypertension in rats. Two groups of rats were studied. One group received the drug, and the other group served as the control group. Hypertension was induced in the rats by exposure to a cold environment. The extent of the induced hypertension in a given rat was measured by monitoring its blood pressure. After 6 weeks of cold exposure, the rats in the sample were tested. The resulting blood pressure data were summarized for the two groups; see the accompanying table. Use these data to determine whether sufficient evidence exists to indicate that rats treated with Propranolol have less hypertension, on the average, than do untreated rats. Use $\alpha = .05$.

Statistic	Group 1 (Received Propranolol)	Group 2 (Control)
Sample size	7	5
Sample mean	129.43	167.60
Sample variance	583.95	249.30

8.46   Refer to the data of Exercise 8.8. Use the MINITAB output shown here to conduct a two-sample t test for $H_0: \mu_1 - \mu_2 = 0$ versus $H_a: \mu_1 - \mu_2 \neq 0$. Give the p-value for your test, and draw conclusions.

```
MTB > READ INTO C1 C2
DATA> 282    284
DATA> 279    285
DATA> 280    286
DATA> 278    277
DATA> 275    283
DATA> END

      5 ROWS READ
MTB > PRINT C1 C2

  ROW     C1      C2

    1     282     284
    2     279     285
    3     280     286
    4     278     277
    5     275     283

MTB > TWOSAMPLE T FOR C1 VS C2;
SUBC> POOLED.

TWOSAMPLE T FOR C1 VS C2
      N       MEAN       STDEV    SE MEAN
C1    5      278.80       2.59       1.2
C2    5      283.00       3.54       1.6

95 PCT CI FOR MU C1 - MU C2: (-8.7, 0.3)

TTEST MU C1 = MU C2 (VS NE): T= -2.14   P=0.064   DF= 8

POOLED STDEV =          3.10

MTB > STOP
```

8.47   Two judges were asked to rate separately the rehabilitative potential of each of 22 inmates. These data appear next.

Inmate	Judge 1	Judge 2	Inmate	Judge 1	Judge 2
1	6	5	12	9	8
2	12	11	13	10	8
3	3	4	14	6	7
4	9	10	15	12	9
5	5	2	16	4	3
6	8	6	17	5	5
7	1	2	18	6	4
8	12	9	19	11	8
9	6	5	20	5	3
10	7	4	21	10	9
11	6	6	22	10	11

Use the accompanying computer output to reach a conclusion about the following.

a   Consider H_0: The distribution of differences is symmetrical about 0 versus H_a: The difference tends to be larger than 0. What is your conclusion? What is the p-value?

b   How would the results of part a compare to those from a paired t test? use the accompanying output to draw conclusions.

OBS	JUDGE-1	JUDGE-2	DIFF
1	6	5	1
2	12	11	1
3	3	4	-1
4	9	10	-1
5	5	2	3
6	8	6	2
7	1	2	-1
8	12	9	3
9	6	5	1
10	7	4	3
11	6	6	0
12	9	8	1
13	10	8	2
14	6	7	-1
15	12	9	3
16	4	3	1
17	5	5	0
18	6	4	2
19	11	8	3
20	5	3	2
21	10	9	1
22	10	11	-1
N=22			

```
                              PAIRED T TEST

                                    STANDARD     STD ERROR
   VARIABLE     LABEL       N      MEAN    DEVIATION    OF MEAN       T    PR>|T|

    DIFF     DIFFERENCE   22   1.09090909  1.47709789  0.31491833  3.46   0.0023
             IN RATINGS
```

```
                            WILCOXON SIGNED RANK TEST
                  STATISTICAL ANALYSIS SYSTEM - NPAR 360 INTERFACE

        WILCOXON MATCHED-PAIRS SIGNED-RANKS TEST
           WITH ONE-TAIL PROBABILITIES OF THIS OR GREATER T

    HIGHER GROUP    LOWER GROUP    N(TOTAL)    N(SIGNED)    WILCOXON T   PROBABILITY

      JUDGE_1        JUDGE_2          22          20          30.00        0.0026
```

8.48 A process of recycled aluminum cans is concerned about the levels of impurities (principally other metals) contained in lots from two sources. Laboratory analysis of sample lots yields the following data (in kilograms of impurities per hundred kilograms of product).

Source 1:	3.8	3.5	4.1	2.5	3.6	4.3	2.1	2.9	3.2	3.7	2.8	2.7
Source 2:	1.8	2.2	1.3	5.1	4.0	4.7	3.3	4.3	4.2	2.5	5.4	4.6

a Use the EXECUSTAT output shown here to test $H_0: \mu_1 - \mu_2 = 0$ against $H_a: \mu_1 - \mu_2 \neq 0$. Give the p-value for the test and draw a conclusion.

b Use the same output to determine whether the t' test would yield the same result. Does your conclusion change?

```
Two Sample Analysis for DATA by SOURCE

                        SOURCE1              SOURCE2

Sample size             12                   12
Mean                    3.26667              3.61667        diff. = -0.35
Variance                0.45697              1.86333        ratio = 0.245243
Std. deviation          0.675995             1.36504

95% confidence intervals
   mu1 - mu2: (-1.26194,0.561937) assuming equal variances
   mu1 - mu2: (-1.28176,0.581758) not assuming equal variances
   variance ratio: (0.0703861,0.85449)


Hypothesis Test - Difference of Means

Null hypothesis: difference of means = 0
Alternative: not equal
Equal variances assumed: no

Computed t statistic = -0.795951
          P value = 0.4376
```

8.49 To compare the performance of microcomputer spreadsheet programs, teams of three students each choose whatever spreadsheet program they wish. Each team is given the same set of standard accounting and finance problems to solve. The time (in minutes) required for each team to solve the set of problems is recorded. The data shown here were obtained for the two most widely used programs; also displayed are the sample means, sample standard deviations, and sample sizes.

Program	Time										$\bar{y}$	s	n
1	39	57	42	53	41	44	71	56	49	63	51.50	10.46	10
2	43	38	35	45	40	28	50	54	37	29			
	36	27	52	33	31	30					38.00	8.67	16

a Use the Minitab output shown here to test $H_0: \mu_1 - \mu_2 = 0$ versus $H_a: \mu_1 - \mu_2 > 0$. Give the p-value for your test, and draw tentative conclusions.

b Additional analyses (shown in the accompanying printout) were done on the sample data. Based on all three analyses, reach a conclusion about the data (that is, make sense of the data).

```
MTB > PRINT C1 C2

ROW   PROGRAM1   PROGRAM2

 1        39         43
 2        57         38
 3        42         35
 4        53         45
 5        41         40
 6        44         28
 7        71         50
 8        56         54
 9        49         37
10        63         29
11                   36
12                   27
13                   52
14                   33
15                   31
16                   30

MTB > DESCRIBE 'PROGRAM1' 'PROGRAM2'

               N      MEAN    MEDIAN    TRMEAN    STDEV    SEMEAN
PROGRAM1      10     51.50     51.00     50.62    10.46      3.31
PROGRAM2      16     38.00     36.50     37.64     8.67      2.17

              MIN       MAX        Q1        Q3
PROGRAM1    39.00     71.00     41.75     58.50
PROGRAM2    27.00     54.00     30.25     44.50


MTB > TWOSAMPLE T 'PROGRAM1' VS 'PROGRAM2';
SUBC> ALTERNATIVE +1.

TWOSAMPLE T FOR PROGRAM1 VS PROGRAM2
            N      MEAN     STDEV    SE MEAN
PROGRAM1   10      51.5      10.5       3.3
PROGRAM2   16     38.00      8.67       2.2

95 PCT CI FOR MU PROGRAM1 - MU PROGRAM2: (5.1, 21.9)

TTEST MU PROGRAM1 = MU PROGRAM2 (VS GT): T= 3.41  P=0.0018  DF= 16
```

```
MTB > TWOSAMPLE T 'PROGRAM1' VS 'PROGRAM2';
SUBC> ALTERNATIVE +1;
SUBC> POOLED.

TWOSAMPLE T FOR PROGRAM1 VS PROGRAM2
                N       MEAN     STDEV    SE MEAN
PROGRAM1    10      51.5     10.5      3.3
PROGRAM2    16     38.00     8.67      2.2

95 PCT CI FOR MU PROGRAM1 - MU PROGRAM2: (5.7, 21.3)

TTEST MU PROGRAM1 = MU PROGRAM2 (VS GT): T= 3.57  P=0.0008  DF=  24

POOLED STDEV =        9.38

MTB > MANN-WHITNEY 'PROGRAM1' VS 'PROGRAM2';
SUBC> ALTERNATIVE +1.

Mann-Whitney Confidence Interval and Test

PROGRAM1   N =  10     Median =       51.00
PROGRAM2   N =  16     Median =       36.50

Point estimate for ETA1-ETA2 is       13.00
95.2 pct c.i. for ETA1-ETA2 is (5.00,22.00)
W = 191.0
Test of ETA1 = ETA2   vs.   ETA1 g.t. ETA2 is significant at 0.0017

MTB > STOP
```

 8.50 Educators compared the exam scores of nursing-degree students with the exam scores of students from diploma and associate degree programs on a state licensing board examination. By random sampling procedures, the educators drew a sample of five from those who completed the nursing-degree program, resulting in a mean score of 400 with a standard deviation of 15. A random sample of five drawn from the associate degree program had a mean score of 370 with a standard deviation of 30. Can the licensing board conclude that the mean score of nursing students who complete the degree program is higher than the mean score of those who complete the associate program? Base your answer on the results of a statistical test. Give the approximate p-value for your test.

 8.51 An educator wants to compare the effects of two different teaching methods. Two classes of students are selected at random; class 1 receives method 1, and class 2 receives method 2. A comprehensive standard examination is administered to both classes to determine the effectiveness of the two methods at the end of the test period. The relevant data are shown here.

Statistic	Class 1	Class 2
Sample size	$n_1 = 64$	$n_2 = 64$
Average test score	$\bar{y}_1 = 88$	$\bar{y}_2 = 80$
Sample variance	$s_1^2 = 56$	$s_2^2 = 56$

Determine a 95% confidence interval for the difference between the two population means, on the basis of the difference between the two sample means. Would a 90% confidence interval be wider?

 8.52 A study was conducted to see whether food prices charged in a ghetto area are higher than thosed charged in a more affluent suburban area. Food prices were obtained from nine stores in each area, and a food price index was computed. The summary results for each area were as follows.

Ghetto Area	Suburban Area
$n_1 = 90$	$n_2 = 9$
$\bar{y}_1 = 11.1$	$\bar{y}_2 = 10.5$
$s_1^2 = 2.5$	$s_2^2 = 1.5$

Conduct a statistical test of $H_0: \mu_1 - \mu_2 = 0$ versus $H_a: \mu_1 - \mu_2 > 0$. Show all steps in the test of hypothesis, and state your conclusion in *nonstatistical terms*. Use $\alpha = .05$. What type of error (Type I or II) could you have made?

8.53 We are given the following data summarizing information on two indpendent samples taken from populations whose variances are known to be equal.

Group	n	$\bar{y}$	s^2
Sample 1	6	30	60
Sample 2	4	20	60

From these data, the following value for t was computed:

$$t = \frac{30 - 20}{7.75\sqrt{\frac{1}{6} + \frac{1}{4}}} = 2.$$

a Show, by computing it, how the 7.75 in the preceding computation was obtained.

b Suppose that the null hypothesis were tested against a two-tailed alternative, with $\alpha = .05$.

  i What would be the rejection region?

  ii Would the null hypothesis be rejected?

c Suppose that the null hypothesis were tested against the one-tailed alternative that the mean of population 1 is larger than the mean of population 2, with $\alpha = .05$.

  i What would be the rejection region for this test?

  ii Would the null hypothesis be rejected?

8.54 To test the research hypothesis that teacher expectation can improve student performance, two groups of 100 students were compared. Teachers of the experimental group were told that their students would show large IQ gains during the test semester, while teachers of the control group were told nothing. At the end of the semester, the students' IQ change scores were calculated, with the following results.

Group	Mean	Standard Deviation	Sample Size
Experimental	16.5	14.2	100
Control	7.0	13.1	100

a Test the null hypothesis that teacher expectation has no effect on mean IQ change scores.

b State your conclusion in two ways:

  i in statistical terms.

  ii in nontechnical terms as you might explain it to an intelligent person who was not familiar with statistical terminology.

8.55 Individuals running for public office must now report the amount of money they spend in each campaign. It has been reported that women candidates usually find it difficult to raise money and therefore spend less in their campaigns than do men candidates. Suppose that the accompanying data represent the campaign expenditures (in thousands of dollars) of a randomly selected group of men and women candidates who have just completed their campaigns for public office. Do the data support the claim that women candidates generally spend less than men candidates in their campaigns for public office?

a   Would you use a one-tailed test or two-tailed test of hypothesis in this case? Why?

	Women	Men
	138	134
	127	137
	134	135
	125	140
		130
		134
Sum	524	810
Mean	131	135

b   State the null and alternative hypotheses in

    i   statistical terms or symbols.

    ii   plain English.

8.56   Refer to Exercise 8.55. Summary data for the two samples are shown in the accompanying MINITAB output, along with the results of a t test for $\mu_1 - \mu_2$. What assumptions must we make in order to run a t test? Which one of these (if any) could cause a problem for these data?

```
MTB > SET INTO C1
DATA> 138 127 134 125
DATA> END
MTB > SET INTO C2
DATA> 134 137 135 140 130 134
DATA> END
MTB > PRINT C1 C2

  ROW     C1      C2

    1     138     134
    2     127     137
    3     134     135
    4     125     140
    5             130
    6             134

MTB > TWOSAMPLE T FOR C1 VS C2;
SUBC> POOLED;
SUBC> ALTERNATIVE -1.

TWOSAMPLE T FOR C1 VS C2
        N       MEAN      STDEV    SE MEAN
C1   4       131.00      6.06       3.0
C2   6       135.00      3.35       1.4

95 PCT CI FOR MU C1 - MU C2: (-10.8, 2.8)

TTEST MU C1 = MU C2 (VS LT): T= -1.36  P=0.11  DF=  8

POOLED STDEV =         4.56

MTB > STOP
```

8.57   The following computer output presents the data of Exercise 8.18 for a t test of $H_0: \mu_1 - \mu_2 = 0$ and a Wilcoxon rank sum test (which is equivalent to the Mann–Whitney test shown here).

```
MTB > NAME C1 'PLUMBER1'
MTB > NAME C2 'PLUMBER2'
MTB > SET C1
MTB > END
MTB > PRINT C1

PLUMBER1
    88.2     94.7    101.8    102.6     89.3     95.7     78.2     80.1     83.9
    86.1     89.4     71.4     92.4     85.3     87.5     94.6     92.7     84.6

MTB > HISTOGRAM C1

Histogram of PLUMBER1    N = 18

Midpoint    Count
    70         1    *
    75         0
    80         2    **
    85         4    ****
    90         5    *****
    95         4    ****
   100         1    *
   105         1    *

MTB > SET C2
MTB > END
MTB > PRINT C2

PLUMBER2
   105.8    117.6    119.5    126.8    108.2    114.7     90.2     95.6    110.1
   115.3    109.6    112.4    104.6    107.2    109.7    102.9     99.1    111.5

MTB > HISTOGRAM C2

Histogram of PLUMBER2    N = 18

Midpoint    Count
    90         1    *
    95         1    *
   100         1    *
   105         4    ****
   110         6    ******
   115         2    **
   120         2    **
   125         1    *
```

```
MTB > TWOSAMPLE 95% CONFIDENCE FOR 'PLUMBER1' AND 'PLUMBER2'

TWOSAMPLE T FOR PLUMBER1 VS PLUMBER2
              N      MEAN      STDEV    SE MEAN
PLUMBER1     18     88.81      7.89       1.9
PLUMBER2     18    108.93      8.73       2.1

95 PCT CI FOR MU PLUMBER1 - MU PLUMBER2: (-25.8, -14.5)

TTEST MU PLUMBER1 = MU PLUMBER2 (VS NE): T= -7.26   P=0.0000   DF=   33

MTB > MANN-WHITNEY 95% CONFIDENCE FOR 'PLUMBER1' AND 'PLUMBER2'

Mann-Whitney Confidence Interval and Test

PLUMBER1    N =   18      Median =        88.75
PLUMBER2    N =   18      Median =       109.65
Point estimate for ETA1-ETA2 is        -20.30
95.2 pct c.i. for ETA1-ETA2 is (-25.40,-14.90)
W = 183.0
Test of ETA1 = ETA2   vs.   ETA1 n.e. ETA2 is significant at 0.0000

MTB > STOP
```

a Compare the results for these two tests, and draw a conclusion about the effectiveness of the dispatcher program.

b Comment on the appropriateness or inappropriateness of the *t* test, based on your findings in Exercise 8.18a and on the output shown here.

c Does it matter which test was used here? Might it be reasonable to run both tests in certain situations? Why?

8.58 Suppose that you are the personnel manager for a company and you suspect a difference exists in the mean length of work time lost due to sickness for two types of employees: those who work at night versus those who work during the day. Particularly, you suspect that the mean time lost for the night shift exceeds the mean time lost for the day shift. To check your theory, you randomly sample the records for ten employees in each shift category and record the number of days each employee has lost due to sickness within the past year. These data are shown next.

Night Shift	Day Shift
15	8
10	9
10	2
7	0
7	10
4	9
9	9
6	7
10	3
12	3

a Would you use a one-tailed test or two-tailed test in your test hypothesis? Why?

b What is the pooled estimate of σ?

c Conduct the statistical test and show all parts of the test:

    i   null hypothesis.

    ii   alternative hypothesis.

    iii   test statistic and computations.

    iv   rejection region.

    v   conclusion.

        a   in statistical terms.

        b   in plain English that nonstatisticians can understand.

8.59 Refer to Exercise 8.58. Based on your decision, could you have made

    a   a Type I error?

    b   a Type II error?

    c   both Type I and Type II errors?

    d   neither error?

    (Answer each question with yes or no.)

8.60 Refer to the data of Exercise 8.6. Use the MINITAB output shown here to determine a 95% confidence interval for $\mu_1 - \mu_2$.

```
MTB > READ INTO C3 C4
DATA> 152     384
DATA> 157     369
DATA> 179     354
DATA> 182     375
DATA> 176     366
DATA> 149     423
DATA> END

        6 ROWS READ
MTB > PRINT C3 C4

  ROW      C3      C4

    1      152     384
    2      157     369
    3      179     354
    4      182     375
    5      176     366
    6      149     423

MTB > TWOSAMPLE T FOR C3 VS C4;
SUBC> POOLED;
SUBC> ALTERNATIVE -1.

TWOSAMPLE T FOR C3 VS C4
        N       MEAN      STDEV    SE MEAN
C3   6        165.8       14.8        6.0
C4   6        378.5       24.0        9.8

95 PCT CI FOR MU C3 - MU C4: (-238.3, -187.1)

TTEST MU C3 = MU C4 (VS LT): T= -18.51   P=0.0000   DF=  10

POOLED STDEV =         19.9

MTB > STOP
```

 8.61 A study was carried out to determine whether nonworking wives from middle-class families have more voluntary association memberships than do nonworking wives from working-class families. A random sample of housewives was obtained, and each was asked for information about her husband's occupation and her own memberships in voluntary associations. On the basis of their husbands' occupations, the women were divided into middle-class and working-class groups, and the mean number of voluntary association memberships was computed for each group.

For the 15 middle-class women, the mean number of memberships per woman was $\bar{y}_1 = 3.4$ with $s_1 = 2.5$. For the 15 working-class wives, $\bar{y}_2 = 2.2$ with $s_2 = 2.8$. Use these data to construct a 95% confidence interval for $\mu_1 - \mu_2$.

8.62 A regional IRS auditor ran a test on a sample of returns that were filed by March 15 to determine whether the average refund for taxpayers is larger this year than it was last year. Sample data are shown here for a random sample of 100 returns for each year.

Statistic	Last Year	This Year
Mean	320	410
Variance	300	350
Sample size	100	100

a In a test of hypothesis, would you use a one-tailed or a two-tailed test? Why?

b What assumptions are required to conduct a t test of $H_0: \mu_1 - \mu_2 = 0$? Do you think the assumptions hold? Why or why not?

8.63 Miss America Pageant officials insist that their pageant is not a beauty contest and that talent is more important than beauty when it comes to success in the pageant. In an effort to evaluate the latter assertion, a random sample of 55 preliminary talent-competition winners and a random sample of 53 preliminary swimsuit-competition winners were taken to see whether there was a significant difference in the mean amount won for the two groups. For the 55 preliminary talent competition winners, the mean amount won was $8645 with a standard deviation of $5829; for the 53 preliminary swimsuit winners, the mean amount won was $9198 with a standard deviation of $8185. Compute a 95% confidence interval for the difference in the mean amount won by the two groups. Does your confidence interval confirm what the pageant officials contend?

8.64 A visitor to the United States from France insisted that recordings made in Europe tend to have selections with longer playing times than do recordings made in the United States. To verify or contradict this contention, a random sample of selections was taken from a group of recordings produced in France and Germany, and another random sample of selections was taken from American-produced recordings. The sample results are given next.

Statistic	Foreign Produced	American Produced
Number in sample	14	14
Mean playing time (in seconds)	207.45	182.54
Standard deviation	41.43	37.32

Do the foreign-produced selections have longer mean playing times? Use $\alpha = .05$.

8.65 A major federal agency located in Washington, D.C., regularly conducts classes in PL/1, a computer programming language used in the programs written within the agency. One week, the course was taught by an individual associated with an outside consulting firm. The following week, a similar course was taught by a member of the agency. The following results were achieved by the classes.

Taught by Outsider	38	42	53	37	36	48	47	47	44
Taught by Staff Member	46	33	38	60	58	52	44	45	51

The values represent scores aggregated over the 1-week course out of a potential maximum of 64. Do the data present sufficient evidence to indicate a difference in teaching effectiveness, assuming that the scores reflect teaching effectiveness? Use $\alpha = .05$.

8.66　Company officials are concerned about the length of time a particular drug retains its potency. A random sample (sample 1) of 10 bottles of the product is drawn from current production and analyzed for potency. A second sample (sample 2) is obtained, stored for one year, and then analyzed. The readings obtained are as follows:

| **Sample 1:** | 10.2 | 10.5 | 10.3 | 10.8 | 9.8 | 10.6 | 10.7 | 10.2 | 10.0 | 10.6 |
| **Sample 2:** | 9.8 | 9.6 | 10.1 | 10.2 | 10.1 | 9.7 | 9.5 | 9.6 | 9.8 | 9.9 |

The data are analyzed by a standard program package (SAS). The relevant output is shown here.

TTEST PROCEDURE

VARIABLE: POTENCY

SAMPLE	N	MEAN	STD DEV	STD ERROR	MINIMUM	MAXIMUM
1	10	10.37000000	0.32335052	0.10225241	9.80000000	10.80000000
2	10	9.83000000	0.24060110	0.07608475	9.50000000	10.20000000

VARIANCES	T	DF	PROB>|T|
UNEQUAL	4.2368	16.6	0.0006
EQUAL	4.2368	18.0	0.0005

FOR H0: VARIANCES ARE EQUAL, F' = 1.81　DF = (9,9)　PROB>F' = 0.3917

a　Identify the sample means and sample standard deviations.

b　Locate the value of the t statistic. Is the pooled-variance t statistic identified as "equal variance" or "unequal variance"?

c　Locate the value of the t' statistic.

d　Why are these two statistics equal in this case?

8.67　Two possible methods for retrofitting jet engines to reduce noise are being considered. Identical planes are fitted with two systems. Noise-recording devices are installed directly under the flight path of a major airport. Each time one of the planes lands at the airport, a noise level is recorded. The data are analyzed by a computer software package (SAS). The relevant output is as follows.

VARIABLE: DBREAD

SYSTEM	N	MEAN	STD DEV	STD ERROR	MINIMUM	MAXIMUM
H	42	100.90476190	2.99438111	0.46204304	95.0000000	110.00000000
R	20	92.50000000	8.19178022	1.83173774	79.0000000	111.00000000

VARIANCES	T	DF	PROB>|T|
UNEQUAL	4.4491	21.5	0.0002
EQUAL	5.9126	60.0	0.0001

a Locate the t statistic.

b Locate the t' statistic.

c Can the research hypothesis of unequal means be supported, using $\alpha = .01$? Does it matter which statistic is used?

8.68 A study was conducted on 16 dairy cattle. Eight cows were randomly assigned to a liquid regimen of water only (group 1); the others received liquid whey only (group 2). In addition, each animal was given 7.5 kg of grain per day and allowed to graze on hay at will. Although no significant differences were observed between the groups in the dairy-milk-production guages, such as milk production and fat content of the milk, the following data on daily hay consumption (in kilograms/cow) were of interest.

Group 1	15.1	14.9	14.8	14.2	13.1	12.8	15.5	15.9
Group 2	6.8	7.5	8.6	8.4	8.9	8.1	9.2	9.5

Use these sample data to test the research hypothesis that there is a difference in mean hay consumption for the two diets. Use $\alpha = .05$.

8.69 Refer to Example 8.10. Suppose that we wish to detect only $\mu_1 - \mu_2 > 0$. Determine the sample size required when $\alpha = .05$ and $\beta \leq .10$ for $\mu_1 - \mu_2 \geq .5$.

8.70 An industrial concern has experimented with several different mixtures of the four components (magnesium, sodium nitrate, strontium nitrate, and a binder) that comprise a rocket propellant. The company has found that two mixtures in particular give higher flare-illumination values than the others. Mixture 1 consists of a blend composed of the proportions .40, .10, .42, and .08, respectively, for the four components of the mixture; mixture 2 consists of a blend that uses the proportions .60, .27, .10, and .05. Twenty different blends (ten of each mixture) are prepared and tested to obtain the flare-illumination values. These data appear below (in units of 1000 candles).

Mixture 1	185	192	201	215	170	190	175	172	198	202
Mixture 2	221	210	215	202	204	196	225	230	214	217

a Plot the sample data. Which test(s) could be used to compare the mean illumination values for the two mixtures?

b Give the level of significance of the test, and interpret your findings.

8.71 Refer to Exercise 8.70. Instead of conducting a statistical test, use the sample data to answer the question, "What is the difference in mean flare illumination for the two mixtures?"

8.72 Refer to Example 8.3. Suppose that the seventh untreated animal died before the study was completed. Analyze the remaining observations to compare the two population means. Assume that $\sigma_1^2 \neq \sigma_2^2$. Give the level of significance for your test.

8.73 Refer to the accompanying computer printout for the statistical test of the data in Exercise 8.11.

a Were there any problems in running a t test?

b Compare these computer results to your calculations for Exercise 8.11.

c Give the value of the test statistic and the level of significance for a t test of the research hypothesis that the experimental mean is greater than the control mean.

Control Group	5.4	6.2	3.1	3.8	6.5	5.8	6.4	4.5	4.9	4.0
Experimental Group	8.8	9.5	10.6	9.6	7.5	6.9	7.4	6.5	10.5	8.3

```
                     TWO SAMPLE T-TEST

                     TTEST PROCEDURE

VARIABLE: LEADPCT

GROUP       N       MEAN      STD DEV    STD ERROR    MINIMUM       MAXIMUM

CONTROL    10   5.06000000   1.18902388  0.37600236   3.10000000    6.50000000
EXPERMT    10   8.56000000   1.47135614  0.46528366   6.50000000   10.60000000


VARIANCES         T       DF     PROB>|T|

UNEQUAL       -5.8507    17.2     0.0001
EQUAL         -5.8507    18.0     0.0001

FOR H0: VARIANCES ARE EQUAL,  F' = 1.53    DF = (9,9)    PROB>F' = 0.5356
```

8.74 A study of anxiety was conducted among residents of a southeastern metropolitan area. Each person selected for the study was asked to check a "yes" or "no" for the presence of each of 12 anxiety symptoms. Anxiety scores ranged from 0 to 12, with higher scores related to the higher perceived presence of any anxiety symptoms. The results for a random sample of 50 residents, categorized by gender, are summarized below. Use these data to test the research hypothesis that the mean perceived anxiety score differs for males and for females. Give the level of significance for your test.

	Sample Size	Mean	Standard Deviation
Female	26	5.26	3.2
Male	24	7.02	3.9

8.75 A clinical trial was conducted to determine the effectiveness of drug A in the treatment of symptoms associated with alcohol withdrawal. A total of 30 patients were treated (under blinded conditions) with drug A and another 30 with an identical-looking placebo. The average symptom score for the two groups after 1 week of therapy was 1.5 and 6.3, respectively. (*Note:* Higher symptom scores indicate more withdrawal "problems.") The corresponding standard deviations were 3.1 and 4.2.

a Compare the mean total symptom scores for the two groups. Give the *p*-value for a two-sample *t* test of $H_0: \mu_1 - \mu_2 = 0$ versus $H_a: \mu_1 - \mu_2 < 0$. Draw conclusions.

b Suppose that the average total symptoms scores for the two groups were 6.8 and 12.2 prior to therapy. How would this affect your conclusions? How could you guard against possible baseline (pretreatment) differences?

8.76 Two analysts, supposedly of identical abilities, each measure the parts per million of a certain type of chemical impurity in drinking water. It is claimed that analyst 1 tends to give higher readings than analyst 2. To test this theory, each of six water samples is divided and then analyzed by both analysts separately. The data are shown in the accompanying table (readings in ppm).

Water Sample	Analyst 1	Analyst 2
1	31.4	28.1
2	37.0	37.1
3	44.0	40.6
4	28.8	27.3
5	59.9	58.4
6	37.6	38.9

a   Is there sufficient evidence to indicate that analyst 1 reads higher on the average than does analyst 2? Give the level of significance for your test.

b   What would be your conclusion based on a Wilcoxon test? Compare your results here to those you got for part a.

8.77   A single leaf was taken from each of 11 different tobacco plants. Each was divided in half; one half was chosen at random and treated with preparation I, and the other half received preparation II. The object of the experiment was to compare the effects of the two preparations of mosaic virus on the number of lesions on the half leaves after a fixed period of time. These data are recorded in the accompanying table. For $\alpha = .05$, use Wilcoxon signed-rank test to examine the research hypothesis that the distributions of lesions are different for the two populations.

Number of Lesions on the Half Leaf

Tobacco Plant	Preparation I	Preparation II
1	18	14
2	20	15
3	9	6
4	14	12
5	38	32
6	26	30
7	15	9
8	10	2
9	25	18
10	7	3
11	13	6

8.78   An investigator plans to compare the mean number of particles of effluent in water collected at two different locations in a water treatment plant. If the standard deviation for particle counts is expected to be approximately 6 for the counts in samples taken at each of the locations, determine the sample sizes required to estimate the mean difference in particles of effluent, using a 99% confidence interval of width 1 (particle).

8.79   The weight gains for $n_1 = n_2 = 8$ rats tested on diets 1 and 2 are summarized here. Set up a statistical test for $\mu_1 - \mu_2$, the difference in average weight gained for the two diets. Use $\alpha = .05$ and draw conclusions.

Statistic	Diet 1	Diet 2
$\sum y$	25	26.2
s	.005	.045
n	8	8

8.80 Refer to the computer simulation in Example 8.5. Another computer simulation study of 1000 samples was run using the same sample sizes but different population standard deviations. The independent normal populations were as follows:

Population 1	$\mu_1 = 100$	$\sigma_1 = 15$	$n_1 = 10$
Population 2	$\mu_2 = 100$	$\sigma_2 = 10$	$n_2 = 10$

For each run of the simulation study, the pooled t test and separate-variance t test were run. The number (%) of times t and t' were rejected at the upper and lower .05 levels is recorded here.

Hypothesis Result	Pooled t Test	Separate-variance t Test
H_0 rejected, H_a: $\mu_1 - \mu_2 < 0$	43 (4.3%)	44 (4.4%)
H_0 rejected, H_a: $\mu_1 - \mu_2 > 0$	48 (4.8%)	46 (4.6%)

a Before reviewing the results of the simulation study, how well do you think the pooled t test will perform, even though one of the assumptions underlying it is obviously not satisfied? Explain.

b Which test performed better? Which would you recommend, and why?

Experiences with Real Data

8.81 The following memorandum opinion on statistical significance was issued by the judge in a trial involving many scientific issues. The opinion has been stripped of some legal jargon and has been taken out of context. Still, it can give us an understanding of how others deal with the problem of ascertaining the meaning of statistical significance. Read this memorandum and comment on the issues raised regarding statistical significance.

Memorandum Opinion

This matter is before the Court upon two evidentiary issues that were raised in anticipation of trial. First, it is essential to determine the appropriate level of statistical significance for the admission of scientific evidence.

With respect to statistical significance, no statistical evidence will be admitted during the course of the trial unless it meets a confidence level of 95%.

Every relevant study before the court has employed a confidence level of at least 95%. In addition, plaintiffs concede that social scientists routinely utilize a 95% confidence level. Finally, all legal authorities agree that statistical evidence is inadmissible unless it meets the 95% confidence level required by statisticians. Therefore, because plaintiffs advance no reasonable basis to alter the accepted approach of mathematicians to the test of statistical significance, no statistical evidence will be admitted at trial unless it satisfies the 95% confidence level.

8.82 Certain baseline determinations were made on 182 patients entered in a study of survival in males suffering from congestive heart failure. When these data were summarized, 88 deaths had been observed. The accompanying table summarizes the baseline data for survivors and nonsurvivors. The variables listed below "Heart rate" are measures of the severity of the heart failure. The arrows to the left of each variable indicates the direction of improvement.

a Discuss these baseline findings (that is, make sense of the data).

b What assumptions have the authors made when doing these t tests?

Variable	Nonsurvivors ($n = 88$)	Survivors ($n = 94$)	t Test p-Value
Age (y)	57 ± 10	56 ± 8	NS
Duration of symptoms (mo)	45 ± 43	39 ± 27	NS
Heart rate (beats/min)	87 ± 15	83 ± 16	NS
↓Mean arterial pressure (mm Hg)	87 ± 13	94 ± 13	< 0.001
↓Left-ventricular filling pressure (mm Hg)	29 ± 7	24 ± 9	< 0.001
↑Cardiac index (l/min/m^2)	2.0 ± 0.7	2.5 ± 0.8	< 0.001
↑Stroke volume (ml/beat)	45 ± 16	59 ± 5	< 0.001
↓Systemic vascular resistance (units)	25 ± 10	21 ± 8	< 0.01
↑Stroke work (g-m)	35 ± 19	56 ± 33	< 0.001

Values are listed as mean ± standard deviation.

Source: American Hospital Assn., Twin Cities Metropolitan Health Board. Used with permission.

 8.83 Hospital administrators studied patterns in length of hospital stay, with particular attention paid to patients having health-maintenance organization (HMO) payment sources versus those with non-HMO payment sources. The accompanying graph shown summarizes the sample data.

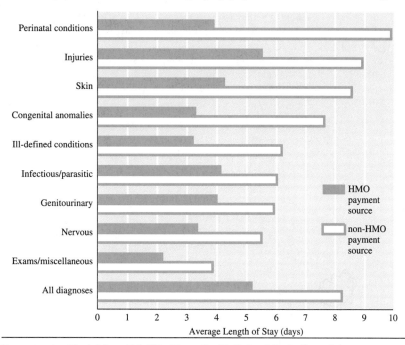

Source: American Hospital Assn., Twin Cities Metropolitan Health Board. Used with permission.

a What general conclusions would you draw from the graph? What additional information would you need to make more definitive statements regarding these results?

b Suppose that, across all diagnoses, the sample statistics were as shown below. Use these data to test $H_0: \mu_1 - \mu_2 = 0$ versus $H_a: \mu_1 - \mu_2 \neq 0$. Give the p-value for your test.

Patient Group	Sample Mean	Sample Size	Sample Standard Deviation
HMO	5.0 (days)	120	1.3
Non-HMO	8.1	130	1.9

8.84 Refer to Exercise 8.83. Run the t' test and compare your results for this test with those you obtained earlier. Which (if either) test is better for these data?

8.85 An abstract for the results of a study of ten congestive heart-failure patients is given here. Read the abstract and try to interpret the results.

Abstract

An experimental compound was studied in ten patients suffering from congestive heart failure. Certain variables were measured at baseline and then four hours after intravenous treatment with the compound. The compound was shown to increase the cardiac index from 11.1 to 34.3% from a baseline average of 2.41 ± 0.49 l/min/m^2 ($p < .01$, heart rate by 6–10% from 72 ± 12 beats/min ($\rho < .02$) and decreased pulmonary capillary wedge pressure by 15.3–24.2% from 18.7($p < .001$).

8.86 Several antidepressant drugs have been studied in the treatment of cocaine abusers. One recent study showed that 20 cocaine abusers who were treated with an antidepressant in an outpatient setting experienced decreases in cravings after two weeks and some reduction in their actual use of cocaine. Comment on these results. Are they compelling? Why or why not?

8.87 In April 1986, the *Australian Journal of Statistics* (Vol. 30, No. 1, pp. 23–44) published the results of a study by S. R. Butler and H. W. Marsh on reading and arithmetic achievement for students from non-English-speaking families. All kindergarten students from seven public schools in Sydney, Australia, were included in the original sample of 392 children. Reading and arithmetic achievement tests were administered at the start of the study during kindergarten and then at years 1, 2, 3, and 6 of primary school.

The table shown on page 355 gives the characteristics of all 286 of the original 392 students who were still available for testing at year 6 ($n = 226$ students from English-speaking families, and $n = 60$ from non-English-speaking families).

a Can you suggest better ways to summarize these baseline characteristics?

b What test(s) may have been used to compare these characteristics?

c What other characteristics could or should have been examined to make a direct comparison of reading and arithmetic achievement?

d What effect (if any) might the attrition rate have had on the study results? Recall that 106 (27%) of the original 392 students were not available for testing at year 6.

| | Group | |
| | English-speaking Family ($n = 226$) | Non-English-speaking Family ($n = 60$) |
Characteristics	$\bar{y}$	$\bar{y}$
Age (in months)	67.17	67.15
Gender (1 = male, 2 = female)	1.50	1.55
Number of children in family	2.54	2.62
Ordinal position in family (1 = oldest child, etc.)	1.89	1.82
Father's occupation (1 = most skilled, 17 = least skilled)	8.26*	11.50
Peabody Picture Vocabulary IQ	99.26*	77.45

*Statistically significant, $p < .01$

8.88 Refer to the Clinical Trials database on your data disk (or in Appendix 1). Use the HAM-D total score data to conduct a statistical test of $H_0: \mu_D - \mu_A = 0$ vs $H_a: \mu_D - \mu_A > 0$; that is, we want to know whether the placebo group (D) has a higher (worse) mean total depression score at the end of the study than does the group receiving treatment A. Use $\alpha = .05$. What are your conclusions?

8.89 Refer to Exercise 8.88. Repeat this same comparison with the placebo group for treatment B, and then again for treatment C. Give the p-value for each of these tests. Which of the three treatment groups (A, B, or C) appears to have the lowest mean HAM-D total score?

8.90 Use the Clinical Trials database to construct a 95% confidence interval for $\mu_D - \mu_A$, based on the HAM-D anxiety score data. What can you conclude about $\mu_D - \mu_A$, based on this interval?

8.91 Refer to the Clinical Trials database on your data disk. Compare the mean ages for treatment groups B and D, using a two-sided statistical test. Set up all parts of the test, using $\alpha = .05$; draw a conclusion. Why might it be important to have patients with similar ages in the different treatment groups in a study of the effects of several drug products on the treatment of depression?

8.92 Refer to Exercise 8.91. What other variables should be comparable among the treatment groups in order to permit us to draw unbiased conclusions about the effectiveness of the drug products for treating depression?

8.93 Refer to the Bonus Level database on your data disk (or in Appendix 1). Compare the mean ages for males and females, using a t test. Does the t' test support these findings? Make sense of the data, based on the results of the two statistical tests.

8.94 Refer again to the Bonus Level database. Run a t test, a t' test, and a Wilcoxon rank sum test to compare bonus levels for males and females. Make sense of the data by discussing your results (in plain English), and draw some conclusions.

9

Inferences About Variances

9.1 Introduction

When people think of statistical inference, they usually think of inferences involving population means. However, the particular population parameter needed to answer an experimenter's practical questions varies from one situation to another, and sometimes a population's variability is more important than its mean. Thus, product quality is often defined in terms of low variability. For example, while the drug manufacturer must certainly be concerned with controlling the mean potency of tablets, it must also attempt to minimize the variation in potency from one tablet to another. Excessive (or inadequate) potency could be very harmful to a patient. Hence, the manufacturer would like to produce tablets with the desired mean potency and with as little variation in potency (as measured by σ or σ^2) as possible.

Inferential problems about a population variance are similar to those for a population mean. We can estimate or test hypotheses about a single population variance, or we can compare two variances.

9.2   Estimation and Tests for a Population Variance

The sample variance

$$s^2 = \frac{\sum (y - \bar{y})^2}{n - 1}$$

point estimate for σ^2
chi-square distribution

can be used for inferences concerning a population variance σ^2. For a random sample of n measurements drawn from a normal population with mean μ and variance σ^2, the value s^2 provides a **point estimate for σ^2**. In addition, the quantity $(n - 1)s^2/\sigma^2$ follows a **chi-square distribution**, with df $= n - 1$ (see Figure 9.1). We do not give the mathematical formula for the chi-square (χ^2, where χ is the Greek letter chi) probability distribution, but instead we list its properties.

FIGURE 9.1
Upper-tail and Lower-tail Values of Chi-Square

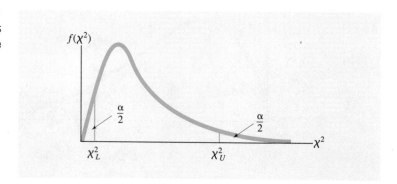

Properties of the Chi-Square Probability Distribution

1   Unlike z and t, values in a chi-square distribution are all positive.

2   The chi-square distribution, unlike the normal and t distributions, is asymmetric.

3   There are many chi-square distributions. We obtain a particular one by specifying the degrees of freedom (df) associated with the sample variances s^2. This quantity is df $= n - 1$. Figure 9.2 shows a chi-square distribution with df $= 4$.

Upper-tail values of the chi-square distribution can be found in Table 4 of Appendix 3. Entries in the table are values of χ^2 that have an area a to the right under the curve. The degrees of freedom are specified in the left column of the table, and values of a are listed across the top of the table. Thus, for df $= 14$, the value of chi-square with an area $a = .10$ to its right under the curve is 21.06 (see Figure 9.3). We can use this information to form a confidence interval for σ^2. Because the chi-square distribution is not symmetrical, the confidence intervals based on this

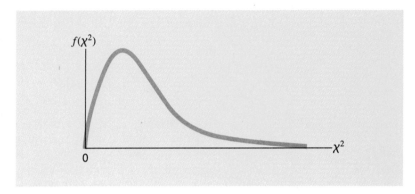

F I G U R E 9.3
Critical Value of the Chi-Square
Distribution for $a = .10$ and
$df = 14$

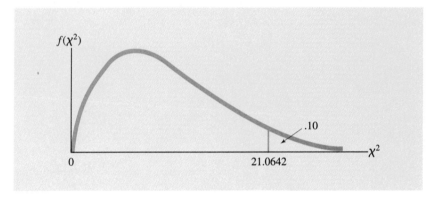

distribution don't have the usual form (estimate $\pm$ error) that we saw for μ and $\mu_1 - \mu_2$. The confidence interval for σ^2 takes the following form:

$$\text{Lower confidence limit} < \sigma^2 < \text{Upper confidence limit.}$$

General Confidence Interval for σ^2 (or σ) with Confidence Coefficient $(1 - \alpha)$

$$\frac{(n - 1)s^2}{\chi_U^2} < \sigma^2 < \frac{(n - 1)s^2}{\chi_L^2},$$

where χ_U^2 is the upper-tail value of chi-square for $df = n - 1$ with area $\alpha/2$ to its right, and χ_L^2 is the lower-tail value with area $\alpha/2$ to its left (see Figure 7.1). We can determine χ_U^2 and χ_L^2 for a specific value of df by obtaining the critical value in Table 4 in Appendix 3 corresponding to $a = \alpha/2$ and $a = 1 - \alpha/2$, respectively.

Note: The confidence interval for σ is found by taking square roots throughout.

EXAMPLE   9.1   The variability in milk production for a 305-day lactation period was observed for a random sample of 15 Holstein cows. Use the milk-yield data in Table 9.1 to estimate σ^2, the population variance of milk yields, using a 95% confidence interval. Then give the corresponding 95% confidence interval for σ.

TABLE 9.1
Milk Production Data (in 1000 pounds), for Example 9.1

12.928	13.812	11.036
12.120	14.358	9.248
14.972	8.998	9.980
14.044	10.620	11.990
14.788	14.744	14.786

Solution   For these data, we find

$$\sum y = 188.424 \qquad \sum y^2 = 2431.470.$$

Substituting into the shortcut formula for s^2 (Chapter 4), we have

$$s^2 = \frac{1}{n-1}\left[\sum y^2 - \frac{(\sum y)^2}{n}\right] = \frac{1}{14}\left[2{,}431.470 - \frac{(188.242)^2}{15}\right] = 4.612.$$

The confidence coefficient for our example is $1 - \alpha = .95$. The upper-tail chi-square value can be obtained from Table 4 in Appendix 3, for df $= n - 1 = 14$ and $a = \alpha/2 = .025$. Similarly, the lower-tail chi-square value is obtained from Table 4 with $a = 1 - \alpha/2 = .975$. Thus,

$$\chi_U^2 = 26.12 \qquad \chi_L^2 = 5.629.$$

The 95% confidence interval is then

$$\frac{14(4.612)}{26.12} < \sigma^2 < \frac{14(4.612)}{5.629}$$

or

$$2.472 < \sigma^2 < 11.471$$

thousand pounds. Thus, we estimate the population variance for milk yields to lie between 2472 and 11,471 pounds.   ▪

By taking square roots of the limits for σ^2, we obtain a 95% confidence interval for σ. From this we obtain

$$\sqrt{2472} < \sigma < \sqrt{11{,}471}$$

or

$$49.72 < \sigma < 107.10.$$

In addition to estimating a population variance, we can construct a statistical test of the null hypothesis that σ^2 equals a specified value, σ_0^2. This test procedure is summarized here.

Statistical Test for σ^2 (or σ)

H_0: $\sigma^2 = \sigma_0^2 (\sigma_0^2$ is specified)

H_a: **1** $\sigma^2 > \sigma_0^2$

     2 $\sigma^2 < \sigma_0^2$

     3 $\sigma^2 \neq \sigma_0^2$

T.S.: $\chi^2 = \dfrac{(n-1)s^2}{\sigma_0^2}$

R.R.: For a specified value of α:

  1 Reject H_0 if χ^2 is greater than χ_U^2 (the upper-tail value for $a = \alpha$ and df $= n - 1$).

  2 Reject H_0 if χ^2 is less than χ_L^2 (the lower-tail value for $a = 1 - \alpha$ and df $= n - 1$).

  3 Reject H_0 if χ^2 is greater than χ_U^2 (based on $a = \alpha/2$ and df $= n - 1$) or less than χ_L^2 (based on $a = 1 - \alpha/2$ and df $= n - 1$).

EXAMPLE 9.2   The manufacturer of a specific pesticide for controlling household bugs claims that its product retains most of its potency for a period of at least 6 months. Specifically, it claims that the drop in potency over the period from 0 to 6 months will vary in the interval from 0% to 8%. To test the manufacturer's claim, a consumer group obtained a random sample of 20 containers of pesticide from the manufacturer. Each can was tested for potency and then stored for a period of 6 months at room temperature. After the storage period, each can was again tested for potency. The drop in potency was recorded for each can, and the sample variance for the drops in potencies was computed to be $s^2 = 6.2$. Use these data to determine whether there is sufficient evidence to indicate that the population of potency drops has more variability than that claimed by the manufacturer. Use $\alpha = .05$.

Solution   The manufacturer has claimed that the population of potency reductions has a range of 8%. Dividing the range by 4, we obtain an approximate population standard deviation of $\sigma = 2\%$ (or $\sigma^2 = 4$).

The appropriate null and alternative hypotheses are

H_0: $\sigma^2 = 4$ (that is, we assume that the manufacturer's claim is correct)

H_a: $\sigma^2 > 4$ (that is, there is more variability than is claimed by the manufacturer)

Using the computed sample variance based on 20 observations, we find that the test statistic and rejection region are as follows:

T.S.: $\chi^2 = \dfrac{(n-1)s^2}{\sigma_0^2} = \dfrac{19(6.2)}{4} = 29.45$

R.R.:   For $\alpha = .05$, we will reject H_0 if the computed value of chi-square is greater than 30.14, as obtained from Table 4 in Appendix 3 for $a = .05$ and df $= 19$.

Conclusion: Since the computed value of chi-square, 29.45, is less than the critical value, 30.14, there is insufficient evidence to support rejecting the manufacturer's claim, based on $\alpha = .05$. However, the consumer group is not prepared to accept H_0: $\sigma^2 = 4$. Since $s^2 = 6.2$ and the p-value of the test is $.05 < p < .10$, it should do additional testing with a larger sample size before reaching a definite conclusion.   ∎

---

normality assumption

The χ^2 methods for making inferences about σ^2 (and σ) are based on the assumption that the population distribution is normal. This **normality assumption** is much more important for making inferences about variances than it is for making inferences about means. The Central Limit Theorem helps greatly in normalizing the sampling distribution of a mean, but there is no comparable theorem for variances. Population nonnormality, in the form of skewness or heavy tails, can have serious effects on the nominal significance and confidence probabilities for a variance (or standard deviation). If a plot of the sample data shows substantial skewness or outliers, the nominal probabilities given by the χ^2 distribution are suspect. Some computationally elaborate inference procedures about a variance (such as the so-called jackknife method) are less sensitive to the normality assumption. These may well replace the χ^2-based methods as computation costs decrease and computer programs become more widely available.

Exercises

Basic Techniques

9.1   Suppose that Y has a χ^2 distribution with 27 df.
   a   Find $P(Y > 46.96)$.
   b   Find $P(Y > 18.11)$.
   c   Find $P(Y < 12.88)$.
   d   What is $P(12.8786 < Y < 46.9630)$?

9.2   A χ^2 distribution has 11 df.
   a   Find $\chi^2_{.025}$.
   b   Find $\chi^2_{.975}$.

9.3   Suppose that Y has a χ^2 distribution with 277 df. Find approximate values for $\chi^2_{.025}$ and $\chi^2_{.975}$.

9.4   A sample of 25 observations is drawn from a normal population with unknown mean μ and variance σ^2. Define

$$\chi^2 = \frac{(n-1)s^2}{\sigma^2}.$$

Find the following probabilities.

a $P(\chi^2 > 12.4)$

b $P(\chi^2 < 36.4)$

c $P(9.89 < \chi^2 < 45.56)$

Applications

9.5   A packaging line fills nominal 32-ounce tomato juice jars with an actual mean of 32.30 ounces. The process should have a standard deviation smaller than .15 ounce per jar (a larger standard deviation leads to too many underweight or overfilled jars). Samples of 61 jars are regularly taken to test the process. One such sample yields a sample mean of 32.28 ounces and a standard deviation of .132 ounce. Does this indicate (using $\alpha = .05$) that $\sigma < .15$? Carry out a formal hypothesis test.

9.6   Suppose that the research hypothesis in Exercise 9.5 is formulated as $\sigma > .15$. Does this reformulation tend to be more or less generous in terms of what sample results cause the packaging line to be shut down for adjustment?

9.7   A certain part for a small assembly should have a diameter of 4000 mm; a maximum standard deviation of .011 mm is allowed by specifications. A random sample of 26 parts shows the following diameters:

3.952	3.978	3.979	3.984	3.987	3.991	3.995	3.997	3.999	3.999	3.999
4.000	4.000	4.000	4.001	4.001	4.002	4.002	4.003	4.004	4.006	4.009
4.010	4.012	4.023	4.041							

a   Calculate the sample mean and sample standard deviation.

b   Can the research hypothesis that $\sigma > .011$ be supported (at $\alpha = .05$) by these data? State all parts of a statistical hypothesis test.

9.8   Calculate 90% confidence intervals for the true variance and for the true standard deviation for the data of Exercise 9.7.

9.9   Plot the data of Exercise 9.7. Does the plot suggest any violation of the assumptions underlying your answers to Exercises 9.7 and 9.8? Would such a violation have a serious effect on the validity of your answers?

9.10   Baseballs vary somewhat in their rebounding coefficient. A "dead ball" has a relatively low rebound, but a "rabbit ball" has a high rebound. A standard test has been developed. A purchaser of large quantities of baseballs requires that the mean value be 85 and that the standard deviation be less than 2 units. A sample of 81 baseballs is tested. The mean value is 84.91 and the standard deviation is 1.80. Can the research hypothesis that $\sigma < 2$ be supported, using $\alpha = .05$? Carry out the steps of a formal hypothesis test.

9.11   Place bounds on the p-value in Exercise 9.10.

9.12   As part of a detailed driver-training program, school officials are requiring teenagers to take a depth-perception test. In one phase of this test, the student is asked to judge the distance between a parked vehicle and a pedestrian stationed a given distance from the student. The following distances (in feet) were reported by 15 driver-education students.

5	8	7	7	10	6	4	11
6	8	4	9	9	6	5	

Use these data to construct a 99% confidence interval for σ^2, the variance of the depth-perception distances.

9.3 Quality Control: *r*-Charts and *s*-Charts

In Chapter 7 we discussed how to construct an use $\bar{y}$-charts to examine and control the mean output from a process. We can construct an *r*-chart or *s*-chart from the samples that are used to construct a $\bar{y}$-chart sample. Of course, the sample range is the difference between the largest and smallest measurements in a sample; hence, it gives a measure of the variability of a process. The smaller the sample ranges are, the smaller the variation in the process output. In contrast, the larger the sample ranges, the more variability in the output of the process.

An *r*-chart is constructed by plotting the individual sample ranges r_i for successive samples of size *n*. The sample size is usually kept small, typically less than 10, because the sample range tends to be more volatile and unreliable for larger sample sizes.

The center line of the *r*-chart is denoted by $\bar{r}$ and is calculated as the average of the sample ranges. The upper and lower control limits for the *r*-chart are given by

$$\text{UCL} = D'_n \bar{r} \quad \text{and} \quad \text{LCL} = D_n \bar{r}$$

where D'_n and D_n are obtained from Table 15 of Appendix 3.

EXAMPLE 9.3   Refer to the colorimeter readings for the 25 samples of Example 7.5. The sample ranges for those data are presented in Table 9.2. Construct an *r*-chart for the data. Plot the sample data on the *r*-chart as well.

Solution   The center line for the *r*-chart is given by

$$\bar{r} = \frac{\sum r_i}{k} = \frac{56.8}{25} = 2.27.$$

For $n = 5$ observations per sample, we find $D'_5 = 2.115$ and $D_5 = 0$, from Table 15 in Appendix 3. Therefore, for $\bar{r} = 2.27$, the upper and lower control limits for the *r*-chart are, respectively,

$$\text{UCL} = D'_n \bar{r} = 2.115(2.27) = 4.80$$

and

$$\text{LCL} = D_n \bar{r} = 0(2.27) = 0.$$

The *r*-chart and the accompanying sample data are shown in Figure 9.4. ∎

Process variability can also be monitored by using an *s*-chart for the sample standard deviations. Although the *s*-chart may be preferable to the *r*-chart (because the sample standard deviation is a better measure of variability than the sample range), the *r*-chart seems to be more popular among production and manufacturing personnel, because the sample range is easily understood and—for small sample sizes—tends to be very comparable to the sample standard deviation. Moreover, *r* is much easier

Sample	Sample Range
1	1.8
2	1.9
3	2.1
4	2.5
5	1.6
6	1.6
7	2.6
8	1.9
9	1.2
10	3.4
11	4.0
12	.7
13	3.8
14	2.2
15	3.5
16	1.4
17	3.2
18	1.2
19	2.8
20	2.8
21	1.5
22	1.3
23	1.8
24	2.0
25	4.0
Total	56.8

FIGURE 9.4
r-Chart for Example 9.3

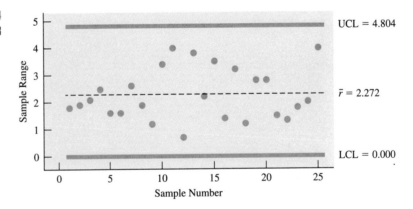

to compute than s, unless the calculations are computer-generated. For these reasons, we do not present the computations for an s-chart; however, the interpretations for an s-chart follow immediately from our previous discussions about $\bar{y}$- and r-charts.

A final comment should be made about the relationship between $\bar{y}$- and r-charts. Since the center line of an r-chart, $\bar{r}$, is used in constructing the limits of a $\bar{y}$-chart, the $\bar{y}$-chart should never be used without first constructing the corresponding r-chart. If the variability of the process is out of control, as determined by the r-chart, then the average of the sample ranges ($\bar{r}$) will not be reliable, and hence the control limits for the $\bar{y}$-chart will be unreliable.

Exercises

Basic Techniques

9.13 Explain the difference between a $\bar{y}$-chart and an r-chart. Give an example of a situation where each might be useful.

9.14 A manufacturing process is monitored for 60 days; on each day a random sample of $n = 6$ items is obtained. If $\bar{r} = 13.0$, determine the center line and the upper and lower control limits for an r-chart.

Application

9.15 Refer to Exercise 7.28. Construct an r-chart for these data. Is the process variability in control, to permit construction of a $\bar{y}$-chart?

9.4   Estimation and Tests for Comparing Two Population Variances

checking equal variance assumption

One major application of a test for the equality of two population variances is for **checking** the validity of the **equal variance assumption** (that is, $\sigma_1^2 = \sigma_2^2$) for a two-sample t test. First we hypothesize two populations of measurements that are normally distributed. We label these populations as 1 and 2, respectively. We are interested in comparing the variance of population 1 (σ_1^2) to the variance of population 2 (σ_2^2).

When independent random samples have been drawn from the respective populations, the ratio

$$\frac{s_1^2/s_2^2}{\sigma_1^2/\sigma_2^2}$$

F distribution

possesses a probability distribution in repeated sampling that is referred to as an **F distribution**. The formula for the probability distribution is omitted here, but we specify its properties next.

Properties of the F Distribution

1   Unlike t or z, but like χ^2, F can assume only positive values.

2   The F distribution, unlike the normal distribution or the t distribution, but like the χ^2 distribution, is nonsymmetrical (see Figure 9.5).

3   There are many F distributions, and each one has a different shape. We specify a particular one by designating the degrees of freedom associated with s_1^2 and s_2^2. We denote these quantities by df_1 and df_2, respectively.

FIGURE 9.5

Distribution of s_1^2/s_2^2, the F Distribution

4   Tail values for the F distribution are tabulated and appear in Tables 5 through 9 of Appendix 3.

Table 5 of Appendix 3 records the upper-tail value of F that has an area equal to .10 to its right (see Figure 9.6). The degrees of freedom for s_1^2, designated by df_1, are indicated across the top of the table, while df_2, the degrees of freedom for s_2^2, appear in the first column to the left. Thus, for $df_1 = 8$ and $df_2 = 10$, the tabulated value is 2.38. Only 10% of the measurements from an F distribution with $df_1 = 8$ and $df_2 = 10$ would exceed 2.38 in repeated sampling.

Table 6 of Appendix 3 gives the upper-tail values for the F distribution that has an area of .05 to its right. Thus, the .05 value of the F distribution with $df_1 = 6$ and $df_2 = 19$ is 2.63 (see Figure 9.7).

A statistical test of the null hypothesis $\sigma_1^2 = \sigma_2^2$ utilizes the test statistic s_1^2/s_2^2. When H_0 is true, s_1^2/s_2^2 follows an F distribution with $df_1 = n_1 - 1$ and $df_2 = n_2 - 1$. If upper-tail and lower-tail values of F were given in Tables 5 through 9 of Appendix 3, we would have no difficulty in performing the test. Unfortunately, only upper-tail values of F are given. To alleviate this situation, we are at liberty to identify either of the two populations as population 1. For a one-tailed alternative hypothesis, the populations are designated 1 and 2 so that H_a is of the form $\sigma_1^2 > \sigma_2^2$. Then the rejection region is located in the upper tail of the F distribution. For a two-tailed alternative, we designated the population with the larger sample variance as population 1. By this convention, we again are only concerned with upper-tail

FIGURE 9.6
Upper-Tail .10 Value for the F
Distribution; $df_1 = 8$ and
$df_2 = 10$

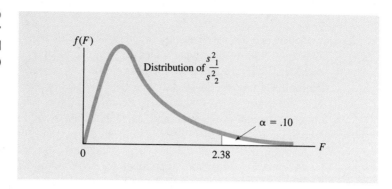

FIGURE 9.7
Critical Value for the F
Distribution; $df_1 = 6$ and
$df_2 = 19$

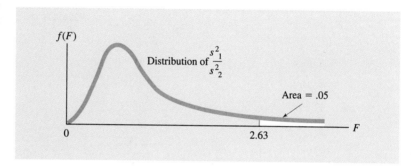

rejection regions. The upper-tail F-value for a two-tailed test can then be obtained from Tables 5 through 9 in Appendix 3.

We summarize the test procedure next.

Statistical Test Comparing σ_1^2 and σ_2^2

H_0: $\sigma_1^2 = \sigma_2^2$

H_a: **1** $\sigma_1^2 > \sigma_2^2$

         2 $\sigma_1^2 \neq \sigma_2^2$

T.S.: $F = \dfrac{s_1^2}{s_2^2}$

R.R.:   For a specified value of α:

1   Reject H_0 if F exceeds the tabulated value of F for $a = \alpha$, $df_1 = n_1 - 1$, and $df_2 = n_2 - 1$.

2   Reject H_0 if F exceeds the tabulated value of F for $a = \alpha/2$, $df_1 = n_1 - 1$, and $df_2 = n_2 - 1$.

Note: Tables 5 through 9 of Appendix 3 give upper-tail values for F corresponding to $a = .10, .05, .025, .01$, and $.005$.

EXAMPLE 9.4   Previously, we discussed an experiment in which company officials were concerned about the length of time a particular drug retained its potency. A random sample of ten bottles was obtained from the production line, and each bottle was analyzed to determine its potency. A second sample of ten bottles was obtained and stored in a regulated environment for one year. Potency readings were then obtained on these bottles at the end of the year, and the sample data were used to place a confidence interval on $\mu_1 - \mu_2$ (the difference in mean potencies for the two time periods).

In order to use t in the confidence interval or in a statistical test, we must draw the samples from normal populations with possibly different means *but* with a common variance. Use the following sample data summary to test the equality of the population variances. Use $\alpha = .05$. Sample 1 data are the readings taken immediately after production, and sample 2 data are the readings taken one year after production. Draw conclusions.

Sample 1: $\bar{y}_1 = 10.37,$ $s_1^2 = 0.105$
Sample 2: $\bar{y}_2 = 9.83,$ $s_2^2 = 0.058$

Solution   The five parts of the statistical test of $H_0: \sigma_1^2 = \sigma_2^2$ are shown here:

$H_0:$ $\sigma_1^2 = \sigma_2^2$

$H_a:$ $\sigma_1^2 \neq \sigma_2^2$

T.S.: $F = \dfrac{s_1^2}{s_2^2} = \dfrac{0.105}{.058} = 1.81$

R.R: For a two-tailed test with $\alpha = .05$, reject H_0 if $F > F_{.025,9,9} = 4.03$ (refer to Table 7 in Appendix 3, with $df_1 = 9$ and $df_2 = 9$). Conclusion: Since 1.81 does not fall in the rejection region, we cannot reject $H_0: \sigma_1^2 = \sigma_2^2$. And since 1.81 is not greater than $F_{.10,9,9} = 2.44$ (see Table 5 in Appendix 3), it appears that the assumption of equality of variances holds for the t-methods used with these data.   ▪

We can now formulate a confidence interval for the ratio σ_1^2/σ_2^2.

General Confidence Interval for σ_1^2/σ_2^2 with Confidence Coefficient $1 - \alpha$)

$$\frac{s_1^2}{s_2^2} F_L < \frac{\sigma_1^2}{\sigma_2^2} < \frac{s_1^2}{s_2^2} F_U$$

If F_{df_1,df_2} represents the $\alpha/2$ upper-tail value of an F distribution with df_1 and df_2 degrees of freedom, and if F_{df_2,df_1} represents the $\alpha/2$ upper-tail value of an F distribution with the degrees of freedom reversed, then

$$F_L = \frac{1}{F_{df_1,df_2}} \quad \text{and} \quad F_U = F_{df_2,df_1},$$

where $df_1 = n_1 - 1$ and $df_2 = n_2 - 1$.

Note: The confidence interval for σ_1/σ_2 is found by taking square roots throughout.

Although our estimation procedure for σ_1^2/σ_2^2 is appropriate for any confidence coefficient $(1 - \alpha)$, Tables 5 through 9 in Appendix 3 allow us to construct 80%, 90%, 95%, 98%, and 99% confidence intervals for σ_1^2/σ_2^2 with particular ease.

E X A M P L E   9.5   The life span of an electrical component was studied under two operating voltages, V_1 and V_2. Ten different components were randomly assigned to each of the two operating voltages. Use the following data to find a 90% confidence interval for σ_1^2/σ_2^2 (the ratio of the variances in life spans for the two populations, populations 1 and 2, corresponding to the components studied under V_1 and V_2, respectively).

Voltage V_1: $n_1 = 10$, $s_1^2 = .51$
Voltage V_2: $n_2 = 10$, $s_2^2 = .20$

Solution   Before constructing our confidence interval, we must obtain F_{df_1,d_2} and F_{df_2,df_1}. For $n_1 = n_2 = 10$, we know that $df_1 = df_2 = 9$; and hence F_{df_1,df_2} and F_{df_2,df_1} are the same. For a 90% confidence interval (that is, $1 - \alpha = .90$), we must look up the .05 F-value based on $df_1 = 9$ and $df_2 = 9$. Table 6 of Appendix 3 indicates that this value is 3.18. The quantities F_L and F_U are

$$F_L = \frac{1}{3.23} \quad \text{and} \quad F_U = 3.18.$$

Substituting into the confidence interval formula, we have

$$\frac{.51}{.20}\left(\frac{1}{3.18}\right) < \frac{\sigma_1^2}{\sigma_2^2} < \frac{.51}{.20}(3.18)$$

$$.80 < \frac{\sigma_1^2}{\sigma_2^2} < 8.11.$$

Therefore, we are 90% confident that the ratio of population variances corresponding to voltages V_1 and V_2 lies in the interval of .80 to 8.11.   ∎

E X A M P L E   **9.6**   Refer to Example 9.5. Suppose that one of the components on V_1 was damaged by the experimenter midway through the test period and had to be removed from the study. Then, with $n_1 = 9$ and $n_2 = 10$, it follows that $\mathrm{df}_1 = 8$ and $\mathrm{df}_2 = 9$. Assuming that s_1^2 and s_2^2 are as given in Example 9.5, set up a 90% confidence interval for σ_1^2/σ_2^2.

Solution   The appropriate .05 F-values can be obtained from Table 6 in Appendix 3:

$$F_{8,9} = 3.23 \qquad F_L = \frac{1}{3.23}$$
$$F_{9,8} = 3.39 \qquad F_U = 3.39.$$

We then have the confidence interval

$$\frac{.51}{.20}\left(\frac{1}{3.23}\right) < \frac{\sigma_1^2}{\sigma_2^2} < \frac{.51}{.20}(3.39)$$
$$.79 < \sigma_1^2/\sigma_2^2 < 8.64. \quad \blacksquare$$

The inferences about σ_1^2/σ_2^2 based on the F distribution are very sensitive to departures from normality by the underlying distributions. The first precaution you should take is to plot the data for each sample separately. At any hint that one or both of the populations may not be normal, be very cautious about the inferences you make on σ_1^2/σ_2^2 using the F distribution; the p-value or confidence coefficient may be substantially different from what you found by using the test or confidence interval based on F.

Several alternative procedures are available and are discussed in detail in other textbooks. For example, the Ansari–Bradley test can be used to compare the variances of two populations that have the same median (that is, the same location). The interested reader is referred to Hollander & Wolfe (1973) for details concerning alternatives to the F methods described in this section.

Exercises

Basic Techniques

9.16   Find the value of F that locates an area a in the upper tail of the F-distribution for these conditions:

a $a = .05$, $\mathrm{df}_1 = 7$, $\mathrm{df}_2 = 12$

b $a = .05$, $\mathrm{df}_1 = 3$, $\mathrm{df}_2 = 10$

c $a = .05$, $\mathrm{df}_1 = 10$, $\mathrm{df}_2 = 20$

d $a = .01$, $\mathrm{df}_1 = 8$, $\mathrm{df}_2 = 15$

e $a = .01$, $\mathrm{df}_1 = 13$, $\mathrm{df}_2 = 25$

9.17   Find approximate values for F_a for these conditions:

a $a = .05$, $\mathrm{df}_1 = 11$, $\mathrm{df}_2 = 24$

b $a = .05$, $\mathrm{df}_1 = 14$, $\mathrm{df}_2 = 14$

c $a = .05$, df$_1 = 35$, df$_2 = 22$

d $a = .01$, df$_1 = 22$, df$_2 = 24$

e $a = .01$, df$_1 = 17$, df$_2 = 25$

(*Note*: Your answers may not agree with those in the back of the book. As long as your answer is close to the recorded answer, it is satisfactory.)

9.18   Random samples of $n_1 = 8$ and $n_2 = 10$ observations were selected from populations 1 and 2, respectively. The corresponding sample variances were $s_1^2 = 7.4$ and $s_2^2 = 12.7$. Do the data provide sufficient evidence to indicate a difference between σ_1^2 and σ_2^2? Test by using $\alpha = .10$. What assumptions have you made?

9.19   An experiment was conducted to determine whether there was sufficient evidence to indicate that data variation within one population—say, population A—exceeded the variation within a second population—population B. Random samples of $n_A = n_B = 8$ measurements were selected from the two populations, and the sample variances were calculated to be

$$s_A^2 = 2.87 \qquad s_B^2 = .91$$

Do the data provide sufficient evidence to indicate that σ_A^2 is larger than σ_B^2? Test by using $\alpha = .05$.

Applications

9.20   A soft-drink firm is debating whether it should invest in a new type of canning machine or continue operating with the machines currently in use. The company has already determined that it will be able to fill more cans per day for the same cost if the new machines are installed. However, an important factor as yet unsolved is the variability of fills. (The company would, of course, prefer the model with the smaller variance in fills.) Let σ_1^2 and σ_2^2 denote the variances for fills from the old model and the new model, respectively. Obtaining samples of fills from the two models and utilizing the test statistics s_1^2/s_2^2, we can set up either a one-tailed or a two-tailed rejection region, using the F distribution.

a   What type of rejection region would be most favored by the manager of the soft-drink company? Why?

b   What type of rejection region would be most favored by the salesperson for the company manufacturing the model presently in use? Why?

9.21   Refer to Exercise 9.20. Suppose that random samples of $n_1 = n_2 = 11$ cans from the two machines are examined to determine the amount of fill (in ounces). The means and variances are

$$\bar{y}_1 = 11.70 \qquad \bar{y}_2 = 11.60$$
$$s_1^2 = .06 \qquad s_2^2 = .022.$$

Do these data present sufficient evidence to indicate less variability of fills for the new model? Use $\alpha = .10$.

9.22   In a gasoline economy study, ten 1-gallon samples of a particular brand of gasoline were used for each of two cars (A and B). Both cars averaged approximately 17 miles per gallon, but the sample standard deviations were .95 and 1.56 for cars A and B, respectively. Use these data to test the hypothesis that the variances in miles per gallon for the two cars are identical. Use $\alpha = .05$.

9.5   Using Computers to Help Make Sense of Data

We can use the Automatic Transmission database to illustrate the use of r and s charts for product variability. Recall that a random sample of five automatic transmissions is obtained from the daily production, and each is tested for internal pressure. This procedure is followed for each of 40 days. Computer software was used to compute the mean, range, and standard deviation of the five measurements for each of the 40 days. Control charts for r and for s have been obtained, and the sample data have been plotted using MINITAB and EXECUSTAT. Examine both sets of output to find the control charts for r and s and the data plots.

MINITAB Output

```
MTB > NAME C1 'DAY' C2 'ITEM1' C3 'ITEM2' C4 'ITEM3' C5 'ITEM4' C6 'ITEM5'
MTB > NAME C7 'MEAN' C8 'RANGE' C9 'STDDEV'
MTB > RMEAN C2 - C6 INTO C7
MTB > RRANGE C2 - C6 INTO C8
MTB > RSTDDEV C2 - C6 INTO C9
MTB > PRINT C1 - C9
```

ROW	DAY	ITEM1	ITEM2	ITEM3	ITEM4	ITEM5	MEAN	RANGE	STDDEV
1	1	6.01	4.46	4.90	3.83	4.61	4.762	2.18	0.79992
2	2	6.06	6.26	5.44	3.86	5.88	5.500	2.40	0.96551
3	3	4.46	6.17	4.07	4.29	4.29	4.656	2.10	0.85760
4	4	5.08	4.68	4.37	4.50	4.40	4.606	0.71	0.29134
5	5	4.11	5.84	5.67	4.55	5.62	5.158	1.73	0.77600
6	6	4.58	5.90	4.35	5.25	4.18	4.852	1.72	0.71314
7	7	6.04	4.45	4.22	5.09	4.68	4.896	1.82	0.71570
8	8	4.98	5.19	5.70	4.91	2.97	4.750	2.73	1.04199
9	9	6.48	5.95	4.53	6.25	6.08	5.858	1.95	0.76842
10	10	5.30	5.98	5.36	3.83	4.56	5.006	2.15	0.82800
11	11	3.56	3.95	6.38	4.90	4.86	4.730	2.82	1.08922
12	12	4.96	6.78	6.56	4.32	5.25	5.574	2.46	1.05843
13	13	4.39	3.16	4.31	4.43	6.33	4.524	3.17	1.13960
14	14	2.88	4.62	5.70	5.77	3.83	4.560	2.89	1.23719
15	15	2.81	4.27	3.19	6.02	5.94	4.446	3.21	1.49954
16	16	2.77	3.20	3.60	5.75	4.57	3.978	2.98	1.19351
17	17	4.88	3.37	4.69	4.02	3.30	4.052	1.58	0.72875
18	18	6.06	4.49	3.40	5.03	6.63	5.122	3.23	1.27741
19	19	6.17	2.64	5.90	4.75	5.22	4.936	3.53	1.39990
20	20	5.85	5.00	3.31	4.58	7.37	5.222	4.06	1.51042
21	21	5.10	5.39	5.37	4.33	7.28	5.494	2.95	1.08698
22	22	7.29	2.77	3.54	7.45	5.14	5.238	4.68	2.12647
23	23	4.44	5.87	5.52	5.03	4.13	4.998	1.74	0.72434
24	24	4.94	5.97	6.30	8.22	4.72	6.030	3.50	1.39435
25	25	5.56	5.59	4.63	3.56	6.84	5.236	3.28	1.22263
26	26	5.48	4.74	6.51	6.76	4.13	5.524	2.63	1.12469
27	27	5.31	6.87	2.82	3.55	3.47	4.404	4.05	1.65927
28	28	3.69	4.01	5.16	3.87	4.93	4.332	1.47	0.66567
29	29	3.32	6.22	2.12	6.01	5.60	4.654	4.10	1.82939
30	30	5.14	6.57	6.37	6.98	6.66	6.344	1.84	0.70812
31	31	4.93	6.02	5.10	5.58	6.62	5.650	1.69	0.68986
32	32	4.93	5.45	2.16	6.25	5.05	4.768	4.09	1.54668
33	33	6.64	7.00	5.39	4.87	6.76	6.132	2.13	0.94195
34	34	7.85	4.97	4.68	5.48	4.07	5.410	3.78	1.45607
35	35	4.02	6.64	7.62	5.91	4.15	5.668	3.60	1.56795
36	36	4.51	5.42	4.81	5.00	3.74	4.696	1.68	0.62812
37	37	3.62	5.41	4.78	1.78	3.88	3.894	3.63	1.38090
38	38	3.06	5.90	6.96	4.96	2.72	4.720	4.24	1.81819
39	39	2.04	3.22	3.76	3.44	6.76	3.844	4.72	1.75479
40	40	4.99	6.20	4.73	3.87	3.79	4.716	2.41	0.98116

MINITAB Output (continued)

```
MTB > NAME C20 'ALLDATA' C21 'ALLDAYS' C22 'SORTDATA' C23 'SORTDAYS'
MTB > STACK C2 - C6 INTO C20
MTB > SET C21
DATA> 5(1:40)
DATA> END
MTB > SORT C20 INTO C22;
SUBC> BY C21.
MTB > SET C23
DATA> (1:40)5
DATA> END
MTB > RCHART C22 SUBGROUPS ARE IN C23;
SUBC> RBAR.
```

```
                        R Chart for SORTDATA
              -
              -
  S   7.50+  -
  a         -
  m         -
  p         ---------------------------------------------UCL=5.901
  l         -
  e   5.00+  -
              -
  R         -                          +
  a         -              +        +   + +  ++      +
  n         -              +     +       ++      + +
  g         -----------+-++++-+--+------------------R̄=2.791
  g   2.50+ +      +     +                   +        +
  e         -+  +    +  ++                +  +
              -     ++        +        +    +  +   +
              -
              -
        0.00+-+-----------------------------------------LCL=0.000
              +---------+---------+---------+---------+
              0        10        20        30        40
                         Sample Number
```

MINITAB Output(Continued)

```
MTB > SCHART C22 SUBGROUPS ARE IN C23;
SUBC> RBAR.
```

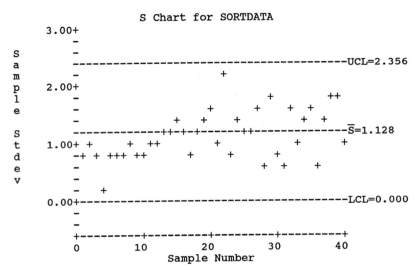

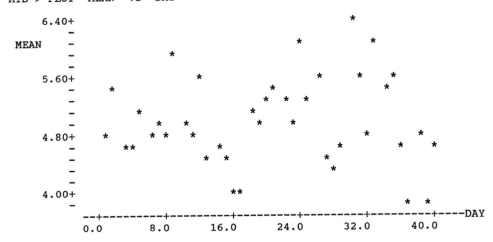

```
MTB > STOP
```

EXECUSTAT Output

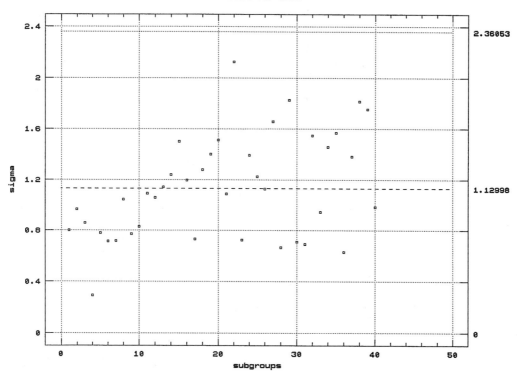

EXECUSTAT Output (continued)

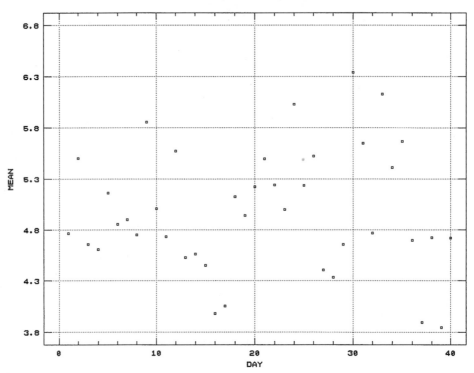

Scatterplot of MEAN vs DAY

There are no obvious ups and downs in the sample data plots; the means seem to stay in the 4.00–6.00 range corresponding to the ideal level of performance. One troubling feature of the data plot is that the means in days 31–40 seem to "jump around." The r and s charts confirm that the variability (ranges and standard deviations) increases over time, but the process is not yet out of control. Higher variability over time means that more transmissions will either be too high or too low in pressure. In seeking a cause, we should bear in mind that the data show no sudden increase in variability, but rather a gradual increase. Thus a factor such as worker complacency or machine wearout might be the cause.

As a second illustration of how computers can help us understand data—as well as some of the assumptions underlying the methods we use for analyzing data—we ran the following simulation to show how sensitive (or not) the chi-square test for population variance is to departures from normality. (In this case, the population of measurements is skewed by the presence of outliers.) A simulation study involved 1000 samples of size 51 each from a moderately outlier-prone population. The population variance was 64.8. A χ^2 test of the variance was performed for each sample. The results are shown in the following output. What do the results indicate about the test?

```
one-tail:

        number of times Ho: variance = 64.8 is rejected in favor of
   alpha      "variance < 64.8"      "variance > 64.8"      total(alpha doubled)
   0.100            205                     289                     494
   0.050            162                     221                     383
   0.025            127                     171                     298
   0.010            106                     111                     217
   0.005             87                      86                     173
```

As can be seen from the computer's output, the "stated" values of α used for testing each sample of 51 values are much smaller than the observed fraction of times H_0 was rejected in the simulation study of 1000. This provides evidence that, for this type of population, the chi-square rejects H_0 at a much higher rate than is specified by α. Consequently, the χ^2 test is quite sensitive to this type of nonnormality.

Summary

In this chapter, we discussed procedures for making inferences involving a population variance and involving the ratio of two population variances. Estimation and statistical tests involving σ^2 make use of the chi-square probability distribution, with df $= n - 1$. Inferences involving the ratio of two population variances utilize an F distribution, with $df_1 = n_1 - 1$ and $df_2 = n_2 - 1$.

The need for inferences involving one or more population variances can be traced to our discussion of numerical descriptive measures of a population in Chapter 4. To describe or make inferences about a population of measurements, we cannot always rely on the mean, since it is strictly a measure of central tendency. Often, in evaluating or comparing the performance of individuals on a psychological test, the consistency of manufactured products emerging from a production line, or the yields of a particular variety of corn, we can gain important information by studying the population variance.

In the next chapter, we consider additional applications that use the chi-square distribution.

Key Terms

point estimate for σ^2

chi-square distribution

normality assumption

checking equal variance assumption

F distribution

Key Formulas

1    $100(1 - \alpha)\%$ confidence interval for σ^2 (or σ):

$$\frac{(n-1)s^2}{\chi_U^2} < \sigma^2 < \frac{(n-1)s^2}{\chi_L^2}$$

or

$$\sqrt{\frac{(n-1)s^2}{\chi_U^2}} < \sigma < \sqrt{\frac{(n-1)s^2}{\chi_L^2}}.$$

2   Statistical test for σ^2:

H_0: $\sigma^2 = \sigma_0^2 (\sigma_0^2$ is specified)

T.S.: $\chi^2 = \dfrac{(n-1)s^2}{\sigma_0^2}$

3   Statistical test for σ_1^2/σ_2^2:

H_0: $\sigma_1^2 = \sigma_2^2$

T.S.: $F = \dfrac{s_1^2}{s_2^2}$

4    $100(1 - \alpha)\%$ confidence interval for σ_1^2/σ_2^2 (or σ_1/σ_2):

$$\frac{s_1^2}{s_2^2} F_L < \frac{\sigma_1^2}{\sigma_2^2} < \frac{s_1^2}{s_2^2} F_U,$$

where

$$F_L = \frac{1}{F_{\mathrm{df}_1, \mathrm{df}_2}} \qquad \text{and} \qquad F_U = F_{\mathrm{df}_2, \mathrm{df}_1}$$

or

$$\sqrt{\frac{s_1^2}{s_2^2} F_L} < \frac{\sigma_1}{\sigma_2} < \sqrt{\frac{s_1^2}{s_2^2} F_U}.$$

Supplementary Exercises

9.23   Two consumer research groups are vying for a large government contract. Since subjective evaluations of consumer products will be made by judges during the study, government officials prefer to award the contract to a company that utilizes judges with consistent ratings (of course, other qualifications are also evaluated before awarding the contract). One measure of consistency is the variability of judges' scores on the same item.

Before issuing the contract, a test is conducted in which 25 judges from each company are asked to rate a single item. The sample variances are given here:

$$\text{company A:}\quad s_1^2 = .50 \qquad \text{company B:}\quad s_2^2 = .15$$

Use these data to test the hypothesis that the variances of the judges' ratings are the same for the two populations. The alternative hypothesis is that the variances are different. Use $\alpha = .10$.

9.24   Refer to Exercise 8.19, in which we were interested in comparing the weights of the combs of roosters fed one of two vitamin-supplemented diets. The Wilcoxon rank sum test was suggested as a test of the hypothesis that the two populations were identical. Would it have been appropriate to run a t test comparing the two population means? Explain.

9.25   A consumer-protection magazine was interested in comparing tires purchased from two different companies, each claiming that its tires would last the same number of miles. A sample of five tires of each brand was obtained and tested under simulated road conditions. The number of miles before significant deterioration in tread was recorded for all tires. The data are given next (in 1000 miles).

Brand I	40.6	35.9	48.5	36.4	38.3
Brand II	40.9	40.2	42.5	39.1	42.6

a   Construct a 98% confidence interval for the ratio of the two population variances.

b   How does the confidence interval change if we're interested in σ_1/σ_2 rather than in σ_1^2/σ_2^2?

9.26   A random sample 20 patients, each of whom has suffered from depression, was selected from a mental hospital, and each patient was administered the Brief Psychiatric Rating Scale. The scale consists of a series of adjectives that the patient scores according to his or her mood. Extensive testing in the part has shown that ratings in certain mood adjectives tend to be similar and hence are grouped together as jointly measuring one or more components of a person's mood. For example, a group consisting of certain adjectives seems to measure depression. Let us suppose that the mean and the standard deviation of the 20 patients in the group are 13.2 and 4.6, respectively.

a   Place a 99% confidence interval on σ^2, the variance of the population of patients' scores from which this sample was drawn.

b   What critical assumption underlies the inference? Do you know whether this assumption is valid for these data?

9.27   Refer to Exercise 9.26. Suppose that extensive testing of a large number of depressed patients throughout the century has indicated that the population standard deviation of scores for the depression adjectives is 5.9. Use the sample data of Exercise 9.26 to test the research hypothesis that the standard deviation for all patients who might be treated for depression in this hospital is less than 5.9. Give a p-value for these data, and draw conclusions.

9.28   A pharmaceutical company manufactures a particular brand of antihistamine tablets. In the quality-control division, certain test are routinely performed to determine whether the product being manufactured meets specific performance criteria prior to release of the product onto the market. In particular, the company requires that the potencies of the tablets lie in the range of 90% to 110% of the labeled drug amount.

a   If the company is manufacturing 25-mg tablets, within what limits must the tablets' potencies lie?

b   A random sample of 30 tablets is obtained from a recent batch of antihistamine tablets. The data for the potencies of the tablets are given below. Is the assumption of normality warranted for inferences about the population variances?

c   Translate the company's 90% to 110% specification on the acceptable range of product potency into a statistical test of the population variance for potencies. Draw conclusions based on $\alpha = .05$.

24.1   27.2   26.7   23.6   26.4   25.2
25.8   27.3   23.2   26.9   27.1   26.7
22.7   26.9   24.8   24.0   23.4   25.0
24.5   26.1   25.9   25.4   22.9   24.9
26.4   25.4   23.3   23.0   24.3   23.8

9.29 A study was conducted to compare the variabilities in strengths of 1-inch-square sections of a synthetic fiber produced under two different procedures. Random samples of nine squares from each process were obtained and tested.

a Plot the data for each sample separately.

b Is the assumption of normality warranted?

c If permissible from part b, use the following data (expressed in psi) to test the research hypothesis that the population variances corresponding to the two procedures are different. Use $\alpha = .10$.

Procedure 1	74	90	103	86	75	102	97	85	69
Procedure 2	59	66	73	68	70	71	82	69	74

9.30 Refer to Example 9.2. Construct a 95% confidence interval for σ^2, and use this interval to help interpret the findings of the consumer group. Does it appear that the test of Example 9.2 had much power to detect an increase in σ^2 of 25% over the claimed value? Explain.

9.31 The risk of an investment is measured in terms of the variance in the return that could be observed. Random samples of 10 yearly returns were obtained from two different portfolios. These yielded the following information.

	Portfolio	
Statistic	1	2
Sample mean return (000)	132	146
Sample variance	10.9	25.6
Sample size	10	10

Does Portfolio 2 have a higher risk? Give a *p*-value for your test.

9.32 Refer to Exercise 9.31. Are there any differences in the average returns for the two portfolios? Indicate the method you used in arriving at a conclusion, and explain why you used it.

9.33 Two different modeling techniques for assessing the resale value of houses were considered. A random sample of 12 existing listings was taken, and each house was valued according to the two techniques. These data are shown here.

Listing	Assessed Value Listing (000) Technique 1	2	Listing	Assessed Value Listing (000) Technique 1	2
1	155	138	7	63	67
2	137	128	8	129	134
3	248	230	9	144	149
4	136	146	10	270	292
5	102	95	11	157	150
6	87	82	12	51	48

a Plot the data. Do the two modeling techniques appear to give similar results?

b Give estimates of the mean and the standard error of the difference between estimates for the two methods.

9.34 Refer to Exercise 9.33. Place a 90% confidence interval on the variance of the difference in estimates. Give the corresponding interval for σ.

9.35 Refer to Exercises 9.33 and 9.34. What is the critical assumption about the sample data? How would you check this assumption? Do the data suggest that the assumption holds? Do you have any cautions about the inferences in Exercise 9.34?

9.36 An important consideration in examining the potency of a pharmaceutical product is the drop in potency for a specific shelf life (time on a pharmacist's shelf). In particular, the variability of these drops in potency is very important. Researchers studied the drops in potency for two different drugs over a 6-month period. These data are summarized in the accompanying table. Suppose that drug 1 is an experimental drug product and that drug 2 is a marketed product. Use a one-tailed test with $\alpha = .01$ to determine whether the data suggest that drug 1 has more variability in potency drop than drug 2.

Statistic	Drug 1	Drug 2
Sample size	10	10
Sample mean	58	56
Sample variance	82	23

9.37 Refer to Exercise 9.36. Would your result have changed if you had used a two-tailed test with $\alpha = .10$? Why might a two-tailed test be important?

9.38 Blood cholesterol levels for randomly selected patients with similar histories were compared for two diets: a low-fat-content diet and a normal diet. The summary data appear in the accompanying table.

Statistic	Low-Fat Content	Normal
Sample size	19	24
Sample mean	170	196
Sample variance	198	435

a Do these data present sufficient evidence to indicate a difference in cholesterol level variabilities for the two diets? Use $\alpha = .10$.

b What other test might be of interest in comparing the two diets?

9.39 Sales from weight-reducing agents marketed in the United States represent sizable chunks of income for many of the companies that manufacture these products. Psychological as well as physical effects often contribute to how well a person responds to the recommended therapy. Consider a comparison of two weight-reducing agents, A and B. In particular, consider the variabilities in the lengths of time people remain on the therapy. A total of 26 overweight males, matched as closely as possible physically, were randomly divided into two groups. Those in group 1 received preparation A, while those assigned to group 2 received preparation B. Use the accompanying summary data to compare the variabilities associated with the lengths of time on therapy. Use a two-tailed test with $\alpha = .10$.

Statistic	Preparation A	Preparation B
Sample size	13	13
Sample mean	25 days	35 days
Sample variance	50	16

9.40   Refer to Exercise 9.39. What might the null and alternative hypotheses have been if preparation A had been a placebo (no active medication) and preparation B a marketed product known to be an effective weight-reducing agent?

9.41   A chemist at an iron ore mine suspects that the variance in the amount (weight, in ounces) of iron oxide per pound of ore tends to increase as the mean amount of iron oxide per pound increases. To test this theory, ten 1-pound specimens of iron ore are selected at each of two locations—one (location 1) containing a much higher mean content of iron oxide than the other (location 2). The amounts of iron oxide contained in the ore specimens are shown in the accompanying table.

Location 1	8.1	7.4	9.3	7.5	7.1	8.7	9.1	7.9	8.4	8.8
Location 2	3.9	4.4	4.7	3.6	4.1	3.9	4.6	3.5	4.0	4.2

Do the data provide sufficient evidence to indicate that the amount of iron oxide per pound of ore is more variable at location 1 than at location 2? Use $\alpha = .05$.

9.42   One index of service quality for telephone reservation systems is the waiting time from the first ring until an agent answers, ready to make reservations. The waiting times for a sample of 30 reservation calls during a period of 1 week showed a mean and a standard deviation of 28.4 and 17.4 seconds, respectively. Use these data to construct a 95% confidence interval for the standard deviation of waiting time.

9.43   A personnel officer was planning to use a t test to compare the mean number of monthly unexcused absences for two divisions of a multinational company, but then she noticed a possible difficulty. The variation in the number of unexcused absences per month seemed to differ for the two groups. As a check, a random sample of 5 months was selected at each division; and for each month, the number of unexcused absences was obtained.

Category A	20	14	19	22	25
Category B	37	29	51	40	26

a   What assumption seemed to bother the personnel officer?

b   Do the data provide sufficient evidence to indicate that the variances differ for the populations of absences for the two employee categories? Use $\alpha = .05$.

9.44   A researcher was interested in weather patterns in Phoenix and Seattle. As part of the investigation, the researcher took a random sample of 20 days in July and observed the daily average temperatures. The data were collected over several years to ensure independence of daily temperatures. The collected data produced the following information.

Statistic	Phoenix Daily Average Temperature	Seattle Daily Average Temperature
Sample size	20	20
Sample mean	95.3	63.3
Sample standard deviation	5.1	7.6

Do the data suggest that there is a difference in the variability of average daily temperatures during July for the two cities? Is there a difference in mean temperatures for the two cities during July? Use $\alpha = .05$ for both tests.

Experiences with Real Data

9.45 Refer to the Clinical Trial database on your data disk (or in Appendix 1) to calculate the sample variances for the anxiety scores within each treatment group. Use these data to run separate tests comparing each of the treatments A, B, and C to the placebo group D. Use two-sided tests, with $\alpha = .05$.

9.46 Do any of the tests in Exercise 9.45 negate the possibility of comparing the treatment means for groups A, B, and C to the treatment mean for the placebo group, using t tests? Explain.

9.47 Use the sleep disturbance scores from the Clinical Trial database to give a 98% confidence interval for σ_B^2/σ_C^2. Do the same for σ_B^2/σ_A^2.

9.48 Refer to the Bonus Level database on your data disk. Construct a 95% confidence interval for σ^2 (and σ), the population variance (and the standard deviation) for bonus level. Graph the data to check any underlying assumptions. Are there any potential problems? What inferences can you make?

9.49 Refer again to the Bonus Level database. Conduct a statistical test to compare the population variances for male and female bonuses. How did you check any assumptions? What sense can you make of these data?

10

Analyzing Count Data

10.1   Introduction

For the inferences discussed in Chapters 7 through 9, we have been concerned primarily with sample data measured on a continuous scale. However, we sometimes encounter situations where levels of the variable of interest are identified by name or rank only and we are interested in the number of observations occurring at each level of the variable. In Chapter 3, we called these types of data *ordinal* (or *count) data*. For example, an item coming off an assembly line may be classified into one of three quality classes: acceptable, second, or reject. Similarly, a traffic study might require a count and classification of the type of transportation used by commuters along a major access road into a city. A pollution study might be concerned with the number of different alga species identified in samples from a lake and the number of times each species is identified. A consumer protection group might be interested in the results of a prescription fee survey to compare prices on a checklist of medications in different sections of a large city.

The data analysis methods in this chapter are appropriate for ordinal and for some discrete data—especially data from a binomial experiment or its extension, the multinomial experiment.

10.2   The Multinomial Experiment

multinomial experiment

The examples in Section 10.1 all exhibit, to a reasonable degree of approximation, the characteristics of a **multinomial experiment**.

The Multinomial Experiment

1   The experiment consists of n identical trials.

2   Each trial results in one of k outcomes.

3   The probability that a single trial will result in outcome i is p_i. $i = 1, 2, \ldots, k$, and remains constant from trial to trial. (*Note:* $\sum_i p_i = 1$.)

4   The trials are independent.

5   We are interested in n_i, the number of trials resulting in outcome i. (*Note:* $\sum_i n_i = n$.)

---

multinomial distribution

The probability distribution for the number of observations resulting in each of the k outcomes, called the **multinomial distribution**, is given by the formula

$$P(n_1, n_2, \ldots, n_k) = \frac{n!}{n_1! n_2! \cdots n_k!} p_1^{n_1} p_2^{n_2} \cdots p_k^{n_k}.$$

Recall from Chapter 5, where we discussed the binomial probability distribution, that

$$n! = n(n-1) \cdots 1$$

and

$$0! = 1$$

We can use the formula for the multinomial distribution to compute the probability of particular events.

EXAMPLE   10.1   Previous breeding experience with a particular herd of cattle suggests that the probability of obtaining one healthy calf from a mating is .83. Similarly, the probability of obtaining zero or two healthy calves is respectively, .15 or .02. If a farmer breeds three dams from the herd, what is the probability of obtaining exactly three healthy calves?

Solution Assuming that the three dams are chosen at random, this experiment can be viewed as a multinomial experiment with $n = 3$ trials and $k = 3$ outcomes. These outcomes are listed below with their corresponding probabilities.

Outcome	Number of Progeny	Probability (p_i)
1	0	.15
2	1	.83
3	2	.02

Notice that outcomes 1, 2, and 3 refer to the events that a dam produces zero, one, or two healthy calves, respectively. Similarly, n_1, n_2, and n_3 refer to the number of dams producing zero, one, or two healthy progeny, respectively. To obtain exactly three healthy progeny, we must observe one of the following possible events.

$$A: \begin{cases} 1 \text{ dam gives birth to } 0 \text{ healthy progeny}: n_1 = 1 \\ 1 \text{ dam gives birth to } 1 \text{ healthy progeny}: n_2 = 1 \\ 1 \text{ dam gives birth to } 2 \text{ healthy progeny}: n_3 = 1 \end{cases}$$

$$B: 3 \text{ dams give birth to } 1 \text{ healthy progeny}: \begin{cases} n_1 = 0 \\ n_2 = 3 \\ n_3 = 0 \end{cases}$$

For event A, with $n = 3$ and $k = 3$,

$$P(n_1 = 1, n_2 = 1, n_3 = 1) = \frac{3!}{1!1!1!}(.15)^1(.83)^1(.02)^1 \simeq .015.$$

Similarly, for event B,

$$P(n_1 = 0, n_2 = 3, n_3 = 0) = \frac{3!}{0!3!0!}(.15)^0(.83)^3(.02)^0 = (.83)^3 \simeq .572.$$

Thus, the probability of obtaining exactly three healthy progeny from three dams is the sum of the probabilities for events A and B—namely, $.015 + .572 \simeq .59$. ▪

10.3 Chi-Square Goodness-of-Fit Test

Our primary interest in the multinomial distribution is as a probability model underlying statistical tests about the probabilities $p_1, p_2, \ldots, p_k$. We hypothesize specific values for the ps and then determine whether the sample data agree with the hypothesized values. One way to test such a hypothesis is to examine the observed number of trials resulting in each outcome and to compare this to the number we would *expect* to result in each outcome. For instance, in our previous example, we gave the probabilities associated with zero, one, and two progeny as .15, .83, and .02. If we

expected number of outcomes

were to examine a sample of 100 mated dams, we would **expect to observe** 15 dams that produce no healthy progeny. Similarly, we would expect to observe 83 dams that produce 1 healthy calf and 2 dams that produce 2 healthy calves.

D E F I N I T I O N   10.1

Expected Number of Outcomes   In a multinomial experiment where each trial can result in one of k outcomes, the **expected number of outcomes** of type i in n trials is np_i, where p_i is the probability that a single trial results in outcome i. ▪

In 1900, Karl Pearson proposed the following test statistic to test the specified probabilities:

χ^2

$$\chi^2 = \sum_i \left[\frac{(n_i - E_i)^2}{E_i} \right],$$

where n_i represents the number of trials resulting in outcome i and E_i represents the number of trials we would expect to result in outcome i when the hypothesized probabilities represent the actual probabilities assigned to each outcome. Frequently,

cell probabilities

we refer to the probabilities $p_1, p_2, \ldots, p_k$ as **cell probabilities,** with one cell corresponding to each of the k outcomes. The observed numbers $n_1, n_2, \ldots, n_k$ corre-

observed cell counts
expected cell counts

sponding to the k outcomes are called **observed cell counts;** and the expected numbers $E_1, E_2, \ldots, E_k$ are termed **expected cell counts.** Suppose that we hypothesize values for the cell probabilities $p_1, p_2, \ldots, p_k$. We can then calculate the expected cell counts by using Definition 10.1 to examine how well the observed data fit, or agree, with what we would expect to observe. Certainly, if the hypothesized values for the p_i are correct, the observed cell counts n_i should not deviate greatly from the expected cell counts E_i and the computed value of χ^2 should be small. Conversely, when one or more of the hypothesized cell probabilities are incorrect, the observed and expected cell counts will differ substantially, making χ^2 large.

chi-square distribution

The distribution of the quantity χ^2 can be approximated by a **chi-square distribution,** provided that the expected cell counts E_i are fairly large (see Figure 10.1).

F I G U R E   10.1
Chi-Square Probability
Distribution for df $= 4$

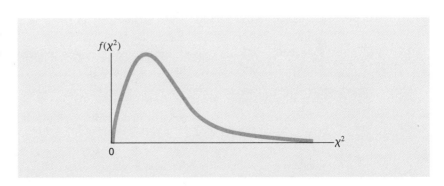

The chi-square goodness-of-fit test based on k specified cell probabilities has $k - 1$ degrees of freedom. Upper-tail values of the test statistic

$$\chi^2 = \sum_i \left[\frac{(n_i - E_i)^2}{E_i} \right]$$

can be found in Table 4 in Appendix 3.

We can now summarize the chi-square goodness-of-fit test with regard to k specified cell probabilities.

Chi-Square Goodness-of-Fit Test

H_0: $p_i = p_{i0}$ for categories $i = 1, \ldots, k$, p_{i0} are specified probabilities or proportions.

H_a: At least one of the cell probabilities differs from the hypothesized value.

$T.S.$: $\chi^2 = \sum_i \left[\frac{(n_i - E_i)^2}{E_i} \right]$, where n_i is the observed number in category i and $E_i = n p_{i0}$ is the expected number under H_0.

$R.R.$: Reject H_0 if χ^2 exceeds the tabulated critical value for $a = \alpha$ and df $= k - 1$.

Some researchers (see, for example, Siegel (1956) and Dixon & Massey (1969)) recommend that all the E_is should be 5 or more before performing this test. This requirement is perhaps too stringent. Cochran (1954) indicates that the approximation should be quite good if no E_i is less than 1 and if no more than 20% of the E_is are less than 5. We recommend applying Cochran's guidelines for determining whether χ^2 can be approximated with a chi-square distribution. We can combine categories if some of the E_is are too small, but care should be taken so that the combination of categories does not change the nature of the hypothesis to be tested.

EXAMPLE   10.2   A test drug is to be compared against a standard drug preparation for patients suffering from high blood pressure. Over many clinical trials at many different locations, patients suffering from comparable hypertension (as measured by the New York Heart Association (NYHA) Classification) have been administered the standard therapy. Responses to therapy for this large patient group were classified into one of four response categories. Table 10.1 lists the categories and percentages of patients treated with the standard preparation who have been classified in each category.

A clinical trial is conducted with a random sample of 200 patients who suffer from high blood pressure. All patients are required to be listed according to the same hypertensive categories of the NYHA Classification as those studied under the standard preparation. Use the sample data in Table 10.2 to test the hypothesis that the cell probabilities associated with the test preparation are identical to those for the standard. Use $\alpha = .05$.

Category	Percentage
Marked decrease in blood pressure	50%
Moderate decrease in blood pressure	25%
Slight decrease in blood pressure	10%
Stationary or slight increase in blood pressure	15%

TABLE 10.2
Sample Data for Example 10.2

Category	Observed Cell Counts
1	120
2	60
3	10
4	10

Solution This experiment possesses the characteristics of a multinomial experiment, with $n = 200$ and $k = 4$ outcomes.

Outcome 1: A person's blood pressure will decrease markedly after treatment with the test drug.

Outcome 2: A person's blood pressure will decrease moderately after treatment with the test drug.

Outcome 3: A person's blood pressure will decrease slightly after treatment with the test drug.

Outcome 4: A person's blood pressure will remain stationary or increase slightly after treatment will the test drug.

The null and alternative hypotheses are then

$$H_0: p_1 = .50, \ p_2 = .25, \ p_3 = .10, \ p_4 = .15$$

and

H_a: At least one of the cell probabilities is different from the hypothesized value.

Before computing the test statistic, we must determine the expected cell numbers. These data are given in Table 10.3.

TABLE 10.3
Observed and Expected Cell
Numbers for Example 10.2

Category	Observed Cell Number (n_i)	Expected Cell Number (E_i)
1	120	$200(.50) = 100$
2	60	$200(.25) = 50$
3	10	$200(.10) = 20$
4	10	$200(.15) = 30$

Since all the expected cell numbers are large, we may calculate the chi-square statistic and compare it to a tabulated value of the chi-square distribution.

$$\chi^2 = \sum_i \left[\frac{(n_i - E_i)^2}{E_i} \right]$$

$$= \frac{(120 - 100)^2}{100} + \frac{(60 - 50)^2}{50} + \frac{(10 - 20)^2}{20} + \frac{(10 - 30)^2}{30}$$

$$= 4 + 2 + 5 + 13.33 = 24.33$$

For the probability of a Type I error set at $\alpha = .05$, we look up the value of the chi-square statistic for $a = .05$ and df $= k - 1 = 3$. The critical value from Table 4 of Appendix 3 is 7.815.

$$\text{R.R.:} \qquad \text{Reject } H_0 \text{ if } \chi^2 > 7.815.$$

Conclusion: Since the computed value of χ^2 is greater than 7.815, we reject the null hypothesis and conclude that at least one of the cell probabilities differs from that specified under H_0. Practically, it appears that a much higher proportion of patients treated with the test preparation fall into the moderate and marked improvement categories. ∎

The assumptions needed for running a chi-square goodness-of-fit test are those associated with a multinomial experiment; the key ones are independence of the trials and constant cell probabilities. Independence of the trials would be violated if, for example, several patients from the same family were included in the sample, since hypertension has a strong hereditary component. The assumption of constant cell probabilities would be violated if the study were conducted over a period of time during which the standards of medical practice shifted, allowing for other "standard" therapies.

The test statistic for the chi-square goodness-of-fit test is the sum of k terms, which is why the degrees of freedom depend on k (the number of categories) rather than on n (the total sample size). However, there are only $k - 1$ degrees of freedom, rather than k of them, because the sum of the $n_i - E_i$ terms must be equal to $n - n = 0$; $k - 1$ of the observed minus expected differences are free to vary, but the last (kth) one is determined by the condition that the sum of the $n_i - E_i$ equals zero.

This goodness-of-fit test has been used extensively over the years to test various scientific theories. Unlike in previous statistical tests, however, the hypothesis of interest here is the null hypothesis, not the research (or alternative) hypothesis. Unfortunately, the logic behind running a statistical test does not hold. In the standard situation where the research (alternative) hypothesis is the one of interest to the scientist, we formulate a suitable null hypothesis and gather data that enable us to reject H_0 in favor of H_a. Thus we "prove" H_a by contradicting H_0.

This is not so with the chi-square goodness-of-fit test. If a scientist has a theory and wants to show that sample data conform to or "fit" that theory, she wants to accept H_0. As our previous work suggests, there is the potential for committing a

Type II error in accepting H_0. Here, as with other tests, calculating β probabilities is difficult. In general, for a goodness-of-fit test, the potential for committing a Type II error is high if n is small or if k (the number of categories) is large. Even if the expected cell counts E_i conform to our recommendations, the probability of a Type II error could be large. Consequently, the results of a chi-square goodness-of-fit test should be viewed suspiciously. Don't automatically accept the null hypothesis as fact just because H_0 was not rejected.

Exercises

Basic Techniques

10.1 List the characteristics of a multinomial experiment.

10.2 How does a binomial experiment relate to a multinomial experiment?

10.3 Determine the rejection region for a chi-square goodness-of-fit test for the following values of $k, n,$ and α.

   a $n = 100, k = 4, \alpha = .05$

   b $n = 500, k = 10, \alpha = .01$

   c $n = 200, k = 8, \alpha = .001$

10.4 Under what conditions is it appropriate to use the chi-square goodness-of-fit test for a multinomial experiment? What qualification(s) might one have to make if the sample data do not yield rejection of the null hypothesis?

10.5 Hypothetical data are presented here. Use these data to run a chi-square goodness-of-fit test with H_0: $p_1 = .2$, $p_2 = .15$, $p_3 = .40$, $p_4 = .15$, and $p_5 = .10$. Use $\alpha = .05$. Do the data fit the hypothesized probabilities?

Category	Observed Cell Number (n_i)
1	60
2	50
3	130
4	40
5	20
Total	300

multiply

10.6 Use the data of Exercise 10.5 to run a chi-square goodness-of-fit test with this new null hypothesis—H_0: $p_1 = .15$, $p_2 = .20$, $p_3 = .45$, $p_4 = .15$, and $p_5 = .05$. Again use $\alpha = .05$. Compare your results to those of Exercise 10.5. How sensitive does this test appear to be for the cell probabilities specified under H_0? What conclusion can be drawn if we do *not* reject H_0?

Applications

10.7 Over the past five years, an insurance company has had a mix of 40% whole life polices, 20% universal life policies, 25% annual renewable-term (ART) policies, and 15% other types of policies. A change in this mix over the long haul could require a change in the commission structure, in reserves, and possibly in investments. A sample of 1000 policies issued over the last few months yielded the results recorded here. Use these data to assess whether there has been a shift from the historical percentages. Give the *p*-value for your test. Which policies (if any) seem to be more popular than before?

Category	Observed Cell Number (n_i)
Whole Life	320
Universal Life	280
ART	240
Other	160
Total	1000

10.8   A work-study program was developed cooperatively by a university and several industries in the surrounding community. Students were to work with industrial sociologists during a three-month internship. Equal numbers of students from the university were sent to a chemical, a textile, and a pharmaceutical industry. Students completing the program were classified according to the industry in which they interned. Consider the following data as a random sample of the many students who could have completed the program. Test the null hypothesis that the probability that a finishing student interned in a pharmaceutical chemical, or textile industry is 1/3. Use $\alpha = .01$, with n_i the number of students in group i finishing the program.

Group	n_i
Pharmaceutical	20
Chemical	13
Textile	30

10.9   An experiment was conducted to determine whether the proportion of mentally ill patients of each social class housed in a county facility agrees with the social class distribution of the county. The observed cell numbers for the 400 patients classified are as follows:

Lower: 215   Upper-middle: 60
Lower-middle: 100   Upper: 25

Use these data to test the null hypothesis

$$p_1 = .25 \quad p_3 = .20$$
$$p_2 = .48 \quad p_4 = .07$$

where the *p*s are the hypothesized proportions of persons in the respective social-class categories in the county. Use $\alpha = .05$ and draw conclusions.

10.10   In previous presidential elections in a given locality, 50% of the registered voters were Republicans, 40% were Democrats, and 10% were registered as independents. Prior to the upcoming election, a random sample of 200 registered voters showed that 90 were registered as Republicans, 80 as Democrats, and 30 as independents. Test the research hypothesis that the distribution of registered voters is different from that in previous election years. Give the *p*-value for your test. Draw conclusions.

10.11   A local doctor suspects that there is a seasonal trend in the occurrence of the common cold. He estimates that 40% of the cases each year occur in the winter, 40% in the spring, 10% in the summer, and 10% in the fall. The accompanying information was collected from a random sample of 1000 cases of patients with the common cold over the past year. Would you agree with the doctor's estimates, based on the sample information? Perform a statistical test, using $\alpha = .05$. Draw conclusions.

Season	Frequency
Winter	374
Spring	292
Summer	169
Fall	165

10.12   Refer to Exercise 10.11. What would the null hypothesis be if the doctor claimed that there are no differences in the percentages of cases over the seasons? Test the hypothesis that there is no seasonal trend in the occurrence of the common cold. Give the level of significance of your test. Do you have any reservations about your conclusion?

10.13   Previous experimentation with a drug developed for the relief of depression was conducted with normal adults who had no signs of depression. We will assume that a large data bank is available from studies conducted with normals and that for all practical purposes, the data bank can represent the population of responses for normals. Each of the adults participating in one of these studies was asked to rate the drug as ineffective, mildly effective, or effective. The percentages of respondents in these categories were 60%, 30%, and 10%, respectively. In a new study of depressed adults, a random sample of 85 adults responded as follows:

Ineffective: 30

Mildly effective: 35

Effective: 20

Is there evidence to indicate a different percentage distribution of responses for depressed adults than for normals? Give the level of significance for your test, and draw conclusions.

10.14   In random sampling, 40 newspaper editors were interviewed to determine their opinions on the degree of future suppression of freedom of the press brought about by recent court decisions. The editors' opinions are summarized below. Use these data to test the null hypothesis that each category is equally preferred. Use $\alpha = .05$. Draw conclusions from these data. What reservation(s), if any, do you have concerning your conclusions?

Degree of Suppression	Frequency
None	8
Very little	8
Moderate	10
Severe	14

10.15   A sample of 125 securities analysts was obtained, and each analyst was asked to select four stocks on the New York Stock Exchange that were expected to outperform the Standard and Poor's Index over a three-month period. One theory suggests that the securities analysts could be expected to do no better than chance and that the number of correct guesses from the four selected should have a multinomial distribution as shown here.

Number Correct	0	1	2	3	4
Multinomial Probabilities (p_i)	.0625	.2500	.3750	.2500	.0625

If the number of correct guesses from the sample of 125 analysts actually had a frequency distribution as shown in the following display, use these data to conduct a chi-square goodness-of-fit test. Use $\alpha = .05$. Draw conclusions.

Number Correct	0	1	2	3	4
Frequency	3	23	51	39	9

10.16   Refer to Exercise 10.15. Suppose that the assumed multinomial probabilities were all .20; that is, suppose that all $p_i = .20$.

a   How would your conclusions change?

b   Can you suggest a problem with using the chi-square goodness-of-fit test, based on the results of part a?

 # 10.4 Inferences About the Binomial Parameter p

The binomial experiment discussed in Chapter 5 is a special case of the multinomial experiment in which each trial results in one of two outcomes, which we label as either a success or a failure. Recall that p is the probability of a success, and $q = 1 - p$ is the probability of a failure. Then the probability distribution for y, the number of successes in n identical trials, is

$$P(y) = \frac{n!}{y!(n - y)!} p^y q^{n-y}.$$

The point estimate of the binomial parameter p is one that we would choose intuitively. In a random sample of n from a population in which the proportion of elements classified as success is p, the best estimate of the parameter p is the sample proportion of successes. If we let y denote the number of successes in the n sample trials, the sample proportion is

$$\hat{p} = \frac{y}{n}.$$

We observed in Section 5.12 that y possesses a mound-shaped probability distribution that can be approximated by using a normal curve when

$$n \geq \frac{5}{\min(p, q)} \quad \text{(or equivalently, } np \geq 5 \text{ and } nq \geq 5\text{).}$$

In a similar way, the distribution of $\hat{p} = y/n$ can be approximated by a normal distribution, with a mean and a standard error as given below.

Mean and Standard Error of $\hat{p}$

$$\mu_{\hat{p}} = p$$

$$\sigma_{\hat{p}} = \sqrt{\frac{pq}{n}}, \quad \text{where} \quad q = 1 - p$$

The normal approximation to the distribution of $\hat{p}$ can be applied under the same condition as that for approximating y by using a normal distribution. In fact, the approximation for both y and $\hat{p}$ becomes more precise for large n. Henceforth, in this text, we will assume that $\hat{p}$ can be adequately approximated by using a normal distribution, and we will base all our inferences on results from our previous study of the normal distribution.

A confidence interval can be obtained for p by using the methods of Chapter 7 for μ, by replacing $\bar{y}$ with $\hat{p}$, and $\sigma_{\bar{y}}$ with $\sigma_{\hat{p}}$. A general $100(1-\alpha)\%$ confidence interval for the binomial parameter is given here.

Confidence Interval for p, with Confidence Coefficient of $(1 - \alpha)$

$$\hat{p} \pm z_{\alpha/2}\sigma_{\hat{p}}$$

where

$$\hat{p} = \frac{y}{n} \quad \text{and} \quad \sigma_{\hat{p}} = \sqrt{\frac{pq}{n}}$$

Note: Since p is unknown, replace p by $\hat{p}$ (and q by $\hat{q}$) in $\sigma_{\hat{p}}$.

EXAMPLE   10.3   Response to an advertising display was measured by counting the number of people who purchased the product out of the total number exposed to the display. If 330 purchased the product out of a total of 870 exposed, estimate the proportion of all persons exposed who will buy the product. Use a 90% confidence interval.

Solution   For these data,

$$\hat{p} = \frac{330}{870} = .38$$

$$\sigma_{\hat{p}} = \sqrt{\frac{(.38)(.62)}{870}} = .016.$$

The confidence coefficient for our example is .90. Recall from Chapter 7 that we can obtain $z_{\alpha/2}$ by looking up the z-value in Table 2 in Appendix 3 corresponding to an area of $(\alpha/2)$. For a confidence coefficient of .90, the z-value corresponding to an area of .05 is 1.645. Hence, the 90% confidence interval on the proportion of persons who will purchase the product after exposure to this display is

$$.38 \pm 1.645(.016) \quad \text{or} \quad .38 \pm .026. \quad \blacksquare$$

The confidence interval for p is based on a normal approximation to a binomial, which is appropriate provided that n is sufficiently large. The rule we've specified is that both np and nq should be at least 5, but since p is the unknown parameter, we'll require that $n\hat{p}$ and $n\hat{q}$ be at least 5. When the sample size is too small and violates this rule, the confidence interval usually will be too wide to be of any use. For example, with $n = 20$ and $\hat{p} = .2$, the rule is not satisfied, since $n\hat{p} = 4$. The 95% confidence interval based on these data would be $.025 < p < .375$, which is practically useless. Very few product managers would be willing to launch a new product if the expected increase in market share was between .025 and .375.

Keep in mind, however, that a sample size that is sufficiently large to satisfy the rule *does not* guarantee that the interval will be informative. It only judges the adequacy of the normal approximation to the binomial—the basis for the confidence level.

Sample-size calculations for estimating p follow very closely the procedures we developed for inferences about μ. In Chapter 7, the required sample size for a $100(1 - \alpha)\%$ confidence interval for p of the form $\hat{p} \pm E$ (where E is specified) is found by solving the expression

$$z_{\alpha/2}\sigma_{\hat{p}} = E$$

for n. This result is shown here.

Sample Size Required for a $100(1 - \alpha)\%$ Confidence Interval for p of the Form $\hat{p} \pm E$

$$n = \frac{z_{\alpha/2}^2 pq}{E^2}, \quad \text{where} \quad q = 1 - p$$

Note: Since p is not known, either substitute an educated guess or use $p = .5$. Using $p = .5$ will generate the largest possible sample size for the specified confidence interval width, $2E$, and will thus give a conservative answer to the required sample size.

E X A M P L E   10.4   A large public opinion polling agency plans to conduct a national survey to determine the proportion of employed adults who fear losing their job within the next year. How many workers must be polled to estimate to within .02, using a 95% confidence interval?

Solution   By design, the agency wants the interval to be of the form $\hat{p} \pm .02$. The sample size necessary to achieve this accuracy is given by

$$n = \frac{z_{\alpha/2}^2}{E^2} pq,$$

where $z_{\alpha/2} = 1.96$ and $E = .02$. If a previous survey has been run recently, we use the sample proportion from that survey to substitute for p; otherwise, we can use $p = .5$. Using $p = .5$, we find that the required sample size is

$$n = \frac{(1.96)^2(.5)(.5)}{(.02)^2} = 2401.$$

That is, 2401 workers must be surveyed to estimate p to within .02.   ▪

A statistical test about a binomial parameter p is very similar to the large-sample test involving a population mean presented in Chapter 7. These results are summarized next, with three different alternative hypotheses along with their corresponding rejection regions. Recall that only one alternative is chosen for a particular problem.

Summary of a Statistical Test for p

H_0: $p = p_0$ (p_0 is specified)

H_a: 1 $p > p_0$

      2 $p < p_0$

      3 $p \neq p_0$

T.S.: $z = \dfrac{\hat{p} - p_0}{\sigma_{\hat{p}}}$

R.R.:   For a probability α of a Type I error,

   1   reject H_0 if $z > z_\alpha$.

   2   reject H_0 if $z < -z_\alpha$.

   3   reject H_0 if $|z| > z_{\alpha/2}$.

Note: Under H_0,

$$\sigma_{\hat{p}} = \sqrt{\frac{p_0 q_0}{n}}, \quad \text{where} \quad q_0 = 1 - p_0.$$

EXAMPLE   10.5   Sports car owners in a town complain that their cars are judged differently from family-style cars at the state vehicle inspection station. Previous records indicate that 30% of all passenger cars fail the inspection on the first time through. In a random sample of 150 sports cars, 60 failed the inspection on the first time through. Is there sufficient evidence to indicate that the percentage of first failures for sports cars is higher than the percentage for all passenger cars? Use $\alpha = .05$.

Solution   The appropriate statistical test is as follows.

H_0: $p = .30$

H_a: $p > .30$

T.S.: $z = \dfrac{\hat{p} - p_0}{\sigma_{\hat{p}}}$

R.R.: For $\alpha = .05$, we will reject H_0 if $z > 1.645$.

Using the sample data, we find that

$$\hat{p} = \frac{60}{150} = .4 \quad \text{and} \quad \sigma_{\hat{p}} = \sqrt{\frac{(.3)(.7)}{150}} = .037.$$

Also,

$$np_0 = 150(.3) = 45$$

and

$$nq_0 = 150(.7) = 105.$$

The test statistic is then

$$z = \frac{.4 - .3}{.037} = 2.7.$$

Since the observed value of z exceeds 1.645, we conclude that sports cars at the vehicle inspection station have a first-failure rate that exceeds .3. However, we must be careful not to attribute this difference to a difference in standards for sports cars and family-style cars. Parallel testing of sports cars versus other cars would have to be conducted to eliminate other sources of variability that might account for the higher first-failure rate for sports cars.   ▪

The z test for p, like the confidence interval for p based on z, depends on the adequacy of the normal approximation to the binomial. When can you use the z test for p? Generally speaking, you should view the results of a z test for p skeptically if either np_0 or nq_0 (where $q_0 = 1 - p_0$) is 2 or less. If both np_0 and nq_0 are at least 5, the z test should be accurate. But for the same sample size n, z tests based on more extreme values of p_0 are less accurate than those based on values of p_0 closer to .5. For example, for $n = 5000$, a test of $H_0\colon p = .001$ (for which $np_0 = 5$) would be much more suspect than would a test of $H_0\colon p = .01$ (for which $np_0 = 50$).

Exercises

Basic Techniques

10.17 Hypothetical sample results from a binomial experiment with $n = 150$ yielded $\hat{p} = .2$.

   a Does this experiment satisfy the sample-test requirement for a confidence interval for p based on z? What sample sizes would be suspect, given the same sample proportion?

   b Construct a 90% confidence interval for p.

10.18 Under what conditions can the formula $\hat{p} \pm z_{\alpha/2}\sigma_{\hat{p}}$ be used to express a confidence interval for p?

10.19 A random sample of 1500 is drawn from a binomial population. Suppose that there are $y = 1200$ successes.

   a Construct a 95% confidence interval for p.

   b Construct a 90% confidence interval for p.

10.20 Refer to the previous exercise. Explain the difference in interpreting the two confidence intervals.

Applications

10.21 Experts have predicted that approximately 1 in 12 tractor-trailer units will be involved in an accident this year. One of the reasons for this is that 1 in 3 tractor-trailer units has an imminently hazardous mechanical condition, probably related to the braking systems on the vehicle. A survey of 50 tractor-trailer units passing through a weighing station confirmed that 19 had a potentially serious braking system problem.

a Do the binomial assumptions hold?

b Can a normal approximation to the binomial be applied here to get a confidence interval for p?

c Give a 95% confidence interval for p using these data. Is the interval informative? What could be done to decrease the width of the interval, assuming that $\hat{p}$ remained the same?

10.22 In a study of self-medication practices, a random sample of 1230 adults completed a survey. Some of the medical conditions that were self-treated are shown in the accompanying table. Summarize the results of this part of the survey, using a 95% confidence interval for each medical condition.

Medical Condition	Home Remedy	% Responding
Sore throat—not related to a cold	Salt water or baking soda mouthwash	30
Burns—other than sunburn	Cold water/butter	28
Overindulgence in alcohol	Homebrew	25
Overweight	Diet	22
Pain associated with injury	Hot or cold compress	21

10.23 In the survey discussed in Exercise 10.22, 441 of the adults reported that they had had a cough or cold recently, and 260 of the respondents said that they had treated the condition with an over-the-counter (OTC) remedy. These data are summarized here.

Survey respondents reporting problem	441
Number of patients using any OTC remedy	260
Patients using specific classes of OTC remedies:	
Adult pain relievers	110
Adult cold caps/tabs	57
Cough remedies	44
Allergy/hay fever remedies	9
Liquid cold remedies	35
Sprays/inhalers	4
Children's pain reliever	22
Cough drops	13
Sore-throat lozenges/gum	9
Children's cold caps/tabs	13
Nose drops	9
Chest rubs/ointments	9
Anesthetic throat lozenges	4
Room vaporizers	4
Other product	4

a How might these data be organized and summarized? Would percentages help? Do the percentages add up to 100%? Why or why not?

b Based on these data, which classes of OTC remedies could be summarized by using a 95% confidence interval for p?

10.24 Many individuals over the age of 40 develop an intolerance for milk and milk-based products. A dairy has developed a line of lactose-free products that are more tolerable to such individuals.

To assess the potential market for these products, the dairy commissioned a market research study of individuals over 40 in its sales area. A random sample of 250 individuals showed that 86 of them suffer from milk intolerance. Calculate a 90% confidence interval for the population proportion that suffers from milk intolerance based on the sample results.

10.25 Shortly before April 15 of the previous year, a team of sociologists conducted a survey to study their theory that tax cheaters tend to allay their guilt by holding certain beliefs. A total of 500 adults were interviewed and asked under what situations they think cheating on an income tax return is justified. The responses include:

56% agree that "other people don't report all their income."

50% agree that "the government is often careless with tax dollars."

46% agree that "cheating can be overlooked if one is generally law abiding."

Assuming that the data are a simple random sample of the population of taxpayers (or tax-nonpayers), calculate 95% confidence intervals for the population proportion that agrees with each statement.

10.26 A national columnist recently reported the results of a survey on marriage and the family. Part of the column has been paraphrased here.

The Ingredients of Marriage

The Gallup people offered respondents a list of well-known ingredients. Here in the United States, such elements as faithfulness, mutual respect, and understanding ranked at the top. These were followed by enough money, same background, good housing, and agreement in politics. Seventy-five percent of the respondents voted for "a good sex life," 59% for children, 52% for common interests, 48% for "living away from in-laws," and 43% for "sharing household chores." (In Germany, by contrast, only 52% voted for a good sex life, and only 19% for sharing household chores.)

a   How could you display the results of this survey in a graph or table?

b   Would you use a confidence interval to convey more information about the "true" percentages expressing an opinion on the various ingredients of a good marriage? Why or why not?

c   What qualms might you have about the way this survey has been reported?

10.27 A substantial part of the U.S. population is "technologically illiterate," according to experts at a National Technological Literacy Conference organized by the National Science Foundation and Pennsylvania State University. At this conference, the results of a national survey of 2000 adults showed that:

- 70% do not understand radiation.

- 40% think space-rocket launchings change the weather and that some unidentified flying objects are actually visitors from other planets.

- More than 80% do not understand how telephones work.

- 75% do not have a clear understanding of what computer software is.

- 72% do not understand the gross national product.

a   How might you display these data in a graph or table? Construct the display.

b   The problem with many newspaper articles that report survey results is that conclusions are given without sufficient details about the study for the reader to assess the data and reach a separate conclusion. What details that are missing here would be necessary for you to reach your own conclusion?

10.28 More and more people are dining out—so say the results of national surveys. Compared to 1978, here are some figures.

Population Meal Eaten Away from Home	1978	Survey Now
Breakfast	3%	5%
Lunch	18%	20%
Dinner	16%	16%

a If the survey data were based on random samples of 1500 adults in 1978 and at present, what conclusions can be drawn for each meal? Is a normal approximation to the binomial valid here?

b Can we conclude from the data shown here that more people are eating out? Why or why not?

10.29 The benign mucosal cyst is the most common lesion of a pair of sinuses in the upper jawbone. In a random sample of 800 males, 35 persons were observed to have a benign mucosal cyst.

a Would it be appropriate to use a normal approximation in conducting a statistical test of the null hypothesis $H_0: p = .096$ (the highest incidence in previous studies among males)? Explain.

b Conduct a statistical test of the research hypothesis $H_a: p < .096$. Use $\alpha = .05$.

10.30 National public opinion polls are based on interviews of as few as 1500 persons in a random sampling of public sentiment toward one or more issues. These interviews are commonly done in person, because mail returns are poor and telephone interviews tend to reach older people, thus biasing the results. Suppose that a random sample of 1500 persons is surveyed to determine the proportion of the adult public that agrees with recent energy conservation proposals.

a If 560 respondents indicate that they favor the policies set forth by the current administration, estimate p, the proportion of adults holding a "favor" opinion. Use a 95% confidence interval. What is the half width of the confidence interval?

b How many persons must be surveyed to establish a 95% confidence interval with a half width of .01?

10.31 A sample of 20 crayfish of all sizes was obtained from a large lake to estimate the proportion of crayfish that exhibit more than 9 units (ppb) of mercury. Of those sampled, 8 exhibited mercury levels exceeding 9 units. Use these data to estimate p, the proportion of all crayfish in the lake whose mercury level exceeds 9, using a 95% confidence interval.

10.32 Refer to Example 10.4. Suppose that a recently done survey resulted in $\hat{p} = .15$. Use this guessed value to compute an appropriate sample size. Comment on the differences between your answer here and your answer in Example 10.4.

10.5 Operating Characteristic Curves and Control Charts for p

Two techniques are particularly appropriate for monitoring product quality as measured by p, the fraction of items that are defective. The first of these, control charts, were initially discussed in Chapter 7, where we presented the $\bar{y}$-chart; in Chapter 9, we discussed the s-chart and r-chart as well. A similar type of chart can be used to monitor p as a measure of the ongoing quality of a manufacturing process. In such situations, we are especially interested in detecting a shift in product quality. The second technique, called **lot acceptance sampling,** provides a means to screen (sample) incoming raw materials or outgoing production from a plant where the product is shipped in large quantities (often called "lots"). We begin by discussing lot acceptance sampling.

lot acceptance sampling

Manufacturers are interested in minimizing not only the amount or proportion of defective raw material used in the production process, but also the proportion of defective finished products shipped from the plant. Thus, they would like to sample, or screen, shipments of raw materials entering the plant and reject shipments (lots) that contain too high a proportion of defectives. Similarly, they must screen the final product to make certain that a shipment does not contain too high a proportion of defectives.

The most obvious type of screen (sampling plan) to employ consists of a careful inspection of each item from the lot. Unfortunately, this screen is both costly and time-consuming. In addition, it remains subject to errors in reporting brought about by human fatigue.

statistical sampling plan

Another type of screen is called a **statistical sampling plan.** Here we obtain a random sample of n items from the lot. Each item of the sample is inspected, and if y, the number of defectives observed in the sample, is less than or equal to some predetermined number a, we accept the lot. Thus, a statistical sampling plan is designated by n, the sample size, and a, the acceptance number. If the lot is accepted ($y \leq a$), we conclude that the proportion of defectives p in the lot is acceptably small. However, if $y > a$, we reject the lot and conclude that p is too large (above an acceptable level of defectives).

operating characteristic (OC) curve

We can characterize the goodness of a particular sampling plan (n, a) by constructing an **operating characteristic (OC) curve.** The OC curve for a sampling plan is a graph displaying the probability of accepting a lot for various values of p, the proportion of defective items in the lot (see Figure 10.2). As you can see, the probability of accepting a lot decreases as the proportion of defectives within the lot increases.

FIGURE 10.2
OC Curve for a Sampling Plan

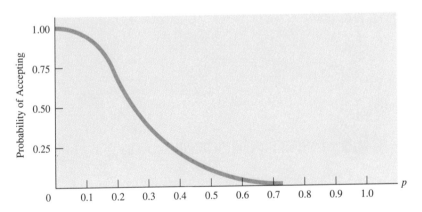

We can construct an OC curve by computing the probability of accepting a lot—namely, $P(y \leq a)$—for several values of p. Consider the sampling plan $n = 4$ and $a = 0$. Here we sample four items and accept the lot if y, the number of defectives, is zero. Hence, we must compute $P(y = 0)$ for $n = 4$ and for different values of p to obtain the OC curve. We use the binomial probability distribution

$$P(y) = \frac{n!}{y!(n-y)!} p^y q^{n-y} \quad \text{where} \quad q = 1 - p.$$

Table 10.4 shows the results of the calculations for $p = .1, .2$, and $.4$. Plotting these three points and connecting them, we have the OC curve as shown in Figure 10.3.

Fraction of Defectives, p	Probability of Accepting, $P(y = 0)$
.1	$\frac{4!}{0!4!}(.1)^0(.9)^4 = .656$
.2	$\frac{4!}{0!4!}(.2)^0(.8)^4 = .410$
.4	$\frac{4!}{0!4!}(.4)^0(.6)^4 = .130$

FIGURE 10.3
OC Curve for the Sampling Plan
$(n = 4, a = 0)$

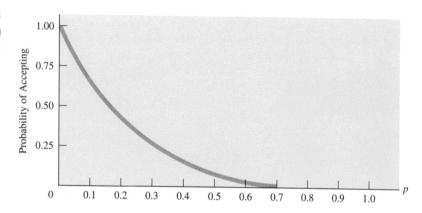

EXAMPLE   10.6   Construct an operating characteristic curve for the sampling plan $(n = 10, a = 1)$.

Solution   The probability of accepting the lot is given by $P(y \leq 1)$. Using the binomial probability distribution, we must calculate $P(y = 0) + P(y = 1)$ for $n = 10$ and for various values of p. For $p = .1$ and $n = 10$,

$$P(y = 0) = \frac{10!}{0!10!}(.1)^0(.9)^{10} = .349$$

$$P(y = 1) = \frac{10!}{1!9!}(.1)^1(.9)^9 = .387$$

Hence, the probability of accepting the lot is $.349 + .387 = .736$. By similar computations for $p = .2$ and $.4$, the probabilities of accepting the lot are $.376$ and $.046$, respectively. Graphing our results, we have the OC curve presented in Figure 10.4.   ∎

Several comments should be made concerning statistical sampling plans. First, each plan (n, a) is unique, so the inspector must choose a sampling plan that possesses

FIGURE 10.4
OC Curve for the Sampling Plan
$(n = 10, a = 1)$ of
Example 10.6

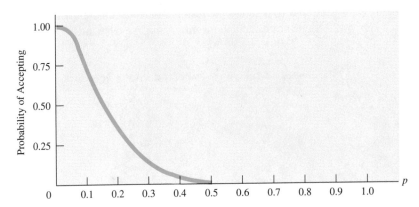

FIGURE 10.4
OC Curve for the Sampling Plan
$(n = 10, a = 1)$ of
Example 10.6

characteristics suitable for his or her particular problem. In general, for a fixed value of a, an increase in n makes the graphed curve of the probability of acceptance drop sharply as p increases; increasing a for a fixed n increases the probability of acceptance for values of p. Second, the OC curve for a particular sampling plan can be thought of as a plot of the probability of a Type II error for the null hypothesis H_0: $p = 0$ for various actual values of p, when the rejection region is $y > a$.

control chart for p

The other method of monitoring product quality makes use of a **control chart for p**. Recall from our previous work that a control chart typically consists of three lines. In a control chart for p, successive sample proportions $\hat{p}$ would be plotted and might appear as shown in Figure 10.5. If one of the sample proportions falls outside either the upper or lower control line, the process is judged to be out of control; that is, the proportion of defectives p in the production has shifted.

FIGURE 10.5
Control Chart for p

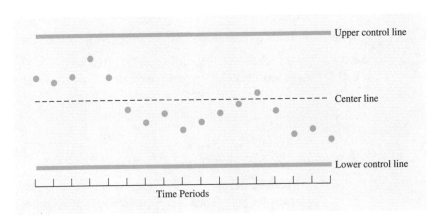

We can compute the three control lines in the following manner. The center line is designated by $\hat{p}_c$. If we obtain k different random samples of n observations each, $\hat{p}_c$ is the sample proportion of defectives for the entire set of kn measurements, or equivalently, $\hat{p}_c$ is the average of the sample proportions computed for the k samples of n measurements.

upper control line (UCL)
lower control line (LCL)

The **upper control line (UCL)** and **lower control line (LCL)** are then

$$\text{UCL} = \hat{p}_c + 3\sqrt{\frac{\hat{p}_c\hat{q}_c}{n}}, \qquad \text{where} \qquad \hat{q}_c = 1 - \hat{p}_c$$

and

$$\text{LCL} = \hat{p}_c - 3\sqrt{\frac{\hat{p}_c\hat{q}_c}{n}}.$$

EXAMPLE 10.7   A pharmaceutical firm has been investigating the possibility of supplying hospital personnel with small disposable vials that can be used to perform many standard laboratory analyses. For a particular analysis, such as blood sugar, the technician would insert a measured amount of fluid (perhaps blood) in an appropriate vial and observe its color when thoroughly mixed with the fluid already stored in the vial. By comparing the optical density of the combined fluid to a color-coded chart, the technician would obtain a reading on the blood sugar level of the patient. Obviously, the system must be tightly controlled to ensure that the vials are correctly sealed and contain the proper amount of fluid prior to shipment to the hospital laboratories. The data in Table 10.5 give the proportion of defectives in 30 different samples (taken from 30 different production hours) of 50 vials each. Use these data to construct the three control lines.

T A B L E 10.5
Thirty Sample Proportions, Each
Based on 50 Observations, for
Example 10.7

Sample	Sample Proportion Defective	Sample	Sample Proportion Defective
1	.18	16	.18
2	.10	17	.14
3	.18	18	.16
4	.16	19	.14
5	.12	20	.12
6	.12	21	.18
7	.16	22	.16
8	.18	23	.18
9	.12	24	.18
10	.12	25	.14
11	.16	26	.16
12	.18	27	.18
13	.12	28	.16
14	.18	29	.18
15	.14	30	.10

Solution   The sum of the 30 sample proportions is 4.58, so

$$\hat{p}_c = \frac{4.58}{30} = .15 \quad \text{and} \quad \hat{q}_c = .85.$$

Substituting into the formulas for the control limits, we have

$$\text{UCL} = .15 + 3\sqrt{\frac{(.15)(.85)}{30}} = .15 + 3(.05) = .30$$

and

$$\text{UCL} = .15 - 3\sqrt{\frac{(.15)(.85)}{30}} = .15 - 3(.05) = 0. \quad \blacksquare$$

We have discussed the binomial probability distribution and various count data problems that utilize the binomial distribution. In the next section, we consider inferences involving two binomial parameters.

Exercises

Basic Techniques

10.33   Sketch the operating characteristic curve for the sampling plan ($n = 5$, $a = 0$).

10.34   Refer to Exercise 10.33. Superimpose the OC curve for the sampling plan with $n = 10$ and $a = 0$ on that for $n = 5$, $a = 0$. What is the effect of increasing the sample size n while holding the acceptance number a constant?

Application

10.35   Refer to the control limits for p obtained in Example 10.7. Suppose that you are now in charge of quality control for the vial-production line. In the next 15 samples of 50 vials, you observe the following numbers of defectives.

Number	1	2	3	4	5	6	7	8	9	10	11	12	13	14	15
Defective	9	8	6	5	2	4	6	8	9	9	12	13	14	10	12

Determine whether the process has remained in control.

10.6   Comparing Two Binomial Proportions

Many practical problems involve comparing two binomial parameters. For example, social scientists may wish to compare the proportions of women who take advantage of prenatal health services for two communities representing different socioeconomic backgrounds. Or the director of marketing may wish to compare public awareness of a new product recently launched and that of a competitor's product.

For comparisons of this type, we assume that independent random samples are drawn from two binomial populations with unknown parameters designated by p_1 and p_2. If y_1 successes are observed for the random sample of size n_1 from population 1 and y_2 successes are observed for the random sample of size n_2 from population 2, then the point estimates of p_1 and p_2 are the observed sample proportions $\hat{p}_1$ and $\hat{p}_2$, respectively:

$$\hat{p}_1 = \frac{y_1}{n_1} \quad \text{and} \quad \hat{p}_2 = \frac{y_2}{n_2}.$$

This notation is summarized here.

Notation for Comparing Two Binomial Proportions

	Population	
	1	2
Population proportion	p_1	p_2
Sample size	n_1	n_2
Number of successes	y_1	y_2
Sample proportion	$\hat{p}_1 = \dfrac{y_1}{n_1}$	$\hat{p}_2 = \dfrac{y_2}{n_2}$

Inferences about two binomial proportions are usually phrased in terms of their difference $p_1 - p_2$, and we use the difference in sample proportions $\hat{p}_1 - \hat{p}_2$ as part of a confidence interval or statistical test. The **sampling distribution for $\hat{p}_1 - \hat{p}_2$** can be approximated by a normal distribution, with mean and standard deviation given by

sampling distribution for $\hat{p}_1 - \hat{p}_2$

$$\mu_{\hat{p}_1 - \hat{p}_2} = p_1 - p_2$$

and

$$\sigma_{\hat{p}_1 - \hat{p}_2} = \sqrt{\frac{p_1 q_1}{n_1} + \frac{p_2 q_2}{n_2}}.$$

This approximation is appropriate if we apply the same requirements to both binomial populations that we used in recommending a normal approximation to a binomial (see Chapter 5). Thus, the normal approximation to the distribution of $\hat{p}_1 - \hat{p}_2$ is appropriate if both np and nq are 5 or more for both populations. Since p_1 and p_2 are never known, you must make your own judgment about the validity of the approximation, using $n\hat{p}$ and $n\hat{q}$ for each sample.

Confidence intervals and statistical tests about $p_1 - p_2$ are straightforward and follow the format we adopted for comparisons using $\mu_1 - \mu_2$. Interval estimation is summarized here. It takes the usual form: point estimate $\pm z$ (standard error).

$100(1 - \alpha)\%$ Confidence Interval for $p_1 - p_2$

$$\hat{p}_1 - \hat{p}_2 \pm z_{\alpha/2}\sigma_{\hat{p}_1-\hat{p}_2},$$

where

$$\sigma_{\hat{p}_1-\hat{p}_2} = \sqrt{\frac{p_1 q_1}{n_1} + \frac{p_2 q_2}{n_2}}$$

Note: Substitute $\hat{p}_1$ and $\hat{p}_2$ for p_1 and p_2 in the formula for $\sigma_{\hat{p}_1-\hat{p}_2}$. When the normal approximation is valid for $\hat{p}_1 - \hat{p}_2$, very little error will result from this substitution.

EXAMPLE 10.8   In a survey to analyze typical funeral expenditures by members of various social classes, a random sample of 162 families from the working (blue-collar) class was taken to determine their funeral expenses for a recent death in the family. Of the 162 families contacted, 61 spent over $800 on the funeral. A similar survey was conducted within the middle/upper classes. Of 189 families contacted, 106 spent more than $800. Estimate $p_1 - p_2$, the difference in the proportions of families who have spent more than $800 on funeral expenses for a recent family death. Use a 95% confidence interval to interpret your findings.

Solution   The point estimate of $p_1 - p_2$ is the difference in sample proportions, $\hat{p}_1 - \hat{p}_2$:

$$\hat{p}_1 - \hat{p}_2 = \frac{61}{162} - \frac{106}{189} = .376 - .561 = -.185.$$

Note also that $n\hat{p}$ and $n\hat{q}$ are 5 or more for both samples, implying that the normal approximation to the binomial is appropriate.

The standard error for $\hat{p}_1 - \hat{p}_2$ is estimated by

$$\sqrt{\frac{p_1 q_1}{n_1} + \frac{p_2 q_2}{n_2}} = \sqrt{\frac{.376(.624)}{162} + \frac{.561(.439)}{189}} = .052.$$

A 95% confidence interval for $p_1 - p_2$ has $z_{\alpha/2} = 1.96$ and is of the form

$$\text{Point estimate} \pm z_{\alpha/2}(\text{standard error}).$$

Substituting into this formula, we have

$$-.185 \pm 1.96(.052) \qquad \text{or} \qquad -.185 \pm .102.$$

This interval indicates that p_2 is larger than p_1; we are 95% confident that the difference in the proportions of families paying more than \$800 per funeral for the working class (p_1) and the middle/upper class (p_2) lies in the interval $-.287$ to $-.083$. ∎

We can readily formulate a statistical test for the equality of two binomial parameters. The test statistic for testing H_0: $p_1 - p_2 = 0$ is a z statistic having the familiar form

$$z = \frac{\text{Point estimate}}{\text{Standard error}} = \frac{\hat{p}_1 - \hat{p}_2}{\sigma_{\hat{p}_1 - \hat{p}_2}}.$$

The standard error is slightly different from what we used for a confidence interval. When H_0 is true, $p_1 = p_2$; we call the common value p. Then

$$\sigma_{\hat{p}_1 - \hat{p}_2} = \sqrt{\frac{p_1 q_1}{n_1} + \frac{p_2 q_2}{n_2}} = \sqrt{pq \left(\frac{1}{n_1} + \frac{1}{n_2} \right)}.$$

The best estimate of p, the proportion of successes common to both populations, is

$$\hat{p} = \frac{\text{Total number of successes}}{\text{Total number of trials}} = \frac{y_1 + y_2}{n_1 + n_2}.$$

We have summarized the test procedure here.

Statistical Test for Comparing Two Binomial Proportions

H_0: $p_1 - p_2 = 0$

H_a: **1** $p_1 - p_2 > 0$

   2 $p_1 - p_2 < 0$

   3 $p_1 - p_2 \neq 0$

T.S.: $z = \dfrac{\hat{p}_1 - \hat{p}_2}{\sigma_{\hat{p}_1 - \hat{p}_2}}$ where $\sigma_{\hat{p}_1 - \hat{p}_2} = \sqrt{pq \left(\dfrac{1}{n_1} + \dfrac{1}{n_2} \right)}$ and p is

   approximated by $\hat{p} = \dfrac{y_1 + y_2}{n_1 + n_2}$

R.R.: For a given value of α,

   1 Reject H_0 if $z > z_\alpha$.

   2 Reject H_0 if $z < -z_\alpha$.

   3 Reject H_0 if $|z| > z_{\alpha/2}$.

Note: $n\hat{p}$ and $n\hat{q}$ must be greater than or equal to 5 for both populations in order for the normal approximation (and hence for this test) to hold.

EXAMPLE 10.9   In a recent survey of county high-school students ($n_1 = 100$ males and $n_2 = 100$ females), 58 of the males and 46 of the females sampled said they consume alcohol on a regular basis. Use the sample data to conduct a test H_0: $p_1 - p_2 = 0$, against the one-sided alternative H_a: $p_1 - p_2 > 0$, that a higher proportion of males than females consume alcohol on a regular basis. Use $\alpha = .05$.

Solution   The four parts of the statistical test are shown here.

$$H_0: \quad p_1 - p_2 = 0$$
$$H_a: \quad p_1 - p_2 > 0$$

T.S.: $z = \dfrac{\hat{p}_1 - \hat{p}_2}{\sigma_{\hat{p}_1 - \hat{p}_2}}$ where $\sigma_{\hat{p}_1 - \hat{p}_2} = \sqrt{pq\left(\dfrac{1}{n_1} + \dfrac{1}{n_2}\right)}$

R.R.:   For $\alpha = .05$, reject H_0 if $z > 1.645$.

From the sample data we find

$$\hat{p}_1 = \frac{58}{100} = .58, \qquad \hat{p}_2 = \frac{46}{100} = .46, \qquad \text{and} \qquad \hat{p} = \frac{58 + 46}{100 + 100} = .52.$$

Notice also that $n\hat{p}$ and $n(1 - \hat{p})$ are 5 or more for both samples, validating the normal approximations to the binomial.

Substituting into the test statistic, we obtain

$$z = \frac{.58 - .46}{\sqrt{.52(.48)\left(\dfrac{1}{100} + \dfrac{1}{100}\right)}} = \frac{.12}{.071} = 1.69.$$

Conclusion: Since $z = 1.69$ exceeds 1.645, we reject H_0: $p_1 - p_2 = 0$; we have shown that proportionately more high-school males than females in the county studied consume alcohol on a regular basis.   ∎

Exercises

Basic Techniques

10.36   A random sample of $n_1 = 1000$ observations was obtained from a binomial population with $p_1 = .4$. Another random sample, independent of the first sample, was selected from a binomial population with $p_2 = .2$. Does the normal approximation hold? Describe the sampling distribution for $\hat{p}_1 - \hat{p}_2$.

10.37   In a study to compare two binomial proportions, $n_1 = 50$, $n_2 = 40$, $y_1 = 20$, and $y_2 = 15$. Use these hypothetical data to construct a 90% confidence interval for $p_1 - p_2$.

10.38 Refer to Exercise 10.37. How large a sample should we take from each population in order to have a 90% confidence interval of the form $\hat{p}_1 - \hat{p}_2 \pm .01$? (*Hint:* Assuming that equal sample sizes will be taken from the two populations, solve the expression

$$z_{\alpha/2}\sigma_{\hat{p}_1 - \hat{p}_2} = .01$$

for n, the common sample size. Use $\hat{p}_1 = .40$ and $\hat{p}_2 = .375$ from Exercise 10.37.)

Applications

10.39 A law student believes that the proportion of registered Republicans in favor of additional tax incentives is greater than the proportion of registered Democrats in favor of such incentives. The student acquired independent random samples of 200 Republicans and 200 Democrats and found 109 Republicans and 86 Democrats in favor of additional tax incentives. Use these data to test: $H_0: p_1 - p_2 = 0$ versus $H_a: p_1 - p_2 > 0$. Give the level of significance for your test.

10.40 In a comparison of the incidence of tumor potential in two strains of rats, 100 rats (50 males, 50 females) were selected from each of two strains and were examined for a period of one year. All the rats were approximately the same age and were housed and fed under comparable conditions. Use the accompanying one-year sample data to construct a 95% confidence interval for the difference in the proportions of rats exhibiting tumor potential for the two strains.

Statistic	Strain A	Strain B
Sample size	100	100
Number exhibiting tumor potential	25	15

10.41 There is a remedy for male pattern baldness—at least that's what millions of males hope, since the FDA approved Upjohn's minoxidil for such treatment. Minoxidil was investigated in a large, 27-center study where patients were randomly assigned to receive topical minoxidil or an identical-looking placebo. Ignoring the center-to-center variation, suppose that the preliminary results were as follows.

Preparation Administered	Sample Size	% with New Hair Growth
Minoxidil	310	32
Placebo	309	20

a Use these data to test $H_0: p_1 - p_2 = 0$ versus $H_a: p_1 - p_2 \neq 0$. Give the *p*-value for your test.

b If you were working for the FDA, what additional information might you want to examine in this study?

10.42 Is cocaine deadlier than heroin? A study reported in the *Journal of the American Medical Association* found that rats with unlimited access to cocaine had poorer health, suffered more behavior disturbances, and died at a higher rate than did a corresponding group of rats given unlimited access to heroin. The death rates after 30 days on the study were as follows.

Drug Made Available	% Dead at 30 Days
Cocaine	90
Heroin	36

a Suppose that 100 rats were used in each group. Conduct a test of $H_0: p_1 - p_2 = 0$ versus $H_a: p_1 - p_2 > 0$. Give the *p*- value for your test.

b What implications are there for human use of the two drugs?

10.7 $r \times c$ Contingency Tables: Chi-Square Test of Independence

In all of our calculations so far in this text, we have assumed that only one measurement is taken on each sampling unit. We might obtain the yield for an acre planted in wheat, a blood pressure reading on a patient who is being administered an anesthetic, or a measurement on the number of potential conflicts at a highway intersection during a one-hour period. However, research problems in the sciences frequently involve more than one variable. If measurements are taken on two (or more) variables for each sampling unit, we say that we have **bivariate** (or **multivariate**) **data.**

bivariate and multivariate data

As with univariate count data, where the data may be summarized in a table, we frequently arrange bivariate data in a two-way table. For example, in a study of the level of public approval for a proposed high-speed bus lane for commuters, the interviewers might also ask individuals information about their occupations. We could then classify each person by his or her opinion about the new lane (favor, do not favor, undecided) and his or her occupation (white-collar worker, blue-collar worker, laborer).

What is the objective of such a classification? In most studies, either we wish to determine whether the two variables are related (dependent) or we wish to predict one variable based on knowledge of the other. This section deals with a test of independence for bivariate count data arranged in a two-way table. The two-way tables are sometimes called **contingency tables** because the alternative hypothesis in our test is that the two variables are dependent (that is, that contingency exists between the two variables).

contingency tables

Suppose that we would like to determine whether the following two variables are dependent: employee classification (staff, faculty, administrator) at a university and opinion about whether the local chapter of the teachers' union should be the sole collective bargaining agent for employee benefits. A random sample of 200 employees is taken from employee records, and each employee is classified according to both variables. The results of the survey appear in Table 10.6. Is there evidence to indicate that a person's opinion concerning collective bargaining depends on his or her employment status? In other words, can we conclude that the two variables are dependent?

TABLE 10.6
Two-way Tabulation of 200 Employees by Employee Classification and Opinion on Collective Bargaining

| Employee Classification | Opinion on Collective Bargaining by Teachers' Union | | | |
	Favor	Do Not Favor	Undecided	Totals
Staff	30	15	15	60
Faculty	40	50	10	100
Administrator	10	25	5	40
Totals	80	90	30	200

independence

To answer this question, we must define the concept of **independence.**

D E F I N I T I O N   10.2

Independent   Two variables that have been categorized in a two-way table are **independent** if the probability that a measurement is classified into a given cell of the table is equal to the probability of its being classified into that row times the probability of its being classified into that column. This must be true for all cells of the table.   ▪

For example, suppose that the probability of selecting a person favoring the teachers' union in the university survey is p_1, the probability of selecting one who does not favor the union for collective bargaining is p_2, and the probability of selecting a person who is undecided is p_3. (*Note:* $p_1 + p_2 + p_3 = 1$.) Similarly, suppose that the probabilities of selecting a staff member, a faculty member, or an administrator are, respectively, p_A, p_B, and p_C (where $p_A + p_B + p_C = 1$). Then the two variables— employee classification and opinion concerning the teachers' union—are independent if the probability of classifying a person into a specific cell of the two-way table is obtained by multiplying the respective row and column probabilities. These ideas are illustrated in Table 10.7.

TABLE 10.7
Cell Probabilities Showing
Independence for the Collective
Bargaining Survey

Employee Classification	Opinion		
	Favor (p_1)	Do Not Favor (p_2)	Undecided (p_3)
Staff (p_A)	$p_A p_1$	$p_A p_2$	$p_A p_3$
Faculty (p_B)	$p_B p_1$	$p_B p_2$	$p_B p_3$
Administrator (p_C)	$p_C p_1$	$p_C p_2$	$p_C p_3$

A test of the independence of two random variables arranged in a two-way table makes use of the test statistic

test statistic

$$\chi^2 = \sum_{i,j} \left[\frac{(n_{ij} - E_{ij})^2}{E_{ij}} \right],$$

where n_{ij} and E_{ij} are, respectively, the observed number and the expected number of measurements falling in the cell for the ith row and the jth column.

D E F I N I T I O N   10.3

Expected Number of Measurements   The **expected number of measurements** E_{ij} falling in the i, j cell (cell of the ith row and jth column of the table) is taken to be

$$E_{ij} = \frac{(\text{Row } i \text{ total})(\text{Column } j \text{ total})}{n},$$

when the two variables are independent. (*Note:* If E_{ij} is not an integer, it should not be rounded to an integer value for the tests that follow.)   ▪

EXAMPLE   10.10   Compute the expected number of measurements falling into each cell of Table 10.6.

Solution   The expected number of measurements falling in the 1, 1 cell (first row, first column is)

$$E_{11} = \frac{\text{(Row 1 total)(Column 1 total)}}{n} = \frac{(60)(80)}{200} = 24.$$

Similarly, the expected number of measurements in the 3, 2 cell is

$$E_{32} = \frac{\text{(Row 3 total)(Column 2 total)}}{n} = \frac{(40)(90)}{200} = 18.$$

These and the remaining cell counts appear in Table 10.8.

TABLE 10.8
Expected Cell Counts for the
Collective Bargaining Survey

Employee Classification	Opinion			
	Favor	Do Not Favor	Undecided	Totals
Staff	24	27	9	60
Faculty	40	45	15	100
Administrator	16	18	6	40
Totals	80	90	30	200

Notice that the expected counts in a row sum to the same row total as do the observed cell counts. The same applies to columns.   ▪

We can now summarize the *chi-square test of independence* for data arranged in a two-way table.

Chi-Square Test of Independence

H_0:   The two variables are independent.

H_a: The two variables are dependent.

T.S.: $\chi^2 = \sum_{i,j} \left[\dfrac{(n_{ij} - E_{ij})^2}{E_{ij}} \right]$

> R.R.: Reject H_0 if χ^2 exceeds the tabulated value of chi-square (Table 4 in Appendix 3) for $a = \alpha$ and df $= (r-1)(c-1)$, where
>
> $$r = \text{Number of rows in the table}$$
> $$c = \text{Number of columns in the table.}$$

The guidelines that Cochran (1954) proposed for the E_{ij}s (see Section 10.2) are still in effect when we use the chi-square test of independence. While agreeing with Cochran, Conover (1971) goes even further, stating that when the E_{ij}s are all of about the same magnitude and both r and c are large, then—even if the E_{ij}s are as small as 1—the approximation by a chi-square distribution will still be good. These guidelines give us a great deal of flexibility in applying the chi-square test without having to collapse some of the categories.

E X A M P L E   10.11   Conduct a chi-square test of independence for the teachers' union data in Table 10.6. Use $\alpha = .05$.

Solution   Using the observed cell counts of Table 10.6 and the expected cell counts of Table 10.8, we can substitute these values into the test statistic:

$$\chi^2 = \sum_{i,j} \left[\frac{(n_{ij} - E_{ij})^2}{E_{ij}} \right]$$

$$= \frac{(30-24)^2}{24} + \frac{(15-27)^2}{27} + \frac{(15-9)^2}{9} + \frac{(40-40)^2}{40} + \frac{(50-45)^2}{45}$$

$$+ \frac{(10-15)^2}{15} + \frac{(10-16)^2}{16} + \frac{(25-18)^2}{18} + \frac{(5-6)^2}{6}$$

$$= 18.2$$

The critical value of χ^2 for $a = .05$ and df $= (r-1)(c-1) = 2(2) = 4$ is 9.488. Since the computed value, 18.2, exceeds 9.488, we reject H_0 and conclude that the two variables are dependent. In particular, we say that the proportion of persons favoring the teachers' union as the collective bargaining agent varies depending on the employee status. From Table 10.6 we see that a much higher proportion of the staff members than of either the faculty or the administrators favor the teachers' union as the collective bargaining agent.   ∎

The objective of this chi-square test is to determine whether the observed dependence is due to random fluctuations. When H_0: independence is rejected in favor of H_a: dependence, we conclude that the two variables that have been summarized in a contingency table are related; that is, an outcome on one variable affects or is affected by an outcome on the other variable. The rejection of H_0 does not, however, indicate the strength or type of the relation between the two variables. As was discussed in

Chapter 4, a percentage comparison can help to identify how the variables are related, and certain "measures of association" (to be presented in the next section) can help to quantify the strength of the relation between the two variables.

The same chi-square test statistic applies to a slightly different sampling procedure. In our discussion of the chi-square goodness-of-fit test, we have implicitly assumed that the data summarized in the contingency table resulted from a single random sample taken from the population of interest. Often, separate random samples are taken from the *subpopulations* defined by the rows (or columns) of the contingency table. For example, the data of Table 10.6 might have resulted from separate random samples of sizes 60 (staff), 100 (faculty), and 40 (administrator) rather than from a single overall sample of 200 individuals. When random samples are taken for categories of the row (or column) variable, the test is called a *test of homogeneity* of the row (or column) distributions. So, in a sampling arrangement organized by rows, the percentage distributions across columns is the same from row to row. This could be seen as a percentage comparison by rows for the sample data. Similarly, in a sampling arrangement organized by columns, the percentage distribution across rows is the same from column to column.

Since the mechanics and conclusions are the same for the chi-square test of independence and for the chi-square test of homogeneity of the distributions, we will not worry about the distinction between the two tests.

Exercises

Basic Techniques

10.43   Data from a random sample of 200 individuals are summarized here in a 2 × 2 contingency table.

   a   Give the null and alternative hypotheses for a test of independence.

   b   Compute the expected cell counts and the value of the χ^2 test statistic.

   c   Based on $\alpha = .05$, determine the rejection region, and draw conclusions about the test of the null hypothesis.

	Column 1	Column 2
Row 1	20	70
Row 2	30	80

10.44   In a 2 × 3 contingency table, what is the minimum number of expected values that need to be computed if the remaining ones are to be computed by subtraction? What is the minimum number of a 3 × 3 table?

Applications

10.45   A survey of student opinion concerning a proposed tuition increase was taken to determine whether student opinion was independent of gender. The results of 300 interviews are recorded in the accompanying table. Run a chi-square test of independence, and give the level of significance for the test results. Draw conclusions.

| | Opinion | | |
Gender of Student	Favor Increase	Oppose Increase	Undecided
Female	91( )†	54( )†	13
Male	59	69	14

†You need to use the formula for computing the expected cell count only for these cells; the remaining cell counts can be computed by subtraction from the appropriate row or column total.

10.46 A scientist was interested in testing the effectiveness of a new drug in controlling worms in the small intestine of sheep. A prestudy test was used to select 40 sheep with approximately the same level of infestation. These sheep were then randomly divided into two groups of 20. Those in the first group were given the drug, those in the second group received no treatment and served as a control group. After two weeks, each of the 40 sheep was examined and classified as either a "responder" or a "nonresponder," depending on the observed worm count. The sample data are summarized here.

Classification	Group 1 (Drug-Tested)	Group 2 (Control)
Responder	15( )	7
Nonresponder	5	13

a Compute the expected cell counts.

b Run a chi-square test of independence, with $\alpha = .05$. State the null hypothesis for this test, and draw appropriate conclusions.

10.47 A carcinogenicity study was conducted to examine the tumor potential of a drug scheduled for initial testing in humans. A total of 300 rats (150 males and 150 females) were studied for a 6-month period. At the beginning of the study, 100 rats (50 males, 50 females) were randomly assigned to the control group, another 100 to the low-dose group, and the remaining 100 (50 males, 50 females) to the high-dose group. On each day of the six-month period, the rats in the control group received an injection of an inert solution, whereas those in the drug groups received an injection of the solution plus drug. The sample data are shown in the accompanying table.

| | Number of Tumors | |
Rat Group	One or More	None
Control	10	90
Low dose	14	86
High dose	19	81

a Give the percentage of rats with one or more tumors for each of the three groups.

b Conduct a chi-square test of independence, with $\alpha = .05$.

c Does there appear to be a drug-related problem regarding tumors for this drug? In other words, as the dose is increased, does there appear to be an increase in the proportion of rats with tumors?

10.48 A total of 210 emphysema patients entering a drug clinic over a one-year period were treated with one of two drugs (either the standard drug, A, or an experimental compound, B) for a period of one week. After this period, each patient's condition was rated as greatly improved, improved, or no change. The sample results follow.

	Patient's Condition		
Therapy	No Change	Improved	Greatly Improved
Standard (A)	20	35	45
Experimental (B)	15	45	50

a Make a percentage comparison for the rows; does there appear to be a difference in the two therapies?

b Run a chi-square test of independence, and draw a conclusion. Use $\alpha = .05$. Does it agree with your speculation in part a?

10.49 Refer to Exercise 10.48. The sum of the expected values must equal $n = 210$; and the expected values will add to the appropriate row and column totals. Determine the minimum number of expected values that need to be computed. How is this number related to the degrees of freedom for the test?

10.50 A university conducted a self-study to satisfy the requirement for accreditation. One aspect of the self-study concerned faculty evaluations. Through the use of student evaluations of instructors, each faculty member was classified both by rank and by ability as a teacher. Use the accompanying results to test the null hypothesis of independence of the two classifications. Use $\alpha = .05$, and draw a conclusion.

	Rank			
Teaching Evaluation	Instructor	Assistant Professor	Associate Professor	Professor
Above average	36	62	45	50
Average	48	50	35	43
Below average	30	13	20	35

10.8 Multiway Contingency Tables

The use of the chi-square probability distribution to analyze count data presented in a two-classification contingency table illustrates the analysis for only one of the many types of classification problems. Many other types of applications are more complicated and hence are omitted from this text. However, the short presentation on contingency tables in Section 10.7 gives you an adequate tool for evaluating and making inferences about count data that are summarized into two classifications. Frequently, sociological studies are summarized in two classification contingency tables for publication in magazines or newspapers. Using what you have learned from Section 10.7, you will be in a position to determine whether two methods of classifying observed events are independent.

Sometimes we are interested in the relationship between two variables while "controlling" for one or more other, related variables. Consider the data of Table 10.9. Suppose that we are interested in the relationship between the two variables "complaint" and "sex." If the third variable, "marital status," is also related to the other variables, we may get more information about the relationship between the two

control variable

variables of interest by using the third variable (marital status) as a **control variable.** Thus we might examine the relationship between a "complaint" and "sex" for each of the three measured levels of marital status.

T A B L E   **10.9**   Drug-related Complaints by Sex and Marital Status

Complaint	Divorced Male	Female	Total	Widowed Male	Female	Total	Married Male	Female	Total	Total Male	Female	Total
Overdose	96	208	304	46	82	128	266	330	596	408	620	1028
Suicide	14	100	114	18	96	114	72	156	228	104	352	456
Psychiatric	38	24	62	22	10	32	116	36	152	176	70	246
Addiction	52	68	120	14	12	26	146	78	224	212	158	370
Total	200	400	600	100	200	300	600	600	1200	900	1200	2100

partial tables

The data of Table 10.9 have been arranged in three **partial tables,** corresponding to the levels of the control variable, in Table 10.10. In examining the relationship between two variables in the presence of a control variable, we could first run separate chi-square tests of independence for each of the partial tables. These results would indicate whether the two variables of interest are independent at each of the indicated levels of the control variable. A combined test of independence of the two variables could be obtained by adding the values of chi-square computed for each of the partial tables. Here we would be testing the independence of the two variables while controlling the influence of a third variable on the two variables being studied. The degrees of freedom of the overall test would be equal to the sum of the degrees of freedom for each of the partial tables (see Table 10.10).

T A B L E   **10.10**
Partial Tables for a Comparison of Drug-related Complaints and Sex in the Presence of the Control Variable, Marital Status

Complaint	Divorced Male	Female	Total	Widowed Male	Female	Total	Married Male	Female	Total
Overdose	96	208	304	46	82	128	266	330	596
Suicide	14	100	114	18	96	114	72	156	228
Psychiatric	38	24	62	22	10	32	116	36	152
Addiction	52	68	120	14	12	26	146	78	224
Total	200	400	600	100	200	300	600	600	1200

E X A M P L E   **10.12**   Use the data of Table 10.10 to test the independence of the two variables "complaint" and "sex," for each of the partial tables. Use $\alpha = .05$ for all tables, and interpret your results.

Solution   Since we already know how to compute expected cell counts by using the formula

$$\frac{(\text{Row total})(\text{Column total})}{n},$$

we have listed the expected cell counts for each of the partial tables in Table 10.11. You should check a few of the cell entries to convince yourself that these are correct.

TABLE 10.11   Expected Cell Counts for the Partial Tables in Table 10.10

Complaint	Divorced			Widowed			Married		
	Male	Female	Total	Male	Female	Total	Male	Female	Total
Overdose	101.3	202.7	304	42.7	85.3	128	298	298	596
Suicide	38.0	76.0	114	38.0	76.0	114	114	114	228
Psychiatric	20.7	41.3	62	10.7	21.3	32	76	76	152
Addiction	40.0	80.0	120	8.6	17.4	26	112	112	224
Total	200	400	600	100	200	300	600	600	1200

Using partial table for "divorced" data in Table 10.10, we have the following test procedure:

H_0:   For the divorced, the two variables are independent.

H_a: For the divorced, the two variables "complaint" and "sex" are dependent.

T.S.:

$$\chi^2 = \sum \frac{(O - E)^2}{E}$$

$$= \frac{(96 - 101.3)^2}{101.3} + \frac{(208 - 202.7)^2}{202.7} + \cdots$$

$$+ \frac{(68 - 80)^2}{80}$$

$$= .2773 + .1386 + \cdots + 1.8000 = 50.2579$$

R.R.:   Reject H_0 if the computed value of χ^2 is greater than 7.81 from Table 4 of Appendix 3, with $a = .05$ and df $= 3$.

Conclusion: We reject the null hypothesis and conclude that, for divorced people, the variables "drug-related complaint" and "sex" are dependent. It appears that a higher percentage of females than males are classified in the suicide category. A clear picture of this trend is seen in Table 10.12, giving a percentage comparison of Table 10.10.

TABLE 10.12
Percentage Comparison of
Drug-related Complaints by Sex,
for the "Divorced" Partial Table

Complaint	Male	Female
Overdose	48%	52%
Suicide	7%	25%
Psychiatric	19%	6%
Addiction	26%	17%
Total	100%	100%
Sample Size	200	400

We can form the same research and null hypotheses for widowed people. The computed value of the test statistic is $\chi^2 = 39.1672$. Again comparing the computed value to 7.81 ($\alpha = .05$, df $= 3$), we reject the null hypothesis of independence of the two variables for widowed people. As with divorced people, a higher percentage of females than males have the drug-related complaint categorized as suicide (see the percentage comparison in Table 10.13). Notice, however, that suicide percentages are much higher for the widowed for both males and females than they are for divorced people. In contrast, the divorced have higher percentages for the addiction complaint.

<table>
<tr><th>TABLE 10.13</th><th>Complaint</th><th>Male</th><th>Female</th></tr>
</table>

TABLE 10.13
Percentage Comparison of
Drug-related Complaints by Sex,
for the "Widowed" Partial Table

Complaint	Male	Female
Overdose	46%	41%
Suicide	18%	48%
Psychiatric	22%	5%
Addiction	14%	6%
Total	100%	100%
Sample Size	100	200

Finally, for married people the computed value of chi-square is $\chi^2 = 100.5678$. Again, this result is highly significant. From the percentage comparison in Table 10.14 and the chi-square test of independence, we see that married females have higher percentages associated with the overdose and suicide complaints than do married males. The married percentages for males and females are very similar to the corresponding percentages for divorced, but the widowed percentages appear to be quite different from the percentages for divorced and married people.

TABLE 10.14
Percentage Comparison of
Drug-related Complaints by Sex,
for the "Married" Partial Table

Complaint	Male	Female
Overdose	44%	55%
Suicide	12%	26%
Psychiatric	19%	6%
Addiction	24%	13%
Total	99%	100%
Sample Size	600	600

In summary, we have seen that sex and drug-related complaints at the hospital emergency room are related at each level of the control variable "marital status." ∎

EXAMPLE 10.13   Combine the separate chi-square tests for the partial tables in Table 10.11 to compute a pooled chi-square. List the parts of the statistical test, and draw conclusions. Give the level of significance of the test.

Solution The five parts of the pooled test of independence are as follows:

H_0: When controlling for marital status, the two variables are independent.

H_a: When controlling for marital status, the two variables "complaint" and "sex" are dependent.

T.S.: Chi-square equals the sum of the computed chi-square test statistics for the partial tables.

R.R.: Since we are asked to give the level of significance for our test, we do not specify a rejection region.

Conclusion: The chi-square test statistic has df $= 9$, the sum of the degrees of freedom for the separate partial tables. Since $\chi^2 = 100.9928$ exceeds the $a = .005$ value of 23.589 in Table 4 of Appendix 3 for df $= 9$, the test is significant at the $p < .005$ level. Indeed, the data indicate a dependence of the two variables when marital status is controlled for. ∎

A word of caution: The pooled chi-square test just performed is *not* equivalent to—and will not necessarily lead to the same conclusion as—the chi-square test of independence between "sex" and "complaint" that *ignores* the control variable "marital status." The results for these two separate tests may lead to quite different conclusions. The pooled chi-square gives us a test of independence between two variables while controlling the influence of a third variable.

To construct a chi-square test of independence that ignores the influence of a third variable, we form a two-way table (by summing over the cells of the third variable) and then perform the chi-square test in the usual way.

10.9 Using Computers to Help Make Sense of Data

The calculations required for a chi-square test of independence are not very difficult; but as the numbers of rows and columns increase, they do become tedious. MINITAB and EXECUSTAT are especially easy to use, but SAS can also be used to conduct a chi-square test of independence. For MINITAB, one enters the cell frequencies and obtains the test results; for SAS, if the data file has not been entered previously, the individual observations (rather than the cell frequencies) must be entered before PROC FREQ can be used to obtain the chi-square test.

Now let's consider an example that involves such output. The market research group of a particular firm conducted a survey in three cities to compare the sales potential of a new soft drink. Each person contacted was asked to try the new drink and to classify it as excellent, satisfactory, or unsatisfactory. The results of the survey are summarized in the accompanying table. MINITAB and EXECUSTAT were used to conduct a chi-square test of independence.

Classification	City 1	City 2	City 3
Excellent	62	51	45
Satisfactory	28	30	35
Unsatisfactory	10	19	20

MINITAB Output

```
MTB > NAME C1 'CITY1' C2 'CITY2' C3 'CITY3'
MTB > READ C1-C3
      3 ROWS READ
MTB > END

MTB > CHISQUARE C1 C2 C3

Expected counts are printed below observed counts

               CITY1      CITY2      CITY3      Total
      1           62         51         45        158
                52.67      52.67      52.67

      2           28         30         35         93
                31.00      31.00      31.00

      3           10         19         20         49
                16.33      16.33      16.33

  Total          100        100        100        300

ChiSq =    1.654 +   0.053 +   1.116 +
           0.290 +   0.032 +   0.516 +
           2.456 +   0.435 +   0.823 = 7.376
df = 4

MTB > STOP
```

EXECUSTAT Output

```
                    Crosstabulation of CITY vs. CLASS
                                                         Row
                    CITY1        CITY2        CITY3      Total

   CLASS1             62           51           45        158
                    52.67        52.67        52.67     158.01

   CLASS2             28           30           35         93
                    31.00        31.00        31.00      93.00

   CLASS3             10           19           20         49
                    16.33        16.33        16.33      48.99

        Column       100          100          100      300.00
        Total        100          100          100      300.00


         Summary Statistics for Crosstabulation

           Chi-square      D.F.          P Value

             7.376          4            0.1173
```

As can be seen from both sets of output, the computed value of chi-square for the test of independence is 7.376. EXECUSTAT indicated the p-value based on df $= 4$ is

.1173. Thus we find insufficient evidence to reject H_0: the hypothesis of independence; there is insufficient evidence to indicate the distribution of classification is different for the three cities.

Summary

This chapter has dealt with categorical data representing the number (frequency) of observations falling into each possible cell. For a single variable, we're interested in the cell counts of the separate categories of the variable. When observations are made on two or more variables, we're interested in the frequencies associated with the cells of the contingency table formed by cross-classifying observations according to the variables of interest.

Categorical data obtained from a single variable arise in a number of practical situations. We discussed a chi-square goodness-of-fit test that is used to test whether the sample frequencies (and percentages) associated with categories of a variable agree with what would be expected according to hypothesized cell percentages. We also examined estimation and test procedures for a binomial proportion p and for comparing two binomial proportions based on independent samples.

Two variable categorical data problems were introduced, each using a chi-square test of independence for data displayed in a $r \times c$ contingency table. We also presented a way to combine information across sets of two-way contingency tables. Finally, it should be stressed that this chapter gave only a brief introduction to problems in the analysis of categorical data. Sequences of courses can be developed at the undergraduate (but more likely the graduate) level.

Key Terms

multinomial experiment

multinomial distribution

expected number of outcomes

χ^2

cell probabilities

observed cell counts

expected cell counts

chi-square distribution

lot acceptance sampling

statistical sampling plan

operating characteristic (OC) curve

control chart for p

upper control line (UCL)

lower control line (LCL)

sampling distribution for $\hat{p}_1 - \hat{p}_2$

bivariate data

multivariate data

contingency tables

independence

test statistic

control variable

partial table

Key Formulas

1   Multinomial distribution:

$$P(n_1, n_2, \ldots, n_k) = \frac{n!}{n_1! n_2! \cdots n_k!} p_1^{n_1} p_2^{n_2} \cdots p_k^{n_k}$$

2   Chi-square goodness-of-fit test:

$$H_0: p_i = p_{i0}$$

$$\text{T.S.:   } \chi^2 = \sum_i \left[\frac{(n_i - E_i)^2}{E_i} \right],$$

where

$$E_i = np_{i0}$$

3   Confidence interval for p:

$$\hat{p} = z_{\alpha/2} \sigma_{\hat{p}},$$

where

$$\hat{p} = \frac{y}{n}$$

and

$$\sigma_{\hat{p}} = \sqrt{\frac{\hat{p}\hat{q}}{n}}$$

4   Sample size required for a $100(1 - \alpha)\%$ confidence interval of the form $\hat{p} \pm E$:

$$n = \frac{z_{\alpha/2}^2 pq}{E^2}$$

(*Hint:* Use $p = .5$ if no estimate is available.)

5   Statistical test for p:

$$H_0: \quad p = p_0$$
$$\text{T.S.:} \quad z = \frac{\hat{p} - p_0}{\sigma_{\hat{p}}},$$

where

$$\sigma_{\hat{p}} = \sqrt{\frac{p_0 q_0}{n}}$$

6   Confidence interval for $p_1 - p_2$:

$$\hat{p}_1 - \hat{p}_2 \pm z_{\alpha/2} \sigma_{\hat{p}_1 - \hat{p}_2},$$

where

$$\sigma_{\hat{p}_1 - \hat{p}_2} = \sqrt{\frac{p_1 q_1}{n_1} + \frac{p_2 q_2}{n_2}}$$

7   Statistical test for $p_1 - p_2$:

$$H_0: p_1 - p_2 = 0$$
$$\text{T.S.:} \ z = \frac{\hat{p}_1 - \hat{p}_2}{\sigma_{\hat{p}_1 - \hat{p}_2}},$$

where

$$\sigma_{\hat{p}_1 - \hat{p}_2} = \sqrt{\hat{p}\hat{q}\left(\frac{1}{n_1} + \frac{1}{n_2}\right)},$$
$$\hat{p} = \frac{y_1 + y_2}{n_1 + n_2}, \quad \text{and} \quad \hat{q} = 1 - \hat{p}$$

8   Chi-square test of independence:

$$\chi^2 = \sum_{i,j} \left[\frac{(n_{ij} - E_{ij})^2}{E_{ij}}\right]$$

where

$$E_{ij} = \frac{(\text{Row } i \text{ total})(\text{Column } j \text{ total})}{n},$$

Supplementary Exercises

10.51 A sociologist studied the relationship between male skin color (light, medium, and dark) and job-mobility orientation (high, medium, and low). Use these data to run a chi-square test of independence. Interpret your results, using $\alpha = .05$.

Job-mobility Orientation	Male Skin Color		
	Light	Medium	Dark
High	35	84	51
Medium	49	78	23
Low	10	13	6

10.52 A study was conducted to determine the relationship between annual income and number of children per family. Compute percentages for each of the income categories listed; then run a chi-square test of independence, and draw conclusions. Use $\alpha = .10$.

Number of Children per Family	Annual Income		
	< $20,000	$20,000–$40,000	>$40,000
≤ 2 children	38	45	22
> 2 children	220	95	30

10.53 A survey of admissions practices at a liberal arts college was conducted to determine whether there appeared to be a difference in the acceptance rates for white and minority (nonwhite) applicants. The results of this survey, which combine information from 4000 applicants, are shown in the accompanying table. Use the accompanying MINITAB computer printout to conduct a chi-square test of independence. Use Table 4 in Appendix 3 to obtain an approximate level of significance for the test, and draw conclusions.

Applicant Accepted?	Applicant		
	Nonwhite	White	Total
Yes	38	126	164
No	362	3474	3836
Total	400	3600	4000

MINITAB Output

```
MTB > NAME C1 'NONWHITE' C2 'WHITE'
MTB > READ C1 C2
      2 ROWS READ
MTB > END
MTB > CHISQUARE C1 C2

Expected counts are printed below observed counts

          NONWHITE     WHITE    Total
      1         38       126      164
             16.40    147.60

      2        362      3474     3836
            383.60   3452.40

Total        400      3600     4000

ChiSq = 28.449 +   3.161 +
           1.216 +   0.135 = 32.961
df = 1

MTB > STOP
```

10.54 A random sample of 145 people with various occupations was taken to investigate the public opinion on police treatment. Each person was asked whether he or she would expect the police to treat him or her as good as, better than, or worse than a common criminal. The table that follows summarizes the results. Is there sufficient evidence to indicate that the expected treatment is independent of occupation? Use $\alpha = .10$.

| | **Expected Treatment** | | | |
Occupation	Better	As Good	Worse	Totals
Unemployed	6	23	11	40
Blue-collar worker	17	30	8	55
White-collar worker	16	28	6	50
Totals	39	81	25	145

10.55 A sociological study was conducted to determine whether there is a relationship between the length of time blue-collar workers remain in their first job and the amount of their education. From union membership records, a random sample of persons was classified. The data are shown in the accompanying table.

| | **Years of Education** | | | |
Years on First Job	0–4.5	4.5–9	9–13.5	13.5
0–2.5	5	21	30	33
2.5–5	15	35	40	30
5–7.5	22	16	15	30
7.5	28	10	8	10

a   Use the accompanying SAS computer output to identify the expected cell numbers.

SAS Output

CATEGORICAL ANALYSES

TABLE OF YRS - JOB BY YRS - ED

YRS - JOB   YRS - ED

FREQUENCY PERCENT ROW PCT COL PCT	0-4.5	4.5-9	9-13.5	13.5	TOTAL
0-2.5	5 1.44 5.62 7.14	21 6.03 23.60 25.61	30 8.62 33.71 32.26	33 9.48 37.08 32.04	89 25.57
2.5-5	15 4.31 12.50 21.43	35 10.06 29.17 42.68	40 11.49 33.33 43.01	30 8.62 25.00 29.13	120 34.48
5-7.5	22 6.32 26.51 31.43	16 4.60 19.28 19.51	15 4.31 18.07 16.13	30 8.62 36.14 29.13	83 23.85
7.5	28 8.05 50.00 40.00	10 2.87 17.86 12.20	8 2.30 14.29 8.60	10 2.87 17.86 9.71	56 16.09
Total	70 20.11	82 23.56	93 26.72	103 29.60	348 100.00

STATISTICS FOR 2-WAY TABLES

STATISTIC	DF	VALUE	PROB
CHI-SQUARE	9	57.830	0.0001
LIKELIHOOD RATIO CHI-SQUARE	9	55.605	0.0001
PHI COEFFICIENT		0.408	
CONTINGENCY COEFFICIENT		0.377	
CRAMER'S V		0.235	

SAS Output (Continued)

CATEGORICAL ANALYSES

OBS	YRS-JOB	YRS-ED	FREQ
1	0-2.5	0-4.5	5
2	0-2.5	4.5-9	21
3	0-2.5	9-13.5	30
4	0-2.5	13.5	33
5	2.5-5	0-4.5	15
6	2.5-5	4.5-9	35
7	2.5-5	9-13.5	40
8	2.5-5	13.5	30
9	5-7.5	0-4.5	22
10	5-7.5	4.5-9	16
11	5-7.5	9-13.5	15
12	5-7.5	13.5	30
13	7.5	0-4.5	28
14	7.5	4.5-9	10
15	7.5	9-13.5	8
16	7.5	13.5	10

N=16

b Test the research hypothesis that the variable "length of time on first job" is related to the variable "amount of education."

c Give the level of significance for the test.

d Draw your conclusions using $\alpha = .05$.

10.56 The personnel department of a large corporation was interested in determining the relationship between performance ratings of recently hired employees and the employees' college grade-point averages. To do this, a random sample of 90 records was obtained, examined, and classified in the following two-way table. Is there sufficient evidence to indicate a relationship between the two variables "performance rating" and "college grade point average"? Use $\alpha = .05$.

	College Grade Point Average		
Performance Rating	A	B	C
Above average	19	8	3
Average	9	12	15
Below average	6	5	13

10.57 Television research to date suggests that "zipping" and "zapping" of commercials by VCR viewers is uncommon. Zipping is the use of the VCR remote control to fast-forward past commercials; zapping is the use of the remote control to change channels when commercials appear. Based on a random sample of 2000 users of VCRs, 66% said that they did not skip the commercials. Obviously, given that more than 30 million households have VCRs, widespread zipping and zapping would have a tremendous impact on the rates charged for TV advertising. Use these data to construct a 95% confidence interval for p, the proportion of VCR users who do not skip commercials.

10.58 Two researchers at Johns Hopkins University have studied the use of drugs among the elderly. Patients in a recent study were asked the extent to which physicians counseled them with regard to their drug therapies. The researchers found the following:

- 25.4% of the patients said that their physicians did not explain what the drug was supposed to do.

- 91.6% indicated that they were not told how the drug might "bother" them.

- 47.1% indicated that their physicians did not ask how the drug "helped" or "bothered" them after therapy was started.

- 87.7% indicated that the drug was not changed after discussion of how the therapy was helping or bothering them.

a Assume that 500 patients were interviewed in this study. Summarize each of these results, using a 95% confidence interval.

b What comments, if any, do you have about the validity of these results?

10.59 People over the age of 40 years tend to notice changes in their digestive systems that alter what and how much they eat. A study was conducted to see whether this observation applies across different ethnic segments of our society. Random samples of Anglo-Saxons, Germans, Latin Americans, Italians, Spaniards, and African-Americans were obtained. The data from this survey are summarized here.

Ethnic Group	Sample Size Responding (60 of Each Group Were Contacted)	Number Reporting Altered Digestive System
Anglo-Saxon	55	7
German	58	6
Latin American	52	34
Italian	54	38
Spanish	30	20
African-American	49	31

a Does it appear that there may be a bias due to the response rates?

b Compare the rates (p_is) for the Anglo-Saxon and German groups, using a 95% confidence interval.

10.60 Refer to Exercise 10.59. There seem to be two distinct rates: those around 12%, and those around 70%. Combine the sample data for the first two groups and for the last four groups. Use these data to test $H_0: p_1 - p_2 = 0$ versus $H_a: p_1 - p_2 < 0$. Here, p_1 corresponds to the population rate for the first combined group, and p_2 is the corresponding proportion for the second combined group. Give the p-value for your test.

10.61 Two sets of 60 ninth-graders were taught algebra I by different methods. The experimental group used self-paced modules developed for use at a computer with a display screen. The control group was given formal lectures by the teachers. At the end of the four-month period, a comprehensive, standardized test was given to both groups. The experimental group had 65% scoring above 80 (out of 100), whereas the control group had just 47% above 80. Use these data to compare the percentages above 80 for the two groups. Give the p-value for your test.

10.62 Refer to Exercise 10.61. How might you have designed this study? What additional data might you want to collect?

10.63 A recent study conducted for a large university examined the effects of intensive behavioral training to change type A habits on the incidence of a second heart attack. A total of 290 patients who had recently had a nonfatal heart attack were studied. Those randomized to group 1 received the intensive behavioral training in addition to routine medical care. The training was designed to help the patient to slow down and be more relaxed. Those assigned to group 2 received only the routine medical care. The data from this five-year study are summarized in the following table. Use these data to test $H_0: p_1 - p_2 = 0$ versus $H_a: p_1 - p_2 < 0$. Give a p-value for your test, and draw a conclusion.

Group	Sample Size	Number of Second Heart Attacks
Group 1	140	17
Group 2	150	29

10.64   A random sample of faculty members of a state university system was polled and classified by university and by which of the three collective bargaining agents (union 101, union 102, union 103) was preferred. The data appear here, and SAS computer output follows.

University	Bargaining Agent		
	101	102	103
1	42	29	12
2	31	23	6
3	26	28	2
4	8	17	37

SAS Output

```
CATEGORICAL ANALYSES

TABLE OF UNIV BY B_AGENT

UNIV                    B_AGENT

FREQUENCY|
PERCENT  |
ROW PCT  |
COL PCT  |       1|       2|       3|  TOTAL
---------+--------+--------+--------+
       1 |     42 |     29 |     12 |     83
         |  16.09 |  11.11 |   4.60 |  31.80
         |  50.60 |  34.94 |  14.46 |
         |  39.25 |  29.90 |  21.05 |
---------+--------+--------+--------+
       2 |     31 |     23 |      6 |     60
         |  11.88 |   8.81 |   2.30 |  22.99
         |  51.67 |  38.33 |  10.00 |
         |  28.97 |  23.71 |  10.53 |
---------+--------+--------+--------+
       3 |     26 |     28 |      2 |     56
         |   9.96 |  10.73 |   0.77 |  21.46
         |  46.43 |  50.00 |   3.57 |
         |  24.30 |  28.87 |   3.51 |
---------+--------+--------+--------+
       4 |      8 |     17 |     37 |     62
         |   3.07 |   6.51 |  14.18 |  23.75
         |  12.90 |  27.42 |  59.68 |
         |   7.48 |  17.53 |  64.91 |
---------+--------+--------+--------+
   Total       107       97       57      261
             41.00    37.16    21.84   100.00
```

```
STATISTICS FOR 2-WAY TABLES

STATISTIC                        DF      VALUE       PROB
----------------------------------------------------------
CHI-SQUARE                        6      75.197     0.0001
LIKELIHOOD RATIO CHI-SQUARE       6      71.991     0.0001
PHI COEFFICIENT                          0.537
CONTINGENCY COEFFICIENT                  0.473
CRAMER'S V                               0.380
```

SAS Output (Continued)

CATEGORICAL ANALYSES

OBS	UNIV	B_AGENT	FREQ
1	1	101	42
2	1	102	29
3	1	103	12
4	2	101	31
5	2	102	23
6	2	103	6
7	3	101	26
8	3	102	28
9	3	103	2
10	4	101	8
11	4	102	17
12	4	103	37

N=12

a Identify the expected cell numbers.

b Use the computer output to determine whether there is sufficient evidence to indicate a difference in the distribution of preference across the four state universities.

c Give the level of significance for the test.

d Draw conclusions.

 10.65 An advertising firm selected to conduct a market awareness study for a brand of house paint obtained information from a national survey of 1500 randomly selected homeowners. Each homeowner selected for the survey was asked if he or she was familiar with a newly marketed line of interior latex paints. If 465 responded affirmatively, use the sample data to test the research hypothesis that the company has reached more than 30% of homeowners with recent advertising. Use $\alpha = .05$.

10.66 Legislators of a particular state were concerned that the enrollment (which affects budget allocations) at a particular university within the state system had been padded by allowing students to overenroll or to enroll for courses that required no academic work. To substantiate their initial findings, they arranged to interview a random sample of 200 graduate students (from the 5000 currently enrolled). If 20 students stated that they had been allowed to pad their enrollments in the past quarter, use these data to construct a 99% confidence interval for p, the proportion of the entire student body with padded enrollments. Interpret your results.

10.67 The relative sensitivities of two fuses were tested under controlled conditions by firing 40 rounds with Type I and 60 rounds with Type II; the firings were conducted in random order. Each round was classified according to whether the fuse functioned or not, with the following results.

Type of Fuse	Functioned	Did Not Function	Total
I	10	30	40
II	40	20	60
Total	50	50	100

a Test the hypothesis that there is no difference between the sensitivities of the two fuses. Use $\alpha = .05$.

b Use the previous data to demonstrate the relationship between the z test and the chi-square test of independence in a 2×2 table. (*Hint:* Compare z^2 and χ^2 for the two tests.)

 10.68 A study was conducted to investigate whether there is any relationship between voting record and education. One hundred five citizens selected at random were interviewed as to how often they vote and what level of formal education they had achieved. The results are shown in the accompanying table. Is there sufficient evidence to indicate a relationship between level of education and frequency of voting? Use $\alpha = .05$.

How Often Do You Vote?

Education	Never	Some Elections	All Elections	Totals
Less than h.s. level	11	12	11	34
High school	7	15	13	35
College	2	20	14	36
Totals	20	47	38	105

10.69 An extension of the traffic study for implementing a priority bus lane involved sampling public opinion about the bus lane during various phases of the study. Three different phases were to be studied. Phase 0 required the bus drivers to use the existing traffic lanes. In Phase 1, bus drivers made use of the exclusive bus lane, but with no preemption of the traffic signals at the intersection. Phase 2 allowed the bus drivers to extend the "green time" on a traffic signal to enable them to pass through before the light changed. Use the accompanying sample data to determine whether the distribution of persons favoring, not favoring, or undecided changes from phase to phase. Use $\alpha = .05$.

Opinion

Phase	Favor	Do Not Favor	Undecided	Totals
0	80	90	30	200
1	60	112	28	200
2	50	125	25	200

10.70 The quality-assurance unit of a large pharmaceutical company was engaged in comparing two new formulations of a drug manufactured in tablet form. Both tablets contained the same amount of active ingredient, but varied in size, shape, and excipient (an inert substance that acts as a vehicle). A random sample of 100 tablets was obtained from a pilot batch for each formulation. The number of tablets classified as acceptable (or not acceptable) with regard to potency is shown below for each formulation. Use these data to place a 95% confidence interval about $p_1 - p_2$, the difference in the proportions of acceptable tablets.

Formulation	Number Acceptable	Number Not Acceptable	Sample Size
1	84	16	100
2	96	4	100

10.71 Refer to Exercise 10.70.

a Run a two-sided statistical test of H_0: $p_1 - p_2 = 0$. Draw conclusions.

b Run a chi-square test of independence for these data. Use $\alpha = .05$. Compare your results here to those in part a.

c Suggest a relationship between the two tests. (*Hint:* Compare z^2 and χ^2 for the two tests.)

10.72 Public concern has been raised about whether there is an association between the use of saccharin as a sweetener and the development of cancer. Suppose that 25 independent, controlled studies have been conducted at elevated doses (greater than or equal to 100 times the normal yearly human intake) in a particular strain of mice. In two of these studies, the treated group of mice had a higher incidence of cancer than in the corresponding control group of mice; no meaningful differences were observed between the treated and controlled groups of mice in the other studies. What conclusions might you draw from these 25 studies? What additional data would you like to see? How do these results extrapolate to the human situation? What error rate should be controlled more carefully?

10.73 A survey of drivers was obtained to compare the proportions who use seat belts regularly for various age categories. These data are shown next. Analyze the data, and draw conclusions. Use $\alpha = .05$.

Regularity of Seat Belt Usage

Age (Years)	Always	Regularly	Sometimes	Never
16–20	1	10	70	19
21–25	4	8	80	8
26–30	8	10	77	5
> 30	15	30	49	6

10.74 Entry-level diastolic blood pressures for a group of 12 hypertensive females are shown in the next table, along with a corresponding diastolic blood pressure following 2 weeks of treatment with an antihypertensive medication. *Note:* The goal of the therapy is to reduce the diastolic blood pressure to 90 mm Hg.

Patient	Entry Level Diastolic BP	BP Following 2 Weeks
1	105	89
2	110	87
3	115	93
4	107	90
5	108	88
6	116	90
7	114	91
8	110	90
9	112	89
10	106	91
11	109	90
12	111	92

a   What percent of the patients were controlled at ≤ 90 mm Hg?

b   Give a p-value for the test $H_0: p = 0$ versus $H_a: p > 0$.

10.75 Refer to Exercise 10.74.

a   Consider using a t-test for $H_0: \mu_d = 0$ versus $H_a: \mu_d > 0$ (see Chapter 8). Might there be a problem with one (or more) of the assumptions? If so, which ones and why?

b   Suggest and perform a suitable test of location for the difference data. Give the p-value, and interpret your findings.

10.76 A poll asked the following question of a random sample of 100 people in various countries: From what you have heard or read, which of these statements comes closest to the way you feel about the United States' involvement in the conflict in Somalia? (1) The United States should increase its support for the U.N. mission there. (2) The United States should carry on its present level of support. (3) The United States should gradually decrease its level of support.

Country	Increase Support	Maintain Present Level	Decrease Support	Don't Know
Argentina	57	6	6	31
Australia	21	43	24	12
Brazil	76	5	5	14
Canada	41	16	23	20
Finland	81	4	5	10
France	72	8	5	15
Great Britain	45	15	15	25
India	66	4	8	22
Sweden	79	10	4	7
Uruguay	62	10	5	23
United States	31	10	53	6
West Germany	58	11	14	17

a   Would it be meaningful to do a percentage comparison of the data?

b   Suggest a statistical test to compare responses across countries.

c   Run the test you selected in part b (using $\alpha = .05$), and draw conclusions.

10.77   Members of a sample of 150 patients suffering from severe acquired immunodeficiency syndrome (AIDS) were treated with an experimental compound. Without treatment, these patients were given little chance to survive beyond the next six months. Baseline characteristics are shown here.

Baseline Characteristics

Gender (% male)	92%
Age in years ($\bar{y} \pm SD$)	32 ± 6
Weight in kg ($\bar{y} \pm SD$)	68 ± 7

For the 120 patients who were treated for more than one week, the 6-month survival rate was 52%. The survival rate for the 30 who did not receive more than one week of therapy was only 17%.

a   Use the survival data to give 95% confidence intervals on 6-month survival rates for patients treated for more than one week and for patients not treated for more than one week.

b   Would it be valid to make a comparison of these two survival rates for the subgroups of $n_1 = 120$ and $n_2 = 30$ patients? Why or why not?

c   Estimate the overall survival rate for the AIDS patients, using a 95% confidence interval. Is this estimate of the survival rate for patients treated with the drug more valid than the one based entirely on patients treated for more than one week?

10.78   As part of a research study conducted at a large behavioral treatment program, 160 subjects trying to quit smoking were randomized to one of four treatment programs.

A: Reading and audiovisual (A/V) material only, no professional contact

B: Reading and A/V material, low professional contact

C: Reading and A/V material, high professional contact

D: High professional contact only

Assessments were made at 4, 12, 26, and 52 weeks, with the results (abstinence rates) shown here.

	Time			
Program	4	12	26	52
A	50%	45%	35%	30%
B	70%	50%	47%	35%
C	95%	75%	60%	40%
D	90%	68%	56%	38%

a Plot the sample data, and draw some tentative conclusions.

b At a given time point, what statistical test would be appropriate for comparing the abstinence rates?

c Run these comparisons for each time point, and draw conclusions. In general, what seems to be the pattern of abstinence?

10.79 A survey was conducted among regular shoppers at three different suburban shopping centers. Subjects were given a questionnaire to be completed in private and deposited in a centrally located, locked collection box. A total of 1000 questionnaires were distributed, and 400 were returned. One of the items on the questionnaire had to do with family income levels. Responses to this question are recorded in the accompanying table.

	Family Income		
Shopping Center	<50,000	50–100,000	>100,000
A	60	25	15
B	66	50	9
C	127	40	8
Total	253	115	32

a Do a percentage comparison of the family income distributions for the three shopping centers. Do the centers appear to have the same distributions?

b Run a test of significance to compare the income distributions. Give a p-value for your test, and draw a conclusion.

c What effect might the response rate have on the conclusions drawn from this survey?

10.80 Three supermarkets have a policy of advertising specials on certain days of the week to attract customers. A customer count is maintained at these supermarkets from 11:00 A.M until 2:00 P.M. on each of the five weekdays. The following data were collected.

	Day of the Week				
Supermarket	Mon.	Tues.	Wed.	Thurs.	Fri.
1	605	650	702	663	568
2	696	741	750	827	663
3	540	668	528	572	516

a Do the data indicate that the choice of a supermarket depends on the day of the week? Use a procedure that has a .01 chance of being in error if this conclusion is drawn. If you decide that choice of a supermarket is related to day of the week, briefly describe the nature of that relationship, based on the sample data.

b Do the data support the hypothesis that the number of customers expected (for all three supermarkets together) is the same for the days Monday through Friday? Use a procedure that has a .05 chance of being in error if it supports the conclusion that different days have different customer counts.

10.81   Student opinion on a resolution presented to the student council was surveyed to determine whether opinion was independent of fraternity and sorority affiliation. Two hundred students were interviewed, and the results were as shown in the accompanying table. In a chi-square test of independence, how many degrees of freedom are there?

Status of Affiliation

Student Opinion	Fraternity	Sorority	Unaffiliated
Favor	40	35	27
Opposed	18	25	55
Total	58	60	82

10.82   Use the data in Exercise 10.81 to determine whether there is sufficient evidence to indicate that student opinion on the resolution is independent of affiliation status (fraternity, sorority, or unaffiliated). Use $\alpha = .05$.

10.83   Refer to Exercise 10.81. Suppose that we introduce a control variable—socioeconomic status of parents (families with incomes of $35,000 or more and families with incomes below $35,000). The partial tables for the data are as given. Run separate chi-square tests of independence to determine whether the variable "student opinion" is independent of affiliation status (fraternity, sorority, unaffiliated), while controlling for socioeconomic status. Interpret your results. Use $\alpha = .05$.

Student Opinion	Income Below $35,000			Income $35,000 or More		
	Fraternity	Sorority	Unaffiliated	Fraternity	Sorority	Unaffiliated
Favor	26	21	11	14	14	16
Opposed	8	15	39	10	10	16
Total	34	36	50	24	24	32

10.84   Use the results of Exercise 10.83 to conduct a pooled chi-square test of independence, while controlling for the socioeconomic status of each student's family. What is the difference between the research hypothesis for this exercise and that for Exercise 10.82? Compare your conclusions here to those you reached in Exercise 10.82.

Experiences with Real Data

10.85   Construct a 4 × 4 contingency table that categorizes patients from the Clinical Trials database by marital status and by treatment group. Compute the percentages of males and females in each group. Do the groups appear to differ in their sex distributions? Test your intuition by using a chi-square test of independence, and give the p-value for your test.

10.86   Refer to the Clinical Trials database, and list other baseline comparisons that could be made among the treatment groups by using a chi-square test of independence.

10.87   Refer to Exercise 10.86. Make the comparisons suggested in that exercise. Do the groups appear to be comparable?

10.88   Refer to the Clinical Trials database.

a Construct a 4×5 contingency table to categorize patients by treatment group and by therapeutic effect.

b Compare expected cell counts. Do the expected cell counts satisfy the criteria for running a chi-square test of independence?

c Run a chi-square test of independence on the table of part a, if appropriate, or on a "collapsed" 4×2 table as shown here.

Treatment Group	Marked/Moderate	Minimal/Unchanged or Worse
A		
B		
C		
D		

10.89 If the table is collapsed, as suggested in Exercise 10.88, what information may be lost?

10.90 Refer to the Crimes database on your data disk (or in Appendix 1). Construct a contingency table with the two variables—religious preference and parents' income—for majors in social sciences and humanities.

a Test the hypothesis that religious preference is independent of parents' income. Use $\alpha = .05$. (*Note:* Combine the two categories of "other" and "Jew.")

b Interpret your findings. What percentage comparison might help you to interpret the data?

10.91 Refer again to the Crimes database. Do the sample data indicate that religious affiliation is independent of major? Explain your findings and how you reached your conclusion.

10.92 For the accompanying news article, comment on the appropriateness of the sample and of the inferences that are drawn. What additional information, if any, would have helped you draw your own inferences?

Study of Divided Families Shows Positive Attitudes

Chicago—A study of divorced mothers and their children has revealed some positive attitudes among members of divided families. Perhaps a broken home is not the psychological disaster for family members that society has long suspected.

The study, involving 20 mothers with one or more children between the ages of 6 and 18, was conducted to determine the basic concerns of divorced mothers and their children. There were 20 mothers and 25 children involved in the study.

All of the women were working full time. Most of them had made plans toward bettering their earning power. The women had been divorced from 3 months to 15 years. The educational level of the women in the study was high, compared to the national average: 12 years to 18 years of education.

A key aim of the study was to determine the feelings of the women and their children about their acceptance in society.

Eighty-six percent of the children felt that at school they were treated the same as children whose parents were married. Children aged 10 through 12 especially preferred that teachers and friends be told about the home situation. They wanted news of the divorce not to come as a surprise to others or to be a source of embarrassment for them.

In general, the children were doing well in school and even excelled in some areas.

Although the trend among most of the women was to socialize mainly with single persons, 80% of them felt accepted in their neighborhoods. Half of them said they felt accepted at church.

Among the children, 91% indicated they were treated no differently at Sunday school. Ninety percent of the sample were active church members.

Most of the women, 85%, said that after their divorces their attitudes toward divorce had shifted from negative to positive. The same proportion saw advantages for their children, in terms of understanding life and people, as a result of the divorce.

10.93 The accompanying article alludes to a study of the effects of orange juice in reducing symptoms of respiratory infections. Examine this article and comment on the published statistics contained in it. Be sure to comment on the adequacy of the sample, the inferences that are drawn, and additional material that might have been included in the article to help you reach an independent conclusion.

Researchers Study Effects of Orange Juice

Gainesville, Fla.—A quart of orange juice, taken every day, has significant effects in reducing the symptoms of respiratory infections caused by the rubella virus, report medical scientists from the University of Florida.

The Florida Study involved 55 human volunteers. Many of those drinking orange juice produced antibodies earlier than those not drinking orange juice. The researchers said that the early antibody production may or may not be the means by which the symptoms were reduced. However, the appearance of this antibody production, they pointed out, may indicate that citrus has a localized effect in stimulating the immune system to fight respiratory infection.

Many of the volunteers who drank fresh frozen orange juice did not experience any sore throat or runny nose. Other volunteers, forbidden to drink orange juice or eat citrus fruit, did show these mild symptoms. Prior to the study all participants were free of infection and had normal antibody levels.

The volunteers were subjected to a weakened, nontransmissible form of the rubella virus. The rubella virus was chosen because it can induce either respiratory or systemic infections, depending on how the virus is administered. Introduced in the nasal passages, the virus causes respiratory infection; introduced under the skin, the virus causes systemic infection.

Half of the 55 volunteers were given the rubella virus by nose drops; the other half received the same virus by injection. Each group was again divided. Half of them were on a quart of orange juice a day; their controls were on regular diets with no orange juice or citrus fruit. Each volunteer was followed for 21 days after the virus was administered.

In the groups who received the virus by nose drops and were not allowed to drink orange juice, 77% showed symptoms of respiratory infection. Among the counterparts who received the same type of virus and drank orange juice, only 27% had respiratory symptoms.

"This represents a significant reduction in symptoms," one researcher said.

Volunteers in the study who received the virus by injection showed no difference in symptoms, whether they were drinking orange juice or not.

6

Step Three: Analyzing Data: Regression, Correlation, and Analysis of Variance

11

Regression and Correlation

11.1   Introduction

We have already discussed (in Chapter 4) some ways to summarize data from more than one variable. In particular, we used the scatter plot as a way to examine the relationship between two quantitative variables based on sample data. In this chapter, we extend the concept of a scatter plot to help describe the relationship between two quantitative variables—in effect extending the data summarization methods of Chapter 4 (step 2 in making sense of data) to a situation involving two quantitative variables. Chapter 12 presents the methods for analyzing data of this type (step 3 in making sense of data). We have grouped these two chapters together because the summarizing techniques and the methods for analyzing these data are topics generally grouped together under the heading of "regression and correlation." We begin with several examples of problems that deal with two (or more) quantitative variables.

A problem of considerable interest to high-school seniors, freshmen entering college, their parents, and university administrators is the expected academic achievement of a particular student after he or she has enrolled in a university. For example, they might wish to estimate what a student's grade-point average (GPA) will be at the end of the freshman year, before the student has been accepted or enrolled in the university. At first glance this task seems difficult.

The statistical approach to this problem is, in many respects, a formalization of the procedure we might follow intuitively. Suppose that data were available giving high-school academic grades, psychological and sociological information, and grades attained at the end of the college freshman year for a large number of students. Then we might categorize the students into groups possessing similar characteristics. For example, highly motivated students who earned a high rank in their high-school class, graduated from a high-school with superior academic standards, and so forth, should achieve, on the average, a higher GPA at the end of their college freshman year than student who lack motivation and who achieved only moderate success in high school. Carrying this line of thought a little farther, we would expect the GPA of a student to be related to many variables that define the individual's psychological and physical characteristics, as well as to the characteristics that define the academic and social environment to which he or she will be exposed. Ideally we would like to obtain a mathematical equation that relates a student's GPA to all these variables, so it could be used for prediction.

You will observe that the problem we have defined has a very general nature. We are interested in some quantitative random variable y that is related to a number of quantitative variables $x_1, x_2, x_3, \ldots$. Generally, the random variable y is called the *dependent variable*, and the x variables are designated as *independent variables*. We are interested in obtaining an equation that relates y to the independent variables. The variable y for our example is the student's grade-point average at the end of the freshman year. The independent variables might be

$x_1 = $ Rank in high-school class

$x_2 = $ Score on a mathematics achievement test

$x_3 = $ Score on a verbal achievement test

and so on. The ultimate objective is to measure $x_1, x_2, x_3, \ldots$ for a particular student, substitute these values into the mathematical equation, and thereby predict the student's grade-point average. In order to accomplish this, we must first determine the related variables $x_1, x_2, x_3, \ldots$, and measure the strength of their relationship to y. Then we must construct a good **prediction equation** that expresses y in terms of the selected independent variables.

prediction equation

Other practical examples of our prediction problem are numerous throughout business, industry, and the sciences. To illustrate this point, let's look at several more examples in the following paragraphs.

Executives of a large oil company would like to relate the performance of a gasoline blend to a number of key ingredients in the gasoline mixture. By varying the proportions of the ingredients in the overall blend and using performance information for these many different blends, we can obtain a prediction equation relating performance to the proportions of the blend ingredients. By substituting different values of the independent variables (proportions of the key ingredients) into the prediction equation, we can determine the proportions that yield the blend with the best gasoline performance.

A biologist would like to relate human physical characteristics such as height, weight, blood pressure, pulse, and age to the amount of secretion from a gland. By observing these variables (characteristics) and the glandular secretion for many different people, we can obtain a prediction equation relating the amount of secretion to these physical characteristics.

A political analyst may wish to predict the success of a candidate in a political primary based on a number of important variables. Success can be measured by the number of votes cast for the candidate. Variables affecting the outcome of the primary might include the amount of money spent on television advertising, the size of the campaign organization, and the amount of money spent advertising in papers and magazines. Ideally the political analyst would like to study the effects of these variables on the outcomes of previous elections, to aid in selecting (predicting) the best strategy for a future campaign.

First we consider the problem of predicting y based on a single independent quantitative variable x. Then we observe that the solution for a *multivariable problem*, where y is related to more than one quantitative independent variable, is based on a generalization of our technique. Since the methodology for the multivariable predictor is fairly complex, we use computer programs to solve these problems.

11.2 Scatter Plots and the Freehand Regression Line

Consider the prediction of a student's GPA at the end of the freshman year based on his or her high-school GPA. The objective is to obtain an equation that will predict the achievement of college freshmen and hence will help admissions officers identify potentially successful students. The GPA data for a sample of 11 students are shown in Table 11.1.

Following the data collection stage, the next step in making sense of the sample data in Table 11.1 is to summarize them, using graphical or numerical methods. In Chapter 4 we introduced the scatter plot as a means for summarizing the relationship between two quantitative variables. Recall that, to construct a scatter plot, we make vertical and horizontal axes of approximately equal length. Generally the independent variable x is labeled along the horizontal axis, and the dependent variable y is labeled along the vertical axis. For our example, the independent variable is the high-school GPA.

Having labeled the axes, we next draw scales along the axes in such a way that all measurements can easily be plotted along the appropriate axis. For our example, both the independent and the dependent variables range between 1 and 4.0. After the axes are drawn, labeled, and scaled, we plot the data of Table 11.1. Each dot on the figure represents the information about one student and can be obtained by plotting the freshman GPA y against the corresponding high-school value x. The dot circled in Figure 11.1 corresponds to student 1. Notice that the dot is placed at a point on the graph corresponding to $x = 2.00$ and $y = 1.60$.

The scatter plot of Figure 11.1 suggests that the freshman GPA, y, increases as the high-school GPA, x, increases. In fact many of the dots seems to be on a straight line. We call a line running through the dots of a scatter plot a *trend line*, or a *regression line*. When the regression line is a straight line, as in Figure 11.2(a) and (b), we say

	TABLE 11.1	
Data for High-school and College GPAs (Based on a 4.0 System)		

Student	High-school GPA (x)	Freshman GPA (y)
1	2.00	1.60
2	2.25	2.00
3	2.60	1.80
4	2.65	2.80
5	2.80	2.10
6	3.10	2.00
7	2.90	2.65
8	3.25	2.25
9	3.30	2.60
10	3.60	3.00
11	3.25	3.10

that there is a linear relationship between x and y. However, not all regression lines are linear. When the trend line is curved as in Figure 11.2(c) and (d) we say that there is a curvilinear relationship between x and y. Trend lines thus provide a further means for summarizing the relationship between two quantitative variables.

FIGURE 11.1
High-school and College GPA Data

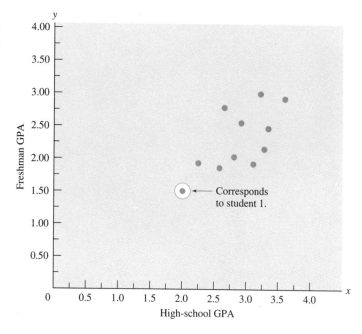

Many methods can be used to obtain a trend line relating y to x. The first—called an "eyeball fit," or a **freehand regression line**—can be obtained by placing a ruler on the graph (Figure 11.1) and moving it about until the distances from the points to the fitted line are minimized. This has been done in Figure 11.3. The resulting

freehand regression line

FIGURE 11.2   Different Types of Regression Lines: (a) Linear Relationship Between x and y; (b) Linear Relationship Between x and y; (c) Curvilinear Relationship Between x and y; (d) Curvilinear Relationship Between x and y

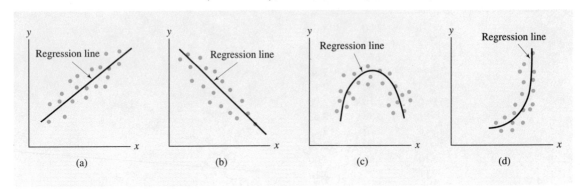

trend line can be used to make a prediction about a freshman's GPA based on high-school achievement. To predict y when $x = 2.5$, refer to the graph and note that the y-coordinate for the point corresponding to $x = 2.5$ is $y = 2.05$ (see the arrows in Figure 11.3).

FIGURE 11.3
Freehand Regression Line for
High-school and College GPA
Data

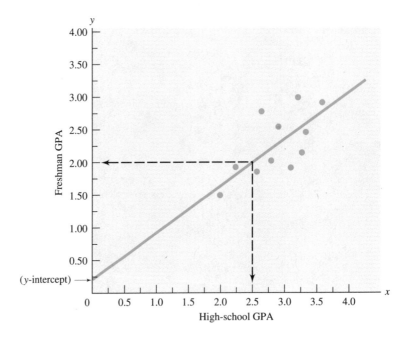

The freehand regression line shown in Figure 11.3 can be represented by a **linear equation** of the form

linear equation

$$y = \beta_0 + \beta_1 x.$$

The two constants, β_0 and β_1, in the equation determine the location and the slope of the line, respectively. The constant β_0 is the y-intercept—that is, the value of y when the line crosses the vertical axis (y-axis). The constant β_1 is the slope of the line—that is, the change in y that corresponds to a one-unit increase in x.

DEFINITION   11.1

y-Intercept β_0   The **y-intercept** β_0 for the straight line

$$y = \beta_0 + \beta_1 x$$

is the value of y at the point where the line crosses the y-axis.   ▪

DEFINITION   11.2

Slope β_1   The **slope** β_1 of the straight line

$$y = \beta_0 + \beta_1 x$$

is the change (increase or decrease) in the value of y for a one-unit increase in x.   ▪

From Figure 11.4 it appears that the y-intercept is approximately $\beta_0 = .20$. When x increases from 0 to 1.0, y increases from about .20 to .95. This change $(.95 - .20 = .75)$ represents the slope (β_1) of the straight line. Thus the equation that corresponds to the freehand regression line of Figure 11.4 is $y = .20 + .75x$. To predict y when $x = 2.5$, we substitute $x = 2.5$ into our prediction equation, $y = .20 + .75x$, and obtain $y = .20 + .75(2.5) = 2.08$, which is about the same as the guessed value of 2.04 (see Figure 11.3).

FIGURE 11.4
Slope β_1 and Intercept β_0 for the Freehand Regression Line

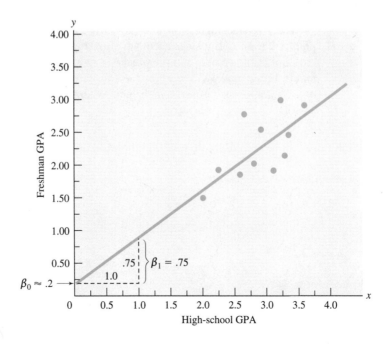

An equation of the form $y = \beta_0 + \beta_1 x$ is called a *deterministic model* because there is no error in reading y. That is, for a given value of the independent variable x, we can predict (determine) y exactly, using the equation $y = \beta_0 + \beta_1 x$. Although deterministic models are simple to use, they are unrealistic in many situations. For example, even though an equation of the form $y = \beta_0 + \beta_1 x$ adequately describes the trend in the GPA data, it cannot be used to predict a student's freshman GPA *exactly*, based on his or her high-school GPA.

A model that allows for the possibility that all observations do not fall on a straight line is the model

$$y = \beta_0 + \beta_1 x + \epsilon$$

where ϵ is a random error. In this model, ϵ represents the difference between a measurement y and a point on the line $\beta_1 + \beta_1 x$. The random error takes into account all unpredictable and unknown factors that are not included in the model. For example, a freshman's GPA could be affected by the student's rank in the graduating high-school class, score on the Scholastic Aptitude Test, motivation, and many other factors. The combined effects of these and other factors not included in the model contribute to ϵ.

One assumption made about the random error is that the average value of ϵ for a given value of x is 0. Thus, since β_0 and β_1 are constant, the average value of y (often called the **expected value of y**) for a fixed value of x is $\beta_0 + \beta_1 x$. This line, denoted by

expected value of y

$$E(y) = \beta_0 + \beta_1 x,$$

is shown in Figure 11.5. The difference between a sample data point and the expected value of y (a point on the line $\beta_0 + \beta_1 x$) is ϵ. The random errors associated with the 11 data points listed in Table 11.1 are also pictured in Figure 11.5.

The freehand regression line provided us with an equation by which we can predict y from x. But since each one of us might determine a different equation for the same data, we need a more precise way to obtain estimates of the constants β_0 and β_1 in the model $y = \beta_0 + \beta_1 x + \epsilon$. The method we will use, called the *method of least squares*, is discussed in the next section.

FIGURE 11.5
The Mean Value of y Plotted for
Various Values of x

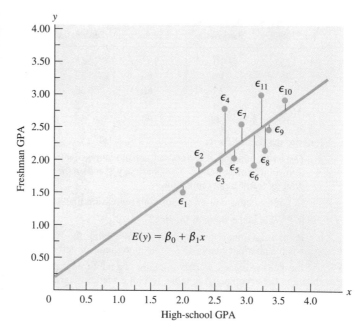

FIGURE 11.5
The Mean Value of y Plotted for
Various Values of x

Exercises

Basic Techniques

11.1 Plot the data shown here in a scatter diagram.

x	5	10	12	15	18	24
y	10	19	21	28	34	40

11.2 Refer to the data of Exercise 11.1.

a Use an eyeball fit to construct a freehand regression line.

b Identify the intercept and slope for your regression line.

c Predict y when $x = 20$.

11.3 Use the equation $y = 1.8 + 2.0x$.

a Predict y when $x = 3$.

b Plot the equation on a graph with the horizontal axis scaled from 0 to 5 and the vertical axis scaled from 1 to 12.

Applications

11.4 Results of a pharmaceutical pricing survey were reported recently, relating the influence of wholesale discounts and other facts on prescription ingredient costs. One problem considered in the study was the relationship between the annual prescription volume y and the percentage of the drug ingredient purchases that were made directly from the drug manufacturers. Suppose that a sample of 10 independent pharmacies yielded the results shown in the accompanying table. Plot these data on a scatter diagram.

Annual Volume (× $1,000), y	Percentage of Purchases, x	Annual Volume (× $1,000), y	Percentage of Purchases, x
25	10	133	63
55	18	90	42
50	25	60	30
75	40	10	5
110	50	100	55

11.5 Refer to the data of Exercise 11.4.

a Use an eyeball fit to determine a freehand regression line for predicting the annual prescription volume of an independent pharmacy based on the percentage of ingredients purchased directly from drug manufacturers.

i What is the y-intercept for your regression line?

ii What is the slope of the line?

b Predict the annual prescription volume, using the freehand regression line from part a, when a pharmacy buys 45% of its ingredients directly from drug manufacturers.

 11.6 Use the data for average teacher's income and public education expenditure per student by state, shown in the accompanying table, to plot a scatter diagram; then draw a freehand regression line.

State	Average Teacher's Income (x)	Public Education Expenditure per Student (y)
Arkansas	$25,200	$2698
Connecticut	$37,300	$7199
Kansas	$27,400	$4404
Maryland	$33,700	$5391
Michigan	$34,400	$4576
Mississippi	$22,000	$2846
Nebraska	$24,200	$3732
New Jersey	$32,900	$7571
Washington	$29,200	$4744
Wisconsin	$31,000	$5117

Source: U.S. Bureau of the Census, *Statistical Abstract of the United States.*

11.7 Refer to Exercise 11.6. Identify the slope and the y-intercept for your regression line. Write the equation corresponding to your regression line, and use it to predict the public education expenditure per student for states with the following average teacher's income.

a $18,500   **b** $41,500

11.8 The data shown in the accompanying table give hospital expenses covered by an insurance carrier and the number of days of hospitalization for a sample of 10 patients.

a Use a scatterplot to summarize the data.

b Construct a freehand regression line relating y (covered expense) to x (days of confinement).

c Identify the y-intercept and the slope for the regression line.

d Predict the hospital expenses ($) covered by an insurance carrier for a patient who is hospitalized for 8 days.

Expense Covered by Insurance (y)	Number of Days Confined (x)
50	1
175	3
180	6
200	7
60	2
140	4
420	12
540	15
170	5
300	9

11.3   Method of Least Squares

method of least squares

The statistical procedure for finding the prediction equation—the **method of least squares**—is, in many respects, an objective way to obtain an eyeball fit to the points. For example, when we fit a line by eye to a set of points, we move the ruler until we think that we have minimized the distance from the points to the fitted line. This same minimizing technique is used in the statistical procedure.

predicted value of y

We denote the **predicted value of y** for a given value of x (obtained from the fitted line) as $\hat{y}$. The prediction equation obtained from the method of least squares is denoted by

$$\hat{y} = \hat{\beta}_0 + \hat{\beta}_1 x,$$

where $\hat{\beta}_0$ and $\hat{\beta}_1$ are *estimates* of the unknown intercept β_0 and of slope β_1 in the linear regression model $y = \beta_0 + \beta_1 x + \epsilon$. The error of prediction (sometimes

residual

called the **residual**) is $y - \hat{y}$, the difference between the actual value of y and what we predict it to be (see Figure 11.6). The method of least squares chooses the prediction equation that minimizes the sum of the squared errors of prediction for all sample measurements. The sum of squared errors (also referred to as the **sum of squares for**

sum of squares for error

error) is denoted by SSE and can be written as

$$\text{SSE} = \Sigma(y - \hat{y})^2,$$

where y is an observed response and $\hat{y}$ is a point on the prediction equation

$$\hat{y} = \hat{\beta}_0 + \hat{\beta}_1 x.$$

Substituting for $\hat{y}$ in SSE, we have

$$\text{SSE} = \Sigma[y - (\hat{\beta}_0 + \hat{\beta}_1 x)]^2.$$

The method of least squares chooses values for the estimates $\hat{\beta}_0$ and $\hat{\beta}_1$ that make SSE a minimum. Deriving these values is beyond the scope of the text, but they can be found by using the formulas that follow.

FIGURE 11.6
Least Squares Fit to the Data in
Table 11.1

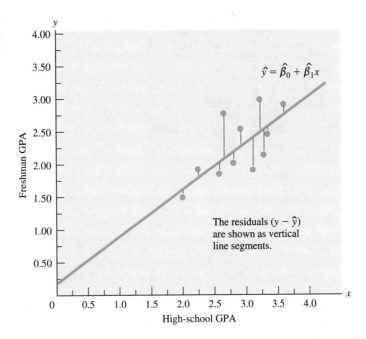

Least Squares Estimates for the Regression Line $\hat{y} = \hat{\beta}_0 + \hat{\beta}_1 x$

$$\hat{\beta}_1 = \frac{S_{xy}}{S_{xx}}$$

and

$$\hat{\beta}_0 = \bar{y} - \hat{\beta}_1 \bar{x},$$

where

$$S_{xx} = \Sigma x^2 - \frac{(\Sigma x)^2}{n}$$

$$S_{xy} = \Sigma xy - \frac{(\Sigma x)(\Sigma y)}{n}$$

and $\bar{y}$ and $\bar{x}$ denote the sample means for the y-values and x-values, respectively; n represents the number of y-values.

The use of these formulas to find $\hat{\beta}_0$, $\hat{\beta}_1$, and the least squares line is illustrated in the next example.

EXAMPLE   11.1   Obtain the regression line $\hat{y} = \hat{\beta}_0 + \hat{\beta}_1 x$ for the GPA data in Table 11.1 by using the method of least squares.

Solution   Calculating the least squares estimates $\hat{\beta}_0$ and $\hat{\beta}_1$ is greatly simplified by using Table 11.2.

TABLE 11.2
Calculations for the GPA Data

x	y	x^2	xy	y^2
2.00	1.60	4.0000	3.2000	2.5600
2.25	2.00	5.0625	4.5000	4.0000
2.60	1.80	6.7600	4.6800	3.2400
2.65	2.80	7.0225	7.4200	7.8400
2.80	2.10	7,8400	5.8800	4.4100
3.10	2.00	9.6100	6.2000	4.0000
2.90	2.65	8.4100	7.6850	7.0225
3.25	2.25	10.5625	7.3125	5.0625
3.30	2.60	10.8900	8.5800	6.7600
3.60	3.00	12.9600	10.8000	9.0000
3.25	3.10	10.5625	10.0750	9.6100

Total	31.70	25.90	93.6800	76.3325	63.5050

Substituting the values from Table 11.2 into the formulas, we obtain

$$S_{xx} = \Sigma x^2 - \frac{(\Sigma x)^2}{n} = 93.68 - \frac{(31.7)^2}{11} = 2.326$$

$$S_{xy} = \Sigma xy - \frac{(\Sigma x)(\Sigma y)}{n} = 76.3325 - \frac{(31.7)(25.9)}{11} = 1.693$$

$$\bar{y} = \frac{\Sigma y}{n} = \frac{25.9}{11} = 2.355$$

$$\bar{x} = \frac{\Sigma x}{n} = \frac{3.17}{11} = 2.882.$$

Hence,

$$\hat{\beta}_1 = \frac{S_{xy}}{S_{xx}} = \frac{1.693}{2.326} = .728$$

$$\hat{\beta}_0 = \bar{y} - \hat{\beta}_1\bar{x} = 2.355 - .728(2.882) = .257.$$

Thus the least squares prediction equation relating the freshman GPA y to the corresponding high-school value x is

$$\hat{y} = .257 + .728x.$$

Notice that the least squares prediction equation is similar to the equation for the freehand line for the same data set, obtained in Figure 11.4.   ▪

Algebraically, it can be shown from a least squares fit of the model $y = \beta_0 + \beta_1 x + \epsilon$ that the difference between an individual y-value (say, y) and the sample mean $\bar{y}$ is

$$y - \bar{y} = (y - \hat{y}) + (\hat{y} - \bar{y}),$$

where $\hat{y}$ is the predicted value of y from the least squares prediction equation. It can also be shown that

$$\Sigma(y - \bar{y})^2 = \Sigma(y - \hat{y})^2 + \Sigma(\hat{y} - \bar{y})^2.$$

While the proof of this equality is beyond the scope of this text, we can obtain an intuitive understanding of this relationship by considering the following situation.

Suppose that we use the model $y = \beta_0 + \epsilon$. In this model, β_0 represents the population mean for the variable y, and intuitively we would estimate its value by using the sample mean $\bar{y}$. (You can confirm this result by using the formula for the estimated intercept $\hat{\beta}_0$ in a linear model.) Since $\hat{y} = \bar{y}$ for this model, the sum of the squared errors of prediction is $\Sigma(y - \bar{y})^2$.

Now suppose that the variable y is related to an independent variable x. From our previous work, we could fit the model $y = \beta_0 + \beta_1 x + \epsilon$ to obtain

$$\hat{y} = \hat{\beta}_0 + \hat{\beta}_1 x.$$

For this model, the sum of the squared prediction errors is

$$\Sigma(y - \hat{y})^2.$$

Figure 11.7 presents two prediction equations: $\hat{y} = \bar{y}$ for the model $y = \beta_0 + \epsilon$, and $\hat{y} = \hat{\beta}_0 + \hat{\beta}_1 x$ for the model $y = \beta_0 + \beta_1 x + \epsilon$. Notice that we can express the distance between an observation y and the sample mean $\bar{y}$ as the sum of two components, $(\hat{y} - \bar{y})$ and $(y - \hat{y})$. The quality $(\hat{y} - \bar{y})$ represent the portion of the overall distance that can be attributed to the independent variable x (through the prediction equation $\hat{y} = \hat{\beta}_0 + \hat{\beta}_1 x$). The quantity $(y - \hat{y})$ represents the portion of the distance between y and $\bar{y}$ that cannot be accounted for by the independent variable x (and that we attribute to error). Combining this information for all sample observations, we can express the total variability in the sample measurements about the sample mean, $\Sigma(y - \bar{y})^2$, called the **sum of squares about the mean**, as the sum of the squared deviations of the predicted values from $\bar{y}$, $\Sigma(\hat{y} - \bar{y})^2$, called the **sum of squares due to regression**, and the sum of the squared error of prediction, $\Sigma(y - \hat{y})^2$, called the **sum of squares for error**. Thus we have

sum of squares about the mean

sum of squares due to regression

sum of squares for error

Sum of squares about the mean = Sum of squares due to regression

+ Sum of squares for error.

There is another way to view this equation. It can be shown that the sample mean $\bar{y}$ is also the average of the fitted values. Therefore, the sum of squares due to regression $\Sigma(\hat{y} - \bar{y})^2$ depicts variability in the fitted values. Similarly, $\Sigma(y - \hat{y})^2$ represents variability in the y-values about the fitted values. As a result, the total variability in the y-values can be written as

FIGURE 11.7
Relationship Between
$\Sigma(y - \bar{y})^2$ and
$\Sigma(y - \hat{y})^2$

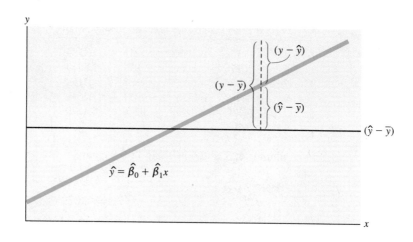

$$\begin{array}{ccccc} \Sigma(y - \bar{y})^2 & = & \Sigma(\hat{y} - \bar{y})^2 & + & \Sigma(y - \hat{y})^2. \\ \text{total variability} & & \text{variability} & & \text{unexplained} \\ \text{in } y\text{-values} & & \text{explained by model} & & \text{variability} \end{array}$$

Figure 11.7 shows how the total variability is partitioned into the two components.

Obviously, if we're interested in predicting y based on the independent variable x, the larger the explained variability is (relative to the unexplained variability), the better the model "fits" the data; and this should lead to more precise predictions of y based on x.

EXAMPLE   11.2   Consider the five data points listed in columns 1 and 2 of Table 11.3.

a Fit the model

$$y = \beta_0 + \beta_1 x + \epsilon.$$

b Verify that

$$\Sigma(y - \bar{y})^2 = \Sigma(y - \hat{y})^2 + \Sigma(\hat{y} - \bar{y})^2.$$

Solution   **a**   For these data, it can be shown that

$$S_{xx} = \Sigma x^2 - \frac{(\Sigma x)^2}{n} = 66 - \frac{(16)^2}{5} = 14.8$$

$$S_{xy} = \Sigma xy - \frac{(\Sigma x)(\Sigma y)}{n} = \frac{145 - (16)(38)}{5} = 23.4$$

$$\hat{\beta}_1 = 1.58$$

$$\hat{\beta}_0 = \bar{y} - \hat{\beta}_1 \bar{x} = 7.6 - 1.58(3.2) = 2.54.$$

b For each x-value, we compute $\hat{y}$ from the least squares prediction equation. We also compute the quantities $(y - \bar{y})$, $(y - \hat{y})$, and $(\hat{y} - \bar{y})$. These quantities are displayed in Table 11.3.

x	y	$\hat{y}$	$y - \bar{y}$	$y - \hat{y}$	$\hat{y} - \bar{y}$
1	4	4.1216	−3.6000	−.1216	−3.4784
2	6	5.7027	−1.6000	.2973	−1.8973
3	7	7.2838	−.6000	−.2838	−.3162
4	9	8.8649	1.4000	.1351	1.2649
6	12	12.0270	4.4000	−.0270	4.4270

From columns 4, 5, and 6 in Table 11.3 we have

$$\Sigma(y - \bar{y})^2 = 37.2000$$
$$\Sigma(y - \hat{y})^2 = .2027$$
$$\Sigma(\hat{y} - \bar{y})^2 = 36.9973$$

Notice that, except for rounding errors,

$$\Sigma(y - \bar{y})^2 = \Sigma(y - \hat{y})^2 + \Sigma(\hat{y} - \bar{y})^2.$$

Intuitively, since the explained variability $\Sigma(\hat{y} - \bar{y})^2 = 36.9973$ accounts for almost all of the total variability $\Sigma(y - \bar{y})^2 = 37.2000$, the model appears to fit the data very well. A scatter plot of y versus x with the prediction equation superimposed shows how tightly the data group about the prediction equation (see Figure 11.8). ▪

FIGURE 11.8
Scatter Plot with Superimposed
Regression Line for the Data of
Table 11.3

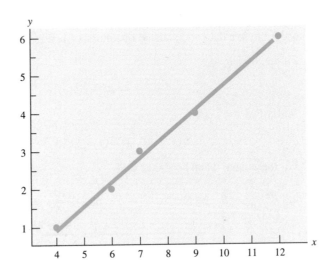

Exercises

Basic Techniques

11.9 Use the accompanying data to determine the least squares prediction equation.

x	1	2	3	4	5
y	2	4	6	7	9

11.10 Examine the accompanying data.

x	1	3	5	7	9
y	1	4	8	9	12

a Determine the least squares prediction equation.

b Use the least squares prediction equation to predict y when $x = 6$.

11.11 Refer to the data of Exercise 11.1. Find the least squares prediction equation, and compare it to the freehand regression line you found in Exercise 11.2.

11.12 A computer solution (using SAS) of the least squares prediction equation for the data of Exercise 11.4 is shown below.

a Plot the data.

b Determine the least squares prediction equation from the output here, and draw the regression line on the data plot in part a.

c Does the prediction equation seem to represent the data adequately?

d Predict y (annual prescription volume) for $x = 35$.

```
OPTIONS NODATE NONUMBER PS=60 LS=78;

DATA ONE;
  INPUT X Y;
  CARDS;
   10     25
   18     55
   25     50
   40     75
   50    110
   63    138
   42     90
   30     60
    5     10
   55    100
  ;

PROC GLM;
  MODEL Y=X/P;
  TITLE1 '    ';
  RUN;
```

General Linear Models Procedure

Number of observations in data set = 10

Dependent Variable: Y

Source	DF	Sum of Squares	Mean Square	F Value	Pr > F
Model	1	13230.96994	13230.96994	162.56	0.0001
Error	8	651.13006	81.39126		
Corrected Total	9	13882.10000			

R-Square	C.V.	Root MSE	Y Mean
0.953096	12.65317	9.021710	71.3000000

Source	DF	Type I SS	Mean Square	F Value	Pr > F
X	1	13230.96994	13230.96994	162.56	0.0001

Source	DF	Type III SS	Mean Square	F Value	Pr > F
X	1	13230.96994	13230.96994	162.56	0.0001

Parameter	Estimate	T for H0: Parameter=0	Pr > |T|	Std Error of Estimate
INTERCEPT	4.697851861	0.79	0.4527	5.95202071
X	1.970477756	12.75	0.0001	0.15454842

Observation	Observed Value	Predicted Value	Residual
1	25.00000000	24.40262942	0.59737058
2	55.00000000	40.16645146	14.83354854
3	50.00000000	53.95979575	-3.95979575
4	75.00000000	83.51696208	-8.51696208
5	110.00000000	103.22173964	6.77826036
6	138.00000000	128.83795046	9.16204954
7	90.00000000	87.45791760	2.54208240
8	60.00000000	63.81218453	-3.81218453
9	10.00000000	14.55024064	-4.55024064
10	100.00000000	113.07412842	-13.07412842

Sum of Residuals	0.00000000
Sum of Squared Residuals	651.13006221
Sum of Squared Residuals - Error SS	0.00000000
First Order Autocorrelation	0.12080212
Durbin-Watson D	1.49533053

11.13   Refer to Exercise 11.6 and the EXECUSTAT output shown here. Identify the least squares prediction equation relating a state's public education expenditure per student to the average teacher's income for the state. Use the least squares equation to predict the per-student expenditure of a state with the following average teacher's income.

a $18,500   **b** $41,500

```
Simple Regression Analysis

Linear model: Y = -3185.29 + 0.269529*X

Table of Estimates
                                    Standard        t           P
                      Estimate        Error       Value       Value
Intercept           -3185.29        1919.33       -1.66      0.1356
Slope                0.269529       0.0637589      4.23       0.0029

R-squared = 69.08%
Correlation coeff. =  0.831
Standard error of estimation = 952.339
Durbin-Watson statistic = 1.56343
Mean absolute error = 595.853
```

Applications

11.14   Family income and annual savings data are displayed here for a sample of nine families.

Annual Savings ($000)	Annual Income ($000)
1	36
2	39
2	42
5	45
5	48
6	51
7	54
8	56
7	59

a   Graph the data, using a scatter plot.

b   Determine an eyeball fit to the data. Predict y (annual savings, in $1000s) based on the annual income of $x = $45,000.

11.15   Refer to Exercise 11.14.

a   Determine the least squares prediction equation.

b   Compute $\Sigma(y - \hat{y})^2$ for the least squares fit and for your eyeball fit. Note that SSE for the least squares fit is less than or equal to SSE for the eyeball fit.

11.16   The following data were obtained in a study of sales volume (per district) as a fraction of the number of client contacts per month.

Sales Volume ($1000), (x)	Average Number of Client Contacts per Month, (y)
15	10
26	15
28	17
30	20
32	23
86	46
109	53
95	48
130	59
160	65

a   Plot the data.

b   Eyeball a linear fit to the data, and guess the value of the intercepts and the value of the slope.

c   Predict sales for $x = 50$.

11.17   Refer to Exercise 11.16.

a   Obtain the linear regression equation $\hat{y} = \hat{\beta}_0 + \hat{\beta}_1 x$, using the method of least squares. Compare this line to the one obtained in Exercise 11.16.

b   Predict sales for $x = 50$, and compare your answer here to the one you gave in Exercise 11.16(c).

 11.18   As one part of a study of commercial bank branches, data are obtained on the number of independent businesses x located in sample ZIP code areas and on the number of bank branches y located in these areas. The commercial centers of cities are excluded.

x	92	116	124	210	216	267	306	378	415	502	615	703
y	3	2	3	5	4	5	5	6	7	7	9	9

$$\Sigma x = 3944 \qquad \Sigma y = 65 \qquad \Sigma xy = 26,208$$
$$\Sigma x^2 = 1,732,524 \quad \Sigma y^2 = 409 \qquad n = 12$$

a   Plot the data. Does a linear equation relating y to x appear plausible?

b   Calculate the regression equation (with y as the dependent variable).

11.4 Correlation

In this section we extend our study of the relationships between two or more variables. Not only might we like to predict the value of one variable (the dependent variable) based on information about one or more independent variables, as we have done in previous sections, but we might also wish to measure the strength of the relationship between these variables.

correlation coefficient

sample correlation
coefficient $\hat{\rho}$

One measure of the strength of the relationship between two variables x and y is called the *coefficient of linear correlation*, or simply the **correlation coefficient**. The stronger the correlation, the better x predicts y. Given n pairs of observations (x_i, y_i), we can compute the **sample correlation coefficient** $\hat{\rho}$ (Greek lower-case rho) as

$$\hat{\rho} = \frac{S_{xy}}{\sqrt{S_{xx}S_{yy}}},$$

where

$$S_{xx} = \Sigma x^2 - \frac{(\Sigma x)^2}{n}$$

$$S_{xy} = \Sigma xy - \frac{(\Sigma x)(\Sigma y)}{n}$$

$$S_{yy} = \Sigma y^2 - \frac{(\Sigma y)^2}{n}$$

We will illustrate the computation of $\hat{\rho}$ with an example and then explain its practical significance.

E X A M P L E   11.3

An engineer is interested in calibrating a flow meter to be used on a liquid soap production line. For the test, the engineer observes 10 meter reading for 10 known flow rates (see Table 11.4).

T A B L E 11.4
Meter Readings for Different
Known Flow Rates

Observed Meter Reading (y)	Actual Flow Rate (x)
1.4	1
2.3	2
3.1	3
4.2	4
5.1	5
5.8	6
6.8	7
7.6	8
8.7	9
9.5	10

a Based on the SAS scatter plot, do the data appear to be linear?

b Compute the sample correlation coefficient $\hat{\rho}$.

Solution **a** The SAS data plot (which appears to be fairly linear) is shown next.

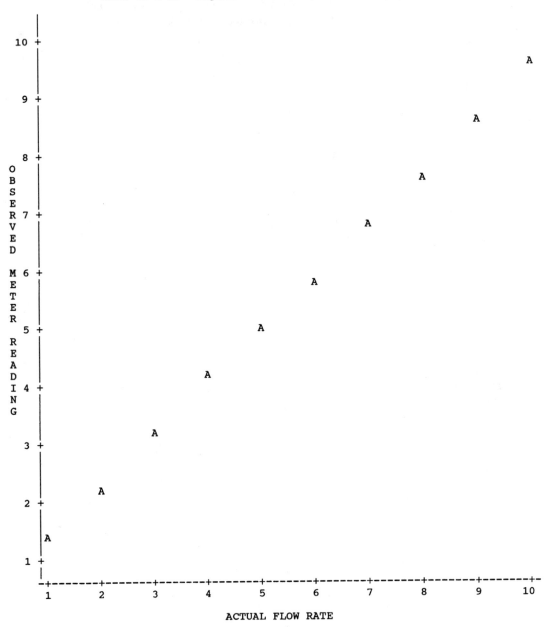

Plot of Y*X.   Legend: A = 1 obs, B = 2 obs, etc.

ACTUAL FLOW RATE

b For these data, it can be shown that $\Sigma x = 55$, $\Sigma x^2 = 385$, $\Sigma y = 54.5$, $\Sigma y^2 = 364.09$, and $\Sigma xy = 374.1$. It follows that

$$S_{yy} = \Sigma y^2 - \frac{(\Sigma y)^2}{n} = 364.09 - \frac{(54.5)^2}{10} = 67.065$$

$$S_{xx} = \Sigma x^2 - \frac{(\Sigma x)^2}{n} = 385 - \frac{(55)^2}{10} = 82.5$$

$$S_{xy} = \Sigma xy - \frac{(\Sigma x)(\Sigma y)}{n} = 374.1 - \frac{55(54.5)}{10} = 74.35.$$

We can substitute these values to obtain

$$\hat{\rho} = \frac{S_{xy}}{\sqrt{S_{xx}S_{yy}}} = \frac{74.35}{\sqrt{82.5(67.065)}} = .9996. \quad \blacksquare$$

How can we interpret this correlation coefficient? First, notice the similarity between the sample correlation coefficient and the slope of the regression line $\hat{y} = \hat{\beta}_0 + \hat{\beta}_1 x$:

$$\hat{\rho} = \frac{S_{xy}}{\sqrt{S_{xx}S_{yy}}} \qquad \text{and} \qquad \hat{\beta}_1 = \frac{S_{xy}}{S_{xx}}$$

In particular, both have the same numerator, and their denominators must always be positive (because they involve the sum of squares of numbers). Thus the correlation coefficient $\hat{\rho}$ has the same sign as $\hat{\beta}_1$, and $\hat{\rho} = 0$ when $\hat{\beta}_1 = 0$. A negative $\hat{\rho}$ implies a negative slope; a positive value implies that y increases as x increases. A slope $\hat{\beta}_1 = 0$ (and hence $\hat{\rho} = 0$) indicates that x is not linearly related to y (see Figure 11.9).

FIGURE 11.9
Interpreting the Correlation Coefficient: (a) Positive Linear Correlation; (b) Negative Linear Correlation; (c) No Linear Correlation

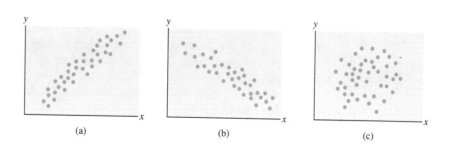

(a)   (b)   (c)

It can be shown that $\hat{\rho}$ must lie in the interval $-1 \leq \hat{\rho} \leq 1$. The implications of various values of $\hat{\rho}$ are indicated in Figure 11.10.

FIGURE 11.10
Possible Values for $\hat{\rho}$, the Sample Correlation Coefficient

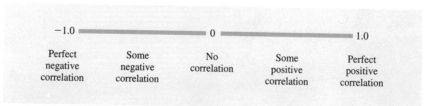

Several misinterpretations of the coefficient of correlation should be noted. First, a correlation coefficient equal to .5 does not mean that the strength of the relationship between y and x is "halfway" between no correlation and perfect correlation. If we designate $S_{yy} = \Sigma(y - \bar{y})^2$ as the total variability of the y-values about their sample mean, it can be shown that an amount equal to $\hat{\rho}^2$ of this total variability can be explained by the variable x. The more closely x and y are linearly related, the more fully the variability in the y-values can be explained by variability in the x-values and the closer $\hat{\rho}^2$ will be to 1. If $\hat{\rho} = .5$, the independent variable x accounts for $\hat{\rho}^2 = .25$ of the total variation in the y-values about $\bar{y}$. The quantity $\hat{\rho}^2$ is called a **coefficient of determination**.

coefficient of determination

Second, y and x could be perfectly related in some way other than linearly, when $\hat{\rho} = 0$ or some very small value. This is shown in the perfect curvilinear fit of Figure 11.11.

F·I G U R E 11.11
Perfect Curvilinear Fit with
$\hat{\rho} = 0$

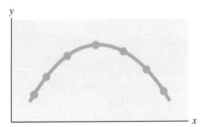

Finally, notice that we cannot add correlations. If the simple linear correlations between y and x_1, y and x_2, and y and x_3 are .1, .3, and .2, respectively, it *does not* follow that x_1, x_2, and x_3 account for $\hat{\rho}_1^2 + \hat{\rho}_2^2 + \hat{\rho}_3^2 = (.1)^2 + (.3)^2 + (.2)^2$ of the variability of the y-values about their sample mean. Indeed, x_1, x_2, and x_3 may be highly correlated and may contribute the same information for predicting y. The relationship between y and several independent variables should not be studied by computing simple correlation coefficients for each of the independent variables. Rather, we should relate y to x_1, x_2, and x_3 by using a single multivariable model. This topic is discussed briefly in the next section.

The ordinary correlation coefficient r assesses the linear association between two variables x and y. In certain situations, the variable y may increase (or decrease) with increases in x—but not necessarily in a linear fashion. When this happens, the correlation coefficient ρ will not depict the full extent of the relation between x and y. One approach is to use the rank correlation coefficient, which measures the *monotonic* association between y and x; that is, the rank correlation coefficient measures whether y increases (or decreases) with x, even when the relation between y and x is not necessarily linear.

The rank order correlation coefficient is easy to calculate. We simply rank all x-values and all y-values separately and then calculate the ordinary correlation coefficient for the ranks. This correlation based on the ranks is called **Spearman's rank order correlation coefficient** ρ_S. The estimate of ρ_S computed from sample data is designated as $\hat{\rho}_S$.

Spearman's rank order correlation coefficient ρ_S

E X A M P L E   11.4   A corporation examined the relationship between profits (in $1000s) and the percentage of operating capacity being used by each of 12 plants (see Table 11.5).

T A B L E 11.5
Relationship Between Profits and Percent of Operating Capacity at 12 Plants

Profits ($1000s), y	% of Operating Capacity, x
2.5	50
6.2	57
3.1	61
4.6	68
7.3	77
4.5	80
6.1	82
11.6	85
10.0	89
14.2	91
16.1	95
19.5	99

   a Based on the MINITAB plot, do the data appear to be linear?

   b Compute the rank correlation coefficient, $\hat{\rho}_S$.

Solution **a** The MINITAB plot is shown next.

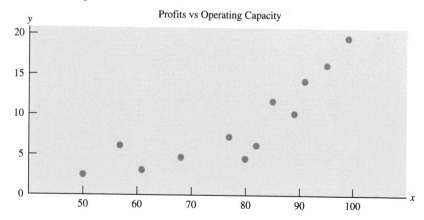

Profits vs Operating Capacity

T A B L E 11.6
Ranks on Profit and on Percent
of Capacity for Example 11.4

Rank on Profits	Rank on % Capacity
1	1
6	2
2	3
4	4
7	5
3	6
5	7
9	8
8	9
10	10
11	11
12	12

b To compute $\hat{\rho}_S$, first we need the ranks (from low to high) separately for each variable. These are shown here in Table 11.6.

   If we let y denote the ranks on profits and x denote the ranks on percent of operating capacity, we can compute $\hat{\rho}_S$ as we would $\hat{\rho}$. To do this, we need S_{yy}, S_{xx}, and S_{xy}.

   For these data, it can be shown that

$$S_{yy} = 143$$
$$S_{xx} = 143$$
$$S_{xy} = 125.$$

Hence, the rank correlation coefficient $\hat{\rho}_S$ is

$$\hat{\rho}_S = \frac{125}{\sqrt{143(143)}} = 0.874.$$

For these data, y increases with x, but the relation between y and x is not linear.   ■

If there are no ties in ranks for either of the two variables, we can use a simpler formula for r_S that makes use of d_i, the difference between the y rank and the x rank on observation i:

$$\hat{\rho}_S = 1 - \frac{6\Sigma d_i^2}{n(n^2 - 1)},$$

where n is the number of x_i, y_i observations.

EXAMPLE 11.5 Compute $\hat{\rho}_S$ for the data of Example 11.4, using the simpler computational formula.

Solution The differences in ranks are shown in Table 11.7.

TABLE 11.7
Differences in the x and y
Ranks from Table 11.6

Rank on Profit	Rank on % Capacity	d_i
1	1	0
6	2	4
2	3	−1
4	4	0
7	5	2
3	6	−3
5	7	−2
9	8	1
8	9	−1
10	10	0
11	11	0
12	12	0

Then $\Sigma d_i^2 = 36$ and

$$\hat{\rho}_S = 1 - \frac{6\Sigma d_i^2}{n(n^2 - 1)}$$

$$= 1 - \frac{6(36)}{12(143)} = .874.$$

Note: This agrees with what we obtained in Example 11.4 except for rounding errors. ▪

Exercises

Basic Techniques

11.19   Plot the sample data shown here, compute the correlation coefficient, and interpret your findings.

x	1	2	3	4	6	9	10
y	2	4	5	7	8	12	13

11.20   Refer to Exercise 11.19. Suppose that the first three y-values are 16, 12, and 10.

   a   Plot the data, and guess a value for $\hat{\rho}$.

   b   Compute the sample correlation coefficient, and compare the computed value to the guessed value.

   c   Why do the correlation coefficients differ for Exercises 11.19 and 11.20?

Applications

11.21   Refer to Example 11.4. Compute the coefficient of determination, and use it to help interpret the value of $\hat{\rho}$.

11.22   An experiment was conducted to investigate the amplitude of the shock wave recorded on sensors placed at different distances from an explosive charge. The charge was to be detonated underground, with three sensors placed at each of the three different distances from the charge, as illustrated in Figure 11.12. The shock-wave amplitudes were recorded and summarized according to the sensors' distance from the explosion. These data are given in the accompanying table.

Distance (x)	5	5	5	10	10	10	15	15	15
Amplitude (y)	8.6	8.2	8.1	5.8	6.2	6.1	5.2	4.8	4.7

FIGURE 11.12
Location of Sensors from the Charge for the Shock-Wave Experiment of Exercise 11.22

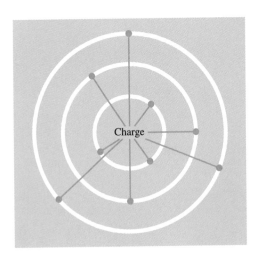

   a   Plot the sample data.

   b   Compute the correlation coefficient, and interpret its value.

11.23 A forester was interested in training an assistant to estimate the timber volume of a standing tree. Having trained her, the forester calibrated the assistant's estimates against known timber volumes. Perhaps a better way to quantify the assistant's estimate would have been to base it on an objective reading, such as the basal area of the tree. If, indeed, volume is related to basal area, the assistant would then have an objective way to estimate the timber volume of a tree. A random sample of 12 trees was obtained. For each tree included in the sample, the basal area x was recorded, along with the cubic-foot volume after the tree was felled. These data appear next.

Tree	1	2	3	4	5	6	7	8	9	10	11	12
Basal Area (x)	.3	.5	.4	.9	.7	.2	.6	.5	.8	.4	.8	.6
Volume (y)	6	9	7	19	15	5	12	9	20	9	18	13

a Plot the data.

b Compute and interpret the correlation coefficient between basal area and timber volume.

11.24 Compute the rank correlation coefficient for the data of Exercise 11.21. What measure seems more appropriate—ρ or ρ_S?

11.25 An equal number of families from eight different cities of various sizes were asked how much money they spend for food, clothing, and housing per year. The city sizes and average family responses are summarized here. (City size is in 1000s; expenditure is in $1000s.)

City Size	30	50	75	100	150	200	175	120
Expenditure	65	77	79	80	82	90	84	81

a Plot the data.

b Compute the correlation coefficient ρ.

11.26 Compute ρ_S for the data of Exercise 11.25. Is the shortcut formula appropriate? Why or why not?

11.5 Multiple Regression

In previous sections of this chapter, we examined methods for obtaining the least squares prediction equation, $\hat{y} = \hat{\beta}_0 + \hat{\beta}_1 x$, based on values from two variables y and x. Now we turn to a more complicated situation in which we collect data on the dependent variable y and on k $(k > 1)$ independent variables $x_1, x_2, \ldots, x_k$. Here we use the sample data to obtain the multiple regression predicting equation

$$\hat{y} = \hat{\beta}_0 + \hat{\beta}_1 x + \hat{\beta}_2 x_2 + \cdots + \hat{\beta}_k x_k.$$

For example, instead of trying to predict a student's GPA y at the end of the student's college freshman year based on the student's corresponding high-school GPA x_1, we may wish to use information on several additional variables, such as rank in high-school class x_2, college board verbal score x_3, and college board math achievement score x_4. To do this, we must examine the records of a sample of n freshmen to obtain information on the variables y, x_1, x_2, x_3, and x_4. Then the method of least squares can be used to obtain the least squares multiple regression equation

$$\hat{y} = \hat{\beta}_0 + \hat{\beta}_1 x_1 + \hat{\beta}_2 x_2 + \hat{\beta}_3 x_3 + \hat{\beta}_4 x_4.$$

Although computational formulas exist for the least squares estimates of the parameters $\beta_0, \beta_1, \beta_2, \ldots, \beta_k$, the formulas are difficult to work with algebraically beyond the simplest case, $\hat{y} = \hat{\beta}_0 + \hat{\beta}_1 x$ (discussed in section 11.3). Rather than spending time developing expressions for the least squares estimates of the parameters in a multiple regression equation, we will make use of available computer software packages to do the work for us for a particular problem

EXAMPLE 11.6 The sales profit y of a product in a sales territory is thought to be related to the total population x_1 of the territory, as well as to the advertising expenditure per person x_2 in the territory. The data from eight different sales territories are shown in Table 11.8. Use these data to find least squares estimates for the coefficients in the multiple regression equation

$$\hat{y} = \hat{\beta}_0 + \hat{\beta}_1 x_1 + \hat{\beta}_2 x_2.$$

Predict sales profit for a territory with a population of $x_1 = 2.8$ and an advertising expenditure of $x_2 = .25$.

TABLE 11.8
Data for Example 11.6

Sales Territory	Sales Profit per Person (Dollars), y	Total Population (Millions), x_1	Advertising per Person (Dollars), x_2
1	3.6	2.4	.16
2	2.5	1.3	.21
3	4.2	5.1	.12
4	4.1	4.9	.14
5	4.0	3.2	.26
6	5.1	6.7	.10
7	4.3	3.2	.41
8	11.5	.7	.11

Solution Since it is important that you be able to locate and identify the least squares estimates from a SAS computer printout, a portion of some sample output follows. The least squares estimates in the printout are shown in color. Notice that the least squares prediction equation, $\hat{y} = \hat{\beta}_0 + \hat{\beta}_1 x_1 + \hat{\beta}_2 x_2$, is

$$\hat{y} = 9.04 - .58x_1 - 11.27x_2.$$

Substituting into this equation, we find that the predicted sales profit per person for a territory with a population of $x_1 = 2.8$ and an advertising expenditure of $x_2 = .25$ per person is

$$\hat{y} = 9.04 - .58(2.8) - 11.27(.25) = \$4.60.$$

SAS Output

```
OPTIONS NODATE NONUMBER PS=60 LS=78;
DATA ONE;
  INPUT Y X1 X2;
  CARDS;
   3.6    2.4    0.16
   2.5    1.3    0.21
   4.2    5.1    0.12
   4.1    4.9    0.14
   4.0    3.2    0.26
   5.1    6.7    0.10
   4.3    3.2    0.41
  11.5    0.7    0.11
PROC GLM;
  MODEL Y=X1 X2/P;
TITLE1 '      ';
RUN;
```

General Linear Models Procedure
Number of observations in data set = 8

Dependent Variable: Y

Source	DF	Sum of Squares	Mean Square	F Value	Pr > F
Model	2	14.97893501	7.48946751	0.98	0.4387
Error	5	38.36981499	7.67396300		
Corrected Total	7	53.34875000			

R-Square	C.V.	Root MSE	Y Mean
0.280774	56.39067	2.770192	4.91250000

Parameter	Estimate	T for H0: Parameter=0	Pr > |T|	Std Error of Estimate
INTERCEPT	9.04432679	2.90	0.0337	3.11645121
X1	−0.58325688	−1.10	0.3218	0.53071152
X2	−11.26824515	−1.09	0.3237	10.29793401

Observation	Observed Value	Predicted Value	Residual
1	3.60000000	5.84159106	−2.24159106
2	2.50000000	5.91976137	−3.41976137
3	4.20000000	4.71752730	−0.51752730
4	4.10000000	4.60881377	−0.50881377
5	4.00000000	4.24816104	−0.24816104
6	5.10000000	4.00968120	1.09031880
7	4.30000000	2.55792427	1.74207573
8	11.50000000	7.39654001	4.10345999

Sum of Residuals	0.00000000
Sum of Squared Residuals	38.36981499
Sum of Squared Residuals - Error SS	0.00000000
First Order Autocorrelation	0.48482122
Durbin-Watson D	0.46055772

The purpose of Example 11.6 is to illustrate that we can rely on available software packages to obtain least squares predicion equations for multiple regression problems. We should realize that the multiple regression equation is often the first step in examining the relationship between a dependent variable y and several independent variables. However, inferences related to multiple regression prediction equations are beyond the scope of this text.   ■

11.6 Using Computers to Help Make Sense of Data

Computer output for data plots and linear regression problems has been used already in this chapter. In this section we provide some MINITAB and SAS programs for your use as well as some EXECUSTAT output.

The data of Table 11.1 were used to illustrate how SAS and MINITAB can solve linear regression and correlation problems. The MINITAB PLOT procedure was used to give a scatter plot for the data. The MINITAB REGRESS procedure computes a number of quantities, including the least squares estimates of β_0 and β_1, $\hat{\rho}^2$ (actually $\hat{\rho}^2 \times 100$, and a table of original values, predicted values, and residuals. Finally, the MINITAB CORRELATION procedure computes the sample correlation coefficient $\hat{\rho}$.

SAS output for the same data is shown here, too. Notice how PROC PLOT, PROC REG, and PROC CORR are used to analyze the linear regression data and to obtain the correlation coefficient. Similarly, EXECUSTAT was used to obtain a data plot, the least squares prediction equation, and the correlation coefficient. You should compare the MINITAB, SAS, and EXECUSTAT outputs to identify where the important results of a regression or correlation analysis are located and how they are labeled. Notice that the least squares prediction equation from all these software systems was computed to be $\hat{y} = .257 + .728x$, and the correlation coefficient was found to be $\hat{\rho} = .699$. These same procedures can then be used to solve other linear regression and correlation problems.

Additional details shown in the regression and correlation output of MINITAB, SAS, and EXECUSTAT will become more familiar to you after you study the inferential methods for linear regression and correlation in Chapter 12. Several other examples of SAS and MINITAB output are shown in the remaining exercises of this chapter.

MINITAB Output

```
MTB > READ INTO C1 C2
DATA> 2.00    1.60
DATA> 2.25    2.00
DATA> 2.60    1.80
DATA> 2.65    2.80
DATA> 2.80    2.10
DATA> 3.10    2.00
DATA> 2.90    2.65
DATA> 3.25    2.25
DATA> 3.30    2.60
DATA> 3.60    3.00
DATA> 3.25    3.10
DATA> END
        11 ROWS READ

MTB > NAME C1 'HS_GPA'
MTB > NAME C2 'FR_GPA'
MTB > PRINT C1 C2
```

MINITAB Output (Continued)

```
ROW   HS_GPA   FR_GPA

  1     2.00     1.60
  2     2.25     2.00
  3     2.60     1.80
  4     2.65     2.80
  5     2.80     2.10
  6     3.10     2.00
  7     2.90     2.65
  8     3.25     2.25
  9     3.30     2.60
 10     3.60     3.00
 11     3.25     3.10
```

MTB > PLOT C2 VS C1

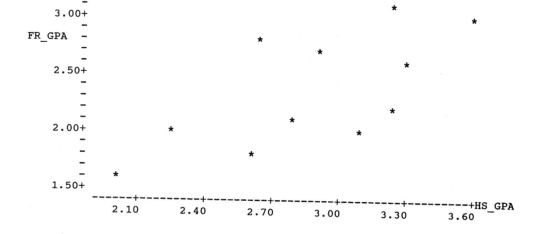

```
MTB > BRIEF 3
MTB > REGRESS C2 ON 1 C1;
SUBC> PREDICT C1.

The regression equation is
FR_GPA = 0.257 + 0.728 HS_GPA

Predictor       Coef        Stdev      t-ratio          p
Constant      0.2568       0.7243         0.35      0.731
HS_GPA        0.7279       0.2482         2.93      0.017

s = 0.3785      R-sq = 48.9%      R-sq(adj) = 43.2%

Analysis of Variance

SOURCE        DF          SS           MS          F          p
Regression     1      1.2327       1.2327       8.60      0.017
Error          9      1.2896       0.1433
Total         10      2.5223
```

MINITAB Output (Continued)

Obs.	HS_GPA	FR_GPA	Fit	Stdev.Fit	Residual	St.Resid
1	2.00	1.600	1.713	0.247	-0.113	-0.39
2	2.25	2.000	1.895	0.194	0.105	0.32
3	2.60	1.800	2.149	0.134	-0.349	-0.99
4	2.65	2.800	2.186	0.128	0.614	1.72
5	2.80	2.100	2.295	0.116	-0.195	-0.54
6	3.10	2.000	2.513	0.126	-0.513	-1.44
7	2.90	2.650	2.368	0.114	0.282	0.78
8	3.25	2.250	2.623	0.146	-0.373	-1.07
9	3.30	2.600	2.659	0.154	-0.059	-0.17
10	3.60	3.000	2.877	0.212	0.123	0.39
11	3.25	3.100	2.623	0.146	0.477	1.37

Fit	Stdev.Fit	95% C.I.		95% P.I.	
1.713	0.247	(1.154,	2.271)	(0.690,	2.735)
1.895	0.194	(1.456,	2.333)	(0.932,	2.857)
2.149	0.134	(1.847,	2.452)	(1.241,	3.058)
2.186	0.128	(1.897,	2.475)	(1.282,	3.090)
2.295	0.116	(2.033,	2.557)	(1.399,	3.191)
2.513	0.126	(2.228,	2.799)	(1.610,	3.416)
2.368	0.114	(2.109,	2.626)	(1.473,	3.262)
2.623	0.146	(2.292,	2.953)	(1.704,	3.541)
2.659	0.154	(2.310,	3.008)	(1.734,	3.584)
2.877	0.212	(2.398,	3.356)	(1.896,	3.859)
2.623	0.146	(2.292,	2.953)	(1.704,	3.541)

```
MTB > CORRELATION COEFFICIENT BETWEEN C1 AND C2

Correlation of HS_GPA and FR_GPA = 0.699

MTB > STOP
```

SAS Output

```
OPTIONS NODATE NONUMBER PS=60 LS=78;
DATA A1;
  INPUT STUDENT HS_GPA FR_GPA;
  LABEL STUDENT='STUDENT'
        HS_GPA='HIGH SCHOOL GPA'
        FR_GPA='FRESHMAN GPA';
  CARDS;
    1    2.00    1.60
    2    2.25    2.00
    3    2.60    1.80
    4    2.65    2.80
    5    2.80    2.10
    6    3.10    2.00
    7    2.90    2.65
    8    3.25    2.25
    9    3.30    2.60
   10    3.60    3.00
   11    3.25    3.10
;
PROC PRINT N;
TITLE1 "DATA FOR HIGH SCHOOL AND COLLEGE GPA'S";
TITLE2 'LISTING OF THE DATA';

PROC PLOT DATA=A1;
  PLOT FR_GPA*HS_GPA;
TITLE2 'EXAMPLE OF PROC PLOT';

PROC REG DATA=A1;
  MODEL FR_GPA=HS_GPA;
TITLE2 'EXAMPLE OF PROC REG';

PROC CORR DATA=A1;
  VAR FR_GPA HS_GPA;
TITLE2 'EXAMPLE OF PROC CORR';

PROC FREQ DATA=A1;
  TABLES FR_GPA*HS_GPA / MEASURES NOPRINT;
TITLE2 'EXAMPLE OF PROC FREQ';
RUN;
```

DATA FOR HIGH SCHOOL AND COLLEGE GPA'S
         LISTING OF THE DATA

OBS	STUDENT	HS_GPA	FR_GPA
1	1	2.00	1.60
2	2	2.25	2.00
3	3	2.60	1.80
4	4	2.65	2.80
5	5	2.80	2.10
6	6	3.10	2.00
7	7	2.90	2.65
8	8	3.25	2.25
9	9	3.30	2.60
10	10	3.60	3.00
11	11	3.25	3.10

N = 11

SAS Output (Continued)

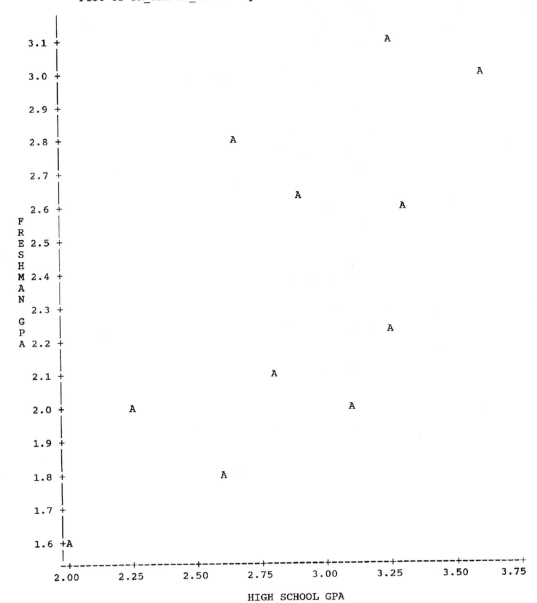

DATA FOR HIGH SCHOOL AND COLLEGE GPA'S
EXAMPLE OF PROC PLOT

Plot of FR_GPA*HS_GPA.   Legend: A = 1 obs, B = 2 obs, etc.

DATA FOR HIGH SCHOOL AND COLLEGE GPA'S
EXAMPLE OF PROC REG

Model: MODEL1
Dependent Variable: FR_GPA   FRESHMAN GPA

Analysis of Variance

Source	DF	Sum of Squares	Mean Square	F Value	Prob>F
Model	1	1.23267	1.23267	8.603	0.0167
Error	9	1.28960	0.14329		
C Total	10	2.52227			

Root MSE	0.37854	R-square	0.4887
Dep Mean	2.35455	Adj R-sq	0.4319
C.V.	16.07681		

Parameter Estimates

Variable	DF	Parameter Estimate	Standard Error	T for H0: Parameter=0	Prob > |T|
INTERCEP	1	0.256809	0.72426138	0.355	0.7311
HS_GPA	1	0.727921	0.24818083	2.933	0.0167

Variable	DF	Variable Label
INTERCEP	1	Intercept
HS_GPA	1	HIGH SCHOOL GPA

DATA FOR HIGH SCHOOL AND COLLEGE GPA'S
EXAMPLE OF PROC CORR

CORRELATION ANALYSIS

2 'VAR' Variables: FR_GPA   HS_GPA

Simple Statistics

Variable	N	Mean	Std Dev	Sum
FR_GPA	11	2.35455	0.50222	25.90000
HS_GPA	11	2.88182	0.48232	31.70000

Simple Statistics

Variable	Minimum	Maximum	Label
FR_GPA	1.60000	3.10000	FRESHMAN GPA
HS_GPA	2.00000	3.60000	HIGH SCHOOL GPA

Pearson Correlation Coefficients / Prob > |R| under Ho: Rho=0 / N = 11

	FR_GPA	HS_GPA
FR_GPA FRESHMAN GPA	1.00000 0.0	0.69908 0.0167
HS_GPA HIGH SCHOOL GPA	0.69908 0.0167	1.00000 0.0

SAS Output (Continued)

DATA FOR HIGH SCHOOL AND COLLEGE GPA'S
EXAMPLE OF PROC FREQ

STATISTICS FOR TABLE OF FR_GPA BY HS_GPA

Statistic	Value	ASE
Gamma	0.547	0.175
Kendall's Tau-b	0.537	0.172
Stuart's Tau-c	0.533	0.170
Somers' D C|R	0.537	0.172
Somers' D R|C	0.537	0.172
Pearson Correlation	0.699	0.144
Spearman Correlation	0.694	0.184
Lambda Asymmetric C|R	0.889	0.105
Lambda Asymmetric R|C	0.889	0.105
Lambda Symmetric	0.889	0.078
Uncertainty Coefficient C|R	0.945	0.035
Uncertainty Coefficient R|C	0.945	0.035
Uncertainty Coefficient Symmetric	0.945	0.023

Sample Size = 11

EXECUSTAT Output

Simple Regression Analysis

Linear model: FR_GPA = 0.256809 + 0.727921*HS_GPA

Table of Estimates

	Estimate	Standard Error	t Value	P Value
Intercept	0.256809	0.724261	0.35	0.7311
Slope	0.727921	0.248181	2.93	0.0167

R-squared = 48.87%
Correlation coeff. = 0.699
Standard error of estimation = 0.378536
Durbin-Watson statistic = 2.52629
Mean absolute error = 0.291256

Correlation Analysis for GPAS

	HS_GPA	FR_GPA
HS_GPA		0.6991
FR_GPA	0.6991	

The table shows estimated product-moment correlation

Correlation Analysis for W147

	HS_GPA	FR_GPA
HS_GPA		0.6941
FR_GPA	0.6941	

The table shows estimated Spearman rank correlation

EXECUSTAT Output (Continued)

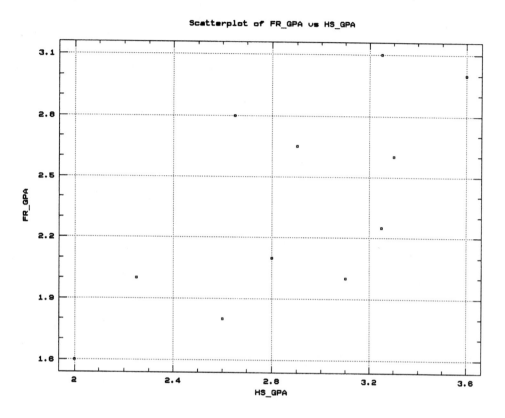

Scatterplot of FR_GPA vs HS_GPA

Summary

In this chapter we presented an introduction to regression and correlation. First we examined the relationship between a dependent variable y and a single independent variable x. A scatter plot was used to provide a graphical display of the data, and the method of least squares was used to obtain the regression equation $\hat{y} = \hat{\beta}_0 + \hat{\beta}_1 x$.

The strength of the linear relationship between y and x can be measured by the sample correlation coefficient $\hat{\rho}$, by the coefficient of determination $\hat{\rho}^2$, or by the rank correlation coefficient $\hat{\rho}_S$.

Fitting multivariable predictors to experimental data is a very powerful and valuable method of inference. The computer prediction of election-eve outcomes uses this technique. Multivariable predictors are also employed in business forecasting, industrial production, medicine, and many areas of science. In this chapter we discussed—and gave an example to illustrate—how to obtain the least squares regression line $\hat{y} = \hat{\beta}_0 + \hat{\beta}_1 x_1 + \hat{\beta}_2 x_2$ when y is related to only two independent variables. Solutions to multiple regression problems for this situation, as well as for situations where we are concerned with more than two independent variables, can conveniently be obtained by using a standard statistical software package. Students are encouraged to pursue with their professors the computer opportunities available at their institution. Additional details on how to obtain solutions to multiple regression problems can be found in some of the references at the end of this book—particularly Mendenhall (1987) and Ott (1993).

Chapter 12 deals with analysis methods (step 3 in making sense of data) for the regression and correlation situations described in this chapter.

Key Terms

prediction equation

freehand regression line

linear equation

expected value of y

method of least squares

predicted value of y

residual

sum of squares for error

sum of squares about the mean

sum of squares due to regression

sum of squares for error

correlation coefficient

sample correlation coefficient $\hat{\rho}$

coefficient of determination

Spearman's rank order correlation coefficient $\hat{\rho}_S$

Key Formulas

1   Least squares prediction equation:

$$\hat{y} = \hat{\beta}_0 + \hat{\beta}_1 x$$

where

$$\hat{\beta}_1 = \frac{S_{xy}}{S_{xx}}, \qquad \hat{\beta}_0 = \bar{y} - \hat{\beta}_1 \bar{x}, \qquad S_{xx} = \Sigma x^2 - \frac{(\Sigma x)^2}{n}$$

and

$$S_{xy} = \Sigma xy - \frac{(\Sigma x)(\Sigma y)}{n}$$

2   Sample correlation coefficient:

$$\hat{\rho} = \frac{S_{xy}}{\sqrt{S_{xx}S_{yy}}}, \qquad -1 \le \hat{\rho} \le 1$$

where

$$S_{yy} = \Sigma y^2 - \frac{(\Sigma y)^2}{n}$$

3   Spearman's rank order correlation coefficient:

$$\hat{\rho}_S = \frac{S_{xy}}{\sqrt{S_{xx}S_{yy}}}, \qquad \text{based on rank.}$$

$$\hat{\rho}_S = 1 - \frac{6\Sigma d^2}{n(n^2 - 1)}, \qquad \text{when there are no ties.}$$

Supplementary Exercises

11.27   An investigator was interested in examining the effect of different doses of a new drug on the pulse rates of human subjects. Four doses of the drug were used in the experiment (1.5, 2.0, 2.5, and 3.0 ml/kg of body weight). Three persons were randomly assigned to each of the four drug doses. After a prestudy pulse rate was recorded for each individual, subjects were injected with the appropriate drug dose. One hour later, pulse rates were again recorded. The changes in pulse rates are listed in the accompanying table.

Change in Pulse Rate (y)	20, 21, 19	16, 17, 17	15, 13, 14	8, 10, 8
Drug Dose (x)	1.5, 1.5, 1.5	2.0, 2.0, 2.0	2.5, 2.5, 2.5	3.0, 3.0, 3.0

a   Plot the sample data.

b   Find the least squares line for these data, using the accompanying SAS output on the following page.

c   Predict the change in pulse rate that would accompany a drug dose of 2.3 ml/kg of body weight. (*Note*: A dose of $x = 2.3$ with no y-value was included as another observation, so the software would compute the predicted value for $x = 2.3$.)

```
OPTIONS NODATE NONUMBER PS=60 LS=78;
DATA RAW;
  INPUT DOSE CHANGE;
  LABEL DOSE = 'DRUG DOSE'
        CHANGE = 'CHANGE IN PULSE RATE';
  CARDS;
  1.5   20
  1.5   21
  1.5   19
  2.0   16
  2.0   17
  2.0   17
  2.3    .
  2.5   15
  2.5   13
  2.5   14
  3.0    8
  3.0   10
  3.0    8
;
PROC PRINT N;
TITLE1 'EFFECT OF DIFFERENT DOSES ON PULSE RATES';
TITLE2 'LISTING OF THE DATA';

PROC PLOT DATA=RAW;
  PLOT CHANGE*DOSE;
TITLE2 'PLOT OF THE DATA';

PROC REG DATA=RAW;
  MODEL CHANGE=DOSE / P;
TITLE2 'EXAMPLE OF PROC REG WITH THE P OPTION (PREDICTED VALUES)';
RUN;
```

```
EFFECT OF DIFFERENT DOSES ON PULSE RATES
        LISTING OF THE DATA

            OBS    DOSE    CHANGE

             1     1.5      20
             2     1.5      21
             3     1.5      19
             4     2.0      16
             5     2.0      17
             6     2.0      17
             7     2.3       .
             8     2.5      15
             9     2.5      13
            10     2.5      14
            11     3.0       8
            12     3.0      10
            13     3.0       8

                  N = 13
```

EFFECT OF DIFFERENT DOSES ON PULSE RATES
PLOT OF THE DATA

Plot of CHANGE*DOSE.   Legend: A = 1 obs, B = 2 obs, etc.

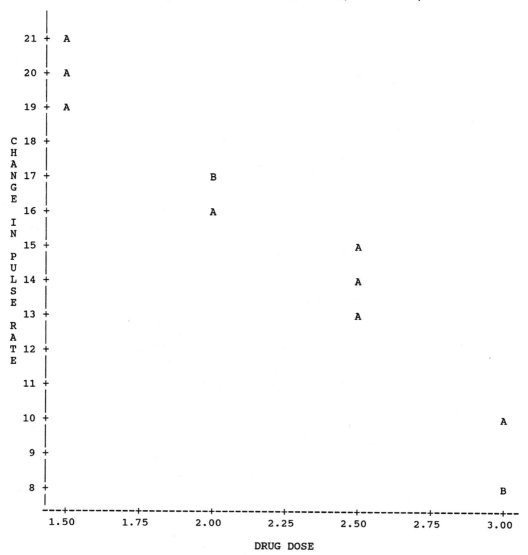

NOTE: 1 obs had missing values.

EFFECT OF DIFFERENT DOSES ON PULSE RATES
EXAMPLE OF PROC REG WITH THE P OPTION (PREDICTED VALUES)

Model: MODEL1
Dependent Variable: CHANGE   CHANGE IN PULSE RATE

Analysis of Variance

Source	DF	Sum of Squares	Mean Square	F Value	Prob>F
Model	1	201.66667	201.66667	168.056	0.0001
Error	10	12.00000	1.20000		
C Total	11	213.66667			

Root MSE	1.09545	R-square	0.9438
Dep Mean	14.83333	Adj R-sq	0.9382
C.V.	7.38502		

Parameter Estimates

Variable	DF	Parameter Estimate	Standard Error	T for H0: Parameter=0	Prob > |T|
INTERCEP	1	31.333333	1.31148770	23.891	0.0001
DOSE	1	-7.333333	0.56568542	-12.964	0.0001

Variable	DF	Variable Label
INTERCEP	1	Intercept
DOSE	1	DRUG DOSE

EFFECT OF DIFFERENT DOSES ON PULSE RATES
EXAMPLE OF PROC REG WITH THE P OPTION (PREDICTED VALUES)

Obs	Dep Var CHANGE	Predict Value	Residual
1	20.0000	20.3333	-0.3333
2	21.0000	20.3333	0.6667
3	19.0000	20.3333	-1.3333
4	16.0000	16.6667	-0.6667
5	17.0000	16.6667	0.3333
6	17.0000	16.6667	0.3333
7	.	14.4667	.
8	15.0000	13.0000	2.0000
9	13.0000	13.0000	-178E-17
10	14.0000	13.0000	1.0000
11	8.0000	9.3333	-1.3333
12	10.0000	9.3333	0.6667
13	8.0000	9.3333	-1.3333

Sum of Residuals   1.953993E-14
Sum of Squared Residuals   12.0000
Predicted Resid SS (Press)   17.7709

11.28   Refer to Exercise 11.27. Calculate the correlation coefficient and the coefficient of determination for the data. Interpret your results.

11.29   A production foreman was concerned about the quality of the outgoing product from his department. He strongly suspected that the percentage of defective items passing through his assembly line during a 30-minute period increased throughout the day. At nine 30-minute periods throughout the day, the assembly line was closely examined to determine the number of defectives being produced. For each of these 30-minute periods, the number of hours that workers had been working (from 8:00 A.M.) was also recorded. The data are given in the accompanying table.

Number of Defectives (y)	Number of Hours x That Workers Are on the Job (x)
13	1.0
14	1.5
16	2.5
14	2.0
15	3.5
20	4.5
18	4.0
18	5.5
20	6.0

Total $\Sigma y = 148$; $\Sigma y^2 = 2490$ $\Sigma x = 30.5$; $\Sigma x^2 = 128.25$; $\Sigma xy = 535.5$

a Plot the data.

b Write a linear model relating the number of defectives, y, to the number of hours on the job, x, for these data.

c Use the method of least squares to fit the model.

d Predict the number of items that would be defective in a 30-minute period if the workers had just completed 5 hours of work.

11.30 Refer to Exercise 11.29. Compute the sample correlation coefficient to measure the strength of the linear relationship between x and y. Interpret your answer.

11.31 A chain of grocery stores conducted a study to determine the relationship between the amount of money x spent on advertising and the weekly volume y of sales. Six different levels of advertising expenditure were tried in random order over a six-week period. The accompanying data were observed (in units of $100).

Weekly Sales Volume (y)	10.2	11.5	16.1	20.3	25.6	28.0
Amount Spent on Advertising (x)	1.0	1.25	1.5	2.0	2.5	3.0

a Plot these data on a scatter diagram.

b Use the method of least squares to fine the regression equation $\hat{y} = \hat{\beta}_0 + \hat{\beta}_1 x$.

c Use the prediction equation of part b to estimate sales volume for an expenditure of $220 in advertising.

11.32 Refer to Exercise 11.31. Compute the correlation coefficient between the sales volume and the advertising volume. Does there appear to be a strong linear relationship between x and y?

11.33 Suppose that the following data were collected on emphysema patients: the number of years x the patient smoked and inhaled, and a physician's evaluation y of the patient's diminution in lung capacity (measured on a scale of 0 to 100). The results for a sample of ten patients appear in the accompanying table. (*Note:* $S_{xx} = 876.9$, $S_{yy} = 2510$, and $S_{xy} = 1148$.)

Patient	Years Smoking (x)	Diminution in Lung Capacity (y)
1	25	55
2	36	60
3	22	50
4	15	30
5	48	75
6	39	70
7	42	70
8	31	55
9	28	30
10	33	35

a Plot the data on a scatter diagram.

b Use the method of least squares to find the regression line $\hat{y} = \hat{\beta}_0 + \hat{\beta}_1 x$.

c Does there appear to be a positive linear relationship between x and y?

d Calculate the correlation coefficient between the variables "lung capacity" and "number of years smoking."

e Predict a person's diminution in lung capacity after 30 years of smoking.

11.34 An experiment was conducted to measure the strength of the linear relationship between two variables: a student's emotional stability (as measured by a guidance counselor's subjective judgment after an encounter session), and the student's score on an achievement test administered to children entering the first grade. The variable of emotional stability was measured on a scale of 0 to 40 (from low to high), and the achievement test was also measured from 10 to 40. Use the accompanying data from a random sample of 15 children to calculate the correlation coefficient. (*Note:* $S_{xx} = 485.33$, $S_{yy} = 522.93$, and $S_{xy} = 316.67$.)

Student	Emotional Stability (x)	Achievement (y)	Student	Emotional Stability (x)	Achievement (y)
1	23	31	9	32	33
2	21	23	10	29	35
3	31	34	11	16	21
4	34	29	12	29	22
5	26	29	13	23	24
6	22	27	14	27	28
7	14	21	15	25	15
8	18	17			

11.35 Earnings from a particular stock for the 7-year period June 1986 to 1992 are listed in the accompanying table. Sketch these seven data points.

Year	1992	1991	1990	1989	1988	1987	1986
Earnings per Share	2.30	1.80	1.50	1.20	1.05	1.10	1.20

11.36 Refer to the sketch in Exercise 11.35.

a Suggest an appropriate multiple regression model for relating earnings per share to the independent variable "year."

b Compute the rank correlation coefficient $\hat{\rho}_S$.

11.37 Yields in bushels of tomatoes are shown here for 12 equal-sized plots, each of which received a different amount of fertilizer.

a Plot the data.

b Based on the plot, specify a linear or multiple regression model that may be appropriate.

c Compute either $\hat{\rho}$ or $\hat{\rho}_S$, depending on the model selected in part b.

Plot	Yield, y (in Bushels)	Amount of Fertilizer, x (in Pounds per Plot)
1	24	12
2	18	5
3	31	15
4	33	17
5	26	20
6	30	14
7	20	6
8	25	23
9	25	11
10	27	13
11	21	8
12	29	18

11.38 A social researcher analyzed 1985 demographic data from 10 nations. These data are shown in the accompanying table. Use the MINITAB output to address parts a through c.

Nation	% Projected Population Increase (y)	Birth Rate (x_1)	Death Rate (x_2)	Life Expectancy (yr.) (x_3)	Per Capita GNP (x_4)
Bolivia	53.2	42	16	51	$510
Cuba	14.9	17	6	73	$1050
Cyprus	14.3	29	9	74	$3720
Egypt	39.3	37	10	57	$700
Ghana	60.1	47	15	52	$320
Jamaica	21.7	28	6	70	$1300
Nigeria	71.6	48	17	50	$760
South Africa	40.1	35	14	54	$2450
South Korea	21.1	23	6	66	$2010
Turkey	36.9	35	10	63	$1230

Source: Population Reference Bureau, *World Population Data Sheet* (Washington, D.C., 1985).

```
MTB > NAME C1 'POPUL' C2 'BIRTH' C3 'DEATH' C4 'LIFE' C5 'GNP'
MTB > PRINT C1 - C5

ROW   POPUL   BIRTH   DEATH   LIFE   GNP

  1    53.2     42      16     51    510
  2    14.9     17       6     73   1050
  3    14.3     29       9     74   3720
  4    39.3     37      10     57    700
  5    60.1     47      15     52    320
  6    21.7     28       6     70   1300
  7    71.6     48      17     50    760
  8    40.1     35      14     54   2450
  9    21.1     23       6     66   2010
 10    36.9     35      10     63   1230

MTB > REGRESS 'POPUL' ON 4 'BIRTH' 'DEATH' 'LIFE' 'GNP'
```

The regression equation is
POPUL = 27.7 + 0.738 BIRTH + 1.46 DEATH − 0.422 LIFE − 0.00406 GNP

Predictor	Coef	Stdev	t-ratio	p
Constant	27.73	40.59	0.68	0.525
BIRTH	0.7384	0.4275	1.73	0.145
DEATH	1.457	1.202	1.21	0.280
LIFE	−0.4224	0.5055	−0.84	0.441
GNP	−0.004056	0.002120	−1.91	0.114

s = 4.910   R-sq = 96.6%   R-sq(adj) = 93.8%

Analysis of Variance

SOURCE	DF	SS	MS	F	p
Regression	4	3377.16	844.29	35.02	0.001
Error	5	120.54	24.11		
Total	9	3497.70			

SOURCE	DF	SEQ SS
BIRTH	1	3097.66
DEATH	1	72.67
LIFE	1	118.56
GNP	1	88.26

```
MTB > REGRESS 'POPUL' ON 4 'BIRTH' 'DEATH' 'LIFE' 'GNP';
SUBC> PREDICT 30 15 65 1100.

The regression equation is
POPUL = 27.7 + 0.738 BIRTH + 1.46 DEATH - 0.422 LIFE - 0.00406 GNP

Predictor       Coef        Stdev       t-ratio         p
Constant        27.73       40.59         0.68        0.525
BIRTH           0.7384      0.4275        1.73        0.145
DEATH           1.457       1.202         1.21        0.280
LIFE           -0.4224      0.5055       -0.84        0.441
GNP            -0.004056    0.002120     -1.91        0.114

s = 4.910       R-sq = 96.6%      R-sq(adj) = 93.8%

Analysis of Variance

SOURCE          DF            SS          MS         F          p
Regression      4         3377.16      844.29     35.02      0.001
Error           5          120.54       24.11
Total           9         3497.70

SOURCE          DF         SEQ SS
BIRTH           1         3097.66
DEATH           1           72.67
LIFE            1          118.56
GNP             1           88.26

    Fit   Stdev.Fit          95% C.I.              95% P.I.
  39.81        7.80    ( 19.75,  59.87)   ( 16.11,  63.51) XX

X  denotes a row with X values away from the center
XX denotes a row with very extreme X values
```

a   Determine the least-square regression line

$$\hat{y} = \hat{\beta}_0 + \hat{\beta}_1 x_1 + \hat{\beta}_2 x_2 + \hat{\beta}_3 x_3 + \hat{\beta}_4 x_4.$$

b   Predict y for birth rate $x_1 = 30$, death rate $x_2 = 15$, life expectancy $x_3 = 65$, and per capita GNP $x_4 = \$1100$.

c   Does the regression equation seem to fit the data?

11.39   Refer to Exercise 11.38.

a   Use the MINITAB computer output given here to find the following least-squares regression equations:

$$\hat{y} = \hat{\beta}_0 + \hat{\beta}_1 x_1$$
$$\hat{y} = \hat{\beta}_0 + \hat{\beta}_1 x_1 + \hat{\beta}_2 x_2$$
$$\hat{y} = \hat{\beta}_0 + \hat{\beta}_1 x_1 + \hat{\beta}_2 x_2 + \hat{\beta}_3 x_3$$

b   Does it appear from Exercise 11.38 and the output here that all four variables are needed to predict y? What about the variables x_1, x_2, and x_3? The variables x_1 and x_2? Might knowledge of x_1 (birth rate) be sufficient? Explain.

```
MTB > REGRESS 'POPUL' ON 1 'BIRTH'

The regression equation is
POPUL = - 25.6 + 1.84 BIRTH

Predictor        Coef       Stdev     t-ratio        p
Constant      -25.563       8.297       -3.08    0.015
BIRTH          1.8441      0.2343        7.87    0.000

s = 7.071       R-sq = 88.6%      R-sq(adj) = 87.1%

Analysis of Variance

SOURCE        DF          SS          MS          F        p
Regression     1      3097.7      3097.7      61.95    0.000
Error          8       400.0        50.0
Total          9      3497.7

Unusual Observations
Obs.   BIRTH      POPUL      Fit Stdev.Fit   Residual    St.Resid
  3     29.0      14.30    27.92      2.54     -13.62       -2.06R

R denotes an obs. with a large st. resid.

MTB > REGRESS 'POPUL' ON 2 'BIRTH' 'DEATH'

The regression equation is
POPUL = - 21.7 + 1.22 BIRTH + 1.60 DEATH

Predictor        Coef       Stdev     t-ratio        p
Constant      -21.740       8.590       -2.53    0.039
BIRTH          1.2213      0.5486        2.23    0.061
DEATH          1.598       1.282         1.25    0.253

s = 6.839       R-sq = 90.6%      R-sq(adj) = 88.0%

Analysis of Variance

SOURCE        DF          SS          MS          F        p
Regression     2      3170.3      1585.2      33.90    0.000
Error          7       327.4        46.8
Total          9      3497.7

SOURCE        DF      SEQ SS
BIRTH          1      3097.7
DEATH          1        72.7

Unusual Observations
Obs.   BIRTH      POPUL      Fit Stdev.Fit   Residual    St.Resid
  3     29.0      14.30    28.06      2.45     -13.76       -2.16R

R denotes an obs. with a large st. resid.
```

```
MTB > REGRESS 'POPUL' ON 3 'BIRTH' 'DEATH' 'LIFE'

The regression equation is
POPUL = 59.3 + 0.916 BIRTH + 0.40 DEATH - 0.944 LIFE

Predictor        Coef        Stdev      t-ratio        p
Constant        59.33        44.54        1.33      0.231
BIRTH          0.9159       0.5013        1.83      0.117
DEATH           0.398        1.283        0.31      0.767
LIFE          -0.9439       0.5114       -1.85      0.114

s = 5.899       R-sq = 94.0%      R-sq(adj) = 91.0%

Analysis of Variance

SOURCE          DF          SS          MS          F          p
Regression       3       3288.9      1096.3      31.50      0.000
Error            6        208.8        34.8
Total            9       3497.7

SOURCE          DF       SEQ SS
BIRTH            1       3097.7
DEATH            1         72.7
LIFE             1        118.6

MTB > STOP
```

11.40   Use the data in the following table and the accompanying MINITAB printout.

a   Locate the regression line, and predict average state and local taxes paid for average personal incomes of $12,000 and $13,500.

b   Does the least squares line seem to fit the data provided in the accompanying computer printout?

c   Locate $\hat{\rho}^2$, and interpret its value.

State	Average Personal Income	Average State and Local Taxes Paid per Capita
Arkansas	$ 9724	$ 771
California	14,344	1337
Connecticut	16,369	1434
Illinois	13,728	1255
Louisiana	10,850	1051
Mississippi	8857	769
New Jersey	15,282	1457
North Dakota	12,461	1110
Oregon	11,582	1229
Oklahoma	11,745	1123

```
MTB > REGRESS 'TAXES' 1 'INCOME';
SUBC> PREDICE 12000;
SUBC> PREDICT 13500.

The regression equation is
TAXES = - 26 + 0.0944 INCOME

Predictor        Coef        Stdev      t-ratio         p
Constant        -26.3        148.5        -0.18     0.864
INCOME        0.09444      0.01169         8.08     0.000

s = 84.91        R-sq = 89.1%      R-sq(adj) = 87.7%

Analysis of Variance

SOURCE          DF           SS           MS          F         p
Regression       1       470263       470263      65.22     0.000
Error            8        57679         7210
Total            9       527942

Unusual Observations
Obs.   INCOME       TAXES        Fit Stdev.Fit   Residual    St.Resid
  9     11582      1229.0     1067.5      28.9      161.5        2.02R

R denotes an obs. with a large st. resid.

     Fit  Stdev.Fit        95% C.I.          95% P.I.
  1106.9       27.5   ( 1043.6, 1170.3)   ( 901.1, 1312.8)

  1248.6       29.3   ( 1181.0, 1316.2)   ( 1041.4, 1455.8)

MTB > STOP
```

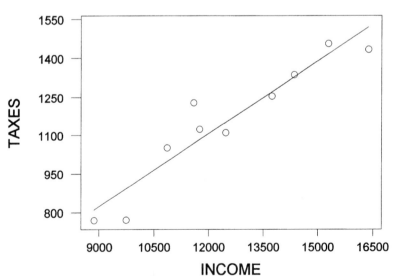

11.41 English, mathematics, and social studies achievement test scores (labeled ENGLACT, MATH-ACT, and SOCSACT, respectively) are displayed for 20 grade-school children in the accompanying MINITAB output. Determine pairwise correlation coefficients for ENGLACT, MATHACT, and SOCSACT.

```
MTB > PRINT C1 - C3

 ROW   ENGLACT   MATHACT   SOCSACT

   1       16        15         8
   2       15        27        24
   3       17        11        19
   4       18        14        25
   5       18         8        12
   6       11        10         7
   7       13        14        11
   8       17         8        23
   9       21        23        26
  10       17        12        13
  11       12        18        11
  12        9        11         5
  13       26        28        26
  14       18        19        15
  15       19        17        14
  16       22        27        27
  17       15        19        11
  18       23        28        21
  19       19        14        13
  20       24        22        15

MTB > CORRELATE 'ENGLACT' 'MATHACT' 'SOCSACT'

           ENGLACT   MATHACT
MATHACT     0.595
SOCSACT     0.668     0.567

MTB > REGRESS 'SOCSACT' 1 'ENGLACT'

The regression equation is
SOCSACT = - 2.41 + 1.07 ENGLACT

Predictor      Coef      Stdev     t-ratio       p
Constant     -2.405      5.049       -0.48     0.640
ENGLACT      1.0689      0.2803       3.81     0.001

s = 5.341      R-sq = 44.7%     R-sq(adj) = 41.6%

Analysis of Variance

SOURCE        DF        SS          MS         F        p
Regression     1      414.72      414.72     14.54    0.001
Error         18      513.48       28.53
Total         19      928.20

Unusual Observations
Obs. ENGLACT     SOCSACT       Fit Stdev.Fit   Residual    St.Resid
   2    15.0       24.00      13.63      1.38      10.37        2.01R

R denotes an obs. with a large st. resid.
```

```
MTB > REGRESS 'SOCSACT' 2 'ENGLACT' 'MATHACT'

The regression equation is
SOCSACT = - 2.77 + 0.820 ENGLACT + 0.274 MATHACT

Predictor        Coef       Stdev     t-ratio          p
Constant       -2.766       4.992       -0.55      0.587
ENGLACT        0.8196      0.3443        2.38      0.029
MATHACT        0.2737      0.2250        1.22      0.240

s = 5.271      R-sq = 49.1%      R-sq(adj) = 43.1%

Analysis of Variance

SOURCE         DF          SS          MS         F          p
Regression      2      455.85      227.92      8.20      0.003
Error          17      472.35       27.79
Total          19      928.20

SOURCE         DF      SEQ SS
ENGLACT         1      414.72
MATHACT         1       41.13

Unusual Observations
Obs.  ENGLACT     SOCSACT      Fit Stdev.Fit  Residual    St.Resid
  8     17.0       23.00     13.36     2.31       9.64       2.03R

R denotes an obs. with a large st. resid.

MTB > STOP
```

11.42   Refer to the computer output of Exercise 11.41.

   a   Identify the least-squares prediction equations

$$\hat{y} = \hat{\beta}_0 + \hat{\beta}_1 x_1$$
$$\hat{y} = \hat{\beta}_0 + \hat{\beta}_1 x_1 + \hat{\beta}_2 x_2,$$

where

$$y = \text{SOCSACT} \qquad x_1 = \text{ENGLACT} \qquad x_2 = \text{MATHACT}.$$

   b   Which regression equation seems to predict y better? Explain.

11.43   Social adjustment and perceived self-image tests were administered to $n = 6$ teenagers who had recently completed a STRAIGHT program (for drug rehabilitation).

   a   Plot the data.

   b   Use these data to compute the correlation coefficient.

   c   Use the coefficient of determination and the data plot to interpret the value of $\hat{\rho}$.

Social Adjustment Score	Perceived Self-image	Social Adjustment Score	Perceived Self-image
55	35	28	18
37	23	52	31
61	42	70	45

11.44   Quantity discount is a usual practice in business; the larger the quantity purchased, the smaller is the unit price. The following data illustrate this phenomenon.

Units Purchased (in 000), x	Unit price (in $00), y
1	100
2	80
3	70
4	60
5	40

Totals	15	350

a   Compute the y-intercept $\hat{\beta}_0$ and the slope $\hat{\beta}_1$.

b   Predict y when $x = 6$. Would you be uncomfortable predicting y when $x = 10$? Why or why not?

11.45   A study was conducted to examine the efficiencies of various manufacturing sites of a large corporation. At each site, the average number of acceptable cartons of manufactured goods per month was recorded, as was the average number of hours of assembly-line operation per month. These data are shown here.

Location	Average Number of Acceptable Cartons (000), y	Average Number of Hours of Line Operation, x
1	12	20
2	11	38
3	15	40
4	16	45
5	20	57
6	18	68
7	22	74
8	26	79
9	20	81
10	21	86
11	27	93
12	32	104
13	33	110
14	34	120
15	31	138

a   Plot the sample data.

b   Obtain a least squares fit to these data, using a linear regression model.

c   Plot the least squares prediction equation on the graph of part a. Does this model appear to fit the data adequately?

11.46   Refer to the data of Exercise 11.45. Suppose that the last four data points (corresponding to locations 12, 13, 14, and 15) were as shown here, rather than as indicated in the previous exercise.

Location	y	x
12	25	104
13	23	110
14	20	120
15	15	138

a   Plot the entire new data set for the 15 locations.

b   Would a linear regression model still provide a good fit to the data?

11.47   A sociologist working for the government of a large city collected data on the number of nonviolent crimes (in 1000s) reported and the number of all crimes over the previous reporting period. Quarterly data are shown here.

Quarter	Nonviolent Crimes	All Crimes
1	7.2	14.1
2	6.4	14.5
3	6.6	13.3
4	7.3	13.6
5	7.5	15.2
6	6.9	15.7
7	7.1	15.3
8	7.4	14.8
9	7.6	16.1
10	7.3	16.6
11	7.1	16.2
12	7.0	15.9

a   Plot the nonviolent crimes data versus quarter. Then plot all the crimes data versus quarter on the same graph.

b   Does there appear to be a relationship between the two crime variables?

c   Compute the correlation coefficient between the number of nonviolent crimes and the number of all crimes.

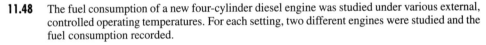

11.48   The fuel consumption of a new four-cylinder diesel engine was studied under various external, controlled operating temperatures. For each setting, two different engines were studied and the fuel consumption recorded.

Observation	External Temperature (°F), x	Fuel Consumption (gallons), y
1	20	25
2	20	26
3	30	28
4	30	27
5	40	32
6	40	35
7	50	42
8	50	46
9	60	55
10	60	53
11	70	55
12	70	57
13	80	60
14	80	58
15	90	61
16	90	58

a Plot the sample data. Do you think a linear regression line will be an adequate model?

b Fit the least squares regression model $y = \beta_0 + \beta_1 x + \epsilon$, and draw the prediction equation on the graph of part a.

c What is the sample correlation coefficient for these data?

11.49 We are given the following scatter diagrams.

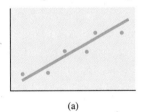
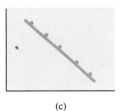

(a)   (b)   (c)

a Which of the following relationships *best* describes diagram (a)?

Strong positive relationship

Strong negative relationship

Rather weak positive relationship

Rather weak negative relationship

Little or no relationship

Perfect positive relationship

Perfect negative relationship

b Which relationship best describes diagram (b)?

c Which relationship best describes diagram (c)?

11.50 A study was conducted in a poverty region to determine the effect that level of education had on level of family income. For ten families in the region, information was collected on the number of grades of school completed by the head of each family and the annual income of the family. Family income is thought to be linearly related to amount of schooling. The data are given next.

Family	Grades of School Completed, x	Family Income (in $000s), y
1	6	21
2	5	19
3	10	31
4	7	25
5	8	28
6	12	33
7	5	20
8	9	29
9	7	22
10	11	32

Summary of data:

$$\Sigma x = 80 \qquad\qquad \Sigma y = 260$$
$$\Sigma x^2 = 694 \qquad \Sigma (y - \bar{y})^2 = 250 \qquad \Sigma (x - \bar{x})(y - \bar{y}) = 113$$

a Compute $\hat{\beta}_1$.

b Explain in plain English the meaning of $\hat{\beta}_0$ and $\hat{\beta}_1$. (Use a sketch if that would help.)

11.51 For the following graphs (a) and (b), write the regression equation for each of the lines, a, b, c, and d.

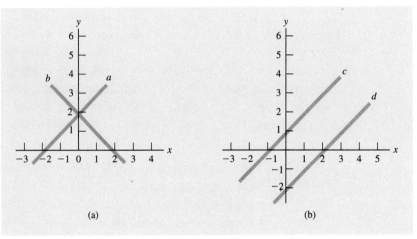

(a)   (b)

11.52 *Fill in the blanks:* The correlation coefficient $\hat{\rho}$ has an upper limit of _____ . If all points are exactly on the straight line, $\hat{\rho}$ will be _____ or _____, depending on whether the relationship is _____ or _____. If the points on the scatter diagram are randomly scattered, $\hat{\rho}$ will be near _____. The better the fit, the _____ the magnitude of $\hat{\rho}$.

11.53 Earnings from a particular stock are listed here for the 8-year period June 1986 to 1993.

Year	1993	1992	1991	1990	1989	1988	1987	1986
Earnings per Share	3.31	2.10	1.60	1.15	1.05	1.00	.99	.97

a Plot the data.

b Use the method of least squares to fit the model $y = \hat{\beta}_0 + \hat{\beta}_1 x + \epsilon$. *Hint:* Use $x =$ (year $- 1985$) to simplify your calculations.

c What earnings per share would you predict for 1994? Do you have reservations about using the prediction equation?

11.54 The accompanying data are mean weights and mean waist sizes for state representatives participating in the Miss America Pageant during the years of 1972–1983. Compute the correlation coefficient between weight and waist size. Weights are measured in pounds and waist sizes in inches.

Year	Mean Weight	Mean Waist Size	Year	Mean Weight	Mean Waist Size
1972	119.4	24.1	1978	116.2	23.9
1973	118.1	24.0	1979	114.7	23.5
1974	118.6	24.0	1980	115.0	23.6
1975	118.0	23.9	1981	116.7	23.7
1976	116.0	23.6	1982	115.1	23.5
1977	115.9	23.6	1983	114.0	23.2

11.55 In the following table, average expenditure per student and average teacher salary for public education in selected cites of Texas are presented. Considering teachers' salaries to be a major cause of expenditures in a school system, determine the linear relationship, using the method of least squares.

Place	Average Expenditure per Student	Average Teacher Salary
Abilene	$1192	$12,486
Amarillo	1342	12,108
Brownsville	988	10,944
Dallas-Fort Worth	1236	12,420
Galveston	1548	13,248
Houston	1332	13,272
Kileen	1118	10,704
Lubbock	1173	11,292
San Antonio	1185	12,252

11.56 **a** Plot the data given in the following table.

b Estimate the association between units sold in April and units sold in December for the salespeople listed.

Salesperson	April Sales	December Sales
Mr. Hinshaw	23	34
Mrs. Fabares	18	22
Ms. Mills	30	36
Mr. Haggar	22	31
Mr. Redburn	19	26
Ms. Blitzer	28	46

11.57 A retailer of satellite dishes would like to know the impact that advertising has on his sales. For six months he records the number of adds run in the newspaper and the number of sales. The results are as follows.

	March	April	May	June	July	August
No. of ads	0	3	5	8	10	9
No. of sales	5	7	12	15	17	15

a Calculate the correlation coefficient for the sample

b Find the regression line equation.

c Predict the number of sales that would be made if the retailer ran six adds.

11.58 The following table lists the distances five trucks have traveled and their maintenance cost. Make a two-dimensional scatter diagram of these data on graph paper.

Distance Traveled (Kilometers)	Maintenance Costs ($)
50,000	9000
35,000	5000
70,000	14,000
43,000	4500
61,000	10,500

11.59 A manufacturing company has developed an aptitude test to be used when hiring new employees. The test is used to predict how well the candidate would perform on the job, if hired.

Let x = score on the aptitude test (range 0 to 50).

Let y = score on job evaluation at the end of one year's employment (range 0 to 10).

To test how well x predicts y, a study was done involving five randomly selected employees after their first year of service. The following sums were obtained:

$$\Sigma x = 150, \quad \Sigma x^2 = 5500, \quad \Sigma y = 25, \quad \Sigma y^2 = 161, \quad \Sigma xy = 930.$$

a Find the equation of the regression line.

b If the company wishes to hire people who will have a predicted job evaluation score of at least $y = 8$, what minimum score x should be used on the aptitude test to give this predicted score?

c Find the correlation coefficient for x and y.

d What percentage of the variation in the job evaluation score y is explained by the aptitude score x?

11.60 Address the following general statements.

a Even if there is a direct cause-and-effect relationship between two variables x and y, their correlation coefficient need not be anywhere near 1 or -1. Give two examples of how this could happen.

b Even if the correlation between two variables x and y is very strong (coefficient near 1 or -1), there need not be any cause-and-effect relationship at all between x and y. Give an example of how this could happen.

11.61 *Modified true/false questions:* Circle T or F. If your answer is false, change the statement so that it becomes true. Only change the words that are underlined.

a **T F** The correlation coefficient for a set of paired data points will always be a number between zero and one (inclusive).

b **T F** If two different scatter diagrams have exactly the same regression lines, they will also have the same correlation coefficient.

c **T F** Truncating the range of x-values in a scatter diagram may cause a significant <u>decrease</u> in the correlation coefficient.

d **T F** If the points of a scatter diagram all lie exactly along a straight line, their correlation coefficient will always be <u>exactly equal to one</u>.

e **T F** If the correlation coefficient for the paired variables x and y is positive, y will tend to <u>decrease</u> as x <u>decreases</u>.

f **T F** If the correlation coefficient for a set of paired variables x and y is zero, there is <u>no relationship</u> between x and y.

11.62 Refer to Exercise 11.48. Since there does not appear to be a simple, suitable transformation on one or both of the variables to linearize the data, rank the data separately and compute the rank order correlation coefficient measure.

11.63 The maximum volume of oxygen uptake (VO_2 max) has been used as a measure of cardiac status in healthy individuals as well as in persons suffering from cardiac-related illnesses (such as congestive heart failure). The VO_2 max readings for 12 healthy adult males following strenuous exercise are recorded here. In general, VO_2 max decreases with any increase in activity level.

Individual	VO_2 Max, y	Duration of Exercise (in minutes), y
1	82	10.0
2	73	9.5
3	68	10.2
4	74	10.5
5	66	11.0
6	63	11.3
7	58	11.6
8	54	12.0
9	56	12.1
10	51	12.5
11	55	12.8
12	44	13.0

a Plot the data.

b Does a linear regression equation seem to be appropriate for these data?

c Fit the data, using the model $y = \beta_0 + \beta_1 x + \epsilon$.

11.64 Refer to Exercise 11.63.

a Obtain the least squares prediction for the same data, but this time assume that the final observation is $x = 13$, $y = 30$.

b How well does a linear regression equation fit these data?

Experiences with Real Data

11.65   Refer to the Bonus Level database on the data disk (or in Appendix 1).

a   Plot the bonus level y versus age for males x.

b   Fit the linear regression $y = \beta_0 + \beta_1 x_1 + \epsilon$, and predict.

11.66   Use the Bonus Level database for females in the data disk.

a   Plot the data.

b   Obtain the least square prediction equation between y (bonus level) and x (age) for females. does the equation appear to fit the data?

c   Refer to Exercise 11.65. Plot the prediction equations for males and females in the same graph. How do they compare?

11.67   Refer to the Bonus Level database.

a   Compute ρ and ρ_s for males. Which seems to be a better measure of the strengths of the relationships between bonus level and age? Explain.

b   Refer to part a. Answer the same questions for the data on females.

11.68   Refer to the Annual Returns database on the data disk (or in Appendix 1).

a   For large capital stocks plot the annual returns y versus year x, for stocks with high book-to-market ratios and for stock with low book-to-market ratios.

b   Comment on the effect of annual returns for this timeframe for each category of book-to-market ratio.

11.69   Refer to Exercise 11.68 and the Annual Returns database.

a   Compute pairwise correlation coefficients for the high, medium, and low book-to-market ratio categories.

b   Interpret your findings.

11.70   Compute the pairwise rank correlation coefficient for the data in Exercise 11.69. Does ρ or ρ_s seem more appropriate for summarizing the relationships between the annual returns for the high, medium, and low book-to-market ratio categories? (*Hint:* Use pairwise scatterplots for the book-to-market ratio categories.)

11.71   Refer to the Clinical Trials database on your data disk (or in Appendix 1) to determine the correlation coefficient between the HAM-D total score and the HAM-D anxiety score.

11.72   Determine the correlation coefficient between the HAM-D total score and the Hopkins OBRIST cluster total.

11.73   For the Clinical Trials data, determine the correlation coefficient between age and daily consumptions of coffee and tea.

12

Inferences Related to Linear Regression and Correlation

12.1   Introduction

In Chapter 11 we focused on methods of summarizing for regression and correlation. As part of our discussion, we gave formulas for finding the least squares estimates for β_0 and β_1 in the linear regression model

$$y = \beta_0 + \beta_1 x + \epsilon.$$

This chapter deals with analysis methods for regression and correlation. We will show how to use the least squares estimates for β_0 and β_1 to make inferences about the relationship between y and x. For example, suppose that x represents the amount of force applied to a 1-foot section of steel and y denotes the corresponding increase in width of the steel sample. Applying the results of Chapter 11, we would obtain a random sample of n observations and complete the least squares estimates for β_0 and β_1 as

$$\hat{\beta}_1 = \frac{S_{xy}}{S_{xx}} \quad \text{and} \quad \hat{\beta}_0 = \bar{y} - \hat{\beta}_1 \bar{x}.$$

Is there a positive linear relationship between x and y? To determine this, we could conduct a statistical test of the null hypothesis $H_0: \beta_1 = 0$ against the alternative hypothesis $H_a: \beta_1 > 0$. The estimate $\hat{\beta}_1$ would be used in this test.

12.2 Inferences About β_0 and β_1

Before we can make any inferences about parameters in the linear regression model, we need to expand on the assumptions we have for the model. Previously we have assumed that the random error term ϵ associated with observation y has an expected value of zero. In addition, we will assume the following:

1   The ϵs are independent of each other.

2   For a given setting of the independent variable x, ϵ is normally distributed, with mean 0 and variance σ_ϵ^2. The variance σ_ϵ^2 is constant for all settings of x.

These two assumptions imply that y_i is normally distributed, with mean $\beta_0 + \beta_1 x_i$ and constant variance $\sigma + \epsilon^2$, and that the ys are independent (see Figure 12.1). For example, y_1 is normally distributed with mean $\beta_0 + \beta_1 x_1$ and variance σ_ϵ^2. Similarly, y_2 is normally distributed with mean $\beta_0 + \beta_1 x_2$ and variance σ_ϵ^2; also, y_1 and y_2 are independent. Under these assumptions, $\hat{\beta}_0$ and $\hat{\beta}_1$ have **sampling distributions** that are normal, with means (called **expected values**) and standard errors as shown here.

sampling distributions for $\hat{\beta}_0$ and $\hat{\beta}_1$

expected values for $\hat{\beta}_0$ and $\hat{\beta}_1$

FIGURE 12.1
Normality Assumption for ϵ in Linear Regression

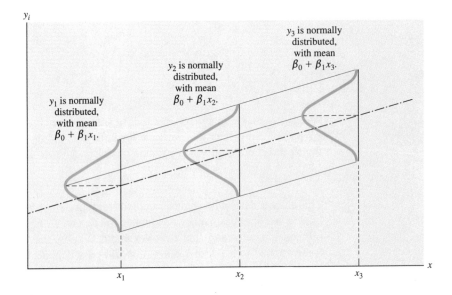

Expected Values and Standard Errors for $\hat{\beta}_0$ and $\hat{\beta}_1$ in Linear Regression

$$\mu_{\hat{\beta}_0} = \beta_0 \qquad\qquad \mu_{\hat{\beta}_1} = \beta_1$$

$$\sigma_{\hat{\beta}_0} = \sigma_\epsilon \sqrt{\frac{1}{n} + \frac{\bar{x}^2}{S_{xx}}} \qquad \sigma_{\hat{\beta}_1} = \frac{\sigma_\epsilon}{\sqrt{S_{xx}}}$$

Based on knowledge of the sampling distributions for $\hat{\beta}_0$ and $\hat{\beta}_1$, we can make inferences (for instance, run a test or construct a confidence interval) about the unknown intercept β_0 and slope β_1.

E X A M P L E   12.1   In examining the weight loss of a compound, a chemist hypothesized that weight loss y (in pounds) is linearly related to the relative humidity x of the room in which the process operates. From a sample of 12 observations, we find $\bar{y} = 4.8$, $\bar{x} = 5.9$, $S_{xy} = 138.2$, $S_{xx} = 126.0$, and $\Sigma x^2 = 544.6$. Compute $\hat{\beta}_0$, $\hat{\beta}_1$ and their standard error.

Solution   For these data,

$$\hat{\beta}_1 = \frac{S_{xy}}{S_{xx}} = 1.09$$

$$\sigma_{\hat{\beta}_1} = \frac{\sigma_\epsilon}{\sqrt{S_{xx}}} = \frac{\sigma_\epsilon}{\sqrt{126}} = \frac{\sigma_\epsilon}{11.225}$$

$$\hat{\beta}_0 = \bar{y} - \hat{\beta}_1 \bar{x} = 4.8 - 1.09(5.9) = -1.63$$

$$\sigma_{\hat{\beta}_0} = \sigma_\epsilon \sqrt{\frac{1}{12} + \frac{(5.9)^2}{126}} = \sigma_\epsilon \sqrt{.3596} = .60\sigma_\epsilon.$$

Before we can make any inferences about β_0 and β_1, we need to compute an estimate of σ_ϵ^2. The estimate we will use is

$$\frac{\text{SSE}}{n-2} = \frac{\Sigma(y - \hat{y})^2}{n-2}.$$

This quantity, which we designate as s_ϵ^2, is based on $n - 2$ degrees of freedom for linear regression problems. ■

Estimates of σ_ϵ^2 and σ_ϵ for Linear Regression

$$s_\epsilon^2 = \frac{\Sigma(y - \hat{y})^2}{n-2} = \frac{\text{SSE}}{n-2}$$

$$s_\epsilon = \sqrt{\frac{SSE}{n-2}}$$

Note: If computer software is not available and calculations are done by hand or with a calculator, use the shortcut formula

$$SSE = S_{yy} - \frac{(S_{xy})^2}{S_{xx}}.$$

E X A M P L E   12.2   The yield per plot in bushels of corn was observed on $n = 10$ plots that had been fertilized to varying degrees. Let the independent variable x denote the amount of fertilizer applied. The data and the coded fertilizer values are recorded in Table 12.1. Use these sample data to obtain the least squares prediction equation $\hat{y}$ for the linear regression model $y = \beta_0 + \beta_1 x + \epsilon$. Calculate estimates of σ_ϵ^2 and σ_ϵ.

TABLE 12.1
Corn Yield Data for Example 12.2

Yield (in Bushels)	Fertilizer (in Pounds per Plot)
12	2
13	2
13	3
14	3
15	4
15	4
14	5
16	5
17	6
18	6

Solution   For these data, we find that

$$n = 10 \qquad \Sigma x = 40$$
$$\Sigma y = 147 \qquad \bar{x} = 4.0$$
$$\bar{y} = 14.7 \qquad \Sigma x^2 = 180$$
$$\Sigma y^2 = 2193 \qquad \Sigma xy = 611$$

Substituting these values into appropriate linear regression formulas, we find the least squares estimates for β_1 and β_0 to be, respectively,

$$\hat{\beta}_1 = \frac{S_{xy}}{S_{xx}} = \frac{611 - \dfrac{40(147)}{10}}{180 - \dfrac{(40)^2}{10}} = \frac{23}{20} = 1.15$$

$$\hat{\beta}_0 = \bar{y} - \hat{\beta}_1 \bar{x} = 14.7 - 1.15(4.0) = 10.10.$$

So the least squares prediction equation is $\hat{y} = 10.10 + 1.15x$.

The estimate for σ_ϵ^2 (and σ_ϵ) requires that we find

$$S_{yy} = \Sigma y^2 - \frac{(\Sigma y)^2}{n} = 2193 - \frac{(147)^2}{10} = 32.10$$

and

$$\text{SSE} = S_{yy} - \frac{(S_{xy})^2}{S_{xx}} = 32.10 - \frac{(23)^2}{20} = 5.65.$$

The estimate for σ_ϵ^2, from the data, is

$$s_\epsilon^2 = \frac{\text{SSE}}{n-2} = \frac{5.65}{8} = 0.71.$$

Hence, $s_\epsilon = 0.84$.   ▪

E X A M P L E   12.3   Refer to Example 12.2 to compute the estimated standard error for $\hat{\beta}_0$ and $\hat{\beta}_1$.

Solution   The formulas for the standard errors are, respectively,

$$\sigma_{\hat{\beta}_0} = \sigma_\epsilon \sqrt{\frac{1}{n} + \frac{\bar{x}^2}{S_{xx}}} \qquad \sigma_{\hat{\beta}_1} = \frac{\sigma_\epsilon}{\sqrt{S_{xx}}}$$

Using $s_\epsilon = 0.84$ as the estimate of σ_ϵ and substituting this into the formulas, we obtain the estimated standard error for $\hat{\beta}_0$ and $\hat{\beta}_1$:

$$s_{\hat{\beta}_0} = 0.84\sqrt{\frac{1}{10} + \frac{4^2}{20}} = 0.84\sqrt{0.90} = 0.80$$

$$s_{\hat{\beta}_1} = \frac{s_\epsilon}{\sqrt{S_{xx}}} = \frac{0.84}{4.47} = 0.19 \quad ▪$$

By substituting s_ϵ for σ_ϵ in the formulas for $\sigma_{\hat{\beta}_1}$ and $\sigma_{\hat{\beta}_0}$, we obtain the estimated standard errors for $\hat{\beta}_1$ and $\hat{\beta}_0$. In practice, since we never know σ_ϵ and hence must always substitute an estimate of its value in the standard error formula, we will drop the word "estimated" and simply call $s_{\hat{\beta}_1}$ and $s_{\hat{\beta}_0}$ the standard errors for $\hat{\beta}_1$ and $\hat{\beta}_0$, respectively.

Using the normality assumption for ϵ, we can construct confidence intervals and statistical tests for β_0 and β_1. If we assume that the ϵ_i from the linear regression model

$$y_i = \beta_0 + \beta_1 x_i + \epsilon_i$$

are normally distributed, we can specify confidence intervals for β_0 and β_1, using the following formula: estimate $\pm t$ standard error.

General Confidence Intervals for β_0 and β_1 in Linear Regression

$$\hat{\beta}_0 \pm ts_\epsilon \sqrt{\frac{1}{n} + \frac{\bar{x}^2}{S_{xx}}} \quad \text{and} \quad \hat{\beta}_1 \pm t \frac{s_\epsilon}{\sqrt{S_{xx}}}$$

where $\quad s_\epsilon = \sqrt{\dfrac{\text{SSE}}{n-2}}$

and t is based on df $= n - 2$.

E X A M P L E 12.4 Use the data from Example 12.2 to develop a 95% confidence interval for β_0 and β_1.

Solution The calculations from Example 12.2 yielded the linear regression equation $\hat{y} = 10.10 + 1.15x$. The $t_{.025}$ value for df $= 8$ is 2.306; $s_\epsilon = 0.84$, $S_{xx} = 20$, and $\bar{x} = 4.0$. Substituting these values into the appropriate formulas, we obtain the 95% confidence intervals shown here:

$$\beta_0 : 10.10 \pm 2.306(0.84)\sqrt{\frac{1}{10} + \frac{4^2}{20}} \quad \text{or} \quad 10.10 \pm 1.84$$

$$\beta_1 : 1.15 \pm 2.306 \frac{(0.84)}{\sqrt{20}} \quad \text{or} \quad 1.15 \pm 0.43$$

In other words, we are 95% confident that the true value of the intercept β_0 lies somewhere in the interval $8.26 \le \beta_0 \le 11.94$. Similarly, we are 95% confident that the true value of the slope β_1 lies somewhere in the interval $0.72 \le \beta_1 \le 1.58$. ▪

E X A M P L E 12.5 A restaurant operating on a "reservations only" basis would like to use the number of advance reservations x to predict the number of dinners y to be prepared. Data on reservations and number of dinners served for one day chosen at random from each week in a 100-week period yielded the following information:

$$\bar{x} = 150 \qquad\qquad \bar{y} = 120$$
$$\Sigma(x - \bar{x})^2 = 90{,}000 \qquad \Sigma(y - \bar{y})^2 = 70{,}000$$
$$\Sigma(x - \bar{x})(y - \bar{y}) = 60{,}000$$

a Find the least squares estimates $\hat{\beta}_0$ and $\hat{\beta}_1$ for the linear regression line $\hat{y} = \hat{\beta}_0 + \hat{\beta}_1 x$.

b Predict the number of means to be prepared if the number of reservations is 135.

c Construct a 90% confidence interval for the slope. Does information on x (number of advance reservations) help in predicting y (number of dinners prepared)?

Solution **a** The least squares estimates are given by

$$\hat{\beta}_1 = \frac{S_{xy}}{S_{xx}} = \frac{60,000}{90,000} = 0.67$$

and

$$\hat{\beta}_0 = \bar{y} - \hat{\beta}_1\bar{x} = 120 - 0.67(150) = 19.50.$$

b The predicted number of means required for the number of advance reservations equal to 135 is

$$\hat{y} = 19.50 + 0.67(135) = 109.95 \qquad \text{or} \qquad 110.$$

c The 90% confidence interval for β_1 uses the formula

$$\hat{\beta}_1 \pm t \text{ standard error,}$$

where the standard error is s_ϵ/S_{xx}.

Although Table 3 in Appendix 3 does not list a t-value for $a = .05$ and df $= 98$, we'll use the t-value for the next higher df (df $= 120$); this value is 1.658.

The standard deviation s_ϵ can be computed by using the summary sample data

$$s_\epsilon^2 = \frac{\text{SSE}}{n-2},$$

where

$$\begin{aligned}
\text{SSE} &= S_{yy} - \hat{\beta}_1 S_{xy} \\
&= 70,000 - 0.67(60,000) \\
&= 29,800.
\end{aligned}$$

Thus

$$s_\epsilon = \sqrt{\frac{29,800}{98}} = \sqrt{304.08} = 17.44,$$

and the 90% confidence interval for β_1 is

$$0.67 \pm 1.658 \frac{(17.44)}{\sqrt{90,000}} \qquad \text{or} \qquad 0.67 \pm .10.$$

Since we are 90% confident that the true value of β_1 lies somewhere in the interval $.57 \le \beta_1 \le .77$, and since $\beta_1 = 0$ does not lie in this interval, it appears that the number of advance reservations is a useful predictor of the number of meals to be prepared in the context of a linear regression model, $y = \beta_0 + \beta_1 x + \epsilon$. ∎

Consider the problem of conducting a statistical test about the slope of a linear regression model. If β_1 is different from zero and the population regression line slopes upward (or downward), as shown in Figure 12.2(a), knowledge of x will help us to predict values of y. When x is large, we predict large values of y; when x is small, we predict small values of y.

FIGURE 12.2
Can x Be Used to Predict y?

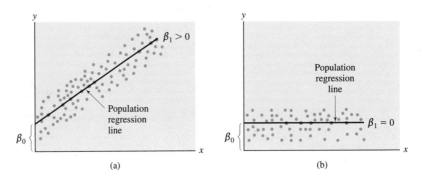

(a)   (b)

However, suppose that the slope of the population regression line is equal to 0 and that the line, therefore, appears as shown in Figure 12.2(b). When $\beta_1 = 0$, knowledge of x will be of no help in predicting y.

In our statistical test for β_1 we ask the question, "Does x contribute information for the prediction of y?" Or in other words, "Does the slope β_1 differ from 0?" Or is it possible the apparent linear arrangement of the data points has occurred solely by chance?

A statistical test of $H_0: \beta_1 = 0$ is summarized next. In order to use this test, we assume that for each value of x there is a population of y-values that can be represented by a normal distribution and that, regardless of the value of x, the population of y-values will have the same variance.

A Statistical Test About the Slope β_1

H_0: $\beta_1 = 0$

H_a: For a one tailed test:

    1 $\beta_1 > 0$

    2 $\beta_1 < 0$

    For a two-tailed test:

    3 $\beta_1 \neq 0$

T.S.: $t = \dfrac{\hat{\beta}_1}{\sqrt{s_\epsilon^2 / S_{xx}}}$ where $s_\epsilon^2 = \dfrac{S_{yy} - \hat{\beta}_1 S_{xy}}{n - 2}$

       and $S_{yy} = \Sigma y^2 - (\Sigma y)^2 / n$

R.R.:   For a given value of α, for df $= (n - 2)$, and for a one-tailed test:

    1 Reject H_0 if $t > t_a$ with $a = \alpha$.

    2 Reject H_0 if $t < -t_a$ with $a = \alpha$.

    For a given value of α, for df $= (n - 2)$, and for a two-tailed test:

    3 Reject H_0 if $|t| > t_a$ with $a = \alpha/2$.

We illustrate the test procedure with an example.

EXAMPLE   12.6   Use the data in Table 11.1 to determine whether there is sufficient evidence to indicate that $\hat{\beta}_1$, the population slope between high-school and college freshman GPAs, is positive. Use $\alpha = .05$.

Solution   The four parts of the statistical test are as follows:

$$H_0: \quad \beta_1 = 0$$
$$H_a: \quad \beta_1 > 0$$

$$\text{T.S.:} \quad t = \frac{\hat{\beta}_1}{\sqrt{s_\epsilon^2/S_{xx}}}$$

R.R.: For $\alpha = .05$ and df $= (n - 2)$, reject H_0 if $t > t_a$, where $a = .05$.

In Example 11.1, we computed

$$\hat{\beta}_1 = .728 \qquad S_{xx} = 2.326 \qquad S_{xy} = 1.693.$$

In order to calculate s_ϵ^2 we must first compute S_{yy}. For the data, we find

$$S_{yy} = \Sigma y^2 - \frac{(\Sigma y)^2}{n} = 63.505 - \frac{(25.9)^2}{11} = 2.522.$$

Substituting, we have

$$s_\epsilon^2 = \frac{S_{yy} - \hat{\beta}_1 S_{xy}}{n - 2} = \frac{2.522 - (.728)(1.693)}{9} = \frac{1.289}{9} = .1432.$$

The test statistic is then

$$t = \frac{.728}{\sqrt{.1432/2.326}} = 2.93.$$

The t-value in Table 3 of Appendix 3 for $a = .05$ and df $= 9$ is 1.833. Since the computed value of t exceeds 1.833, we reject H_0 and conclude that the slope β_1 for the linear regression model $y = \beta_0 + \beta_1 x + \epsilon$) is positive, and hence that the variable x (high-school GPA) is useful in predicting y (a college freshman's GPA). ∎

EXAMPLE   12.7   The laboratory of a hospital participating in the clinical trial of an antibiotic drug had to be validated to see that the laboratory personnel could accurately assay blood samples "spiked" with fixed amounts of the antibiotic. The validation consisted of the following experiment. Thirteen spiked samples (with amounts known only to the study investigator) were sent to the laboratory to be assayed for the amount of the antibiotic present. The results of the validation experiment are shown in Table 12.2. (*Note:* The spiked samples with known amounts added were supplied in blind fashion to the laboratory.) The amounts found are the assay results found by the hospital laboratory.

TABLE 12.2
Amounts of Antibiotic for
Example 12.7

Amount Added (μg/ml), x	Amount Found (μg/ml), y
0	0
5	4.5
5	5.0
5	4.8
10	8.9
10	8.9
10	8.9
20	17.0
20	18.2
20	15.4
40	32.6
40	36.1
40	31.5

a   Plot the sample data.

b   Fit these data to a linear regression model.

c   Test the null hypothesis $H_0: \beta_1 = 1$ versus $H_a: \beta_1 \neq 1$. Give the p-value for your test.

Solution   a   A plot of the sample data is shown in Figure 12.3.

b   For the data, it can be shown that $\hat{\beta}_1 = .822$, with a standard error of $s_\epsilon / \sqrt{S_{xx}} = .024$; the estimate of β_0 is $\hat{\beta}_0 = .529$, with a standard error of .537. The linear regression equation is then

$$\hat{y} = .529 + .822x.$$

FIGURE 12.3

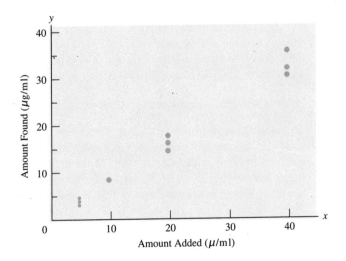

c   The statistical test for $H_0\colon \beta_1 = 1$ is a slight variation of the test for $H_0\colon \beta_1 = 0$. The test statistic for this variation is

$$t = \frac{\hat{\beta}_1 - \beta_{10}}{s_\epsilon/\sqrt{S_{xx}}},$$

where β_{10} is the hypothesized value of β_1 under H_0. The test of $H_0\colon \beta_1 = 1$ is shown here:

$H_0\colon\ \beta_1 = 1$
$H_a\colon\ \beta_1 \neq 1$
T.S.: $t = \dfrac{.822 - 1}{.024} = -7.42$
R.R.: Based on df $= 11$ and Table 3 of Appendix 3, the p-value for the test result is $p < .001$. ∎

Although used less often in experimental situations, a statistical test about β_0 follows the same format as that for β_1. The details are shown here.

A Statistical Test About the Intercept β_0

$H_0\colon\ \ \beta_0 = 0$
$H_a\colon\ \ $ For a one-tailed test:

  1   $\beta_0 > 0$
  2   $\beta_0 < 0$

For a two-tailed test:

  3   $\beta_0 \neq 0$

T.S.: $t = \dfrac{\hat{\beta}_0}{s_\epsilon\sqrt{\dfrac{1}{n} + \dfrac{\bar{x}^2}{S_{xx}}}}$ where $s_\epsilon^2 = \dfrac{S_{yy} - \hat{\beta}_1 S_{xy}}{n - 2}$

and $S_{yy} = \Sigma y^2 - (\Sigma y)^2/n$

R.R.: For a given value of α, for df $= (n - 2)$, and for a one-tailed test:

  1   Reject H_0 if $t > t_a$ with $a = \alpha$.
  2   Reject H_0 if $t < -t_a$ with $a = \alpha$.

For a given value of α, for df $= (n - 2)$, and for a two-tailed test:

  3   Reject H_0 if $|t| > t_a$ with $a = \alpha/2$.

Exercises

Basic Techniques

12.1   Consider the data shown here.

  a   Compute Σx, Σy, Σx^2, Σy^2, and Σxy.

  b   Use these computations to find least squares estimates for β_0 and β_1, and an estimate of σ_ϵ^2.

  c   Give the standard errors for $\hat{\beta}_0$ and $\hat{\beta}_1$.

x	y
1	9
1	10
2	10
2	11
3	12
3	12
4	11
4	13
5	14
5	15

12.2   Refer to Exercise 12.1.

  a   Give all parts for a two-tailed statistical test of $H_0: \beta_1 = 0$ based on $\alpha = .05$.

  b   Conduct the test and draw a conclusion.

  c   Give the p-value for the test in part b.

12.3   A manufacturer wishes to examine the relationship between different concentrations of pectin (0%, 1.5%, and 3% by weight) on the firmness of canned sweet potatoes after storage in a controlled 25°C environment. The sample data for six cans are repeated here.

Firmness (y)	50.5	46.8	62.3	67.7	80.1	79.2
Concentration of Pectin (x)	0	0	1.5	1.5	3.0	3.0

  a   Obtain the least squares estimates for the parameters in the model $y = \beta_0 + \beta_1 x + \varepsilon$.

  b   Obtain an estimate of σ_ϵ^2.

  c   Give the standard error of $\hat{\beta}_1$.

12.4   Refer to Exercise 12.3. Perform a statistical test of the null hypothesis that there is no linear relationship between the concentration of pectin and the firmness of canned sweet potatoes after 30 days of storage at 25°C. Give the p-value for this test, and draw conclusions.

Applications

12.5   A biologist is interested in studying the growth rate of a bacteria culture over a period of time. In a laboratory experiment, five different bacterial cultures were chosen. One culture was randomly selected and assigned to an incubation time of 1 hour; one to an incubation time of 3 hours; and one each to the incubation times 5, 7, and 9 hours. The growth rate y was measured on each culture after the required incubation period. Let x denote the incubation time.

  a   Use the sample data of the accompanying table and the method of least squares to obtain the regression line

$$\hat{y} = \hat{\beta}_0 + \hat{\beta}_1 x.$$

b   Conduct a test of significance to determine whether there is a linear relationship between the mean growth rate and time. Use $\alpha = .05$.

Incubation Time (x)	Growth Rate (y)
1	10.0
3	10.3
5	12.2
7	12.6
9	13.9

12.6   An experiment was conducted to examine the relationship between the weight gain of chickens whose diets were supplemented by different amounts of amino acid lysine and the amount of lysine ingested. Since the percentage of lysine is known and we can monitor the amount of feed consumed, we can determine the amount of lysine eaten. A random sample of twelve 2-week-old chickens was selected for the study. Each was caged separately and allowed to eat at will from feed composed of a base supplemented with lysine. The sample data summarizing weight gains and amounts of lysine eaten over the test period are given here. (In the data, y represents weight gain in grams, and x represents the amount of lysine ingested in grams.)

a   Plot the data in a scatter diagram. Does a linear model seem appropriate?

b   Fit the linear regression model $y = \beta_0 + \beta_1 x + \epsilon$.

Chick	y	x	Chick	y	x
1	14.7	.09	7	17.2	.11
2	17.8	.14	8	18.7	.19
3	19.6	.18	9	20.2	.23
4	18.4	.15	10	16.0	.13
5	20.5	.16	11	17.8	.17
6	21.1	.23	12	19.4	.21

12.7   Refer to Exercise 12.6.

a   Compute an estimate of σ_ϵ^2.

b   Identify the standard error of $\hat{\beta}_1$.

c   Conduct a statistical test of the research hypothesis that for this diet preparation and length of study, there is a direct (positive) linear relationship between weight gain and the amount of lysine eaten.

12.8   Refer to Exercise 12.6. Use the sample data to construct a 95% confidence interval for the intercept β_0. Does the interval include zero as a possible value? Does the validation experiment conform to theory in regard to the intercept?

12.9   Refer to Exercises 11.6, 11.7, and 11.13. Conduct a statistical test to show that the slope of the population regression line relating public education expenditure per student for a state to the average salary for teachers in the state is positive. Use $\alpha = .05$.

12.10   Research in dentistry over the past 20 years has indicated that plaque from different locations in the mouth can differ in chemical composition. Since the quantity of plaque at a given site might be quite small, it is necessary to have a sensitive procedure to permit study of the chemical composition of plaque. One such procedure relates the DNA content (one important chemical component) of plaque to the weight of plaque.

In order to study the relationship between weight of plaque and DNA content, ten male volunteers (ages 18–20) were selected at random from a group of volunteers. Over a four-day period each person consumed his normal diet, supplemented by 30 grams of sucrose per day. No tooth brushing was allowed. The four-day accumulation of plaque for each person was weighed and analyzed for DNA content. These sample data are summarized in the accompanying table.

Person	Plaque Weight (mg) x	DNA (μg) y
1	42.7	260
2	52.3	303
3	24.6	175
4	33.4	214
5	41.8	226
6	36.7	246
7	27.0	181
8	47.3	251
9	31.4	154
10	33.9	247

a Graph these data in a scatter plot.

b Use the EXECUSTAT computer output shown here to determine the least squares regression line.

Simple Regression Analysis for W161

Linear model: Y = 62.8061 + 4.38949*X

Table of Estimates

	Estimate	Standard Error	t Value	P Value
Intercept	62.8061	35.6907	1.76	0.1165
Slope	4.38949	0.938097	4.68	0.0016

R-squared = 73.24%
Correlation coeff. = 0.856
Standard error of estimation = 24.8794
Durbin-Watson statistic = 2.17969
Mean absolute error = 17.3348

Analysis of Variance

Source	Sum of Squares	D.F.	Mean Square	F-Ratio	P Value
Model	13552.2	1	13552.2	21.89	0.0016
Error	4951.86	8	618.983		
Total (corr.)	18504.1	9			

X	Y	STRESID	FITS	RESIDS
42.7	260	0.42423	250.237	9.7628
52.3	303	0.56462	292.376	10.6237
24.6	175	0.20571	170.787	4.2125
33.4	214	0.19640	209.415	4.5850
41.8	226	-0.87484	246.287	-20.2867
36.7	246	0.93645	223.900	22.0997
27.0	181	-0.01491	181.322	-0.3223
47.3	251	-0.90031	270.429	-19.4289
31.4	154	-2.02882	200.636	-46.6360
33.9	247	1.51177	211.610	35.3903

12.11   Refer to Exercise 12.10. If the least squares prediction equation is $\hat{y} = 62.81 + 4.39x$, test to see whether there is a significant linear relationship between the DNA content and the plaque weight; that is, use the sample data to test whether the population slope β_1 differs from zero. Use $\alpha = .05$. (*Note:* $S_{yy} = 18{,}504.1$, $S_{xy} = 3{,}087.43$, and $S_{xx} = 703.37$.)

12.12   Use the output of Exercise 12.10 to compare the results of your test in Exercise 12.11.

12.3 Inferences About $E(y)$

The methods of previous sections can be expanded to include inferences about the average value of y for a given setting of the independent variable. For example, in evaluating the effects of different levels of advertising expenditure x on sales y, we may wish to estimate the average sales per month for a given level of expenditure x. The estimate of $E(y)$ for a specific setting of x can be obtained by evaluating the prediction equation

$$\hat{y} = \hat{\beta}_0 + \hat{\beta}_1 x$$

sampling distribution for $\hat{y}$

at that setting. It can be shown that, in repeated sampling at a particular setting of x, the **sampling distribution for $\hat{y}$** has a mean

$$E(y) = \beta_0 + \beta_1 x$$

and standard error given by

$$s_\epsilon \sqrt{\frac{1}{n} + \frac{(x - \bar{x})^2}{S_{xx}}}.$$

Again assuming that the ϵ_is are normally distributed, a general confidence interval for $E(y)$ is given by the formula that follows.

Confidence Interval for $E(y)$

$$\hat{y} \pm ts_\epsilon \sqrt{\frac{1}{n} + \frac{(x - \bar{x})^2}{S_{xx}}},$$

where

$$s_\epsilon^2 = \frac{\text{SSE}}{n - 2}$$

and the t-value is based on df $= n - 2$.

E X A M P L E   12.8   Use the data of Example 12.2 to give a 90% confidence interval for the mean corn yield when 5 lb of fertilizer is applied to a plot.

Solution   The prediction equation in Example 12.2 was

$$\hat{y} = 10.10 + 1.15x,$$

where x is the amount of fertilizer applied. For our example we have $x = 5$, so

$$\hat{y} = 10.10 + 1.15(5) = 15.85$$

The standard error of $\hat{y}$ can be computed by using $S_{xx} = 20$, $s_\epsilon = 0.84$, $\bar{x} = 4$, and $n = 10$. The t-value in Table 3 of Appendix 3 for $a = .05$ and df $= n - 2 = 8$ is 1.86. Hence, the appropriate confidence interval for the average corn yield per plot when 5 lb of fertilizer is applied is

$$15.85 \pm 1.86(.84)\sqrt{\frac{1}{10} + \frac{(5-4)^2}{20}} \quad \text{or} \quad 15.85 \pm .61;$$

that is, the interval extends from 15.24 to 16.46.   ▪

E X A M P L E   12.9   In Example 12.8 we constructed a 90% confidence interval for the mean corn yield when 5 lb of fertilizer is applied. Use the same sample data to construct a 90% confidence interval on $E(y)$ for any specific value of fertilizer in the range from 2 to 6. Graph your results.

Solution   Using the results from Example 12.8, we can compute that $\hat{y} = 10.10 + 1.15x$, $s_\epsilon = .84$, and

$$\sqrt{\frac{1}{n} + \frac{(x - \bar{x})^2}{S_{xx}}} = \sqrt{.1 + \frac{(x - 4)^2}{20}}.$$

Our 90% confidence interval for $E(y)$, then, is of the form

$$\hat{y} \pm 1.86(.85)\sqrt{.1 + \frac{(x - 4)^2}{20}}.$$

All we need to do is substitute a specific value of x in this form to determine a confidence interval. For fertilizer settings of 2, 3, 4, 5, and 6, the 90% confidence limits are given in Table 12.3.

T A B L E 12.3
Confidence Interval Estimates for Different Fertilizer Settings x for Example 12.9

x	90% Confidence Interval
2	11.54 to 13.26
3	12.94 to 14.16
4	14.21 to 15.19
5	15.24 to 16.46
6	16.14 to 17.86

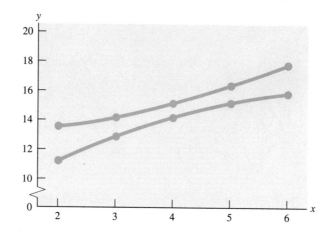

Plotting the endpoints of the confidence intervals and connecting the points, we get the general 90% confidence interval on $E(y)$ for any value of x between 2 and 6. The graph in Figure 12.4 displays 90% confidence bands for $E(y)$. Notice how the width of the confidence interval (the vertical distance between the two curves of the graph) varies for different values of x. Since the confidence width is narrower for the "central" values of x, it follows that $E(y)$ is estimated more precisely for values of x in the center of the experimental region. The widening of the gap between the bands at the extremities of the experimental region indicates that it would be unwise to extrapolate [try to estimate $E(y)$] beyond the region of experimentation.   ▪

A statistical test about $E(y)$ for a given setting of the independent variable x in linear regression can also be formulated using the test procedure shown here.

Statistical Test About $E(y)$

H_0: $E(y) = \mu_0$ (where μ_0 is specified)

H_a: For a one-tailed test:

    1 $E(y) > \mu_0$

    2 $E(y) < \mu_0$

    For a two-tailed test:

    3 $E(y) \neq \mu_0$

T.S.: $t = \dfrac{\hat{y} - \mu_0}{s_\epsilon \sqrt{\dfrac{1}{n} + \dfrac{(x - \bar{x})^2}{S_{xx}}}}$

where $\quad s_\epsilon^2 = \dfrac{S_{yy} - \hat{\beta}_1 S_{xy}}{n-2}$

R.R.: For a given value of α, for df $= n - 2$, and a one-tailed test:

1   Reject H_0 if $t > t_a$,   with $a = \alpha$.

2   Reject H_0 if $t < -t_a$,   with $a = \alpha$.

For a given value of α, for df $= n - 2$, and a two-tailed test:

3   Reject H_0 if $|t| > t_a$,   where $a = \alpha/2$.

E X A M P L E   12.10   An experiment was run to examine the growth rate of a particular type of bacteria. The growth rate y was determined for two different cultures at five equally spaced time intervals (1, 2, 3, 4, and 5 hours past culture seeding), as shown in Table 12.4.

T A B L E 12.4
Bacterial Growth Rate for Two
Cultures at Five Hourly Time
Intervals, for Example 12.10

Sample	Time 1	2	3	4	5
Culture 1	8.0	9.0	9.1	10.2	10.4
Culture 2	8.5	9.2	9.3	9.8	10.1

a   Use the SAS computer output that follows on the next page to determine the least squares fit to the linear regression model $y = \beta_0 + \beta_1 x + \epsilon$.

b   Conduct a test of $H_0: E(y) = 9.5$ when $x = 3.5$. Use $\alpha = .05$ for a two-tailed test.

Solution   a   By examining the computer output, we see that the least squares fit is

$$\hat{y} = 7.89 + 0.49x,$$

where y is growth rate and x is time.

b   The parts of the statistical test are shown here:

H_0: $E(y) = 9.5$

H_a: $E(y) \neq 9.5$

T.S.: $t = \dfrac{\hat{y} - \mu_0}{s_\epsilon \sqrt{\dfrac{1}{n} + \dfrac{(x - \bar{x})^2}{S_{xx}}}}$

R.R.: For $\alpha = .05$ and df $= n - 2 = 8$, the desired t-value from Table 3 of Appendix 3 is 2.306; so we will reject H_0 if $|t| > 2.306$.

SAS Output

Linear Regression
General Linear Models Procedure

Dependent Variable: RATE   GROWTH RATE

Source	DF	Sum of Squares	Mean Square	F Value	Pr > F
Model	1	4.80200000	4.80200000	70.88	0.0001
Error	8	0.54200000	0.06775000		
Corrected Total	9	5.34400000			

R-Square	C.V.	Root MSE	RATE Mean
0.898578	2.780858	0.260288	9.36000000

Source	DF	Type I SS	Mean Square	F Value	Pr > F
TIME	1	4.80200000	4.80200000	70.88	0.0001

Source	DF	Type III SS	Mean Square	F Value	Pr > F
TIME	1	4.80200000	4.80200000	70.88	0.0001

Parameter	Estimate	T for H0: Parameter=0	Pr > |T|	Std Error of Estimate
INTERCEPT	7.890000000	40.87	0.0001	0.19303497
TIME	0.490000000	8.42	0.0001	0.05820223

Observation	Observed	Predicted Residual	Lower 95% CLM Upper 95% CLM
1	8.00000000	8.38000000	8.05123956
		-0.38000000	8.70876044
2	8.50000000	8.38000000	8.05123956
		0.12000000	8.70876044
3	9.00000000	8.87000000	8.63753127
		0.13000000	9.10246873
4	9.20000000	8.87000000	8.63753127
		0.33000000	9.10246873
5	9.10000000	9.36000000	9.17019007
		-0.26000000	9.54980993
6	9.30000000	9.36000000	9.17019007
		-0.06000000	9.54980993
7 *	.	9.60500000	9.40367617
		.	9.80632383
8	10.20000000	9.85000000	9.61753127
		0.35000000	10.08246873
9	9.80000000	9.85000000	9.61753127
		-0.05000000	10.08246873
10	10.40000000	10.34000000	10.01123956
		0.06000000	10.66876044
11	10.10000000	10.34000000	10.01123956
		-0.24000000	10.66876044

SAS Output (Continued)

General Linear Models Procedure

* Observation was not used in this analysis

```
Sum of Residuals                              -0.00000000
Sum of Squared Residuals                       0.54200000
Sum of Squared Residuals - Error SS            0.00000000
Press Statistic                                0.88400885
First Order Autocorrelation                   -0.20885609
Durbin-Watson D                                2.04501845
```

Using the least squares equation, we find that the predicted value of y when $x = 3.5$ is

$$\hat{y} = 7.89 + 0.49(3.5) = 9.605.$$

It can be shown that $S_{xx} = 20$ and $s_\epsilon = 0.26$ (see ROOT MSE in the output). When we substitute these values into the test statistic, we find that

$$t = \frac{9.605 - 9.5}{0.26\sqrt{\dfrac{1}{10} + \dfrac{(3.5 - 3)^2}{20}}},$$

which simplifies to $t = 1.20$. Since $t = 1.20$ does not exceed $t = 2.306$, we do not have sufficient evidence to conclude that $E(y)$ differs from 9.5 when $x = 3.5$. ▪

12.4 Predicting y for a Given Value of x

In Section 12.3 we were concerned with estimating the expected value of y for a given value of x. But suppose that, after obtaining a least squares prediction equation for the general linear model, we would like to predict the actual value of y (say, the next measurement) for a given value of the independent variable x. This problem differs from the problem discussed in the previous section in that we do not want to estimate the average value of y for a given value of x; rather, we wish to predict what a particular observation will be for that same setting of x.

We still use the least squares equation $\hat{y}$ as our predictor, but the corresponding interval about the observation y is called a *prediction interval*. (Prediction intervals are constructed about variables, whereas confidence intervals are constructed about parameters.)

General $100(1 - \alpha)\%$ Prediction Interval

$$\hat{y} \pm t_{\alpha/2} s_\epsilon \sqrt{1 + \frac{1}{n} + \frac{(x - \bar{x})^2}{S_{xx}}},$$

where

$$s_\epsilon^2 = \frac{\text{SSE}}{n - 2}$$

and $t_{\alpha/2}$ is based on df $= n - 2$.

Notice the similarity between the confidence interval for $E(y)$ and the prediction interval for the variable y. The only difference is that the prediction interval has a 1 added to the quantity under the square root sign. This makes the interval wider, to account for the fact that it predicts a variable (future value of y) rather than a constant $E(y)$.

E X A M P L E   12.11   Use the data of Example 12.2 to predict the actual crop yield for a plot fertilized with 5 lb of fertilizer. Place a 90% prediction interval about the actual value of y.

Solution   From our previous work in Example 12.8, we know that the predicted value of y (using $\hat{y}$) at $x = 5$ is $\hat{y} = 15.85$. Moreover, for $x = 5$, $\bar{x} = 4$, $n = 10$, and $S_{xx} = 20$,

$$\frac{1}{n} + \frac{(x - \bar{x})^2}{S_{xx}} = .15.$$

The corresponding t-value for $a = .05$ and df $= n - 2 = 8$ is 1.86. Hence, the 90% prediction interval is

$$15.85 \pm 1.86(.84)\sqrt{1 + .15} \qquad \text{or} \qquad 15.85 \pm 1.68;$$

that is, the interval extends 14.17 to 17.53.

Notice that this interval is almost three times wider than the corresponding confidence interval for $E(y)$ in Example 12.8. This is to be expected, since here we are placing an interval about a quantity that may vary, whereas in Example 12.8 we were placing an interval about $E(y)$, which cannot vary. Since both intervals are 90% intervals, the prediction interval must be wider to encompass y in the same fraction of intervals (.90) in repeated sampling.   ▪

E X A M P L E   12.12

a   Refer to the data of Example 12.2. Construct a general 90% prediction interval for y when x takes the values 2, 3, 4, 5, and 6.

b   Graph your results to show a 90% prediction band for y when $2 \le x \le 6$.

c   On the graph of part b, superimpose the graph of the 90% confidence band for $E(y)$.

Solution   a   From previous calculations, $\hat{y} = 10.10 + 1.15x$, $s_\epsilon = .84$, and

$$\sqrt{1 + \frac{1}{n} + \frac{(x - \bar{x})^2}{S_{xx}}} = \sqrt{1 + \frac{1}{10} + \frac{(x - 4)^2}{20}}.$$

Hence a general 90% prediction interval is of the form

$$\hat{y} \pm 1.85(.84)\sqrt{1.1 + \frac{(x - 4)^2}{20}}.$$

Substituting the values $x = 2, 3, 4, 5$, and 6 into this form, we obtain the intervals given in Table 12.5.

TABLE 12.5
Prediction Interval Estimates for Different Fertilizer Settings x for Example 12.12

x	90% Prediction Interval
2	10.618 to 14.182
3	11.874 to 15.226
4	13.061 to 16.339
5	14.174 to 17.526
6	15.218 to 18.782

b and c   Plotting the endpoints of the prediction intervals and connecting the dots, we obtain the 90% prediction bands, shown by the solid lines in Figure 12.5. The dotted lines indicate the corresponding 90% confidence bands for $E(y)$. Notice that the prediction bands are wider than the corresponding confidence bands, reflecting the fact that we are predicting the value of a random variable rather than estimating a parameter.

FIGURE 12.5
90% Prediction and Confidence Bands for y and $E(y)$ for Example 12.12

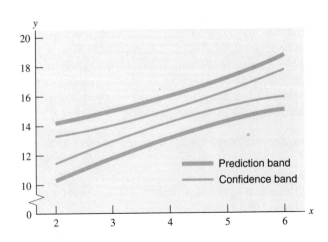

Studying Figure 12.5 and the formulas for these confidence and prediction intervals should suggest factors that may influence the precision of our confidence
intervals for $E(y)$ and our prediction intervals for y. First, the plus or minus terms in
the confidence interval and prediction interval involve $t_{\alpha/2}$, s_ϵ, n, $(x - \bar{x})^2$, and S_{xx}.
Forgetting about $t_{\alpha/2}$ and s_ϵ for the moment, we observe that the width of these intervals will decrease as n increases, $(x - \bar{x})^2$ decreases, and S_{xx} increases. Obviously, if
we take more observations, n increases. The quantity $(x - \bar{x})^2$ can be made smaller
by making predictions at values of x closer to the mean of the x-values $(\bar{x})$. This is
shown in Figure 12.5, where the confidence and prediction intervals are wider for
values of x farther away from the center of the region $\bar{x} = 4$. Finally, we can increase
$S_{xx} = \Sigma(x_i - \bar{x})^2$ and hence improve the confidence and prediction intervals based
on our model by increasing the spread of the x-values in our sample. This is certainly
an important point when the experimenter has control of the x-values. However, we
can reach a point of diminishing returns. The width of the confidence interval for
$E(y)$ and the prediction interval for y are adequate measures of precision, assuming
that the model adequately fits the data. If the x-values are spread too far, making
S_{xx} quite large (say, the values $x = 1, 5, 6, 7$, and 11 in Example 12.12), a linear
regression model may no longer adequately describe the relation between x and y,
thus rendering invalid the confidence interval for $E(y)$ and the prediction interval for
y discussed in this section.

It is important to understand the factors affecting the precision of the confidence
and prediction intervals based on a linear regression model. Clearly, these methods
should not be applied blindly. Plot the data; see how well the model fits the data; plot
the confidence and prediction intervals to see the penalty associated with predictions
away from $\bar{x}$; and so on.   ▪

Exercises

Basic Techniques

12.13   Refer to Exercise 12.6. Estimate the mean weight gain for chickens fed a diet supplemented
with lysine if .19 grams of lysine were ingested over a study period of the same duration. Use
a 95% confidence interval.

12.14   Refer to Exercise 12.13. Construct a 95% prediction interval for the weight gain of a chick
chosen at random and observed to ingest .19 grams of lysine. Compare this interval to the
confidence interval of Exercise 12.13.

Application

12.15   The extent of disease transmission is affected greatly by the viability of infectious organisms
suspended in the air. Because of the infectious nature of the disease under study, the viability
of these organisms must be studied in an airtight chamber. One way to do this is to disperse
an aerosol cloud, prepared from a solution containing the organisms, into the chamber. The
biological recovery at any particular time is the percentage of the total number of organisms
suspended in the aerosol that remain viable. The accompanying data are the biological recovery

percentages computed from 13 different aerosol clouds. For each of the clouds, recovery percentages were determined at different times.

Cloud	Time (in Minutes), x	Biological Recovery, %
1	0	70.6
2	5	52.0
3	10	33.4
4	15	22.0
5	20	18.3
6	25	15.1
7	30	13.0
8	35	10.0
9	40	9.1
10	45	8.3
11	50	7.9
12	55	7.7
13	60	7.7

For the least squares equation $\hat{y} = \hat{\beta}_0 + \hat{\beta}_1 x$, estimate the mean log biological recovery percentage at 30 minutes, using a 95% confidence interval.

12.16   Using the data of Exercise 12.15, construct a 95% prediction interval for the log biological recovery percentage at 30 minutes. Compare your result to the confidence interval on $E(y)$ of Exercise 12.15.

 12.17   A chemist is interested in determining the weight loss y of a particular compound as a function of the amount of time the compound is exposed to the air. The data in the following table give the weight losses associated with $n = 12$ settings of the independent variable, exposure time.

Weight Loss y (in Pounds),	Exposure Time (in Hours)
4.3	4
5.5	5
6.8	6
8.0	7
4.0	4
5.2	5
6.6	6
7.5	7
2.0	4
4.0	5
5.7	6
6.5	7

a   Find the least squares prediction equation for the model

$$y = \beta_0 + \beta_1 x + \epsilon.$$

b   Test $H_0: \beta_1 = 0$; give the p-value for $H_a: \beta_1 > 0$; and draw conclusions.

12.5 Statistical Tests About ρ and ρ_S

Before concluding our discussion of linear regression, we present a test of significance for ρ. We may be interested in a test of the null hypothesis $H_0: \rho = 0$ (that is, there is no linear correlation between x and y). Although the results and conclusions drawn from this test are identical to those for $H_0: \beta_1 = 0$, we present the test of ρ for completeness. The assumptions required for this test are identical to those for a test of $H_0: \beta_1 = 0$.

Statistical Test About ρ

$H_0:$ $\rho = 0$

$H_a:$ For a one-tailed test:

    1 $\rho > 0$

    2 $\rho < 0$

    For a two-tailed test:

    3 $\rho \neq 0$

T.S.: $\dfrac{\hat{\rho}}{\sqrt{\dfrac{1 - \hat{\rho}^2}{n - 2}}}$

R.R.:   For a specified value of α, for df $= (n - 2)$, and a one-tailed test:

    1 Reject H_0 if $t > t_a$,   where $a = \alpha$.

    2 Reject H_0 if $t < -t_a$,   where $a = \alpha$.

    For a specified value of α, for df $= (n - 2)$, and a two-tailed test:

    3 Reject H_0 if $|t| > t_a$,   where $a = \alpha/2$.

In a similar fashion, when dealing with ranks of two variables, we can test whether there is a relationship (positive or negative) between two variables by using the following procedure.

Large-sample Rank Correlation Test for ρ_S

$H_0:$ $\rho_S = 0$ (that is, no association exists between the ranks x and y).

$H_a:$ For a one-tailed test:

    1 $\rho_S > 0$

    2 $\rho_S < 0$

For a two-tailed test:

3 $\rho_S \neq 0$

T.S.: $z = \hat{\rho}_S \sqrt{n-1}$

R.R.: For $\alpha = .05(.01)$ and for a one-tailed test:

    1 Reject H_0 if $z > 1.645$ (2.33).

    2 Reject H_0 if $z < -1.645$ (-2.33).

For $\alpha = .05(.01)$ and for a two-tailed test:

    3 Reject H_0 if $|z| > 1.96$ (2.58).

Note: The number n of pairs of ranks must be 10 or more.

E X A M P L E   12.13   Each of 10 judges was asked to rate an experimental mattress according to both firmness and comfort, with scores to be assigned in the range from 0 to 7.0. Higher scores indicate greater firmness or greater comfort. The results of the study are presented in the second and fourth columns of Table 12.6, and the ranked data for firmness and comfort are given in the third and sixth columns of the table.

T A B L E 12.6
Results of the Experimental
Mattress Study for Example
12.13

Judge	Firmness	Rank (x)	Comfort	Rank (y)
1	2.5	2	5.0	7
2	3.0	4	4.7	6
3	5.0	8	3.0	1.5
4	4.0	7	4.2	4
5	3.5	6	4.5	5
6	2.0	1	3.0	1.5
7	3.3	5	5.9	9
8	5.2	9	5.5	8
9	2.8	8	8.2	3
10	5.8	10	6.1	10

  a   Calculate the rank correlation coefficient.

  b   Use these data to test the hypothesis that there is a positive association between pairs of firmness scores and comfort scores for the experimental mattress data. Use the rank correlation coefficient in your test, and set $\alpha = .05$.

Solution   It is convenient to construct a table of ranks, squares, and crossproducts to aid in our calculations (see Table 12.7).

  a   Using the data shown in the preceding table, we can determine the values of S_{xx}, S_{yy}, and S_{xy}, and hence we can obtain the value of $\hat{\rho}_S$, by using these values:

$$S_{xx} = \Sigma x^2 - \frac{(\Sigma x)^2}{n} = 385 - \frac{(55)^2}{10} = 82.5$$

TABLE 12.7
Computations for Example
12.13

Judge	Firmness	Rank (x)	Comfort	Rank (y)	x^2	y^2	xy
1	2.5	2	5.0	7	4	49	14
2	3.0	4	4.7	6	16	36	24
3	5.0	8	3.0	1.5	64	2.25	12
4	4.0	7	4.2	4	49	16	28
5	3.5	6	4.5	5	36	25	30
6	2.0	1	3.0	1.5	1	2.25	1.5
7	3.3	5	5.9	9	25	81	45
8	5.2	9	5.5	8	81	64	72
9	2.8	3	3.2	3	9	9	9
10	5.8	10	6.1	10	100	100	100
Total		55		55	385	384.5	335.5

$$S_{yy} = \Sigma y^2 - \frac{(\Sigma y)^2}{n} = 384.5 - \frac{(55)^2}{10} = 82.0$$

$$S_{xy} = \Sigma xy - \frac{(\Sigma x)(\Sigma y)}{n} = 335.5 - \frac{(55)(55)}{10} = 33.0.$$

Thus, for

$$\hat{\rho}_S = \frac{S_{xy}}{\sqrt{S_{xx}S_{yy}}},$$

we have

$$\hat{\rho}_S = \frac{33}{\sqrt{(82.5)(82)}} = .40.$$

b   For $H_0: \rho_S = 0$ and $H_a: \rho > 0$, the test statistic is

$$z = \hat{\rho}_S\sqrt{n-1} = .40\sqrt{10-1} = .40(3) = 1.20.$$

The rejection region, for $\alpha = .05$, is

R.R.: Reject H_0 if $z > 1.645$.

Since $z = 1.20$, z does not fall in the rejection region, and we have insufficient evidence to conclude that there is a positive association between the ranks of firmness and comfort. Consequently, we have insufficient information to suggest that the actual firmness and comfort scores are related.   ∎

12.6 Using Computers to Help Make Sense of Data

The MINITAB, SAS, and EXECUSTAT outputs shown in Chapter 11, those shown here for the corn yield data of Table 12.1, and the output for Exercises 12.34 and 12.37 amply illustrate how useful computer programs can be at supplying the necessary calculations for the descriptive and inferential problems of linear regression and correlation presented in Chapters 11 and 12.

The MINITAB, SAS, and EXECUSTAT outputs that follow show the calculations necessary for drawing inferences about β_0 and β_1 based on the corn yield data of Table 12.1. Check these results against the hand-computed calculations we have done on these data throughout the chapter, and familiarize yourself with where to locate the computations in MINITAB, SAS, and EXECUSTAT needed for inferences in linear regression.

MINITAB Output

```
MTB > READ INTO C1 C2
DATA> 12    2
DATA> 13    2
DATA> 13    3
DATA> 14    3
DATA> 15    4
DATA> 15    4
DATA> 14    5
DATA> 16    5
DATA> 17    6
DATA> 18    6
DATA> END
     10 ROWS READ
MTB > NAME C1 'YIELD'
MTB > NAME C2 'FERTLIZR'
MTB > PLOT C1 VS C2


        18.0+                                          *
            -
   YIELD   -                                          *
            -
            -
        16.0+                                  *
            -
            -                          2
            -
            -
        14.0+                *                  *
            -
            -    *           *
            -
            -
        12.0+    *
            -
          --------+---------+---------+---------+---------+--------FERTLIZR
               2.40      3.20      4.00      4.80      5.60
```

MINITAB Output (continued)

```
MTB > REGRESS C1 ON 1, C2;
SUBC> DW.

The regression equation is
YIELD = 10.1 + 1.15 FERTLIZR

Predictor       Coef       Stdev     t-ratio        p
Constant      10.1000      0.7973      12.67      0.000
FERTLIZR       1.1500      0.1879       6.12      0.000

s = 0.8404      R-sq = 82.4%     R-sq(adj) = 80.2%

Analysis of Variance

SOURCE        DF          SS         MS        F         p
Regression     1        26.450     26.450    37.45    0.000
Error          8         5.650      0.706
Total          9        32.100

Unusual Observations
Obs.FERTLIZR      YIELD     Fit Stdev.Fit  Residual   St.Resid
  7     5.00     14.000    15.850    0.325    -1.850     -2.39R

R denotes an obs. with a large st. resid.

Durbin-Watson statistic = 2.30

MTB > STOP
```

SAS Output

```
OPTIONS NODATE NONUMBER PS=60 LS=78;
DATA RAW;
  INPUT YIELD FERTLIZR @@;
  LABEL YIELD = 'YIELD (BUSHELS)'
        FERTLIZR = 'FERTILIZER (POUNDS/PLOT)';
  CARDS;
  12 2 13 2 13 3 14 3 15 4 15 4 14 5 16 5 17 6 18 6
;
PROC PRINT N;
TITLE1 'RELATIONSHIP BETWEEN YIELD AND THE AMOUNT OF FERTILIZER';
TITLE2 'LISTING OF THE DATA';

PROC REG;
  MODEL YIELD=FERTLIZR;
TITLE2 'EXAMPLE OF PROC REG';
RUN;
```

SAS Output (continued)

RELATIONSHIP BETWEEN YIELD AND THE AMOUNT OF FERTILIZER
LISTING OF THE DATA

OBS	YIELD	FERTLIZR
1	12	2
2	13	2
3	13	3
4	14	3
5	15	4
6	15	4
7	14	5
8	16	5
9	17	6
10	18	6

N = 10

RELATIONSHIP BETWEEN YIELD AND THE AMOUNT OF FERTILIZER
EXAMPLE OF PROC REG

Model: MODEL1
Dependent Variable: YIELD   YIELD (BUSHELS)

Analysis of Variance

Source	DF	Sum of Squares	Mean Square	F Value	Prob>F
Model	1	26.45000	26.45000	37.451	0.0003
Error	8	5.65000	0.70625		
C Total	9	32.10000			

Root MSE	0.84039	R-square	0.8240
Dep Mean	14.70000	Adj R-sq	0.8020
C.V.	5.71692		

Parameter Estimates

Variable	DF	Parameter Estimate	Standard Error	T for H0: Parameter=0	Prob > |T|
INTERCEP	1	10.100000	0.79726094	12.668	0.0001
FERTLIZR	1	1.150000	0.18791620	6.120	0.0003

Variable	DF	Variable Label
INTERCEP	1	Intercept
FERTLIZR	1	FERTILIZER (POUNDS/PLOT)

EXECUSTAT Output

Simple Regression Analysis

Linear model: YIELD = 10.1 + 1.15*FERTLIZR

Table of Estimates

	Estimate	Standard Error	t Value	P Value
Intercept	10.1	0.797261	12.67	0.0000
Slope	1.15	0.187916	6.12	0.0003

R-squared = 82.40%
Correlation coeff. = 0.908
Standard error of estimation = 0.840387
Durbin-Watson statistic = 2.29912
Mean absolute error = 0.56

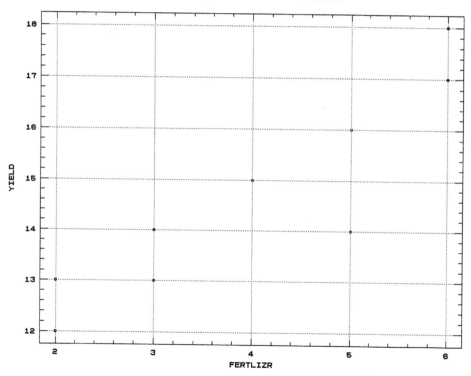

Scatterplot of YIELD vs FERTLIZR

We can also use computer software systems to generate confidence intervals for $E(y)$ and prediction intervals for y. The following MINITAB, SAS, and EXECU-STAT outputs provide a complete analysis of the weight loss data of Exercise 12.17. Locate the least squares prediction equation and the t test (with p-value) for $H_0: \beta_1 = 0$ versus $H_a: \beta_1 > 0$. Compare these results to those you obtained in Exercise 12.17. Then locate the 95% confidence limits for $E(y)$ and the 95% prediction limits for y for x-values (exposure times) of 4, 5, 6, and 7 hours. Notice that the prediction limits for y are not part of the SAS output—only confidence limits for $E(y)$.

MINITAB Output

```
MTB > READ INTO C1 C2
DATA> 4.3   4
DATA> 5.5   5
DATA> 6.8   6
DATA> 8.0   7
DATA> 4.0   4
DATA> 5.2   5
DATA> 6.6   6
DATA> 7.5   7
DATA> 2.0   4
DATA> 4.0   5
DATA> 5.7   6
DATA> 6.5   7
DATA> END
     12 ROWS READ
MTB > NAME C1 'WT_LOSS'
MTB > NAME C2 'TIME'

MTB > REGRESS C1 ON 1 C2 STRES C3 PRED C4;
SUBC> RESIDS C5;
SUBC> PREDICT 4;
SUBC> PREDICT 5;
SUBC> PREDICT 6;
SUBC> PREDICT 7.

The regression equation is
WT_LOSS = - 1.73 + 1.32 TIME

Predictor      Coef       Stdev     t-ratio       p
Constant     -1.733       1.165       -1.49    0.168
TIME          1.3167      0.2076       6.34    0.000

s = 0.8041     R-sq = 80.1%     R-sq(adj) = 78.1%

Analysis of Variance

SOURCE       DF         SS          MS         F        p
Regression    1      26.004      26.004     40.22    0.000
Error        10       6.465       0.647
Total        11      32.469

Unusual Observations
Obs.    TIME    WT_LOSS     Fit  Stdev.Fit  Residual   St.Resid
  9     4.00     2.000    3.533     0.388    -1.533     -2.18R

R denotes an obs. with a large st. resid.

    Fit   Stdev.Fit      95% C.I.           95% P.I.
  3.533      0.388    ( 2.668,  4.399)   ( 1.543,  5.523)
  4.850      0.254    ( 4.283,  5.417)   ( 2.971,  6.729)
  6.167      0.254    ( 5.600,  6.733)   ( 4.287,  8.046)
  7.483      0.388    ( 6.618,  8.349)   ( 5.493,  9.473)
```

MINITAB Output (Continued)

```
MTB > NAME C3 'STRES' C4 'PREDS' C5 'RESIDS'
MTB > PRINT C1 - C5

ROW   WT_LOSS   TIME     STRES      PREDS     RESIDS

  1      4.3      4     1.08898    3.53333    0.76667
  2      5.5      5     0.85213    4.85000    0.65000
  3      6.8      6     0.83028    6.16667    0.63333
  4      8.0      7     0.73388    7.48333    0.51667
  5      4.0      4     0.66286    3.53333    0.46667
  6      5.2      5     0.45884    4.85000    0.35000
  7      6.6      6     0.56809    6.16667    0.43333
  8      7.5      7     0.02367    7.48333    0.01667
  9      2.0      4    -2.17796    3.53333   -1.53333
 10      4.0      5    -1.11433    4.85000   -0.85000
 11      5.7      6    -0.61179    6.16667   -0.46667
 12      6.5      7    -1.39673    7.48333   -0.98333

MTB > STOP
```

SAS Output

```
OBS     WT_LOSS     TIME

  1       4.3         4
  2       5.5         5
  3       6.8         6
  4       8.0         7
  5       4.0         4
  6       5.2         5
  7       6.6         6
  8       7.5         7
  9       2.0         4
 10       4.0         5
 11       5.7         6
 12       6.5         7
 13        .          4
 14        .          5
 15        .          6
 16        .          7

            N = 16
```

General Linear Models Procedure

Number of observations in data set = 16

NOTE: Due to missing values, only 12 observations can be used in this
      analysis.

```
                    General Linear Models Procedure

Dependent Variable: WT_LOSS    WEIGHT LOSS (LBS)
                                     Sum of          Mean
Source                   DF          Squares         Square    F Value      Pr > F

Model                     1       26.00416667     26.00416667    40.22      0.0001
Error                    10        6.46500000      0.64650000
Corrected Total          11       32.46916667

               R-Square              C.V.        Root MSE        WT_LOSS Mean
               0.800888           14.59701       0.804052         5.50833333


Source                   DF        Type I SS     Mean Square    F Value      Pr > F
TIME                      1      26.00416667     26.00416667      40.22      0.0001

Source                   DF      Type III SS     Mean Square    F Value      Pr > F
TIME                      1      26.00416667     26.00416667      40.22      0.0001


                                            T for H0:     Pr > |T|    Std Error of
Parameter               Estimate         Parameter=0                    Estimate

INTERCEPT            -1.733333333           -1.49        0.1677        1.16518239
TIME                  1.316666667            6.34        0.0001        0.20760539


        Observation            Observed           Predicted        Lower 95% CLM
                                                  Residual         Upper 95% CLM

             1               4.30000000          3.53333333         2.66793184
                                                 0.76666667         4.39873483
             2               5.50000000          4.85000000         4.28346174
                                                 0.65000000         5.41653826
             3               6.80000000          6.16666667         5.60012840
                                                 0.63333333         6.73320493
             4               8.00000000          7.48333333         6.61793184
                                                 0.51666667         8.34873483
             5               4.00000000          3.53333333         2.66793184
                                                 0.46666667         4.39873483
             6               5.20000000          4.85000000         4.28346174
                                                 0.35000000         5.41653826
             7               6.60000000          6.16666667         5.60012840
                                                 0.43333333         6.73320493
             8               7.50000000          7.48333333         6.61793184
                                                 0.01666667         8.34873483
             9               2.00000000          3.53333333         2.66793184
                                                -1.53333333         4.39873483
            10               4.00000000          4.85000000         4.28346174
                                                -0.85000000         5.41653826
            11               5.70000000          6.16666667         5.60012840
                                                -0.46666667         6.73320493
            12               6.50000000          7.48333333         6.61793184
                                                -0.98333333         8.34873483
            13     *             .               3.53333333         2.66793184
                                                      .             4.39873483
            14     *             .               4.85000000         4.28346174
                                                      .             5.41653826
            15     *             .               6.16666667         5.60012840
                                                      .             6.73320493
            16     *             .               7.48333333         6.61793184
                                                      .             8.34873483

        * Observation was not used in this analysis
```

SAS Output (Continued)

```
Sum of Residuals                            -0.00000000
Sum of Squared Residuals                     6.46500000
Sum of Squared Residuals - Error SS          0.00000000
Press Statistic                             10.03092643
First Order Autocorrelation                  0.60849016
Durbin-Watson D                              0.54253674
```

EXECUSTAT Output

Simple Regression Analysis

Linear model: WT_LOSS = -1.73333 + 1.31667*TIME

Table of Estimates

	Estimate	Standard Error	t Value	P Value
Intercept	-1.73333	1.16518	-1.49	0.1677
Slope	1.31667	0.207605	6.34	0.0001

R-squared = 80.09%
Correlation coeff. = 0.895
Standard error of estimation = 0.804052
Durbin-Watson statistic = 0.542537
Mean absolute error = 0.638889

Table of Predicted Values

Row	TIME	Predicted WT_LOSS	95.00% Prediction Limits Lower	Upper	95.00% Confidence Limits Lower	Upper
1	4	3.53333	1.54372	5.52294	2.66794	4.39873
2	5	4.85	2.97101	6.72899	4.28346	5.41654
3	6	6.16667	4.28768	8.04565	5.60013	6.7332
4	7	7.48333	5.49372	9.47294	6.61794	8.34873

Refer to the user's manual for MINITAB, SAS, and EXECUSTAT if you want to become more acquainted with the details of these two software systems. Both systems can do much, much more than we have illustrated in this text.

Summary

The material in this chapter follows closely from that presented in Chapter 11, where we dealt with linear regression and correlation in a descriptive sense; there, we discussed the notion of relating a response y to a variable by using a regression equation, we presented an objective way to estimate the intercept and slope of the linear regression model $y = \beta_0 + \beta_1 x + \epsilon$, and we discussed the correlation coefficient and interpreted it as a measure of the strength of the *linear* relationship between two variables y and x. In this chapter we dealt with statistical inferences involving linear regression and correlation. First we presented the formula for computing s_ϵ^2, the estimate of σ_ϵ^2 (the error variance in linear regression). Using this estimate, we presented a statistical test and confidence interval for the slope β_1 and for $E(y)$, the expected value of y. All of these procedures followed the general format for a confidence interval or statistical test based on t.

Inferences based on multiple regression are beyond the scope of this text. For additional material on multiple regression, see Ott (1993) and Hildebrand & Ott (1991).

Key Terms

sampling distributions for $\hat{\beta}_0$ and $\hat{\beta}_1$

expected values for $\hat{\beta}_0$ and $\hat{\beta}_1$

sampling distributions for $\hat{y}$

Key Formulas

1 Expected value and standard error of the sampling distribution for $\hat{\beta}_1$ in linear regression:

$$\text{Expected value: } \beta_1$$
$$\text{Standard error: } \frac{\sigma_\epsilon}{\sqrt{S_{xx}}}$$

2 Estimate of σ_ϵ^2 (or σ_ϵ):

$$s_\epsilon^2 = \frac{\text{SSE}}{n-2} \quad \text{where} \quad \text{SSE} = \Sigma(y - \hat{y})^2 = S_{yy} - \frac{S_{xy}^2}{S_{xx}}$$

$$s_\epsilon = \sqrt{\frac{\text{SSE}}{n-2}}$$

3 General confidence interval for β_1:

$$\hat{\beta}_1 \pm t \frac{s_\epsilon}{\sqrt{S_{xx}}}$$

4 Statistical test for β_1:

$$H_0: \beta_1 = 0$$

$$\text{Test statistic: } t = \frac{\hat{\beta}_1}{\sqrt{s_\epsilon^2 / S_{xx}}}$$

5 Expected value and standard error of the sampling distribution for $\hat{\beta}_0$ in linear regression:

$$\text{Expected value: } \beta_0$$

$$\text{Standard error: } \sigma_\epsilon \sqrt{\frac{1}{n} + \frac{\bar{x}^2}{S_{xx}}}$$

6 General confidence interval for β_0:

$$\hat{\beta}_0 \pm t s_\epsilon \sqrt{\frac{1}{n} + \frac{\bar{x}^2}{S_{xx}}}$$

7 Statistical test for β_0:

$$H_0: \beta_1 = 0$$

$$\text{Test statistic: } t = \frac{\hat{\beta}_0}{s_\epsilon \sqrt{\frac{1}{n} + \frac{\bar{x}^2}{S_{xx}}}}$$

8 General confidence interval for $E(y)$:

$$\hat{y} \pm t s_\epsilon \sqrt{\frac{1}{n} + \frac{(x - \bar{x})^2}{S_{xx}}}$$

9 Statistical test for $E(y)$:

$$H_0: E(y) = \mu_0$$

$$\text{Test statistic: } t = \frac{\hat{y} - \mu_0}{s_\epsilon \sqrt{\frac{1}{n} + \frac{(x - \bar{x})^2}{S_{xx}}}}$$

10 General prediction interval for y:

$$\hat{y} \pm t s_\epsilon \sqrt{1 + \frac{1}{n} + \frac{(x - \bar{x})^2}{S_{xx}}}$$

11 Statistical test for ρ:

$$H_0: \rho = 0$$

$$\text{Test statistic } t = \frac{\hat{\rho}}{\sqrt{\frac{1 - \hat{\rho}^2}{n - 2}}}$$

12 Statistical text for ρ_S:

$$H_0: \rho_S = 0$$

$$\text{Test statistic: } z = \hat{\rho}_S \sqrt{n - 1}$$

Supplementary Exercises

12.18   Refer to the pulse rate data of Exercise 11.27. Use the data reproduced here to test the research hypothesis that a change in pulse rate y is linearly related to drug dose x. Use $\alpha = .05$.

Change in Pulse Rate (y)	20, 21, 19	16, 17, 17	15, 13, 14	8, 10, 8
Drug dose (x)	1.5, 1.5, 1.5	2.0, 2.0, 2.0	2.5, 2.5, 2.5	3.0, 3.0, 3.0

12.19   Refer to Exercise 12.18. Suppose that you had been asked to test the research hypothesis $H_a: \rho \neq 0$. What would your conclusion have been?

12.20   Crude birth rates and infant mortality rates for each of nine nations are shown here.

Nation	Crude Birth Rate	Infant Mortality Rate
Benin	49	149
Canada	15	12
Chile	22	38
Dominican Republic	37	96
Guyana	28	46
Haiti	42	130
Hong Kong	17	13
Japan	14	8
Mexico	33	70

a   Plot the sample data, using crude birth rate as x and infant mortality rate as y.

b   Compute the sample correlation coefficient.

12.21   Refer to Exercise 12.20. Assume that these data represent a random sample from all the nations of the world.

a   Use these data to test $H_0: \rho = 0$ versus $H_a: \rho \neq 0$.   **b**   Draw conclusions.

12.22   Given the following data relating x (gross margin) to y (total sales), answer parts a through c.

Gross Margin (in $1000s), x	Sales (in $1000s), y
15	20
19	38
20	25
22	28
24	31
27	40
23	35
21	24
20	25
5	7

a   Plot the data.

b   Compute the least squares estimates of β_0 and β_1 for the model $y = \beta_0 + \beta_1 x + \epsilon$.

c   Predict y when $x = 25$.

12.23   Refer to Exercise 12.22. Give a 95% confidence interval for β_1.

12.24   Refer to Exercise 12.22. Use these data to construct a 95% confidence interval for $E(y)$ when $x = 25$.

12.25 Refer to Exercise 12.22. Can you compute a confidence interval for $E(y)$ when $x = 40$? Any problems?

12.26 The amount of heat loss was examined for a new brand of thermal panes for different controlled outdoor temperatures. Nine panes were tested, with the results shown here.

Outdoor Temperature (in °F)	Heat Loss
20	86
20	80
20	77
40	68
40	74
40	65
60	43
60	33
60	38

  a Plot the data, with outdoor temperature on the x-axis and heat loss on the y-axis.

  b Use these data to obtain the prediction equation $\hat{y} = \hat{\beta}_0 + \hat{\beta}_1 x$.

12.27 Refer to Exercise 12.26. Use the sample data to test $H_0: \beta_1 = 0$ versus $H_a: \beta_1 < 0$. Set $\alpha = .05$.

12.28 Refer to Exercise 12.26. Conduct a statistical test of $H_0: E(y) = 55$ when $x = 50$, using a two-tailed test with $\alpha = .05$.

12.29 Earlier we examined the relationship between annual income and annual savings.

Annual Savings (in $1000s)	Annual Income (in $1000s)
1	36
2	39
2	42
5	45
5	48
6	51
7	54
8	56
7	59

  a Compute the sample correlation coefficient, and interpret its value.

  b Conduct a statistical test of $H_0: \rho = 0$ versus $H_a: \rho > 0$. Use $\alpha = .05$. Draw conclusions.

12.30 Refer to Example 11.2. Use the sample data shown here to conduct a two-tailed statistical test of $H_0: \rho = 0$, with $\alpha = .05$. Draw conclusions.

Observed Meter Reading (y)	Actual Flow Rate (x)
1.4	1
2.3	2
3.1	3
4.2	4
5.1	5
5.8	6
6.8	7
7.6	8
8.7	9
9.5	10

12.31 Refer to Exercise 12.30. Give a 95% confidence interval for $E(y)$ when $x = 5$. How would this interval change if we used a 90% confidence coefficient? Would a 95% confidence interval for $E(y)$ at $x = 9$ be wider or narrower than the corresponding interval at $x = 5$?

 12.32 In an earlier exercise, we computed the correlation coefficient between profit (y) and percent of operating capacity for a sample of twelve plants. If $\hat{\rho} = .81$, conduct a statistical test of $H_0: \rho = 0$ versus $H_a: \rho > 0$, using $\alpha = .05$.

 12.33 The advertising expense data from an earlier exercise are shown here. Use these data to give a 90% confidence interval for β_1, the slope of the linear regression model, $y = \beta_0 + \beta_1 x + \epsilon$.

Weekly Sales Volume (y)	Amount Spent on Advertising (x)
10.2	1.0
11.5	1.25
16.1	1.5
20.3	2.0
25.6	2.5
28.0	3.0

12.34 Refer to Exercise 12.33 and the SAS output shown here to find a 95% confidence interval for $E(y)$ when $x = 1.75$. Notice that confidence limits for $E(y)$ are given for all x-values in the sample data as well as for the specified x-value, 1.75. Locate the confidence limits for $E(y)$ when $x = 1.25$.

```
OPTIONS NODATE NONUMBER PS=60 LS=78;
DATA RAW;
  INPUT SPENT VOLUME @@;
  LABEL SPENT = 'AMOUNT SPENT ON ADVERTISING'
        VOLUME = 'WEEKLY SALES VOLUME';
  CARDS;
1.00    10.2
1.25    11.5
1.50    16.1
1.75     .
2.00    20.3
2.50    25.6
3.00    28.0
;
PROC PRINT N;
TITLE1 'RELATIONSHIP BETWEEN ADVERTISING EXPENSE';
TITLE2 'AND WEEKLY SALES VOLUME';
TITLE3 'LISTING OF DATA';

PROC REG;
  MODEL VOLUME = SPENT / P CLM;
TITLE3 'REGRESSION ANALYSIS';
RUN;
```

RELATIONSHIP BETWEEN ADVERTISING EXPENSE
AND WEEKLY SALES VOLUME
LISTING OF DATA

OBS	SPENT	VOLUME
1	1.00	10.2
2	1.25	11.5
3	1.50	16.1
4	1.75	.
5	2.00	20.3
6	2.50	25.6
7	3.00	28.0

N = 7

RELATIONSHIP BETWEEN ADVERTISING EXPENSE
AND WEEKLY SALES VOLUME
REGRESSION ANALYSIS

Model: MODEL1
Dependent Variable: VOLUME   WEEKLY SALES VOLUME

Analysis of Variance

Source	DF	Sum of Squares	Mean Square	F Value	Prob>F
Model	1	261.96637	261.96637	190.453	0.0002
Error	4	5.50196	1.37549		
C Total	5	267.46833			

Root MSE	1.17281	R-square	0.9794
Dep Mean	18.61667	Adj R-sq	0.9743
C.V.	6.29980		

Parameter Estimates

Variable	DF	Parameter Estimate	Standard Error	T for H0: Parameter=0	Prob > |T|
INTERCEP	1	1.003509	1.36312865	0.736	0.5025
SPENT	1	9.393684	0.68067860	13.800	0.0002

Variable	DF	Variable Label
INTERCEP	1	Intercept
SPENT	1	AMOUNT SPENT ON ADVERTISING

RELATIONSHIP BETWEEN ADVERTISING EXPENSE
AND WEEKLY SALES VOLUME
REGRESSION ANALYSIS

Obs	Dep Var VOLUME	Predict Value	Std Err Predict	Lower95% Mean	Upper95% Mean	Residual
1	10.2000	10.3972	0.764	8.2755	12.5189	-0.1972
2	11.5000	12.7456	0.640	10.9673	14.5239	-1.2456
3	16.1000	15.0940	0.543	13.5876	16.6005	1.0060
4	.	17.4425	0.486	16.0923	18.7926	.
5	20.3000	19.7909	0.486	18.4407	21.1410	0.5091
6	25.6000	24.4877	0.640	22.7094	26.2660	1.1123
7	28.0000	29.1846	0.903	26.6771	31.6920	-1.1846

```
Sum of Residuals              3.197442E-14
Sum of Squared Residuals            5.5020
Predicted Resid SS (Press)         16.2666
```

12.35   The earnings per share data from Exercise 11.53 are shown here.

Year	1993	1992	1991	1990	1989	1988	1987	1986
Earnings per Share	2.30	1.80	1.50	1.20	1.05	1.00	.99	.97

Conduct a statistical test of $H_0: \beta_1 = 0$ versus $H_a: \beta_1 > 0$, using $\alpha = .05$. Recall that we let $x = (\text{year} - 1985)$.

12.36   Refer to Exercise 12.35. Discuss the results of a statistical test of $H_0: \rho = 0$ versus $H_a: \rho > 0$.

12.37   A study was conducted to investigate the effect of different levels of nitrogen on the yield of lettuce plants. Simple data are shown here.

Nitrogen (lb per Plot), x	Yield (Emergent Stalks per Plot), y
2.0	21
2.8	18
3.0	24
3.6	26
4.1	32
4.3	29

Use the SAS output shown on the next pages to do the following.

a   Observe the sample data.

b   Locate the least squares estimates of β_0 and β_1.

c   Predict the yield when 4 lb of nitrogen is applied to a plot.

```
OPTIONS NODATE NONUMBER PS=60 LS=78;
DATA RAW;
  INPUT NITROGEN YIELD @@;
  LABEL NITROGEN = 'NITROGEN (POUNDS PER PLOT)'
        YIELD = 'EMERGENT STALKS PER PLOT';
  CARDS;
2.0    21
2.8    18
3.0    24
3.6    26
4.0    .
4.1    32
4.3    29
;
PROC PRINT N;
TITLE1 'RELATIONSHIP BETWEEN DIFFERENT LEVELS';
TITLE2 'OF NITROGEN AND YIELD OF LETTUCE PLANTS';
TITLE3 'LISTING OF DATA';

PROC PLOT;
  PLOT YIELD*NITROGEN;
TITLE3 'PLOT OF THE DATA';

PROC REG;
  MODEL YIELD = NITROGEN / P;
TITLE3 'REGRESSION ANALYSIS';
RUN;

     RELATIONSHIP BETWEEN DIFFERENT LEVELS
  OF NITROGEN AND YIELD OF LETTUCE PLANTS
            LISTING OF DATA

        OBS    NITROGEN    YIELD

         1       2.0        21
         2       2.8        18
         3       3.0        24
         4       3.6        26
         5       4.0         .
         6       4.1        32
         7       4.3        29

              N = 7
```

RELATIONSHIP BETWEEN DIFFERENT LEVELS
OF NITROGEN AND YIELD OF LETTUCE PLANTS
REGRESSION ANALYSIS

Model: MODEL1
Dependent Variable: YIELD   EMERGENT STALKS PER PLOT

Analysis of Variance

Source	DF	Sum of Squares	Mean Square	F Value	Prob>F
Model	1	95.00266	95.00266	10.271	0.0327
Error	4	36.99734	9.24934		
C Total	5	132.00000			

Root MSE	3.04127	R-square	0.7197
Dep Mean	25.00000	Adj R-sq	0.6496
C.V.	12.16509		

Parameter Estimates

Variable	DF	Parameter Estimate	Standard Error	T for H0: Parameter=0	Prob > |T|
INTERCEP	1	8.412234	5.32261107	1.580	0.1892
NITROGEN	1	5.026596	1.56841626	3.205	0.0327

Variable	DF	Variable Label
INTERCEP	1	Intercept
NITROGEN	1	NITROGEN (POUNDS PER PLOT)

RELATIONSHIP BETWEEN DIFFERENT LEVELS
OF NITROGEN AND YIELD OF LETTUCE PLANTS
REGRESSION ANALYSIS

Obs	Dep Var YIELD	Predict Value	Residual
1	21.0000	18.4654	2.5346
2	18.0000	22.4867	-4.4867
3	24.0000	23.4920	0.5080
4	26.0000	26.5080	-0.5080
5	.	28.5186	.
6	32.0000	29.0213	2.9787
7	29.0000	30.0266	-1.0266

Sum of Residuals   3.552714E-15
Sum of Squared Residuals   36.9973
Predicted Resid SS (Press)   102.0687

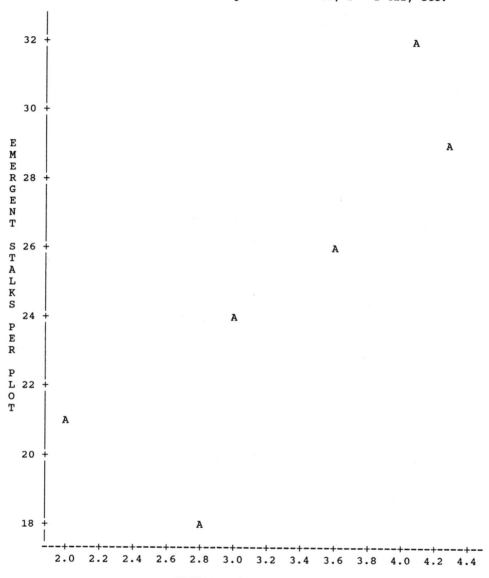

RELATIONSHIP BETWEEN DIFFERENT LEVELS
OF NITROGEN AND YIELD OF LETTUCE PLANTS
PLOT OF THE DATA

Plot of YIELD*NITROGEN.   Legend: A = 1 obs, B = 2 obs, etc.

NOTE: 1 obs had missing values.

12.38   Refer to Exercise 12.37. Use the sample data to construct a 90% confidence interval for β_1.

12.39   The efficiency of a set of production workers was observed, after which the workers attended a one-week efficiency improvement course. The efficiency of the workers after the course was observed. The following data show the before-the-course rating and the after-the-course rating for the workers. Is there a linear relationship between the before-the-course ratings and the after-the-course ratings (that is, can one predict what the after-the-course efficiency rating will be, based on the before-the-course efficiency rating)?

Rating Before the Course	Rating After the Course
32	44
30	52
37	53
44	63
43	44
33	50
36	45
46	53
30	48

Test your linear relationship, using a two-tailed test with $\alpha = .05$. If a before-the-course rating was 35, predict what the after-the-course rating will be.

12.40   In a preliminary study to appraise the behavior of professional football teams, an analyst wanted to determine whether there was an association between points scored in the first half by winning teams and points scored in the second half by the same teams. A random sample of 12 games was taken, with the following results.

Points Scored in First Half	Points Scored in Second Half	Points Scored in First Half	Points Scored in Second Half
12	3	28	5
17	7	13	7
10	10	14	3
17	10	6	16
17	7	19	10
14	3	10	17

Based on $\alpha = .05$, is there an association (positive or negative) between points scored in the first half by winning teams and points scored in the second half by winning teams?

12.41   The numbers of residential customers (in thousands) and of kilowatt-hours (in millions) for Public Service Company of New Mexico for recent years are listed next. Treating kilowatt-hours of electricity as a function of the number of residential customers, determine the linear relationship and predict the value for kilowatt-hours, given that the number of residential customers is 100,000. Using $\alpha = .01$, determine whether the slope is greater than 0.

Year	Kilowatt Hours	Customers	Year	Kilowatt Hours	Customers
1988	1135	91	1983	957	40
1987	1105	81	1982	917	32
1986	1090	73	1981	875	29
1985	1068	66	1980	828	23
1984	1001	51	1979	786	21

12.42 The thermal pollution of automobiles was studied for 1987 and 1988 car models. The data relating automobile weight to BTU (in 1000s) per vehicle mile are shown in the accompanying table below.

a Compute s_ϵ^2.

b Give a 95% confidence interval for β_1 and discuss the meaning of your finding.

Weight (in 1000 lb)	BTU per Vehicle Mile (in 1000s)
x	y
1.8	4
2.6	5.2
4.2	8.5
5.0	11.6
4.8	10.1
3.4	6.3

12.43 Labor data (in terms of man-hours) are presented here for the number of orders processed per month by a large manufacturing center.

Month	Orders processed (x)	Man-Hours Required to Process Orders (y)
1	3000	8000
2	3400	9200
3	4000	10,000
4	2800	7500
5	2000	5800
6	1700	5000
7	1400	4400
8	1300	3700
9	1000	3100
10	600	2220
11	1500	4100
12	2200	5500
13	3300	8100
14	3600	9400
15	4100	10,600
16	3200	7900

a Examine the plot of orders versus months. Are there cyclical patterns?

b Judging from the following plot of orders (x) versus man-hours (y), and ignoring the apparent cyclical effect of part a, what regression model might adequately describe the data?

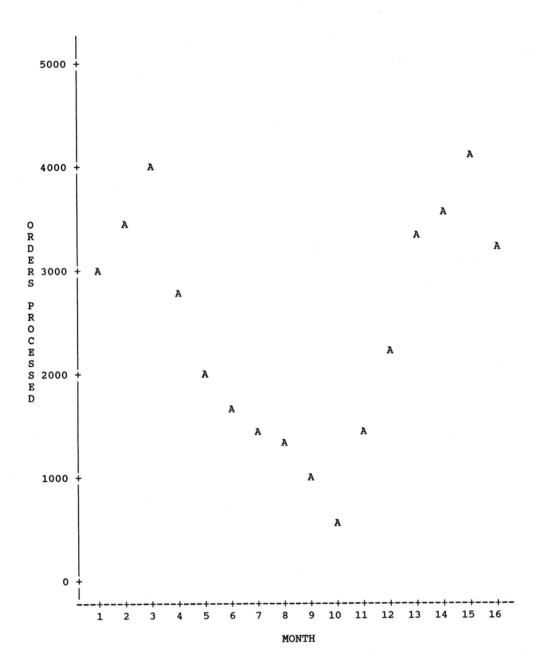

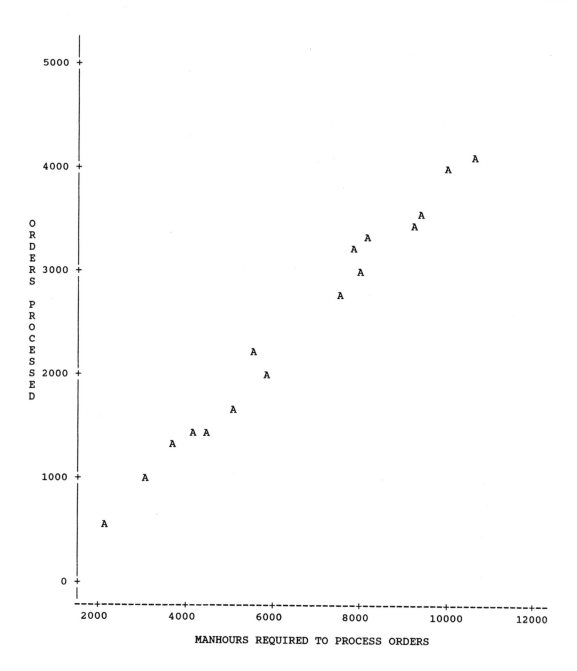

MANHOURS REQUIRED TO PROCESS ORDERS

12.44   Refer to Exercise 12.43. Fit the linear regression model $y = \beta_0 + \beta_1 x + \epsilon$, and draw conclusions about the slope and the intercept.

12.45   A random sample of 14 pharmacies was used to examine the relation between sales volume and profit before tax (PBT). These data are shown next.

Pharmacy	Sales Volume (in $1000s), x	PBT (in $1000s), y
1	38	1.3
2	20	2.1
3	48	2.2
4	44	2.6
5	56	3.3
6	39	4.0
7	65	4.1
8	84	4.2
9	82	5.5
10	105	5.7
11	126	7.0
12	52	7.5
13	80	7.7
14	101	7.9

a   Plot the sample data.

b   Calculate the sample correlation coefficient.

c   Is there a significant linear trend between sales volume (x) and PBT (y)?

12.46   Ten pairs of identical twins participated in an experiment to investigate two methods of teaching children to read music. One child from each pair was taught to read music by method A; the other received instruction according to method B. The results of an examination given at the end of a six-week training period are presented in the accompanying table.

Teaching Method	Pair 1	2	3	4	5	6	7	8	9	10
A	75	80	67	73	93	88	70	95	84	92
B	73	76	65	70	95	82	65	85	83	95

a   Replace the scores for the children taught by method A with ranks. Do the same for the scores from method B. Calculate the rank correlation coefficient, to measure the strength of the relationship between methods A and B.

b   Conduct a test of the alternative hypothesis $H_a: \rho_S \neq 0$. Use $\alpha = .05$.

12.47   The following data show actual population changes for the years of 1970–1979 and predicted population changes for the years of 1980–1989 for selected cities. The numerical values show percentages.

Area	Actual Percentage Change, 1970–1979	Predicted Percentage Change, 1980–1989
Las Vegas	69.1	31.3
Phoenix	55.6	23.7
Orlando	53.3	16.1
San Diego	37.0	22.6
Austin	47.8	28.4
Tampa	42.4	19.0
Sacramento	25.3	12.2
Seattle	12.4	20.1
Tulsa	23.6	21.8
Salt Lake City	32.6	21.5
Pittsburgh	−5.8	−6.2
Boston	−4.8	−1.4

Using Spearman's rank correlation and the MINITAB output shown here, test the hypothesis that there is no correlation in the ranking of the cities for the actual percentage changes and for the predicted percentage changes. Use $\alpha = .05$.

```
MTB > READ INTO C1 C3
DATA> 69.1    31.3
DATA> 55.6    23.7
DATA> 53.3    16.1
DATA> 37.0    22.6
DATA> 47.8    28.4
DATA> 42.4    19.0
DATA> 25.3    12.2
DATA> 12.4    20.1
DATA> 23.6    21.8
DATA> 32.6    21.5
DATA> -5.8    -6.2
DATA> -4.8    -1.4
DATA> END
     12 ROWS READ
MTB > LET C2 = RANKS (C1)
MTB > LET C4 = RANKS (C3)
MTB > PRINT C1-C4

 ROW      C1     C2      C3     C4

   1    69.1     12    31.3     12
   2    55.6     11    23.7     10
   3    53.3     10    16.1      4
   4    37.0      7    22.6      9
   5    47.8      9    28.4     11
   6    42.4      8    19.0      5
   7    25.3      5    12.2      3
   8    12.4      3    20.1      6
   9    23.6      4    21.8      8
  10    32.6      6    21.5      7
  11    -5.8      1    -6.2      1
  12    -4.8      2    -1.4      2

MTB > STOP
```

12.48   Two students at a university were asked to rate rock artists on their competence, with 1 being assigned to the lowest-ranked artist and 10 being assigned to the highest-ranked artist. The ratings of the two students were as shown here.

Rock Artist Number	1	2	3	4	5	6	7	8	9	10
First student ranks	2	6	10	8	7	1	9	5	8	4
Second student ranks	5	4	1	9	3	2	8	6	10	7

Test the hypothesis that there is no association between the rankings of the two students. Use Spearman's rank correlation procedures when making your test. Set $\alpha = .05$.

12.49   Interest rates charged for home mortgages have, in general, declined over recent months. With the apparent favorable influence for new home building, the data shown here are the prevailing mortgage interest rates and the number of housing starts in a midwestern city over a period of 18 months.

Month	Interest Rate (x)	Number of Housing Starts (y)
1	10.5	360
2	10.3	340
3	10.6	370
4	11.4	360
5	11.8	330
6	11.3	300
7	11.0	290
8	10.5	340
9	10.2	360
10	10.0	370
11	9.8	380
12	9.8	390
13	9.9	375
14	10.0	350
15	10.0	345
16	9.9	360
17	9.8	380
18	9.7	395

a   Plot the data.

b   Use these data to obtain a linear regression equation.

c   Is the slope significantly different from 0?

12.50   Refer to Exercise 12.49. Predict the number of housing starts for interest rates of 10.2% and 9.5%. Do you predict that the prevailing interest rate will increase or decrease next month (month 19)?

Experiences with Real Data

12.51 Refer to the Bonus Level database on the data disk (or in Appendix 1) and the data plots from Exercises 11.65 and 11.66.

    a Test $H_0: \beta_1 = 0$ for the linear regression equation relating bonus level (y) to age (x) for males. Draw conclusions.

    b Repeat part a for females.

12.52 Use the Bonus Level data for males.

    a Construct and plot 95% confidence bands for $E(y)$.

    b Calculate 95% prediction bands for y, and superimpose those limits on the graph of part a.

    c Referring to parts a and b, distinguish between a confidence interval on $E(y)$ and a prediction limit for y. Use plain English.

12.53 Refer to the Annual Returns database on the data disk (or in Appendix 1) and the data plots of Exercise 11.68.

    a Does it appear that a linear regression equation is appropriate to relate the annual returns $E(y)$ to year (x) for large capital stocks for any of the three book-to-market ratio categories. Why or why not?

    b How would you test your conclusion in part a? Do you have any concerns about the underlying assumption(s)?

12.54 Repeat Exercise 12.53 using the data for small capital stocks in the Annual Returns database.

12.55 Refer to the large and small capital stocks in the Annual Returns database.

    a Compute the correlation coefficient behavior of the annual returns for large and small stocks for each of the three book-to-market ratio categories.

    b Give the p-value for $H_0: \rho = 0$ versus $H_a: \rho \neq 0$ for each category. Make sense of the data.

12.56 Refer to Exercise 12.55.

    a Compute the rank correlation coefficient for each book-to-market category.

    b Discuss whether ρ or ρ_S is a more appropriate measure for comparing the annual returns data for large and small stocks. Do tests of $H_0: \rho_S = 0$ versus $H_a: \rho_S \neq 0$ yield the same general conclusions as those observed in Exercise 12.55?

12.57 Refer to Exercise 11.71 to conduct a two-sided test of $H_0: \rho = 0$ for the HAM-D total score and the HAM-D anxiety score data from the Clinical Trials database on the data disk (or in Appendix 1). Give the p-value for these data.

12.58 Refer to Exercise 11.72 and the Clinical Trials database. Is there a significant positive correlation between the HAM-D total score and the Hopkins OBRIST cluster total? Use $\alpha = .05$ for your test.

12.59 Refer to Exercise 11.73 and the Clinical Trials database. Can age be used to predict daily consumption of coffee and tea? Explain.

13

Analysis of Variance

13.1   Introduction

We discussed methods for comparing two population means in Chapter 8. Very often, however, the two-sample problem greatly simplifies what we encounter in real life; that is, frequently we want to compare more than two population means.

For example, suppose that we wanted to compare the mean incomes of steelworkers for three different ethnic groups—say black, white, and Hispanic Americans—in a certain city. Independent random samples of steelworkers would have to be selected from each of the three ethnic groups (the three populations). We would first have to consult the personnel files of the steel companies in the city, list steelworkers in each ethnic group, and select a random sample from each. On the basis of the three sample means, we would try to determine whether the population mean incomes differed and, if so, by how much. In fact, the sample means most likely would differ, but this would not necessarily imply a difference in mean income for the three ethnic groups. Even if the population mean incomes were identical, the sample means most probably would differ. How then, could you decide whether the differences among the sample means were large enough to imply a difference among the corresponding population means? We answer this question by using a technique known as an analysis of variance.

13.2   The Logic Behind an Analysis of Variance

The reason we call the method an analysis of variance can be seen more easily with an example. Assume that we want to compare three population means based on three samples of five observations each. The data for the three samples are shown in Table 13.1. Do the data present sufficient evidence to indicate a difference among the three population means? A brief visual analysis of the data in Table 13.1 leads us to a rapid, intuitive "yes." A glance at each of the three samples indicates that there is very little variation within each sample; that is, the variation in the measurements within each sample is very small. In contrast, the spread or variation among the sample means is so large in comparison to the within-sample variation that we intuitively conclude that a real difference does exist among the population means.

TABLE 13.1
A Comparison of Three Population Means (Small Amount of Within-Sample Variation)

Sample 1	Sample 2	Sample 3
29.0	25.1	20.1
29.2	25.0	20.0
29.1	25.0	19.9
28.9	24.9	19.8
28.8	25.0	20.2
$\bar{y}_1 = 29.0$	$\bar{y}_2 = 25.0$	$\bar{y}_3 = 20.0$
$s_1 = .16$	$s_2 = .07$	$s_3 = .16$

Now let's see how our intuition works when a larger within-sample variation is present (see Table 13.2). In this case the variation among the sample means is not large relative to the variation within the samples. Hence it would be difficult to conclude that the samples were drawn from populations with different means.

TABLE 13.2
A Comparison of Three Population Means (Large Amount of Within-Sample Variation)

Sample 1	Sample 2	Sample 3
29.0	33.1	15.2
14.2	7.4	39.3
45.1	17.6	14.8
48.9	44.2	25.5
7.8	22.7	5.2
$\bar{y}_1 = 29.0$	$\bar{y}_2 = 25.0$	$\bar{y}_3 = 20.0$
$s_1 = 18.19$	$s_2 = 14.18$	$s_3 = 12.96$

The variations in the observations for the two sets of data (Tables 13.1 and 13.2) are shown graphically in Figure 13.1. The strong evidence to indicate a difference in population means for the data of Table 13.1 is apparent in Figure 13.1(a). The lack of evidence to indicate a difference in population means for the data of Table 13.2 is indicated by the overlapping data points for the samples in Figure 13.1(b).

F I G U R E 13.1   Dot Diagrams for the Data of Tables 13.1 and 13.2

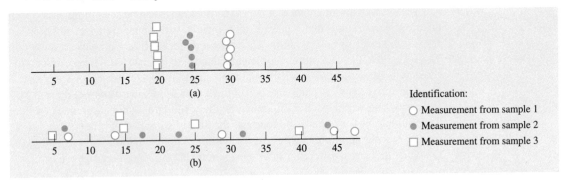

(a)

(b)

Identification:

○ Measurement from sample 1

● Measurement from sample 2

☐ Measurement from sample 3

analysis of variance

Figure 13.1 and the data of Tables 13.1 and 13.2 indicate very clearly, then, what we mean by an **analysis of variance**. All differences in sample means are judged to be statistically significant or not statistically significant by comparison to a measure of the random variation within the population data. Recall that we will measure the variability of the population data by the population standard deviation σ (or equivalently, by the population variance σ^2).

13.3 A Statistical Test About More Than Two Population Means: An Example of an Analysis of Variance

Earlier (in Chapter 8), we presented a method for testing the equality of two population means based on independent random samples from the two populations. We hypothesized two normal populations (1 and 2), with means denoted by μ_1 and μ_2, respectively. To test the null hypothesis that $\mu_1 = \mu_2$, independent random samples of sizes n_1 and n_2 were drawn from the two populations. The sample data were then used to compute the value of the test statistic

$$t = \frac{\bar{y}_1 - \bar{y}_2}{s_p\sqrt{\dfrac{1}{n_1} + \dfrac{1}{n_2}}},$$

where

$$s_p^2 = \frac{(n_1 - 1)s_1^2 + (n_2 - 1)s_2^2}{(n_1 - 1) + (n_2 - 1)} = \frac{(n_1 - 1)s_1^2 + (n_2 - 1)s_2^2}{n_1 + n_2 - 2}$$

is the pooled estimate of the common population variance σ^2. The rejection region for a specified value of α (the probability of a Type I error) was then found by using Table 3 in Appendix 3.

Now suppose that we wish to extend this method to test the equality of more than two population means. The preceding test procedure applies only to two means and therefore is inappropriate. Hence we must employ a more general method of data analysis known as the analysis of variance. We illustrate its use with the following example.

Students from five different campuses throughout the country were surveyed to determine their attitudes toward industrial pollution. Each student was asked a specific set of questions and then given a total score for the interview (see Table 13.3). Suppose that nine students were surveyed at each of the five campuses and that we want to compare the average student scores for the five campuses.

TABLE 13.3
Summary of Sample Results for
Five Populations

Statistic	Population (Campus)				
	1	2	3	4	5
Sample mean	$\bar{y}_1$	$\bar{y}_2$	$\bar{y}_3$	$\bar{y}_4$	$\bar{y}_5$
Sample variance	s_1^2	s_2^2	s_3^2	s_4^2	s_5^2

If we are interested in testing the equality of the population means (that is, $\mu_1 = \mu_2 = \mu_3 = \mu_4 = \mu_5$), we might be tempted to perform all possible pairwise comparisons of the population means. Hence if we assume that the five distributions are approximately normal, with a common variance σ^2, we might conduct ten t tests comparing the five means, two at a time, as listed in Table 13.4 (see Section 8.2).

TABLE 13.4
All Possible Null Hypotheses for
Comparing Two Means from Five
Populations

$\mu_1 = \mu_2$	$\mu_1 = \mu_4$	$\mu_2 = \mu_3$	$\mu_2 = \mu_5$	$\mu_3 = \mu_5$
$\mu_1 = \mu_3$	$\mu_1 = \mu_5$	$\mu_2 = \mu_4$	$\mu_3 = \mu_4$	$\mu_4 = \mu_5$

One obvious disadvantage to this test procedure is that it is tedious and time-consuming. But the more important and less apparent disadvantage of running multiple t tests to compare means is that the probability of incorrectly rejecting at least one of the hypotheses increases as the number of t tests increases. Thus, although the probability of a Type I error may be fixed at $\alpha = .05$ for each individual test, the probability of incorrectly rejecting H_0 on at least one of these test is larger than .05. In other words, the combined probability of a Type I error for the set of ten hypotheses is larger than the value .05 set for each individual test. Indeed, it could be as large as .40.

What we need, then, is a single test of the hypothesis "all five population means are equal" that is less tedious than the individual t tests and that can be performed with a specified probability of a Type I error (say, $\alpha = .05$). First, we assume that the five sets of measurements are normally distributed, with means given by $\mu_1, \mu_2, \mu_3, \mu_4$, and μ_5 and with a common variance σ^2. Consider the quantity

$$s_W^2 = \frac{(n_1 - 1)s_1^2 + (n_2 - 1)s_2^2 + (n_3 - 1)s_3^2 + (n_4 - 1)s_4^2 + (n_5 - 1)s_5^2}{(n_1 - 1) + (n_2 - 1) + (n_3 - 1) + (n_4 - 1) + (n_5 - 1)}$$

$$= \frac{(n_1 - 1)s_1^2 + (n_2 - 1)s_2^2 + (n_3 - 1)s_3^2 + (n_4 - 1)s_4^2 + (n_5 - 1)s_5^2}{n_1 + n_2 + n_3 + n_4 + n_5 - 5}.$$

Notice that this quantity is merely an extension of

$$s_p^2 = \frac{(n_1 - 1)s_1^2 + (n_2 - 1)s_2^2}{n_1 + n_2 - 2},$$

which is used as the estimator of the common variance for two populations in a test of the hypothesis $\mu_1 = \mu_2$ (Section 8.4). Thus, s_W^2 represents a pooled estimate of the common variance σ^2 and measures the **variability** of the observations **within** the five **populations**. (The subscript W refers to the within-population variability.)

variability within populations

Next we consider a quantity that measures the variability between or among the population means. If the null hypothesis $\mu_1 = \mu_2 = \mu_3 = \mu_4 = \mu_5$ is true, the populations are identical, with mean μ and variance σ^2. Drawing single samples from the five populations is then equivalent to drawing five different samples from the same population. What kind of variation might be expected for these sample means? If the variation is too great, we would reject the hypothesis that $\mu_1 = \mu_2 = \mu_3 = \mu_4 = \mu_5$.

To assess the variation from sample mean to sample mean, we need to know the distribution of the mean of a sample of nine observations in repeated sampling. From the Central Limit Theorem (Chapter 6), we know that the sampling distribution of $\bar{y}$ based on nine observations has a mean of $\mu_{\bar{y}} = \mu$ and a variance of $\sigma_{\bar{y}}^2 = \sigma^2/9$. Since we have five different samples of nine observations each, we can estimate the variance $\sigma^2/9$ of the sampling distribution by computing the sample variance for the five sample means:

$$\text{Sample variance (of the means)} = \frac{\sum \bar{y}^2 - (\sum \bar{y})^2/5}{5 - 1}.$$

Here, we merely consider the $\bar{y}$s as a sample of five observations and calculate the sample variance. This quantity is an estimate of $\sigma^2/9$; so 9 times the sample variance of the means is an estimate of σ^2. We designate this estimate of σ^2 by s_B^2, where the subscript B denotes a measure of the **variability among** (between) the **sample means**. For this problem,

variability among sample means

$$s_B^2 = 9 \times (\text{Sample variance of the means}).$$

Under the null hypothesis that all five population means are identical, we have two estimates of σ^2—namely, s_W^2 and s_B^2. Suppose that the ratio s_B^2/s_W^2 is used as a test statistic to test the hypothesis that $\mu_1 = \mu_2 = \mu_3 = \mu_4 = \mu_5$. What would the distribution of this quantity be if we were to repeat the experiment over and over again, each time calculating s_B^2 and s_W^2? As you might surmise from Chapter 9, s_B^2/s_W^2 follows an F distribution, with degrees of freedom that can be shown to be $df_1 = 4$ for s_B^2 and $df_2 = 40$ for s_W^2. The proof of these remarks is beyond the scope of this text; however, we make use of the result for testing the null hypothesis $\mu_1 = \mu_2 = \mu_3 = \mu_4 = \mu_5$.

The test statistic used to test equality of population means is

$$F = \frac{s_B^2}{s_W^2}.$$

When the null hypothesis is true, both s_B^2 and s_W^2 are estimates of σ^2, and F can be expected to assume a value near $F = 1$. When the hypothesis of equality is false, s_B^2 tends to be larger than s_W^2, due to the differences among the population means. Hence we reject the null hypothesis in the upper tail of the distribution of $F = s_B^2/s_W^2$. For α (the probability of a Type I error) equal to .10, .05, .025, .01, or .005, we can locate the rejection region for the one-tailed test by using Tables 5 through 9 in Appendix 3, with $df_1 = 4$ and $df_2 = 40$. Thus, for $\alpha = .05$, the tabulated value of F is 2.61 (see Figure 13.2). If the calculated value of $F = s_B^2/s_W^2$ falls in the rejection region, we conclude that at least one of the population means differs from the others.

FIGURE 13.2
Rejection Region of F for
$\alpha = .05$, $df_1 = 4$, and
$df_2 = 40$

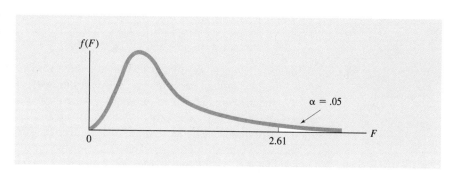

This procedure can be generalized, with only slight modification in the formulas, to test the equality of k (where k is an integer equal to or greater than two) population means from normal populations, with a common variance σ^2. Independent random samples of size $n_1, n_2, \ldots, n_k$ are drawn from the respective populations. Then we compute the sample means and variances. The null hypothesis $\mu_1 = \mu_2 = \cdots = \mu_k$ is tested against the alternative hypothesis that at least one of the population means differs from the others. The test procedure is given next.

An Analysis of Variance for Testing the Equality of k Population Means

H_0: $\mu_1 = \mu_2 = \cdots = \mu_k$

H_a: At least one of the population means is different from the others.

T.S.: $F = \dfrac{s_B^2}{s_W^2},$

where

$$s_W^2 = \frac{(n_1 - 1)s_1^2 + (n_2 - 1)s_2^2 + \cdots + (n_k - 1)s_k^2}{n_1 + n_2 + \cdots + n_k - k}$$

$$s_B^2 = \frac{\sum n_i \bar{y}_i^2 - (\sum n_i \bar{y}_i)^2 / n}{k - 1}$$

and $n = n_1 + n_2 + \cdots + n_k$.

R.R.:   Reject H_0 if $F > F_{\alpha, df_1, df_2}$ where $df_1 = (k - 1)$, and $df_2 = (n_1 + n_2 + \cdots + n_k - k)$. See Tables 5–9 of Appendix 3 corresponding to $\alpha = .10, .05, .025, .01,$ and $.005$.

Note: When $n_1 = n_2 = \cdots = n_k$, the formula for s_B^2 simplifies to $s_B^2 = n' \left[\dfrac{\sum \bar{y}^2 - (\sum \bar{y})^2 / k}{k - 1} \right]$, where n' is the common sample size.

We illustrate these ideas in an example.

E X A M P L E   13.1   Several psychologists were interested in studying the effect of anxiety on learning, as measured by student performance on a series of tests. On the basis of a prestudy test, 27 students were classified into one of three anxiety groups. Group 1 students were those who scored extremely low on a scale measuring anxiety. Those placed in group 3 were students who scored extremely high on the anxiety scale. The remaining students were placed in group 2. The results of the prestudy anxiety test indicated that 6 students were in group 1, 12 were in group 2, and 9 were in group 3.

Following the prestudy assignment of students to groups, the same battery of tests was given to each of the 27 students. The sample mean and sample variance of these test scores (based on a total of 100 points) are summarized in Table 13.5 for each group. Use the sample data to test the hypothesis that the average test scores for low-, middle-, and high-anxiety students are identical (that is, that anxiety has no effect on a student's performance on this battery of tests). Use $\alpha = .01$.

T A B L E 13.5
Summary of Test Scores for
Example 13.1

Group 1 (Low)	Group 2 (Medium)	Group 3 (High)
$n_1 = 6$	$n_2 = 12$	$n_3 = 9$
$\bar{y}_1 = 88$	$\bar{y}_2 = 82$	$\bar{y}_3 = 78$
$s_1^2 = 10.1$	$s_2^2 = 14.8$	$s_3^2 = 13.9$

Solution   First we hypothesize a population for each anxiety group that corresponds to all possible test scores for students who could have been included in the study. We assume that the measurements in each population are approximately normally distributed, with the mean of population 1 equal to μ_1, the mean of 2 equal to μ_2, and the mean of 3 equal to μ_3. In addition, we assume that the populations have a common variance σ^2. From these populations, independent random samples of size $n_1 = 6$, $n_2 = 12$, and $n_3 = 9$ students were obtained and assigned to the respective groups.

To test the null hypothesis of equality of the population means, $\mu_1 = \mu_2 = \mu_3$, we first compute s_W^2 and s_B^2. We calculate s_W^2 directly from the sample data:

$$
\begin{aligned}
s_W^2 &= \frac{(n_1 - 1)s_1^2 + (n_2 - 1)s_2^2 + (n_3 - 1)s_3^2}{n_1 + n_2 + n_3 - 3} \\
&= \frac{5(10.1) + 11(14.8) + 8(13.9)}{6 + 12 + 9 - 3} = \frac{324.5}{24} = 13.52.
\end{aligned}
$$

However, before obtaining s_B^2, we must first compute $\sum n_i \bar{x}_i$ and $\sum n_i \bar{x}_i^2$. From Table 13.5 we determine that

$$
\sum n_i \bar{y}_i = 6(88) + 12(82) + 9(78) = 2214
$$
$$
\sum n_i \bar{y}_i^2 = 6(88)^2 + 12(82)^2 + 9(78)^2 = 181{,}908
$$

Hence, for $k = 3$ and $n = n_1 + n_2 + n_3 = 27$, we have

$$
\begin{aligned}
s_B^2 &= \frac{\sum n_i \bar{y}_i^2 - (\sum n_i \bar{y}_i)^2/n}{k - 1} \\
&= \frac{181{,}908 - (2214)^2/27}{2} = \frac{181{,}908 - 181{,}548}{2} = 180.
\end{aligned}
$$

The test statistic for the null hypothesis $\mu_1 = \mu_2 = \mu_3$ is

$$
F = \frac{s_B^2}{s_W^2} = \frac{180.0}{13.52} = 13.31.
$$

Using the probability of a Type I error, $\alpha = .01$, we can locate the upper-tail rejection region for this one-tailed test by using Table 8 of Appendix 3, with

$$
\mathrm{df}_1 = k - 1 = 2
$$
$$
\mathrm{df}_2 = n_1 + n_2 + n_3 - 3 = 24.
$$

The tabulated value of F is 5.61. Since the observed value of F is greater than 5.61, we reject the hypothesis of equality of the population means; that is, we conclude that at least one of the means is different from the rest. Although the F test does not tell us which of the population means are different, the sample data suggest that anxiety has a detrimental effect on a student's performance in the battery of tests.   ■

After completing the F test in an analysis of variance, the results of a study are usually summarized in an **analysis of variance table**. The format of an analysis of variance (ANOVA) table shown in Table 13.6.

analysis of variance table

TABLE 13.6
Example of an ANOVA Table

Source	Sum of Squares	Degrees of Freedom	Mean Square	F Test
Between samples	SSB	$k-1$	s_B^2	s_B^2/s_W^2
Within samples	SSW	$n-k$	s_W^2	
Total	SSB + SSW	$n-1$		

The analysis of variance table lists the sources of variability in the first column. The second column lists quantities called sums of squares. For our purposes, the sums of squares between samples (SSB) and within samples (SSW) are the numerators of the expressions for s_B^2 and s_W^2, respectively. Therefore,

$$SSB = \sum n_i \bar{y}_i^2 - \left(\sum n_i \bar{y}_i\right)^2/n$$
$$SSW = (n_1 - 1)s_1^2 + (n_2 - 1)s_2^2 + \cdots + (n_k - 1)s_k^2.$$

mean square

The third column of the ANOVA table gives the degrees of freedom associated with each source of variability, and the fourth column displays the **mean squares**. Historically, s_B^2 and s_W^2 have become known as mean squares; s_B^2 is called the mean square between samples, and s_W^2 is called the mean square within samples. Notice that a mean square is merely a sum of squares divided by its degrees of freedom. Finally, the last column of the ANOVA table gives the value of the F statistic for the analysis of variance.

E X A M P L E   13.2   Construct an analysis of variance table for the data of Example 13.1.

Solution   Using the previous results, we obtain the ANOVA table shown as Table 13.7.   ▪

TABLE 13.7
ANOVA Table for the Data of Example 13.1

Source	Sum of Squares	Degrees of Freedom	Mean Square	F Test
Between groups	360	2	180	13.31
Within groups	324.5	24	13.52	
Total	684.5	26		

13.4 Checking on the Equal Variance Assumption

The assumption of equal population variance, like the assumption of normality in populations, has been made in several places in this text, such as for the t test comparing two population means and for the analysis of variance F test in a completely randomized design.

Let us now consider an experiment that compares t population means based on independent random samples from each of the populations. Again we assume that we are dealing with normal populations with a common variance σ_ϵ^2 and possible different means. If there were just two populations of interest, we could verify the assumption of equality of the two population variances by using the F test of Chapter 9. However, with $t > 2$, rather than making all pairwise F tests, we seek a single test that can be used to verify our assumption about the equality of the population variances.

Hartley's test

Hartley's test is the one test we will use in this text for the null hypothesis

$$H_0: \sigma_1^2 = \sigma_2^2 = \cdots = \sigma_t^2.$$

This test represents a logical extension to the F test for $t = 2$. If s_i^2 denotes the sample variance computed from the ith sample, the test statistic is

F_{max}

$$F_{\textbf{max}} = \frac{s_{max}^2}{s_{min}^2},$$

where s_{max}^2 and s_{min}^2 are the largest and smallest of the s_i^2s, respectively. Critical values of the F_{max} test statistic were tabulated by H. O. Hartley (see Table 12 of Appendix 3). The test procedure is summarized here.

Hartley's Test for Homogeneity of Population Variances

H_0: $\sigma_1^2 = \sigma_2^2 = \cdots = \sigma_t^2$ (i.e., homogeneity of variances)

H_a: Not all population variances are the same.

T.S.: $F_{max} = \dfrac{s_{max}^2}{s_{min}^2}$

R.R.: For a specified value of α, reject H_0 if F_{max} exceeds the tabulated F-value (Table 12 of Appendix 3) for $a = \alpha, t$, and $df_2 = n - 1$, where n is the number of observations in each sample.

Theoretically we require all the sample sizes to be the same. In practice, if the sample sizes are nearly equal, the largest n_i can be used for running the test of homogeneity. This procedure will slightly increase the probability of a Type I error beyond the nominal value α.

Hartley's test has some drawbacks, and therefore most practitioners do not routinely run it. For one thing, the test is extremely sensitive to departures from normality. So, in checking one assumption (constant variance), we must be very careful about departures from another analysis of variance assumption (normality of the populations). Fortunately, as noted in Chapter 8, the assumption of homogeneity (equality) of population variances is less critical when the sample sizes are substantially different. When the sample sizes are nearly equal, the variances can be markedly different, and the p-values for an analysis of variance will still be only mildly distorted. Thus we recommend that Hartley's test be used only for the more extreme cases. In these extreme situations where homogeneity of the population variances is a problem, a transformation of the data may help to stabilize the variances. Then inferences can be made from an analysis of variance.

E X A M P L E   **13.3**   The mean dissolved oxygen contents (in ppm) of three different lakes were to be compared based on independent random samples of 10 observations taken from the center of each lake at a depth of 1 foot. The sample data are given in Table 13.8.

T A B L E **13.8**
Mean Dissolved Oxygen Contents (in ppm) of Three Lakes, Example 13.3

Lake 1	Lake 2	Lake 3
0	1	14
2	3	26
1	4	25
3	6	18
1	8	19
2	7	22
3	5	21
4	3	16
1	4	20
5	5	30
$\bar{y} = 2.2$	$\bar{y} = 4.6$	$\bar{y} = 21.1$
$s = 1.55$	$s = 2.07$	$s = 4.84$

Run a test of the equality of the population variances. Use Hartley's test, with $\alpha = .05$.

Solution   The F test for the equality of population variances has

$$F_{max} = \frac{(4.84)^2}{(1.55)^2} = 9.75.$$

The critical value of F_{max} for $a = .05$, $t = 3$, and $df_2 = 9$ is 5.34. Since F_{max} is greater than 5.34, we reject the hypothesis of homogeneity of the population variances.   ▪

13.5   Using Computers to Help Make Sense of Data

Calculations for an analysis of variance can be greatly simplified by using available software. Outputs from MINITAB, SAS, and EXECUSTAT illustrate how these software systems can be used to run an analysis of variance for the data of Table 13.1. Familiarize yourself with how each software system displays the pertinent calculations for an analysis of variance. These same programs can be used for other data sets that you may need to analyze.

MINITAB Output

```
MTB > READ  INTO  C1-C3
DATA> 29.0    25.1    20.1
DATA> 29.2    25.0    20.0
DATA> 29.1    25.0    19.9
DATA> 28.9    24.9    19.8
DATA> 28.8    25.0    20.2
DATA> END
      5 ROWS READ
MTB > PRINT C1-C3

  ROW      C1      C2      C3

    1    29.0    25.1    20.1
    2    29.2    25.0    20.0
    3    29.1    25.0    19.9
    4    28.9    24.9    19.8
    5    28.8    25.0    20.2

MTB > AOVONEWAY C1-C3

ANALYSIS OF VARIANCE
SOURCE      DF        SS        MS         F         p
FACTOR       2  203.3333  101.6667   5545.41     0.000
ERROR       12    0.2200    0.0183
TOTAL       14  203.5533

                                    INDIVIDUAL 95 PCT CI'S FOR MEAN
                                    BASED ON POOLED STDEV
  LEVEL      N      MEAN     STDEV   ----+---------+---------+---------+--
  C1         5    29.000     0.158                                   (*
  C2         5    25.000     0.071                          *)
  C3         5    20.000     0.158   (*
                                    ----+---------+---------+---------+--
POOLED STDEV =    0.135           21.0      24.0      27.0      30.0
MTB > STOP
```

SAS Output

```
OPTIONS NODATE NONUMBER PS=60 LS=78;
DATA RAW;
  INPUT SAMPLE RESPONSE @@;
  CARDS;
1   29.0   1   29.2   1   29.1   1   28.9   1   28.8
2   25.1   2   25.0   2   25.0   2   24.9   2   25.0
3   20.1   3   20.0   3   19.9   3   19.8   3   20.2
;
PROC PRINT N;
TITLE1 'A COMPARISON OF THREE POPULATION MEANS';
TITLE2 'LISTING OF THE DATA';

PROC ANOVA;
  CLASS SAMPLE;
  MODEL RESPONSE = SAMPLE;
  MEANS SAMPLE;
TITLE2 'EXAMPLE OF PROC ANOVA';
RUN;
```

A COMPARISON OF THREE POPULATION MEANS
LISTING OF THE DATA

OBS	SAMPLE	RESPONSE
1	1	29.0
2	1	29.2
3	1	29.1
4	1	28.9
5	1	28.8
6	2	25.1
7	2	25.0
8	2	25.0
9	2	24.9
10	2	25.0
11	3	20.1
12	3	20.0
13	3	19.9
14	3	19.8
15	3	20.2

N = 15

A COMPARISON OF THREE POPULATION MEANS
EXAMPLE OF PROC ANOVA

Analysis of Variance Procedure

Dependent Variable: RESPONSE

Source	DF	Sum of Squares	Mean Square	F Value	Pr > F
Model	2	203.3333333	101.6666667	5545.45	0.0001
Error	12	0.2200000	0.0183333		
Corrected Total	14	203.5533333			

R-Square	C.V.	Root MSE	RESPONSE Mean
0.998919	0.548922	0.135401	24.6666667

SAS Output (Continued)

Source	DF	Anova SS	Mean Square	F Value	Pr > F
SAMPLE	2	203.3333333	101.6666667	5545.45	0.0001

Level of SAMPLE	N	----------RESPONSE---------- Mean	SD
1	5	29.0000000	0.15811388
2	5	25.0000000	0.07071068
3	5	20.0000000	0.15811388

EXECUSTAT Output

Oneway ANOVA

Source of Variation	Sum of Squares	D.F.	Mean Square	F-Ratio	P Value
SAMPLE	203.333	2	101.667	5545.45	0.0000
Error	0.22	12	0.0183333		
Total (corr.)	203.553	14			

Table of Means

SAMPLE	Sample Size	Sample Mean	Standard Error	Estimated Effect
1	5	29	0.060553	4.33333
2	5	25	0.060553	0.333333
3	5	20	0.060553	-4.66667
Overall	15	24.6667	0.0349603	

Summary

The test procedure described in this chapter is called an analysis of variance because testing the null hypothesis of equality of the means relies on a test statistic composed of a measure of variability between populations, s_B^2, and a measure of variability within populations, s_W^2. The rejection region can be fixed so that the probability α of a Type I error is some specified value (our tables allow α to be either .05 or .01). The conclusions that are drawn relate to all the means and not to individual ones. Thus, if the test statistic falls in the rejection region, we conclude that at least one of the means differs from the others. However, we cannot specify exactly which means differ from the others based on an ANOVA. In this text, following an ANOVA in which we reject H_0, we have asked you to look at the data and determine which means appear to be different. This will have to suffice for now. Formal procedures for comparing pairs of means following rejection of H_0 in an ANOVA are discussed in Ott (1993).

The analysis of variance is analogous to the use of a floodlight. Large objects (the differences among means) may be recognizable, but smaller objects (differences between individual means) may not. In contrast, the *t*-test is more like a spotlight, which has a smaller field of vision but provides a more intensified light. Thus we are better able to detect individual differences by using the Student's *t* test.

We conclude from this discussion that we should perform an analysis of variance to indicate overall differences. If a significant value of the *F*-statistic is obtained, we can run a few *t* tests to detect individual differences. Thus, in the study of attitudes toward industrial pollution on five campuses, we might want to compare the mean total scores for campuses 1 and 5, 2 and 5, and so forth. However, we should limit ourselves to a small number of comparisons. Otherwise we increase the risk of detecting a difference that does not exist.

In this chapter we have presented an explanation and an example of an analysis of variance. We needed such a procedure because we were unable to test the equality of more than two population means by using a single test statistic. The analysis of variance is a very general test procedure and can be applied to solve many different problems. However, because of the complexity of the subject, we will go no further. Several useful references are Hicks (1973), Mendenhall (1968), and Ott (1993), for those interested in extending their knowledge of analysis of variance.

Key Terms

analysis of variance

variability within populations

variability among sample means

analysis of variance table

mean square

Hartley's test

F_{max}

Key Formulas

Analysis of variance *F* test:

$$F = \frac{s_B^2}{s_W^2}$$

where

$$s_W^2 = \frac{(n_1 - 1)s_1^2 + (n_2 - 1)s_2^2 + \cdots + (n_k - 1)s_k^2}{n_1 + n_2 + \cdots + n_k - k}$$

and

$$s_B^2 = \frac{\sum n_i \bar{y}_i^2 - (\sum n_i \bar{y}_i)^2 / n}{k - 1}$$

Supplementary Exercises

13.1   Examine the logic behind an analysis of variance.

13.2   Elasticity readings for random samples of size 6 drawn from three different plastic processes (A, B, and C) are given in the accompanying table.

	Process A	Process B	Process C
	4.2	5.6	3.2
	1.1	5.1	2.5
	3.7	4.4	2.9
	2.6	4.2	3.6
	2.1	4.2	3.2
	3.7	5.1	4.1
Sample mean	2.90	4.77	3.25
Sample variance	1.39	.34	.31

Perform an analysis of variance for the experiment to determine whether there are differences in the mean elasticity readings among the three plastic processes. Use $\alpha = .05$.

13.3   Refer to Exercise 13.2. Use the sample data and the results of the analysis of variance to construct a 95% confidence interval for $(\mu_B - \mu_A)$. (*Hint*: Use s_W^2 for the pooled sample variance.)

13.4   The length of life of an electronics component was to be studied under five different operating voltages: V_1, V_2, V_3, V_4, and V_5. Ten different components were randomly assigned to each of the five operating voltages. A summary of the resulting lengths of life is recorded for each of the five groups in the accompanying table. Perform an analysis of variance to test the hypothesis that $\mu_1 = \mu_2 = \mu_3 = \mu_4 = \mu_5$. Use $\alpha = .05$.

Voltage V_1	Voltage V_2	Voltage V_3	Voltage V_4	Voltage V_5
$n_1 = 10$	$n_2 = 10$	$n_3 = 10$	$n_4 = 10$	$n_5 = 10$
$\bar{y}_1 = 3.2$	$\bar{y}_2 = 3.8$	$\bar{y}_3 = 4.1$	$\bar{y}_4 = 4.0$	$\bar{y}_5 = 3.7$
$s_1^2 = .46$	$s_2^2 = .51$	$s_3^2 = .39$	$s_4^2 = .20$	$s_5^2 = .28$

13.5   Refer to Exercise 13.4. Suppose that two components assigned to voltage V_1 and three assigned to voltage V_3 were found to be damaged prior to experimentation. Assuming that no new components can be found to replace the damaged ones, run an analysis of variance to test the equality of the means, using the revised data in the accompanying table. Use $\alpha = .05$.

Voltage V_1	Voltage V_2	Voltage V_3	Voltage V_4	Voltage V_5
$n_1 = 8$	$n_2 = 10$	$n_3 = 7$	$n_4 = 10$	$n_5 = 10$
$s_1^2 = .59$	$s_2^2 = .51$	$s_3^2 = .45$	$s_4^2 = .20$	$s_5^2 = .28$

13.6   A clinical pharmacology lab conducted an experiment to compare two pain-relieving drugs. Six subjects were allotted at random to drug A and eight to drug B. The accompanying figures give the number of hours of pain relief provided by each drug.

	Drug A		Drug B
	4		9
	5		6
	5		4
	7		8
	5		6
	4		9
			8
			6
Total	30	Total	56
Mean	5	Mean	7
Variance	1.20	Variance	3.14

Perform an analysis of variance and construct an ANOVA table. Draw conclusions, using $\alpha = .05$. Do the drugs differ with regard to the number of hours of pain relief they provide?

13.7   Refer to Exercise 13.6.

a   Since only two drugs were tested, compare the mean hours of relief provided by drugs A and B, using a t test with $\alpha = .05$ (two-sided). Draw a conclusion.

b   Did your conclusion in part a agree with the one you made in Exercise 13.6? They should agree.

13.8   A clinical psychologist wanted to compare three methods for reducing hostility levels in university students. A certain psychological test (HLT) was used to measure the degree of hostility. High scores on this test indicate great hostility. Eleven students obtaining high, nearly equal scores were used in the experiment. Five were selected at random from among the eleven problem cases and treated by method A. Three were taken at random from the remaining six students and treated by method B. The other three students were treated by method C. All treatments continued throughout a semester. Each student was given the HLT test at the end of the semester, with the score results as shown in the accompanying table. Give the level of significance for a test of the null hypothesis, "There is no difference in mean HLT test score for the three methods after treatment." (*Hint*: Use Tables 5–9 of Appendix 3 to give an approximate level of significance, such as $p > .05$.)

	Method A	Method B	Method C
	80	70	63
	92	81	76
	87	74	70
	83		
	78		
Sample mean	84.0	75.0	69.7

13.9   Refer to Exercise 13.8. Use the sample data to construct a 99% confidence interval for $(\mu_A - \mu_B)$.

13.10   Three different methods of instruction in speed-reading were to be compared with respect to the mean level of comprehension. A total of 13 students volunteered for the study: 4 were randomly assigned to instructional procedure 1; 4 to procedure 2; and 5 to procedure 3. After a one-week training period, all students were asked to read an identical passage on a film, which was delivered at the rate of 300 words per minute. Students were then asked to answer questions on the film passage. Comprehension grades on these questions are listed in the accompanying table. Perform an analysis of variance to test the hypothesis that all three instructional groups have the same average level of comprehension. Use $\alpha = .05$.

	Procedure 1	Procedure 2	Procedure 3
	82	71	91
	80	79	93
	81	78	84
	83	74	90
			88
Sample mean	81.5	75.5	89.2

13.11 An experiment was conducted to compare the effectiveness of three mouthwashes (A, B, and C) in the treatment of morning halitosis. Although ads were run in the local newspapers at the study site, only 16 people responded and qualified for entrance into the study. Of these 16, 4 did not complete the study; consequently, only the results for the 12 who completed the study are shown in the computer printout that follows. Scores in the data represent results for individual participants on a pleasurable–nonpleasureable scale. (Higher scores imply greater pleasure.) Use the computer printout to respond to the following statements.

a Identify the three sample means and sample standard deviations.

b Compute the standard error of the mean for mouthwash B. Compare this value to the value in the EXECUSTAT output.

c State the results of an F test on the equality of the population mean scores.

d Based on the analysis of variance and the magnitudes of the sample means, which mouthwashes appear to differ?

WASHA	WASHB	WASHC
59	53	57
66	56	61
58	50	72
61		66
		68

Summary Statistics Mouthwash Data

	WASHA	WASHB	WASHC
Sample size	4	3	5
Mean	61	53	64.8
Std. deviation	3.55903	3	5.89067
Std. error	1.77951	1.73205	2.63439

Oneway ANOVA for Mouthwash Data

Source of Variation	Sum of Squares	D.F.	Mean Square	F-Ratio	P Value
WASH	262.117	2	131.058	6.05506	0.0216
Error	194.8	9	21.6444		
Total (corr.)	456.917	11			

13.12   A psychologist used a class of 88 students as subjects for an experiment to study the effects of preconditioning on accuracy of recall of details of observed events. The professor divided the group at random into four equal groups of 22. The students in group A were the controls and were not given any preconditioning; those in group B were privately told to watch for something unusual during the next Wednesday morning's class, but were not told what the unusual event would be; those in group C were privately told that, between 9:25 and 9:30, the professor's lecture would be interrupted by a messenger, who would enter the classroom, hand the professor an envelope, and depart quietly; those in group D were given the same information given to group C, but were also privately told that they would later be asked to describe the messenger's personal appearance in detail. Subjects were asked not to discuss the matter with anyone. A questionnaire was administered to all subjects after the event, in which they were asked to describe certain physical characteristics, items of clothing, and general appearance of the messenger. The resulting scores, on a scale ranging from 0 to 100, were analyzed by the analysis of variance. Mean scores for the four groups were as follows:

Group	Mean Score	Sample Variance
A	80	21
B	83	25
C	85	30
D	92	26

Complete the following analysis of variance table for the preceding data.

Source	Sums of Squares	Degrees of Freedom	Mean Squares	F test
Between groups				
Within groups				
Total	17,676			

13.13   Refer to Exercise 13.12.

a   What hypothesis is tested by the F test in Exercise 13.12?

b   What is your conclusion from the F test? Give the table F-value.

c   The preceding type of analysis requires random samples. What populations are sampled here? Are the samples really random? How might nonrandomness affect your conclusions?

d   One quantity in the ANOVA table is closely related to the s_p^2 that one computes in the t test. Give the numerical value of the quantity.

e   Compute the 95% confidence interval for the mean difference in groups C and D (i.e., $\mu_C - \mu_D$). Is the difference significant?

13.14   The Agricultural Experiment Station of a university tested two different herbicides and their effects on crop yield. Out of 90 acres set aside for the experiment, herbicide 1 was used on a random sample of 30 acres; herbicide 2 was used on a second random sample of 30 acres; and the remaining 30 acres were used as a control. At the end of the growing season, the yields (in bushels per acre) were as follows.

Test Group	Sample Mean	Sample Standard Deviation
Herbicide 1	90.2	6.5
Herbicide 2	89.3	7.8
Control 3	85.0	7.4

Use these data to conduct a one-way analysis of variance. Use $\alpha = .05$. Do any of the yields differ? If so, which ones?

 13.15 A horticulturist was investigating the phosphorus content of tree leaves from three different varieties of apple trees—A, B, and C. Random samples of five leaves from each of the three varieties were analyzed for phosphorus content, as recorded in the accompanying table. Use these data to test the hypothesis of equality of the mean phosphorus levels for the three variables. Use $\alpha = .05$.

	Variety A	Variety B	Variety C
	.35	.65	.60
	.40	.70	.80
	.58	.90	.75
	.50	.84	.73
	.47	.79	.66
Sample mean	.46	.78	.71
Sample variance	.008	.010	.006

 13.16 Refer to the data of Exercise 13.15. Construct a 95% confidence interval for the difference in mean phosphorus content for varieties A and C.

 13.17 An experiment was conducted to compare the effect of three different paints on the corrosion of pipes. A long pipe was cut into 12 segments, which were randomly assigned to one of the three paints so that each paint would be used on four segments. The segments were painted and allowed to weather for a period of six months. The accompanying corrosion readings were than obtained. Is there sufficient evidence to indicate a difference in the mean levels of corrosion for the three paints? Use an analysis of variance with $\alpha = .01$. Summarize your findings in an ANOVA table.

	Paint 1	Paint 2	Paint 3
	10.1	13.4	12.7
	11.4	12.9	11.9
	12.1	13.3	12.5
	10.8	13.1	12.3
Sample mean	11.10	13.18	12.35
Sample variance	.727	.049	.117

13.18 To compare the water-repellent properties of four different chemical coatings, the following tests were performed. Twelve different fabric samples were obtained from the same bolt of material, with three samples randomly assigned to each of the four chemical groups (A, B, C, D). Each of the samples was then treated with the assigned chemical coating. Following the chemical treatment, a fixed amount of water was applied to the fabric and the amount of moisture penetration was recorded. These data are recorded in the computer printout that follows at the top of page 578. Use the information from the accompanying EXECUSTAT output to respond to the following statements.

a Give the sample mean and sample standard deviation for the amount of moisture penetration for each chemical coating.

b Interpret the results of an analysis of variance to compare the mean moisture penetrations for the four chemicals.

	GROUPA	GROUPB	GROUPC	GROUPD
	10.1	11.4	9.9	12.1
	12.2	12.9	12.3	13.4
	11.9	12.7	11.4	12.9

Summary Statistics for Group Data

	GROUPA	GROUPB	GROUPC	GROUPD
Sample size	3	3	3	3
Mean	11.4	12.3333	11.2	12.8
Std. deviation	1.13578	0.814453	1.21244	0.655744
Std. error	0.655744	0.470225	0.7	0.378594

Oneway ANOVA for Group Data

Source of Variation	Sum of Squares	D.F.	Mean Square	F-Ratio	P Value
GROUP	5.2	3	1.73333	1.79931	0.2252
Error	7.70667	8	0.963333		
Total (corr.)	12.9067	11			

13.19   Use the data of Exercise 13.18 to conduct an analysis of variance, using a computer program available to you. Compare your results to those in the output of Exercise 13.18.

13.20   To minimize the loss in efficiency caused by breakdowns in machinery, production records of each machine in a manufacturing plant must be closely monitored so that operators can anticipate when equipment is run-down and in need of repair. The data in the accompanying table give the production records for four different machines, based on the outputs (in hundreds of pounds) from random samples of five shifts over the past week.

	Machine			
	1	2	3	4
	26.2	20.6	30.7	32.1
	32.0	26.4	35.2	34.7
	34.1	25.1	36.3	35.5
	33.6	24.9	31.9	36.8
	35.6	24.3	30.4	33.3
Sample mean	32.30	24.26	32.90	34.48
Sample variance	13.28	4.77	7.24	3.38

a   Perform an analysis of variance to determine whether the mean outputs differ for the four machines. Use $\alpha = .05$.

b   Which machine appears to be least productive?

13.21 Refer to the data of Exercise 13.20.

    a Construct a 95% confidence interval for $(\mu_4 - \mu_2)$.

    b Construct a 95% confidence interval for $(\mu_4 - \mu_1)$.

13.22 Sustained-release drugs are now marketed by many pharmaceutical companies. Even though single-dose, nonsustained-release capsules usually get more drug into the bloodstream quicker, sustained-release preparations supposedly achieve and maintain a more even level of release of the drug over a longer period of time. Consider a study comparing three different drug preparations—A, B, and C. Equivalent doses of the three preparations were placed in mixtures of fluids similar to the gastric juices of the stomach and observed until 50% of the drug product was released from the capsule formulation. The release times (in minutes) appear in the accompanying table.

	Preparation	
A	**B**	**C**
15	38	19
24	33	21
20	39	27
16	31	22
18	26	24
19	29	18

Sample mean   18.67   32.67   21.83

    a Which preparation would you suspect is the sustained-release formulation?

    b Perform an analysis of variance to test the equality of the population mean release times for the three preparations. Use $\alpha = .05$. Summarize your findings in an ANOVA table.

13.23 Students in an environmental engineering class were assigned the project of comparing the mean dissolved oxygen content of samples drawn from four different locations of a lake. The four locations were the center of the lake, the north and south edges, and a spot midway between the center and the east side of the lake. At each location, five different vial samples were drawn and subsequently analyzed for dissolved oxygen content (in parts per million). Use the accompanying table of data to determine whether there is sufficient evidence to indicate a difference in the mean dissolved oxygen content of samples from the four locations. Use $\alpha = .05$.

	Location		
1	**2**	**3**	**4**
4.6	6.7	6.4	5.8
4.8	6.2	6.3	5.3
4.3	6.4	6.6	5.7
4.9	6.5	6.7	5.2
4.7	6.3	6.5	5.0

Sample mean   4.66   6.42   6.50   5.40
Sample variance   .05   .04   .03   .12

13.24 A computer printout for the data of Exercise 13.23 is shown on page 580. Compare the EXECUSTAT computer results to those you obtained by hand in Exercise 13.23.

13.25 A survey was conducted to examine changes in prices (over the last month) of items included in a typical market basket. Six grocery stores in each of four geographic locations were sampled. The data shown in the accompanying table correspond to increases in the price of lettuce over the past month for the sampled stores. Use the data to perform an analysis of variance. Draw appropriate conclusions.

EXECUSTAT Output for Exercise 13.24

```
        LOC1    LOC2    LOC3    LOC4
        4.6     6.7     6.4     5.8
        4.8     6.2     6.3     5.3
        4.3     6.4     6.6     5.7
        4.9     6.5     6.7     5.2
        4.7     6.3     6.5     5.0
```

Summary Statistics for Location Data

	LOC1	LOC2	LOC3	LOC4
Sample size	5	5	5	5
Mean	4.66	6.42	6.5	5.4
Std. deviation	0.230217	0.192354	0.158114	0.339116
Std. error	0.102956	0.0860233	0.0707107	0.151658

Oneway ANOVA for Location Data

Source of Variation	Sum of Squares	D.F.	Mean Square	F-Ratio	P Value
LOCATION	11.6095	3	3.86983	67.3014	0.0000
Error	0.92	16	0.0575		
Total (corr.)	12.5295	19			

Geographic Location

1	2	3	4
10.1	15.3	11.8	16.8
11.3	14.8	12.6	9.2
8.2	10.4	14.2	17.5
8.7	9.3	13.9	18.2
12.1	10.7	8.9	10.9
10.4	15.6	7.5	14.5

a   Give the sample mean and sample standard deviation for lettuce-price increases at each of the four locations.

b   Identify all parts of a statistical test about the equality of the four population means.

c   State conclusions, using the results of an analysis of variance.

d   Which means appear to differ?

13.26   Construct a 99% confidence interval for $(\mu_4 - \mu_1)$ for the data of Exercise 13.25.

13.27   Use a computer program available to you to perform an analysis of variance for the data of Exercise 13.22. Compare the computer results to those you obtained by hand in Exercise 13.22.

Class Exercise

13.28   Every consumer has been faced with the problem of assessing value when either buying or selling goods or services. Frequently, when a major purchase or sale is contemplated, one can enlist the services of a professional appraiser to help assess value. For example, in making arrangements to move furniture and belongings from an apartment or home to another location, one could enlist (purchase) the services of a moving company. A standard procedure is to solicit appraisals from two or three companies before reaching a decision. Similarly, in trying to sell an automobile to a used car dealer, each dealer visited would be asked to give an appraisal of the worth of the car. For those who have had occasion to encounter different appraisers, of the worth of the car. For those who have had occasion to encounter different appraisers, it is interesting to observe the difference in values assigned to the goods or services. Some appraisers may tend to overvalue worth; others may tend to undervalue worth; and still others may tend to vary up and down with no apparent pattern.

Set up an experiment to compare the mean appraised values of three or more appraisers of a specific item. For example, to illustrate differences in assessing worth, you could choose three bicycle shops in town. With the help of other persons in your class, take the same five bicycles (at different times) to the bicycle shops for an assessment of worth. Try to enlist the services of the same appraiser from each shop for all bicycles. As another example, you may wish to work with three or more real estate agents on multiple listings to obtain appraisals on the same houses. For the experiment you choose, answer the following items.

a   Perform an analysis of variance to compare the mean appraised value. Use $\alpha = .05$.

b   Do any of the appraisers appear to overestimate (underestimate)?

c   Which appraisers appear to agree?

13.29   A supermarket manager was interested in the effect of location in his market on the sales of Diet Coke. To measure the effect, displays of Diet Coke were located near the front entrance for one six-week period, near the dairy and meat sections in the rear of the market for another six-week period, and on a shelf along with other soft drinks for a third six-week period. The number of six-packs sold each week in the three locations over the relevant six-week periods are shown here.

Near the Front Entrance	Near the Meat Counter	On the Soft Drink Shelf
38	30	25
44	41	18
58	43	26
51	48	30
43	43	27
54	40	31

Perform an analysis of variance test with $\alpha = .01$. What can you conclude?

13.30   At a large university in the midwest, a course in beginning statistics was taught in a large lecture section, with 250 students seated in rows A through L. In an effort to assess the row effect on student performance, random samples were taken of students in rows A, D, H, and L; the scores shown in the accompanying table were recorded.

Row A	Row D	Row H	Row L
58	60	48	52
47	55	44	49
53	49	36	17
50	51	43	32
64	42	23	28
48	36	26	23

Perform an analysis of variance to determine whether the mean scores for the students are related to seating locations in mass lecture classes. Set $\alpha = .05$. Does the performance of students in any particular row seem to be inferior?

13.31 Ten firms were selected from four different sales-volume categories to see whether the ratios of cash flow to debt were the same for the different categories. The sample data are represented in the following table.

Statistic	Small	Medium	Large	Very Large
Number	10	10	10	10
Mean cash flow to debt ratio	.245	.259	.274	.201
Standard deviation	.0418	.0229	.0317	.0239

Perform an analysis of variance test for equality of means, using $\alpha = .05$. What conclusions can be drawn?

13.32 An experiment similar to that described in Exercise 13.14 was conducted to compare the tar content y of five different brands of cigarettes. Sample data are shown here.

Brand	$\bar{y}$ (mg)	s	n_i
1	10.6	2.4	10
2	11.2	2.5	10
3	10.5	1.8	10
4	11.8	1.9	10
5	13.7	2.2	10

a Based on your intuition, is there sufficient evidence to indicate any differences among the mean contents of the five brands?

b Run an analysis of variance to confirm or reject your conclusion of part a.

13.33 The number of units of production was recorded for a random sample of ten hourly periods from the three bottling assembly lines of a plant. These data are shown next.

Assembly Line		
1	2	3
290	258	249
265	276	257
286	277	264
275	243	266
288	248	278
250	259	273
279	265	281
294	282	254
285	275	261
293	268	265

a Plot the data separately for each line. Are there any obvious differences?

b Identify the means and standard deviations for the three lines, using the accompanying SAS computer output.

c Use the output to construct an analysis of variance table. Draw conclusions based on the analysis of variance.

```
              LISTING OF DATA

     OBS    LINE    UNITS

      1      1      290
      2      1      265
      3      1      286
      4      1      275
      5      1      288
      6      1      250
      7      1      279
      8      1      294
      9      1      285
     10      1      293
     11      2      258
     12      2      276
     13      2      277
     14      2      243
     15      2      248
     16      2      259
     17      2      265
     18      2      282
     19      2      275
     20      2      268
     21      3      249
     22      3      257
     23      3      264
     24      3      266
     25      3      278
     26      3      273
     27      3      281
     28      3      254
     29      3      261
     30      3      265

           N = 30
```

MEANS BY LINE

VARIABLE	MEAN	STANDARD DEVIATION

```
------------------------------------ LINE=1 ------------------------------------
           ----------------------------------
           UNITS   280.5000000   13.8984412
           ----------------------------------

------------------------------------ LINE=2 ------------------------------------
           ----------------------------------
           UNITS   265.1000000   12.9995726
           ----------------------------------

------------------------------------ LINE=3 ------------------------------------
           ----------------------------------
           UNITS   264.8000000   10.2610374
           ----------------------------------
```

Analysis of Variance Procedure

Dependent Variable: UNITS

Source	DF	Sum of Squares	Mean Square	F Value	Pr > F
Model	2	1612.466667	806.233333	5.17	0.0125
Error	27	4207.000000	155.814815		
Corrected Total	29	5819.466667			

R-Square	C.V.	Root MSE	UNITS Mean
0.277082	4.620896	12.48258	270.133333

Source	DF	Anova SS	Mean Square	F Value	Pr > F
LINE	2	1612.466667	806.233333	5.17	0.0125

13.34 An experiment was conducted to compare the number of major defectives observed along each of five production lines in which changes were being instituted. Production was monitored continuously during the period of changes, and the number of major defectives was recorded per day for each line. These data are shown here.

Production Line				
1	2	3	4	5
34	54	75	44	80
44	41	62	43	52
32	38	45	30	41
36	32	10	32	35
51	56	68	55	58

Compute $\bar{y}$ and s^2 for each sample. Does there appear to be a problem with nonconstant variances?

 13.35 The yields of corn, in bushels per plot, were recorded for four different varieties of corn—A, B, C, and D. In a controlled greenhouse environment, each variety was randomly assigned to 8 of 32 plots available for the study. The yields are listed here.

A	2.5	3.6	2.8	2.7	3.1	3.4	2.9	3.5
B	3.6	3.9	4.1	4.3	2.9	3.5	3.8	3.7
C	4.3	4.4	4.5	4.1	3.5	3.4	3.2	4.6
D	2.8	2.9	3.1	2.4	3.2	2.5	3.6	2.7

Perform an analysis of variance on these data, and draw your conclusions. Use $\alpha = .05$.

 13.36 An experiment was conducted to test the effects of five different diets on turkeys. Six turkeys were randomly assigned to each of the five diet groups and were fed for a fixed period of time.

Group	Weight Gained (lb)
Control diet	4.1, 3.3, 3.1, 4.2, 3.6, 4.4
Control diet + level 1 of additive A	5.2, 4.8, 4.5, 6.8, 5.5, 6.2
Control diet + level 2 of additive A	6.3, 6.5, 7.2, 7.4, 7.8, 6.7
Control diet + level 1 of additive B	6.5, 6.8, 7.3, 7.5, 6.9, 7.0
Control diet + level 2 of additive B	9.5, 9.6, 9.2, 9.1, 9.8, 9.1

  a Plot the data separately for each sample.

  b Compute $\bar{y}_i$ and s_i for each sample.

  c Is there any evidence of unequal variances? If not, run an analysis of variance, and draw conclusions.

Experiences with Real Data

13.37 Refer to the Bonus Level database on the data disk (or in Appendix 1). Run an analysis of variance, using $\alpha = .05$, to compare genders. Draw a conclusion based on the analysis of variance.

13.38   Refer to the Crimes database on the data disk (or in Appendix 1). Compare the four different departmental majors with respect to y (the number of acts regarded as a crime). Use an analysis of variance based on $\alpha = .05$. Did you check the underlying assumptions? If so, how?

13.39   Refer to the Crimes database. Compare the four departmental majors with respect to age. Is an analysis of variance appropriate? Why or why not? Draw conclusions.

13.40   Refer to the Annual Returns database on the data disk (or in Appendix 1). Compare the annual returns among the three categories of book-to-market ratios for large capital stocks. Make sense of the data, and report your results in plain English.

13.41   Refer to Exercise 13.40. Make sense of the annual returns data for small capital stocks, as you did for large capital stocks.

13.42   Refer to Clinical Trials database on the data disk (or in Appendix 1). In previous chapters, we have made pairwise comparisons among the means of the four treatment groups from the Clinical Trials database. Now an analysis of variance can be used to compare all four at once. Use the sample data on the HAM-D total score to compare the four treatment groups with an analysis of variance. Construct an ANOVA table to summarize the results of your analysis. Which means (if any) appear to differ?

13.43   Repeat the work of Exercise 13.42, using the Hopkins OBRIST cluster total. Do you get the same results?

13.44   Refer to the Clinical Trials database. "Baseline" comparisons are run in most clinical trials to examine the comparability of the treatment groups *prior* to receiving the assigned medication. Why might this be important? What baseline comparison could we make for the Clinical Trials database?

13.45   Refer to the Clinical Trials database. Run an analysis of variance to compare the ages of the patients in the four treatment groups at the start of the study. Are the four groups comparable with respect to age? Summarize your results in an ANOVA table.

13.46   Refer to the Clinical Trials database. Compare the number of tablets or capsules taken for the four treatment groups, using an analysis of variance. Did patients in the four groups appear to take approximately the same number of tablets? Explain.

Step Four: Communicating Results

14

Communicating the Results of Analyses

14.1   Introduction

In Chapter 1, we introduced the subject of statistics as a study of making sense of data, and we identified the four major components of making sense of data: data gathering, data summarization, data analysis, and communicating results. In this chapter, we deal with methods of communicating the results of statistical analyses. But rather than tell you what to write—which, of course, depends on the particular problem being discussed, the intended audience, and the form of the communication—we consider some important elements of a statistical report. We also discuss some of the potential pitfalls to effective communication. Finally, we discuss how to document analyses to ensure reproducibility of results at a later date.

14.2   Good Communication Is Not Easy

The final step in the process of making sense of data consists of communicating results. How might one communicate the results of a study or survey? The list of possibilities is almost endless, including the many forms of verbal and written communication. For example, written communication within a company can vary from an informal short note or memo to a formal project report (see Figure 14.1).

FIGURE 14.1
Forms of Written and Verbal Communication

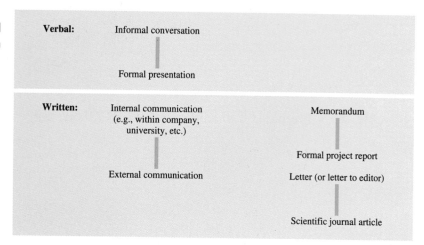

Verbal: Informal conversation

Formal presentation

Written: Internal communication
(e.g., within company,
university, etc.)

Memorandum

Formal project report

External communication   Letter (or letter to editor)

Scientific journal article

Communicating the results of a statistical analysis in concise, unambiguous terms is difficult. In fact, descriptions of most things are difficult. Try, for example, to describe the person sitting next to you in terms so precise that a stranger could select the individual from a group of others having similar physical characteristics. It is not an easy task. Fingerprints, voiceprints, and photographs—all pictorial descriptions—are the most precise methods of human identification. Describing a set of measurements is also difficult; but like describing a person, it can be accomplished more easily by using graphic or pictorial methods.

Cave drawings convey to us scattered bits of information about the life of prehistoric people. Similarly, extensive knowledge about the ancient lives and cultures of the Babylonians, Egyptians, Greeks, and Romans has been gleaned from their drawings and sculpture. Art thus conveys a picture of various life-styles, history, and culture. Not surprisingly, using graphs and tables in conjunction with a written description can help to convey the meaning of a statistical analysis.

In reading the results of a statistical analysis and in communicating the results of our own analyses, we must be careful not to distort them in our presentation of the data and results. You have undoubtedly heard the expression, "It is easy to lie with statistics." The idea is *not* new. The famous British statesman Benjamin Disraeli is quoted as saying, "There are three kinds of lies: lies, damned lies, and statistics." Where do things go wrong?

First of all, distortions of truth can occur only when we communicate. And since communication can be accomplished with graphs, pictures, sound, aroma, taste, words, numbers, or any other means devised to reach our senses, distortions can occur when we use any one or any combination of these methods of communication.

In this respect, statements that we make could be misleading to others if we omit something in our explanation of the data-gathering stage or of the analyses we performed. For example, we might fail to explain clearly the meaning of a numerical statement. Or we might omit some background information that is necessary for a clear interpretation of the results. Even a correct statement may produce a distorted impression if the reader lacks knowledge of elementary statistics. Thus a very clear expression of an inference based on a 95% confidence interval is meaningless to a person who has not been exposed to the introductory concepts of statistics.

Now let's look at some potential hurdles to effective communication that must be carefully considered when we present the results of a statistical analysis—or when we try to interpret what someone else has presented.

14.3   Communication Hurdles: Graphical Distortions

Pictures can easily distort the truth. The marketing of many products, including soft drinks, beers, cosmetics, clothing, and automobiles, involves the use of attractive, youthful models. The not-so-subtle impression we are left with is that, by using the product, we, too, will look like these models. Have you ever "stepped back" from one of these commercials and wondered how the commercial message relates to the quality and usefulness of the product? Or how you are being misled by a commercial? Try it sometime. Mail-order catalog sketches of products are frequently more attractive than the real thing, but we usually take this type of distortion for granted.

Statistical pictures are the histograms, frequency polygons, pie charts, and bar graphs of Chapter 3. These drawings or displays of numerical results are difficult to combine with sketches of lovely women or handsome men and hence are secure from the most common form of visual distortion. But other distortions are possible, such as by shrinking or sketching one of the axes. The idea behind these distortions is that shallow and steep slopes are commonly associated with small and large increases, respectively.

For example, suppose that the values of a leading consumer price index over the first six months of the year were 160, 165, 178, 189, 196, and 210. We might show the upward movement of this consumer price index by using the frequency polygon of Figure 14.2. In this graph, the increase in the index is evident, but it does not appear to be very great. On the other hand, we could present the same sample data in a much different light as shown in Figure 14.3. For this graph, the vertical axis is stretched and does not include zero. Note the impression of a substantial rise that is

FIGURE 14.2
Changes in a Consumer Price Index

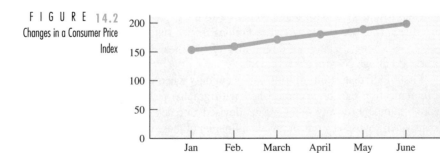

FIGURE 14.3
Changes in a Consumer Price Index

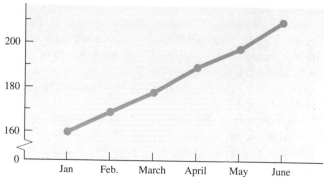

indicated by the steeper slope. Another way to achieve the same effect—decreasing or increasing a slope—is to stretch or shrink the horizontal axis.

When we present data in the form of bar graphs, histograms, frequency polygons, or other figures, we must be careful not to shrink or stretch axes, because this would catch most readers off guard. Increases or decreases in responses should be judged large or small depending on the arbitrary importance to the observer of the change, not on the slopes shown in graphic representations. In reality, most people look only at the slopes in the "pictures."

14.4 Communication Hurdles: Biased Samples

One of the most common statistical distortions occurs because the experimenter unwittingly (or sometimes knowingly) samples the wrong population; that is, he or she draws the sample from a set of measurements that is not the proper population of interest.

For example, suppose that we want to assess the reaction of taxpayers to a proposed park and recreation center for children. A random sample of households is selected, and interviewers are sent to those households in the sample. Unfortunately, no one is at home in 40% of the sample households, so we randomly select and substitute other households in the city to make up the deficit. The resulting sample is selected from the wrong population, and the sample is therefore said to be biased.

The specified population of interest in the household survey is the complete set of all households in the city. In contrast, the sample was drawn from a much smaller population or subset of this group—the set of householders who were at home when the sample was taken. It is possible that the fractions of householders favoring the park in these two populations are equal, in which case no damage is done by confining the sampling to those at home. But it is much more likely that those at home have more small children than the population as a whole, and this group is likely to yield a higher fraction in favor of the park. Thus we have a biased sample, because it is loaded in favor of families with small children; we unwittingly selected the sample only from a special subset of the population of interest.

Biased samples frequently result from surveys that utilize mailed questionnaires. In a sense, the investigator lets the selection and number of the sampling units depend on the interests, available time, and various other personal characteristics of the individuals who receive the questionnaires. Extremely busy and energetic people may drop the questionnaires into the nearest wastebasket; likewise, you rarely hear from low-energy people who are uninterested or who are engrossed with other activities. Most often the respondents are activists—those who are highly in favor, those who are opposed, or those who have something to gain from a certain outcome of the survey.

Although numerous newscasters and analysts utilize election results as an expression of public opinion on major issues, voting results represent a biased sample of public opinion. People who actually vote represent much less than half of all eligible voters; they are individuals who desire to exercise their rights and responsibilities as citizens or who have been specially motivated to participate. The resultant subset of voters does not represent the interests and opinions of all eligible voters in the country.

Sampling the wrong population also occurs when people attempt to extrapolate experimental results from one population to another. Numerous experimental results have been published about the effect of various products (for example, saccharin) in inducing cancer in moles, rats, beagles, and so forth. These results are often taken to imply that humans have a high risk of developing cancer after frequent or extended exposure to the product. But such inferences are not always justified, because the experimental results were not obtained on humans. It is quite possible that humans are capable of resisting much higher doses than rats, or perhaps humans may be completely resistant for some reason.

Another common error in drawing inferences about the carcinogenic character of various products relates to their level of risk in the form in which they would actually be encountered in the real world. For instance, the dose of saccharin that produces cancer in some lab rats might turn out to require ingestion by the rats of several quarts of diet soda per day, if the dose were converted into an amount of the substance as it actually occurs in a food product. In other words, what is being consumed by the lab rat is *not* a product that one (whether a rat or a person) would consume in its test-administered form; rather it is an ingredient in some much larger formulation—such as aspartame in a diet soda or red dye #2 in a jar of maraschino cherries—of which (in some cases) ludicrously large quantities would have to be consumed to achieve an equivalent dosage. It seems fair to ask whether drinking 6 quarts of diet soda every day wouldn't ruin a rat's kidneys before it had a chance to cause cancer.

Drug induction of cancer in small mammals *does* indicate a need for concern and caution by humans, but it does not prove that the drug is definitely harmful to humans. We are not criticizing experimentation in various species of animals, because it is frequently the only way to obtain any information about potential toxicity in human beings. We simply point out that the experimenter is knowingly sampling a population that is only similar (and quite likely not too similar) to the one of interest, and that the tests are conducted under conditions different from those under which the risk would be encountered in real life.

Engineers also test "rats" instead of "humans." Rats in this context are miniature models or pilot plants of a new engineering system. Experiments on the models occasionally yield result that differ substantially from the results of the larger, real systems. So again we see an instance of sampling from the wrong population—but again it is the best the engineer can do because of the economics of the situation. Funds are not usually available to test a number of full-scale models prior to production.

Many other examples could be given of biased samples or of sampling from the wrong populations. The point is that when we communicate the results of a study or survey we should be clear about how the sample was drawn and whether it was *randomly* selected from the population of interest. If this information is not given in the published results of a survey or experiment, the reader should take the inferences with a grain of salt.

14.5   Communication Hurdles: What's the Sample Size?

Distortions can occur when the sample size is not discussed. For example, suppose that a survey indicates that approximately 75% of a sample favor a new high-rise building complex. Further investigation might reveal that the investigators sampled only four people. When three out of the four favored his project, he decided to stop the survey. Of course, we exaggerate with this example; but we could also have revealed inconclusive results based on a sample of 25, even though many buyers would consider this sample size to be large enough. Very large samples are required to achieve adequate information in sampling binomial populations.

Fortunately, many publications now provide more information about the sample size and how opinion surveys are conducted. Fifteen years ago, news accounts rarely identified how many people were sampled, much less how they were sampled. Things are different now. In fact, sometimes the media have gone too far in an attempt to be completely open about how a survey was done. A case in point is the following article from the *Wall Street Journal* (September 23, 1988). How many readers understand much more than the number of persons sampled and the approximate plus or minus (confidence interval)? A person would have to be well trained in statistics and survey sampling to interpret what was done in this case. Again, the moral of the story is, try to communicate in unambiguous terms.

How Poll Was Conducted

The *Wall Street Journal*/NBC News poll was based on nationwide telephone interviews conducted last Friday through Monday with 4159 adults age 18 or older. There were 2630 likely voters.

The sample was drawn from a complete list of telephone exchanges, chosen so that each region of the country was represented in proportion to its population. Households were selected by a method that gave all telephone numbers, listed and unlisted, a proportionate chance of being included. The results of the survey were weighted to adjust for variations in the sample relating to education, age, race, sex, and region.

Chances are 19 of 20 that, if all adults in the United States had been surveyed using the same questionnaire, the findings would differ from these poll results by no more than two percentage points in either direction. The margin of error for subgroups may be larger.

14.6 The Statistical Report

Now that you have seen various ways to communicate the results of a statistical analysis and some of the more common hurdles to effectively communicating these results, let us address the content of a statistical report that would appear as part of an internal project report in a book or scientific journal article. Obviously companies and journals do not all abide by the same outline for a statistical report; however, based on what we present, you should have sufficient material to rearrange some sections and delete others to satisfy specific requirements.

A statistical report of the results of a study or experiment should clearly reflect all stages of making sense of data, much as we have emphasized these stages throughout the textbook. A general outline for a statistical report is shown in Table 14.1, along with a brief description of the content for each major section of the report.

T A B L E 14.1
General Outline for a Statistical Report

Outline	Stage of Making Sense of Data	Description
Summary		Sometimes this section comprises the abstract of a journal article or scientific paper. It is a short summary of the study results, preferably one page or less.
Introduction	Data-gathering stage	The introduction briefly summarizes the background and rationale for the study (survey) run.
Study design and procedures	Data-gathering stage	This section identifies the study (survey) objective, the study (survey) design, and a summary of procedures for conducting the study (survey), including details about the data-gathering stage.
Descriptive statistics	Data summarization stage	This section includes the main descriptive techniques (such as histograms, scatter plots, means, and standard errors)

(Continued)

Outline	Stage of Making Sense of Data	Description
Discussion		used to summarize the study (survey) data. Do the results confirm or contradict previous study (survey) results? If they contradict previous results, a viable explanation for the difference in results should be offered. Recommendations for further studies or surveys should be made, when appropriate.
Data listings (optional)		Sometimes it is appropriate to provide listings of the data on which the summaries and analyses are based. This section could also include computer output for some (all) of the statistical analyses done in the data-analysis stage of the study.
Statistical methodology	Data analysis stage	This section describes the methods used to analyze the study (survey) data—specifically, the statistical tests and estimation procedures used to address the objectives of the study (survey), including those on which the major conclusions are drawn. References for the tests and estimations procedures may also be included, especially if the intended audience may not be familiar with the techniques.
Results and conclusions	Data-analysis stage	In this section, the authors address the main results of the statistical analyses and the conclusions that can be drawn from these analyses in light of the study (survey) objectives.
Discussion		This section provides an opportunity to interpret the results of the statistical analyses and to put the conclusions in the context of previous studies (surveys).

14.7 Documentation and Storage of Results

The final part of the cycle of data processing, analysis, and summarization consists of documenting and storing results. For formal statistical analyses that are subject to careful scrutiny by others, it is important to provide detailed documentation for all data processing and statistical analyses so that the data trail is clear and the database or workfiles are readily accessible. Then the reviewer can follow what has been done, redo it, or extend the analyses. Before we list the elements of a documentation and storage file, we'll discuss several different categories of analyses.

primary analyses
backup analyses

Primary analyses are those used to address the objectives of the study and the analyses on which conclusions are drawn. **Backup analyses** include alternative methods for examining the data that confirm the results of the primary analyses; they may also include new statistical methods that are not as readily accepted as the more standard methods. Several guidelines for analyses are presented here.

Guidelines for Preliminary, Primary, and Backup Analyses

1   Analyses should be performed with software that has been extensively tested.

2   Computer output should be labeled to reflect which study is analyzed and what subjects (animals, patients, etc.) are used in the analysis, and should include a brief description of the analysis preferred. For example, TITLE statements in SAS are very helpful.

3   Variable labels and value labels (e.g., 0 = none, 1 = mild) should appear on the output.

4   A list of the data used in each analyses should be provided.

5   The output for all analyses should be checked *carefully*. Did the job run successfully? Are the samples sizes, means, and degrees of freedom correct? Other checks may be necessary as well.

6   All preliminary, primary, and backup analyses that provide the informational base from which study conclusions are drawn should be saved.

The elements of a documentation and storage file depend on the particular setting in which you work. The contents for a general documentation storage file are shown next.

Contents of a Study Documentation and Storage File

1   Statistical report

2   Study description

3   Random code (used to assign subjects to treatment groups)

4   Important correspondence

5   File creation information

6   Preliminary, primary, and backup analyses

7   Raw data source

8   Data management sheet, including the log and information on the storage of the data files

The major purpose of the documentation and storage file is to provide a clear data and analysis "trail" for your own use or for someone else's use, should there be a need to revisit the data. For any given situation, you must ask yourself whether such documentation is necessary and, if so, how detailed it must be. A good test of the completeness and understandability of your documentation is to ask a colleague who is unfamiliar with your project but knowledgeable in your field to try to reconstruct and even redo the primary analyses you did. If he or she can navigate through your documentation trail, you have done the job.

Summary

In this chapter, we have discussed how to present the results of a statistical analysis of data, and we have considered some of the problems that can stand in the way of effectively communicating these results to the intended audience. Such obstacles or hurdles include graphical distortions, biased sampling, and failure to include a discussion of the sample size and sampling technique. With some understanding of these obstacles, we can better critique and understand communications aimed at us and also do a better job of communicating the results of our analyses to others.

The final topic in this chapter was the documentation and storage of results. Having completed your analyses, drawn conclusions, and communicated these results to the intended audience, you may be tempted to postpone or eliminate the documentation and storage of results. However, it is worth your time to assess the potential for revisiting your analyses in the future and to determine what steps should be taken to facilitate this process (for you or for others).

Finally, we should put our statistical analyses in the context of the practical problem(s) being addressed. A report of statistical analyses is not necessarily the answer to an important question; it is only *part* of the answer. For example, we may demonstrate that a tablet delivers a drug more quickly than a capsule, but this is not the only consideration in the decision to market the drug in capsule or tablet form. Such factors as cost, palatability, and stability must also be considered. Some of the relevant analyses that address these considerations are not statistical.

Supplementary Exercise

Class Project

Have each person in the class choose a commercial he or she has heard or seen lately and critique it for possible distortions. Make suggestions for improvement with regard to clarity of message, etc. (*Note*: We are not asking you to improve it from a commercial standpoint.) Present these findings.

Databases

Murder Rate Database
and Murder Rate Sample Means Database

Murder rates (per 100,000 inhabitants) are listed for 90 cities from the South, North and West

South	Rate	North	Rate	West	Rate
Atlanta, GA	14	Albany, NY	2	Bakersfield, CA	21
Augusta, GA	13	Allentown, PA	2	Boise, ID	3
Baton Rouge, LA	10	Atlantic City, NJ	10	Colorado Springs, CO	6
Beaumont, TX	12	Canton, IL	6	Denver, CO	9
Birmingham, AL	16	Chicago, IL	14	Eugene, OR	2
Charlotte, NC	14	Cincinnati, OH	6	Fresno, CA	20
Chattanooga, TN	10	Cleveland, OH	16	Honolulu, HI	8
Columbia, SC	10	Detroit, MI	16	Kansas City, MO	15
Corpus Christi, TX	16	Evansville, IN	8	Lawton, OK	6
Dallas, TX	18	Grand Rapids, MI	6	Los Angeles, CA	23
El Paso, TX	12	Johnstown, PA	8	Modesto, CA	10
Fort Lauderdale, FL	17	Kankakee, IL	6	Oklahoma City, OK	12
Greensboro, NC	8	Kenosha, WI	4	Oxnard, CA	7
Jackson, MS	17	Lancaster, PA	2	Pueblo, CO	4
Knoxville, TN	7	Lansing, MI	3	Sacramento, CA	9
Lexington, KY	6	Lima, OH	6	St. Louis, MO	15
Lynchburg, VA	12	Madison, WI	2	Salinas, CA	8
Macon, GA	8	Mansfield, OH	3	Salt Lake City, UT	5
Memphis, TN	20	Milwaukee, WI	6	San Diego, CA	10
Monroe, LA	13	Newark, NJ	11	San Francisco, CA	12
Nashville, TN	14	Paterson, NJ	9	San Jose, CA	8
Newport News, VA	11	Philadelphia, PA	12	Seattle, WA	7

(continued)

South	Rate	North	Rate	West	Rate
Orlando, FL	10	Pittsfield, MA	1	Sioux City, IA	1
Richmond, VA	12	Racine, WI	5	Spokane, WA	4
Roanoke, VA	10	Rockford, IL	5	Stockton, CA	18
San Antonio, TX	18	South Bend, IN	11	Tacoma, WA	5
Shreveport, LA	20	Springfield, IL	6	Topeka, KS	11
Washington, DC	11	Syracuse, NY	3	Tucson, AZ	9
Wichita Falls, KS	13	Vineland, NJ	9	Vallejo, CA	6
Wilmington, DE	8	Youngstown, OH	7	Waco, TX	15

Source: Department of Justice, *Uniform Crime Reports for the United States: 1980* (Washington, DC: U.S. Government Printing Office, 1980), pp. 60–86.

Means of 50 samples of size 5 selected from the Murder Rates Database

12.6	9.5	12.1	6.7	9.6
9.5	13.1	15.6	9.4	9.3
14.9	8.2	12.0	9.1	8.0
11.9	7.1	8.5	10.3	11.0
11.0	9.2	7.9	5.1	9.6
10.7	8.3	6.8	11.3	9.2
9.7	8.7	12.2	12.0	10.3
10.5	9.6	8.3	9.1	7.4
8.2	10.9	14.0	8.3	11.2
7.8	9.4	9.4	5.9	10.3

Clinical Trial Database

The data presented here are from a clinical trial that was conducted to compare the safety and efficacy of three different compounds (A, B, and C) and a placebo (D) in the treatment of patients who exhibited characteristic signs and symptoms of depression. Certain predrug (baseline) determinations were made on each of the 100 patients to determine suitability for the study. Then each patient who qualified for entrance into the study was assigned at random to one of the four treatment groups and was dispensed medication for the duration of the study. Neither the investigator nor the patient knew which medication had been assigned.

At the end of the study, scores on numerous anxiety and depression scales were made. Data descriptions for the variables measured are shown next.

Variable Descriptions

Following is a list of the variables and codes that are used in the accompanying output:

PATIENT   = patient number

AGE   = age (yrs)

MAR_STAT = marital status
   1) single
   2) married
   3) separated or divorced
   4) widowed

COFF_TEA = coffee/tea consumption (cups/day)

TOBACCO = tobacco consumption
   0) none
   1) <1 pack daily
   2) 1 pack daily
   3) >1 pack daily

ALCOHOL = alcohol consumption
   0) none
   1) social drinker (<1 drink weekly)
   2) social drinker (1 to 2 drinks weekly)
   3) 1 to 2 drinks most days
   4) 3 or more drinks most days

TRT_EMOT = previous treatment for emotional problems
   1) psychiatrist
   2) non-psychiatrist physician
   3) both
   4) other

HOSPITAL = hospitalization for emotional problems
   0) no
   1) yes

PSY_DIAG = psychiatric diagnosis
   1) major depressive disorder, single episode
   2) major depressive disorder, recurrent episode
   3) bipolar affective disorder
   4) chronic depressive disorder
   5) atypical depressive disorder
   6) adjustment disorder with depressed mood

ANXIETY   = HAM-D anxiety score
RETARDTN = HAM-D retardation score
SLEEP   = HAM-D sleep disturbance score
TOTAL   = HAM-D total score
OBRIST   = HOPKINS OBRIST cluster total
APPETITE   = appetite disturbance score
CHANGED   = how much has the patient changed
          1) very much improved
          2) much improved
          3) minimally improved
          4) no change
          5) minimally worse
          6) much worse
          7) very much worse
THER_EFF   = therapeutic effect
          1) marked
          2) moderate
          3) minimal
          4) unchanged
          5) worse
ADV_EFF   = adverse effects
          1) none
          2) does not significantly interfere with patient's functioning
          3) significantly interferes with patient's functioning
          4) nullifies therapeutic effect
TREATMNT = drug treatment group
          A
          B
          C
          D-Placebo (control) group

    Additional cross tabulations were generated for the clinical trial data, using SAS. These tabulations should be helpful in solving some of the chapter exercises that refer to this database.

DATA SET

PATIENT	AGE	MAR_STAT	COFF_TEA	TOBACCO	ALCOHOL	TRT_EMOT	HOSPITL	PSY_DIAG	ANXIETY	RETARDED	SLEEP	TOTAL	OBRIST	APPETITE	CHANGED	THER_EFF	ADV_EFF	TREATMNT
1	23	2	3	1	0	0	0	1	0.33	0.75	0.00	16	56	91.0	4	4	1	D
2	18	1	0	2	0	1	0	2	0.33	1.25	0.33	12	57	42.5	3	3	1	A
3	36	2	2	2	2	1	0	2	0.50	0.25	0.33	6	40	91.0	1	1	2	B
4	51	4	5	3	0	0	0	1	0.17	0.75	0.67	6	39	61.0	2	1	1	A
5	24	1	6	2	1	1	0	2	1.00	1.00	1.00	13	49	1.5	3	3	1	B
6	59	4	3	1	0	1	0	4	0.33	1.50	0.00	11	40	72.0	2	2	1	A
7	56	1	2	0	0	1	1	4	1.67	1.75	0.00	21	44	7.5	2	2	1	B
8	70	4	1	0	0	1	0	2	0.50	1.75	0.00	12	39	92.5	2	2	2	A
9	30	3	4	3	2	1	0	2	0.83	1.00	0.67	15	49	2.0	2	2	4	D
10	55	4	2	0	2	0	0	1	0.33	1.00	1.00	11	44	31.0	2	2	1	D
11	40	2	4	2	0	1	0	2	0.83	1.50	1.33	23	79	92.0	4	4	1	C
12	61	2	2	0	1	1	1	2	0.50	0.75	0.00	8	30	1.0	2	2	2	C
13	64	2	3	2	0	1	0	2	0.33	1.50	0.00	9	48	11.5	2	2	3	A
14	19	1	10	2	2	1	0	1	0.50	0.75	0.00	7	42	91.5	2	2	1	B
15	46	3	2	0	1	1	0	1	0.17	0.25	0.00	4	35	92.0	1	1	1	A
16	36	2	10	3	0	1	0	2	0.50	1.50	0.00	9	42	72.0	2	2	1	C
17	30	2	2	0	2	0	0	1	0.00	0.50	0.67	4	35	41.0	2	1	1	B
18	34	2	8	3	1	1	0	1	0.50	1.00	0.33	14	56	91.5	3	3	1	D
19	28	1	2	0	1	0	0	1	0.67	1.00	0.33	12	43	72.5	1	2	2	B
20	33	3	3	2	1	1	0	2	0.67	0.75	0.33	8	39	93.0	2	2	2	C

(continued)

DATA SET (continued)

PATIENT	AGE	MAR_STAT	COFF_TEC	TOBACCO	ALCOHOL	TRT_PAT	HOSPITAL	PSY_DIAG	ANXIETY	RETARD	SLEEP	TOTAL	APPROB	PERCENT	CHANGED	THER_EFF	ADV_EFF	TREATMNT
21	51	3	7	3	2	1	1	5	0.83	2.00	0.00	21	99	42.5	4	4	2	A
22	51	2	0	0	0	1	0	4	0.83	1.00	0.00	12	68	61.0	3	3	3	C
23	54	2	3	0	0	0	1	2	0.83	1.00	1.00	16	49	11.0	3	3	2	B
24	35	3	5	0	2	1	0	4	0.67	1.50	0.00	11	42	61.0	2	2	3	A
25	46	2	3	0	1	0	1	2	0.67	2.00	1.33	17	63	61.5	4	4	1	D
26	34	2	7	0	0	1	1	2	0.33	0.25	0.33	6	51	12.5	2	2	2	C
27	27	3	2	1	2	1	0	5	0.83	0.75	0.00	11	58	11.0	2	2	2	A
28	23	2	0	1	0	0	0	1	0.33	0.75	0.00	6	47	91.5	2	2	2	C
29	35	3	6	0	3	1	0	5	0.33	0.50	0.00	9	68	1.0	3	2	2	B
30	19	1	2	0	0	0	0	4	0.67	0.50	0.00	11	47	2.5	3	3	2	B
31	40	3	3	0	3	0	0	1	1.00	1.75	0.00	17	71	62.0	4	4	1	D
32	52	2	5	0	0	1	1	2	0.83	1.75	0.67	19	41	91.5	3	3	2	B
33	51	3	6	0	0	1	0	2	0.83	1.00	1.67	20	63	11.0	3	4	2	D
34	34	3	0	0	2	1	0	4	0.17	0.50	0.33	5	37	13.0	2	2	1	C
35	59	2	4	2	0	1	0	1	0.67	2.50	0.00	17	54	62.5	3	3	2	C
36	31	3	2	0	1	1	1	5	0.50	1.75	0.33	12	51	11.5	3	2	2	C
37	54	2	10	0	2	1	0	2	1.17	1.25	1.33	22	96	92.5	5	5	1	A
38	63	4	2	0	2	1	0	1	0.33	0.25	0.33	7	58	32.5	2	2	2	B
39	34	2	1	2	2	0	1	1	0.50	0.25	0.33	27	90	92.0	5	5	1	D
40	30	1	1	0	1	1	0	1	1.00	0.50	0.67	13	59	33.0	2	2	1	A
41	32	3	2	3	1	1	0	2	0.67	1.25	0.67	14	58	61.0	2	2	2	C
42	21	1	2	0	3	1	0	2	0.83	1.00	1.00	20	60	41.5	3	3	2	B
43	42	2	1	2	1	1	1	2	1.00	1.50	1.00	24	85	62.0	3	2	1	B
44	60	2	0	3	0	1	0	4	0.17	0.75	0.33	5	39	1.5	1	1	1	A
45	53	2	2	0	1	1	0	2	0.67	1.00	0.67	10	38	31.0	2	2	2	B
46	54	4	4	0	1	1	0	4	1.50	1.75	0.00	14	42	93.5	2	2	1	C
47	38	2	2	1	0	1	0	2	1.50	1.75	1.00	24	85	15.0	3	4	2	D
48	41	2	4	0	0	1	1	2	0.33	0.75	0.00	11	47	2.5	2	2	1	A
49	32	3	0	0	1	1	0	4	1.00	1.00	1.00	20	35	42.5	3	2	2	B
50	43	2	4	0	0	1	0	4	0.83	1.25	1.33	21	44	31.5	4	4	1	B

DATA SET (continued)

PATIENT	AGE	MAR_STAT	COFF_TEA	TOBACCO	ALCOHOL	TRT_EMOT	HOSPITAL	PSY_DIAG	ANXIETY	RETARDTN	SLEEP	TOTAL	OBRIST	APPETITE	CHANGED	THER_EFF	ADV_EFF	TREATMNT
51	51	2	1	0	1	1	0	2	0.83	2.25	0.00	20	80	61.0	5	5	1	A
52	23	2	0	1	3	1	0	2	1.33	1.25	0.67	20	39	31.5	3	3	1	A
53	55	2	2	0	0	1	0	1	0.83	1.25	0.00	16	52	12.5	3	3	1	C
54	45	1	3	3	0	1	0	2	0.33	1.25	0.00	14	46	41.5	2	2	1	C
55	30	2	1	0	0	0	0	1	1.17	1.75	0.00	17	64	2.0	3	3	2	C
56	53	4	1	3	4	0	0	1	0.83	0.50	0.00	19	82	62.0	3	4	2	D
57	45	1	3	1	3	0	0	4	0.83	0.50	0.00	8	40	2.0	1	2	1	A
58	48	2	10	1	2	1	0	4	0.33	1.00	0.00	8	32	41.0	2	2	2	B
59	49	1	4	0	3	1	0	5	0.67	1.75	0.33	16	68	12.0	2	2	2	A
60	55	2	6	0	2	0	1	4	0.50	1.00	0.00	9	42	91.0	2	2	2	A
61	33	2	1	0	1	1	0	2	1.17	2.25	2.00	32	112	42.0	4	4	1	D
62	27	1	1	2	3	1	0	5	0.17	0.00	0.00	3	34	72.5	1	1	2	C
63	30	1	2	1	3	1	0	4	0.67	0.00	0.00	6	37	73.5	1	1	2	A
64	35	2	4	3	0	1	0	2	0.50	1.50	0.33	16	43	11.0	3	3	1	A
65	55	2	4	0	0	1	0	2.	1.00	2.00	0.33	24	37	92.0	3	3	1	B
66	22	3	0	1	1	1	0	2	1.00	1.50	0.33	20	46	11.5	3	3	1	D
67	37	3	1	2	0	1	0	2	0.50	0.75	0.00	11	32	41.5	2	2	1	C
68	49	2	6	3	0	1	0	2	0.50	0.75	0.00	13	54	2.5	2	1	1	C
69	21	1	0	1	0	1	0	2	1.17	2.50	2.00	34	74	11.5	5	5	1	A
70	33	3	1	3	2	1	0	1	0.17	0.50	0.00	8	34	42.5	1	1	1	C
71	35	3	10	2	0	1	0	2	1.17	2.50	0.00	24	39	92.5	4	4	2	B
72	39	3	0	0	0	1	0	2	0.50	1.25	0.00	22	66	41.0	2	3	1	D
73	34	1	0	0	0	1	0	2	0.33	0.75	0.00	14	48	71.0	2	2	1	C
74	53	3	3	1	0	1	0	2	0.50	1.00	0.00	12	39	43.5	2	2	1	A
75	34	2	0	0	0	1	0	2	0.33	0.50	0.00	10	36	3.0	2	2	1	A
76	35	2	2	0	1	1	0	2	0.83	1.00	0.33	14	35	92.0	2	2	1	B
77	32	2	0	0	0	1	0	2	0.50	1.00	0.00	12	39	91.0	2	2	1	B
78	43	2	2	2	0	1	0	2	0.17	0.50	0.67	5	57	41.5	2	2	1	C
79	64	2	3	0	0	0	1	1	1.33	1.50	1.67	26	42	72.5	5	5	1	D
80	31	3	2	2	0	1	0	2	0.17	0.00	0.00	2	35	43.0	2	1	1	B

DATA SET (continued)

PATIENT	AGE	MAR_STAT	COFF_TEA	TOBBACCO	ALCOHOL	TRT_EMOT	HOSPITAL	PSY_DIAG	ANXIETY	RETARDED	SLEEP	TOTAL	OBRIST	APPETITE	CHANGED	THER_EFF	ADV_EFF	TREATMNT
81	41	2	2	2	1	0	0	1	0.83	1.25	0.00	16	66	41.5	3	3	1	C
82	53	2	2	0	0	1	0	2	1.67	1.50	1.67	23	47	32.0	6	5	4	D
83	61	2	3	2	1	1	0	2	0.83	1.25	0.00	15	43	31.0	3	2	1	C
84	36	2	2	2	0	1	0	4	1.00	1.75	0.00	16	63	42.0	3	3	1	C
85	29	1	0	0	0	0	1	1	0.33	0.25	0.00	5	53	12.0	1	1	1	A
86	25	2	8	3	2	1	0	1	1.17	1.25	0.00	14	80	91.0	4	4	4	A
87	55	2	1	0	0	1	0	2	0.67	0.50	2.00	16	68	31.0	2	2	2	D
88	34	2	3	0	0	0	0	1	0.67	0.00	0.00	5	31	1.0	1	1	2	C
89	20	1	0	0	0	1	0	2	0.50	0.75	1.67	15	36	43.0	3	2	1	B
90	33	2	1	2	2	1	1	2	0.83	1.25	0.67	16	63	61.5	2	2	1	B
91	44	2	0	0	0	1	0	2	0.33	1.75	1.00	16	55	42.5	4	4	1	B
92	58	2	3	0	0	1	1	2	1.17	1.75	1.00	21	75	11.5	3	4	1	D
93	46	4	3	0	2	1	0	1	0.83	1.25	1.00	16	83	13.5	3	3	2	D
94	31	2	2	0	1	1	0	4	1.17	1.50	0.00	21	65	61.5	4	4	2	D
95	29	3	2	2	2	0	0	1	0.50	1.00	0.67	20	92	32.5	4	4	2	D
96	50	3	3	0	2	1	0	1	0.83	0.25	0.67	18	50	91.0	2	1	2	D
97	27	2	0	0	0	1	0	2	0.67	1.00	2.00	17	64	62.5	3	3	2	D
98	31	3	5	0	1	1	0	2	0.50	0.75	0.67	19	74	61.5	3	3	1	D
99	69	4	4	1	2	1	0	1	1.67	2.25	1.33	26	87	63.5	5	5	3	D
100	41	2	0	0	0	1	0	2	0.00	0.50	0.00	3	37	92.5	2	2	1	D

CLINICAL TRIAL—FREQUENCY COUNTS BY TREATMENT

TABLE OF TREATMNT BY MAR_STAT

TREATMNT	MAR_STAT	MARITAL STATUS			
FREQUENCY	1	2	3	4	TOTAL
A	7	10	5	3	25
B	7	13	4	1	25
C	3	15	6	1	25
D	0	13	8	4	25
TOTAL	17	51	23	9	100

CLINICAL TRIAL—FREQUENCY COUNTS BY TREATMENT

TABLE OF TREATMNT BY TOBACCO

TREATMNT	TOBACCO	TOBACCO DAILY CONSUMPTION			
FREQUENCY	0	1	2	3	TOTAL
A	11	7	2	5	25
B	17	1	7	0	25
C	10	1	9	5	25
D	16	4	2	3	25
TOTAL	54	13	20	13	100

CLINICAL TRIAL—FREQUENCY COUNTS BY TREATMENT

TABLE OF TREATMNT BY ALCOHOL

TREATMNT	ALCOHOL	HISTORY OF ALCOHOL USE				
FREQUENCY	0	1	2	3	4	TOTAL
A	12	3	6	4	0	25
B	11	6	6	2	0	25
C	15	7	2	1	0	25
D	10	6	7	1	1	25
TOTAL	48	22	21	8	1	100

CLINICAL TRIAL—FREQUENCY COUNTS BY TREATMENT

TABLE OF TREATMNT BY TRT_EMOT

TREATMNT	TRT_EMOT		PREVIOUS TRT FOR EMOTIONAL PROBLEMS
FREQUENCY	0	1	TOTAL
A	4	21	25
B	4	21	25
C	4	21	25
D	8	17	25
TOTAL	20	80	100

CLINICAL TRIAL—FREQUENCY COUNTS BY TREATMENT

TABLE OF TREATMNT BY HOSPITAL

TREATMNT	HOSPITAL		EVER HOSPITALIZED FOR EMOTIONAL PROBLEMS
FREQUENCY	0	1	TOTAL
A	21	4	25
B	20	5	25
C	22	3	25
D	21	4	25
TOTAL	84	16	100

CLINICAL TRIAL—FREQUENCY COUNTS BY TREATMENT

TABLE OF TREATMNT BY PSY_DIAG

TREATMNT	PSY_DIAG		PSYCHIATRIC DIAGNOSIS		
FREQUENCY	1	2	4	5	TOTAL
A	5	11	6	3	25
B	4	15	5	1	25
C	7	12	4	2	25
D	11	13	1	0	25
TOTAL	27	51	16	6	100

CLINICAL TRIAL—FREQUENCY COUNTS BY TREATMENT

TABLE OF TREATMNT BY CHANGED

TREATMNT	CHANGED	HOW MUCH HAS PATIENT CHANGED?					
FREQUENCY	1	2	3	4	5	6	TOTAL
A	5	12	3	2	3	0	25
B	2	10	10	3	0	0	25
C	3	13	8	1	0	0	25
D	0	6	9	6	3	1	25
TOTAL	10	41	30	12	6	1	100

CLINICAL TRIAL—FREQUENCY COUNTS BY TREATMENT

TABLE OF TREATMNT BY THER_EFF

TREATMNT	THER_EFF	THERAPEUTIC EFFECT				
FREQUENCY	1	2	3	4	5	TOTAL
A	5	12	3	2	3	25
B	3	13	6	3	0	25
C	4	14	6	1	0	25
D	1	4	6	10	4	25
TOTAL	13	43	21	16	7	100

CLINICAL TRIAL—FREQUENCY COUNTS BY TREATMENT

TABLE OF TREATMNT BY ADV_EFF

TREATMNT	ADV_EFF	SEVERITY OF ADVERSE EFFECTS			
FREQUENCY	1	2	3	4	TOTAL
A	16	6	2	1	25
B	13	12	0	0	25
C	14	10	1	0	25
D	13	9	1	2	25
TOTAL	56	37	4	3	100

CLINICAL TRIAL—MEANS BY TREATMENT

Variable	Label	N	Mean	Standard Deviation
				TREATMNT=A
Age		25	42.2400000	14.6380782
Coff_tee	Coffee or tea cups/day	25	2.9600000	2.7306898
Anxiety	Anxiety/somatization factor	25	0.6264000	0.3481245
Retardtn	Retardation factor	25	1.1300000	0.6419372
Sleep	Sleep disturbance factor	25	0.2664000	0.4906193
Total	Total score	25	12.7200000	6.7485801
Obrist	Obrist cluster total	25	53.1600000	19.0999127
Appetite	Appetite disturbance	25	42.3400000	33.8334524
Changed	How much has patient changed?	25	2.4400000	1.2609520
				TREATMNT=B
Age		25	37.6000000	12.8452326
Coff_tee	Coffee or tea cups/day	25	3.2000000	3.0550505
Anxiety	Anxiety/somatization factor	25	0.6928000	0.3595775
Retardtn	Retardation factor	25	1.0500000	0.5863020
Sleep	Sleep disturbance factor	25	0.5332000	0.4911867
Total	Total score	25	14.0400000	6.4838260
Obrist	Obrist cluster total	25	45.9600000	12.6671491
Appetite	Appetite disturbance	25	50.0200000	32.0898738
Changed	How much has patient changed?	25	2.5600000	0.8205689
				TREATMNT=C
Age		25	40.5200000	10.9549076
Coff_tee	Coffee or tea cups/day	25	2.6000000	2.3452079
Anxiety	Anxiety/somatization factor	25	0.6000000	0.3329289
Retardtn	Retardation factor	25	1.0400000	0.6110101
Sleep	Sleep disturbance factor	25	0.1596000	0.3205137
Total	Total score	25	11.4800000	4.9843087
Obrist	Obrist cluster total	25	48.8800000	12.9077000
Appetite	Appetite disturbance	25	44.3000000	31.6497762
Changed	How much has patient changed?	25	2.2800000	0.7371115
				TREATMNT=D
Age		25	42.0800000	12.9193653
Coff_tee	Coffee or tea cups/day	25	2.4800000	1.9390719
Anxiety	Anxiety/somatization factor	25	0.8400000	0.4243819
Retardtn	Retardation factor	25	1.1500000	0.5728220
Sleep	Sleep disturbance factor	25	0.8536000	0.7015155
Total	Total score	25	19.2000000	5.7445626
Obrist	Obrist cluster total	25	66.6800000	18.6049277
Appetite	Appetite disturbance	25	49.5600000	29.5967059
Changed	How much has patient changed?	25	3.3600000	1.1135529

Minimum Value	Maximum Value	Std Error Of Mean	Sum	Variance
18.0000000	70.0000000	2.92761564	1056.00000	214.27333
0.0000000	10.0000000	0.54613796	74.00000	7.45667
0.1700000	1.3300000	0.06962490	15.66000	0.12119
0.0000000	2.5000000	0.12838743	28.25000	0.41208
0.0000000	2.0000000	0.09812387	6.66000	0.24071
4.0000000	34.0000000	1.34971602	318.00000	45.54333
35.0000000	99.0000000	3.81998255	1329.00000	364.80667
1.5000000	92.5000000	6.76669048	1058.50000	1144.70250
1.0000000	5.0000000	0.25219040	61.00000	1.59000
19.0000000	63.0000000	2.56904652	940.00000	165.00000
0.0000000	10.0000000	0.61101009	80.00000	9.33333
0.0000000	1.6700000	0.07191551	17.32000	0.12930
0.0000000	2.5000000	0.11726039	26.25000	0.34375
0.0000000	1.6700000	0.09823733	13.33000	0.24126
2.0000000	24.0000000	1.29676521	351.00000	42.04000
32.0000000	85.0000000	2.53342982	1149.00000	160.45667
1.0000000	92.5000000	6.41797476	1250.50000	1029.76000
1.0000000	4.0000000	0.16411378	64.00000	0.67333
23.0000000	61.0000000	2.19098152	1013.00000	120.01000
0.0000000	10.0000000	0.46904158	65.00000	5.50000
0.1700000	1.5000000	0.06658578	15.00000	0.11084
0.0000000	2.5000000	0.12220202	26.00000	0.37333
0.0000000	1.3300000	0.06410273	3.99000	0.10273
3.0000000	23.0000000	0.99686174	287.00000	24.84333
30.0000000	79.0000000	2.58154000	1222.50000	166.60900
1.0000000	93.5000000	6.32995524	1107.50000	1001.70833
1.0000000	4.0000000	0.14742230	57.00000	0.54333
22.0000000	69.000000	2.58387306	1052.00000	166.910000
0.0000000	8.000000	0.38781439	62.00000	3.760000
0.0000000	1.670000	0.08487638	21.00000	0.180100
0.2500000	2.250000	0.11456400	29.75000	0.328125
0.0000000	2.000000	0.14030310	21.34000	0.492124
3.0000000	32.000000	1.14891253	480.00000	33.000000
37.0000000	112.000000	3.72098553	1667.00000	346.143333
2.0000000	92.500000	5.91934118	1239.00000	875.965000
2.0000000	6.000000	0.22271057	84.00000	1.240000

Insurance Claims Database

A division of a large insurance company was considering expanding its sales efforts for automobile insurance in the Midwest region of the country. As part of its study, the Claims Department collected data on the size of claims for collision damage in the Midwest over the past year. The data (in $1000s) are shown for a key city in the Midwest region.

INSURANCE CLAIMS SIZES (in $1000s)

(1) 20.5	(19) 4.6	(37) 2.6	(55) 3.3	(73) 4.2	(91) 15.2
(2) 4.8	(20) 0.8	(38) 4.3	(56) 0.7	(74) 8.9	(92) 5.7
(3) 5.3	(21) 19	(39) 10.1	(57) 14.4	(75) 6.5	(93) 1
(4) 4.6	(22) 7.9	(40) 6.2	(58) 7.4	(76) 8.7	(94) 13.1
(5) 0.7	(23) 0.9	(41) 7.4	(59) 2.7	(77) 1.1	(95) 20.9
(6) 1.1	(24) 4	(42) 8.5	(60) 2.7	(78) 3.5	(96) 19.4
(7) 5.9	(25) 0.7	(43) 2.7	(61) 1.2	(79) 3	(97) 1.6
(8) 0.7	(26) 12.9	(44) 14.2	(62) 4.9	(80) 1.1	(98) 2
(9) 4.6	(27) 3.2	(45) 1.3	(63) 26.6	(81) 4.2	(99) 1
(10) 7.7	(28) 2.5	(46) 1.2	(64) 10.1	(82) 4.9	(100) 7.6
(11) 0.7	(29) 1.1	(47) 8.3	(65) 11	(83) 2.5	(101) 16.8
(12) 5.8	(30) 3.7	(48) 5.2	(66) 2.2	(84) 0.9	(102) 1.6
(13) 4.7	(31) 4.2	(49) 7.1	(67) 3.1	(85) 1.1	(103) 0.8
(14) 0.9	(32) 1.3	(50) 1.9	(68) 0.7	(86) 7.7	(104) 3
(15) 1.2	(33) 8.3	(51) 11.3	(69) 6.1	(87) 4.7	(105) 1.9
(16) 0.7	(34) 0.7	(52) 1.8	(70) 7.7	(88) 3.6	(106) 22.6
(17) 2.3	(35) 3.5	(53) 8.8	(71) 0.8	(89) 5.3	(107) 4.2
(18) 2	(36) 1.3	(54) 7.3	(72) 6.5	(90) 5.8	(108) 1.4

(109) 3.6	(127) 0.7	(145) 4.3	(163) 3.5	(181) 0.8
(110) 1.6	(128) 0.9	(146) 7.1	(164) 8.5	(182) 0.7
(111) 6.1	(129) 3.5	(147) 1.3	(165) 1.4	(183) 13.3
(112) 7.2	(130) 0.9	(148) 7.3	(166) 11.4	(184) 1.7
(113) 7.1	(131) 1.1	(149) 5	(167) 1.2	(185) 4.1
(114) 4.4	(132) 33.7	(150) 3	(168) 1.9	(186) 0.8
(115) 2.9	(133) 7.4	(151) 0.9	(169) 6.7	(187) 20.7
(116) 3	(134) 0.9	(152) 2.9	(170) 1.8	
(117) 0.7	(135) 2.1	(153) 6	(171) 1.8	
(118) 4.4	(136) 1.8	(154) 8.2	(172) 3.6	
(119) 1.9	(137) 2.5	(155) 5.7	(173) 3.5	
(120) 0.9	(138) 12.9	(156) 3.3	(174) 8.8	
(121) 1.3	(139) 11.5	(157) 3.7	(175) 4.6	
(122) 9.7	(140) 3.8	(158) 4.6	(176) 2.6	
(123) 2	(141) 13.2	(159) 0.7	(177) 1.9	
(124) 3.4	(142) 1.6	(160) 2.1	(178) 17.4	
(125) 2.3	(143) 0.8	(161) 8.6	(179) 1.2	
(126) 0.8	(144) 4.7	(162) 3.3	(180) 4.4	

Automatic Transmission Database

The internal fluid pressure of automatic transmissions for automobiles is critical to the performance of the transmission (and hence of the car). A manufacturer of automatic transmissions randomly samples five transmissions from each day of production and checks the internal fluid pressure in each one. The results of forty consecutive days of testing the sampled transmissions is shown here. A value of 55 (in standardized units) is considered ideal for internal fluid pressure. Too low pressure results in sluggish performance of the transmission, with a value of 50 or less yielding poor performance. Too high levels of internal fluid pressure results in "jumpy" overperformance; a value of 60 or more gives poor performance.

AUTOMATIC TRANSMISSION VALUES

DAY	item1	item2	item3	item4	item5	mean
1	56.01	54.46	54.90	53.83	54.61	54.762
2	56.06	56.26	55.44	53.86	55.88	55.500
3	54.46	56.17	54.07	54.29	54.29	54.656
4	55.08	54.68	54.37	54.50	54.40	54.606
5	54.11	55.84	55.67	54.55	55.62	55.158
6	54.58	55.90	54.35	55.25	54.18	54.852
7	56.04	54.45	54.22	55.09	54.68	54.896
8	54.98	55.19	55.70	54.91	52.97	54.750
9	56.48	55.95	54.53	56.25	56.08	55.858
10	55.30	55.98	55.36	53.83	54.56	55.006
11	53.56	53.95	56.38	54.90	54.86	54.730
12	54.96	56.78	56.56	54.32	55.25	55.574
13	54.39	53.16	54.31	54.43	56.33	54.524
14	52.88	54.62	55.70	55.77	53.83	54.560
15	52.81	54.27	53.19	56.02	55.94	54.446
16	52.77	53.20	53.60	55.75	54.57	53.978
17	54.88	53.37	54.69	54.02	53.30	54.052
18	56.06	54.49	53.40	55.03	56.63	55.122
19	56.17	52.64	55.90	54.75	55.22	54.936
20	55.85	55.00	53.31	54.58	57.37	55.222
21	55.10	55.39	55.37	54.33	57.28	55.494
22	57.29	52.77	53.54	57.45	55.14	55.238
23	54.44	55.87	55.52	55.03	54.13	54.998
24	54.94	55.97	56.30	58.22	54.72	56.030
25	55.56	55.59	54.63	53.56	56.84	55.236
26	55.48	54.74	56.51	56.76	54.13	55.524
27	55.31	56.87	52.82	53.55	53.47	54.404
28	53.69	54.01	55.16	53.87	54.93	54.332
29	53.32	56.22	52.12	56.01	55.60	54.654

(**continued**)

AUTOMATIC TRANSMISSION VALUES (continued)

DAY	item1	item2	item3	item4	item5	mean
30	55.14	56.57	56.37	56.98	56.66	56.344
31	54.93	56.02	55.10	55.58	56.62	55.650
32	54.93	55.45	52.16	56.25	55.05	54.768
33	56.64	57.00	55.39	54.87	56.76	56.132
34	57.85	54.97	54.68	55.48	54.07	55.410
35	54.02	56.64	57.62	55.91	54.15	55.668
36	54.51	55.42	54.81	55.00	53.74	54.696
37	53.62	55.41	54.78	51.78	53.88	53.894
38	53.06	55.90	56.96	54.96	52.72	54.720
39	52.04	53.22	53.76	53.44	56.76	53.844
40	54.99	56.20	54.73	53.87	53.79	54.716

Violent Crime Rates Database

The Violent Crime Rates Database gives the rate of violent crimes for each of 30 cities selected at random from the South, North, and West. Rates shown represent the number of violent crimes (murder, forcible rape, robbery and aggravated assault) per 100,000 inhabitants during the reporting period.

VIOLENT CRIME RATES PER 100,000 INHABITANTS

South	Rate	North	Rate	West	Rate
Albany, GA	876	Allentown, PA	189	Abilene, TX	570
Anderson, SC	578	Battle Creek, MI	661	Albuquerque, NM	928
Anniston, AL	718	Benton Harbor, MI	877	Anchorage, AK	516
Athens, GA	388	Bridgeport, CT	563	Bakersfield, CA	885
Augusta, GA	562	Buffalo, NY	647	Brownsville, TX	751
Baton Rouge, LA	971	Canton, OH	447	Denver, CO	561
Charleston, SC	698	Cincinnati, OH	336	Fresno, CA	1,020
Charlottesville, VA	298	Cleveland, OH	526	Galveston, TX	592
Chattanooga, TN	673	Columbus, OH	624	Houston, TX	814
Columbus, GA	537	Dayton, OH	605	Kansas City, MO	843
Dothan, AL	642	Des Moines, IA	496	Lawton, OK	466
Florence, SC	856	Dubuque, IA	296	Lubbock, TX	498
Fort Smith, AR	376	Gary, IN	628	Merced, CA	562
Gadsden, AL	508	Grand Rapids, MI	481	Modesto, CA	739
Greensboro, NC	529	Janesville, WI	224	Oklahoma City, OK	562
Hickory, NC	393	Kalamazoo, MI	868	Reno, NV	817
Knoxville, TN	354	Lima, OH	804	Sacramento, CA	690
Lake Charles, LA	735	Madison, WI	210	St. Louis, MO	720
Little Rock, AR	811	Milwaukee, WI	421	Salinas, CA	758

South	Rate	North	Rate	West	Rate
Macon, GA	504	Minneapolis, MN	435	San Diego, CA	731
Monroe, LA	807	Nassau, NY	291	Santa Ana, CA	480
Nashville, TN	719	New Britain, CT	393	Seattle, WA	559
Norfolk, VA	464	Philadelphia, PA	605	Sioux City, IA	505
Raleigh, NC	410	Pittsburgh, PA	341	Stockton, CA	703
Richmond, VA	491	Portland, ME	352	Tacoma, WA	809
Savannah, GA	557	Racine, WI	374	Tucson, AZ	706
Shreveport, LA	771	Reading, PA	267	Victoria, TX	631
Washington, DC	685	Saginaw, MI	684	Waco, TX	626
Wilmington, DE	448	Syracuse, NY	685	Wichita Falls, TX	639
Wilmington, NC	571	Worcester, MA	460	Yakima, WA	585

Note: Rates represent the number of violent crimes (murder, forcible rape, robbery, and aggravated assault) per 100,000 inhabitants, rounded to the nearest whole number.
Source: Department of Justice, Uniform Crime Reports for the United States, 1990.

Crimes Perception Database

A random sample of students majoring in four different departments at a west coast state university were asked: "Which of the following acts do you personally think should be publicly regarded as crimes?" The acts presented were: aggravated assault, armed robbery, arson, atheism, automobile theft, burglary, civil disobedience, communism, drug addiction, embezzlement, forcible rape, gambling, homosexuality, land fraud, masturbation, Nazism, payola (kickbacks), price fixing, prostitution, sexual abuse of children, sexual discrimination, shoplifting, strip mining, treason, and vandalism.

Variable Descriptions

Each student is is identified with a number. Descriptions for the variables measured are as follows:

STUDENT:	student number
CRIMES:	number of acts regarded as a crime
COLLEGE:	year in college (freshman, sophomore, junior, or senior)
SEX:	sex (M or F)
AGE:	age (years)
MAR_STAT:	marital status (single or married)
RELIGION:	religious preference (Protestant, Catholic, Jewish, other, or none)
RACE:	race (black, white, or other)
INCOME:	income of parents (low, medium, or high)
MAJOR:	major in college (business, humanities, natural sciences, social sciences)

CRIMES PERCEPTION DATA

STUDENT	CRIMES	COLLEGE	SEX	AGE	MAR—STAT	RELIGION	RACE	INCOME	MAJOR
001	25	Junior	F	26	Single	Protestant	Other	High	SocSciences
002	09	Sophomore	F	20	Single	Other	Other	High	SocSciences
003	18	Sophomore	F	21	Single	Catholic	Other	Medium	SocSciences
004	18	Junior	M	30	Married	Catholic	White	Medium	SocSciences
005	18	Freshman	F	31	Married	None	White	Low	SocSciences
006	20	Junior	M	20	Single	Catholic	Other	High	SocSciences
007	17	Senior	M	28	Married	Other	White	Medium	SocSciences
008	16	Junior	F	26	Married	Other	White	Medium	SocSciences
009	17	Senior	F	21	Single	Other	White	Medium	SocSciences
010	16	Freshman	M	24	Single	Protestant	White	Medium	SocSciences
011	18	Senior	M	21	Single	Jewish	White	Medium	SocSciences
012	12	Senior	M	27	Married	Protestant	White	High	SocSciences
013	18	Sophomore	M	27	Married	Protestant	Black	High	SocSciences
014	17	Junior	F	26	Married	None	White	Medium	SocSciences
015	15	Junior	M	20	Single	None	White	Medium	SocSciences
016	22	Sophomore	F	29	Single	None	Black	High	SocSciences
017	18	Junior	F	32	Single	None	Black	Low	SocSciences
018	17	Junior	F	19	Single	Protestant	Black	Medium	SocSciences
019	11	Sophomore	F	30	Single	None	Other	Low	SocSciences
020	18	Junior	F	21	Single	None	Black	Medium	SocSciences
021	16	Junior	M	24	Married	None	White	High	SocSciences
022	17	Junior	F	31	Married	None	White	High	SocSciences
023	15	Senior	M	32	Single	None	White	High	SocSciences
024	18	Junior	M	20	Single	None	White	Low	SocSciences
025	14	Freshman	F	18	Single	Protestant	White	Medium	SocSciences
026	17	Junior	F	21	Married	Protestant	White	High	SocSciences
027	17	Senior	M	29	Single	None	Black	High	SocSciences
028	19	Senior	F	19	Single	Protestant	White	Medium	SocSciences
029	16	Senior	F	33	Married	Protestant	White	High	SocSciences
030	21	Junior	F	21	Single	Catholic	Black	High	SocSciences
031	16	Senior	M	32	Married	Protestant	White	Low	SocSciences
032	14	Senior	F	23	Single	None	White	High	SocSciences
033	20	Senior	M	26	Single	Jewish	White	High	SocSciences
034	16	Senior	M	23	Single	Catholic	White	High	SocSciences
035	12	Senior	F	22	Single	None	White	High	SocSciences
036	16	Senior	M	25	Married	Protestant	White	Medium	SocSciences
037	15	Junior	M	19	Single	Other	White	High	SocSciences
038	16	Senior	F	35	Married	Other	Black	Medium	SocSciences

STUDENT	CRIMES	COLLEGE	SEX	AGE	MAR—STAT	RELIGION	RACE	INCOME	MAJOR
039	19	Junior	F	21	Single	Protestant	Black	Medium	SocSciences
040	21	Senior	F	34	Married	Catholic	Black	Medium	SocSciences
041	18	Senior	F	22	Married	Catholic	Other	Medium	SocSciences
042	12	Junior	M	33	Married	Protestant	Black	Medium	Humanities
043	19	Sophomore	F	18	Single	None	White	Medium	Humanities
044	18	Senior	F	32	Married	None	White	Low	Humanities
045	14	Senior	M	25	Single	None	White	Medium	Humanities
046	20	Senior	F	30	Married	Catholic	White	Medium	Humanities
047	16	Senior	F	22	Single	Other	White	High	Humanities
048	12	Sophomore	F	18	Single	Protestant	Other	High	Humanities
049	18	Senior	F	20	Single	Catholic	White	Medium	Humanities
050	21	Sophomore	F	21	Single	Catholic	White	Medium	Humanities
051	16	Senior	F	22	Single	Protestant	White	High	Humanities
052	12	Junior	F	23	Single	Other	Black	Low	Humanities
053	19	Senior	F	32	Married	Protestant	White	High	Humanities
054	21	Junior	F	22	Married	Other	White	Medium	Humanities
055	15	Junior	M	31	Married	None	Other	Low	Humanities
056	18	Senior	F	24	Married	Jewish	White	High	Humanities
057	18	Senior	F	33	Married	Protestant	White	High	Humanities
058	20	Sophomore	M	31	Married	Protestant	White	Medium	Humanities
059	16	Junior	F	32	Married	None	White	High	Humanities
060	13	Freshman	F	18	Single	None	White	High	Humanities
061	16	Senior	F	30	Married	Catholic	Other	High	Humanities
062	20	Senior	F	27	Married	Protestant	Black	Medium	Humanities
063	12	Junior	M	28	Married	None	White	Low	Humanities
064	20	Senior	F	24	Single	Jewish	White	High	Humanities
065	17	Junior	M	37	Married	Protestant	White	Low	Humanities
066	21	Senior	M	35	Married	Other	White	High	Humanities
067	17	Junior	M	22	Single	None	White	High	Humanities
068	18	Sophomore	M	19	Single	None	White	Low	Humanities
069	17	Junior	M	21	Single	None	White	High	Humanities
070	18	Junior	F	21	Single	Other	White	Medium	Humanities
071	17	Senior	M	23	Single	None	White	High	Humanities
072	18	Junior	F	24	Single	Protestant	Other	Low	Humanities
073	13	Junior	M	22	Married	None	White	Medium	Humanities
074	25	Senior	F	32	Married	Protestant	Other	Medium	Humanities
075	20	Junior	F	20	Single	Protestant	White	High	Humanities
076	20	Senior	F	21	Single	None	White	Medium	Humanities

(continued)

STUDENT	CRIMES	COLLEGE	SEX	AGE	MAR—STAT	RELIGION	RACE	INCOME	MAJOR
077	17	Senior	F	25	Single	Catholic	Other	Medium	Humanities
078	15	Junior	F	24	Married	Catholic	Other	Medium	Humanities
079	14	Senior	M	23	Single	Other	Other	High	NatSciences
080	16	Senior	F	26	Married	None	White	Low	NatSciences
081	18	Senior	F	22	Married	None	White	High	NatSciences
082	09	Sophomore	F	18	Single	Other	Other	Medium	NatSciences
083	18	Sophomore	F	20	Single	None	Black	Medium	NatSciences
084	17	Sophomore	M	18	Single	None	Other	Medium	NatSciences
085	15	Senior	M	27	Married	Catholic	Other	Low	NatSciences
086	19	Senior	F	35	Single	None	White	Medium	NatSciences
087	14	Junior	F	20	Single	None	White	Medium	NatSciences
088	15	Sophomore	F	19	Single	None	White	High	NatSciences
089	20	Sophomore	F	26	Single	Catholic	White	Medium	NatSciences
090	15	Senior	F	24	Married	Catholic	Other	Medium	NatSciences
091	17	Junior	F	20	Single	None	White	Medium	NatSciences
092	24	Junior	M	27	Single	Protestant	White	Low	NatSciences
093	15	Junior	F	20	Single	Protestant	White	Medium	NatSciences
094	19	Senior	M	23	Single	Catholic	White	Medium	NatSciences
095	18	Junior	F	19	Single	None	Other	High	NatSciences
096	16	Junior	F	23	Single	Other	White	High	NatSciences
097	19	Senior	M	24	Married	Other	Other	Medium	NatSciences
098	15	Senior	F	29	Single	None	White	High	NatSciences
099	17	Senior	F	20	Single	Catholic	Black	Medium	NatSciences
100	21	Senior	M	38	Married	Protestant	White	High	NatSciences
101	14	Junior	M	22	Married	None	White	High	NatSciences
102	11	Senior	M	26	Married	None	White	Medium	NatSciences
103	20	Senior	F	27	Married	Protestant	White	Medium	NatSciences
104	15	Senior	M	23	Single	Protestant	White	High	NatSciences
105	11	Senior	F	28	Single	Other	Black	Medium	NatSciences
106	11	Senior	M	29	Married	None	Black	Medium	NatSciences
107	17	Junior	M	21	Single	Other	Other	Medium	NatSciences
108	16	Junior	F	19	Single	Protestant	White	High	NatSciences
109	15	Sophomore	M	17	Single	Protestant	Other	High	NatSciences
110	16	Senior	M	29	Married	None	White	High	NatSciences
111	15	Senior	M	28	Married	None	Other	High	NatSciences
112	18	Senior	F	21	Single	Catholic	White	Medium	NatSciences
113	18	Junior	F	20	Single	Other	Other	High	NatSciences
114	15	Senior	F	20	Single	Catholic	Black	High	NatSciences
115	15	Senior	F	23	Single	Other	Other	Low	NatSciences

STUDENT	CRIMES	COLLEGE	SEX	AGE	MAR—STAT	RELIGION	RACE	INCOME	MAJOR
116	19	Senior	M	22	Single	Catholic	Other	Medium	Business
117	12	Senior	M	20	Single	Protestant	White	Medium	Business
118	22	Senior	F	23	Married	Catholic	White	High	Business
119	23	Senior	M	22	Single	Catholic	White	Medium	Business
120	18	Freshman	F	18	Single	Catholic	Black	High	Business
121	21	Junior	M	20	Single	Other	White	Medium	Business
122	16	Sophomore	M	21	Single	None	White	High	Business
123	23	Sophomore	F	19	Single	Other	Black	High	Business
124	16	Senior	F	22	Single	Catholic	White	Low	Business
125	13	Sophomore	M	20	Single	Catholic	Other	Medium	Business
126	18	Senior	M	34	Married	None	White	High	Business
127	14	Junior	F	20	Single	Other	Black	Low	Business
128	15	Sophomore	M	19	Single	Catholic	Other	Low	Business
129	16	Senior	M	28	Married	None	White	High	Business
130	16	Freshman	F	18	Single	None	White	High	Business
131	17	Junior	M	29	Married	Protestant	White	High	Business
132	23	Junior	M	34	Married	Protestant	White	Medium	Business
133	15	Senior	F	33	Married	Protestant	White	Medium	Business
134	20	Junior	F	22	Married	Other	White	High	Business
135	15	Senior	F	38	Married	None	White	Medium	Business
136	20	Senior	M	20	Single	Protestant	White	Low	Business
137	19	Sophomore	M	18	Single	Protestant	Other	High	Business
138	13	Senior	M	21	Married	None	White	High	Business
139	14	Senior	F	28	Married	None	White	Medium	Business
140	16	Senior	M	20	Single	None	White	Medium	Business
141	22	Senior	M	21	Single	Protestant	White	Medium	Business
142	17	Senior	F	23	Single	None	White	Medium	Business
143	17	Senior	F	30	Married	Protestant	White	High	Business
144	16	Junior	M	27	Married	None	Black	Medium	Business
145	16	Senior	M	23	Single	Other	White	High	Business
146	21	Senior	M	26	Single	Catholic	White	High	Business
147	23	Senior	F	22	Single	Other	White	Medium	Business
148	11	Junior	M	20	Single	None	White	High	Business
149	17	Junior	F	21	Single	None	White	Medium	Business
150	13	Senior	F	20	Single	Catholic	White	High	Business
151	15	Senior	F	21	Single	None	Other	Medium	Business
152	16	Senior	M	23	Single	Protestant	Other	High	Business
153	17	Senior	M	27	Married	Catholic	White	High	Business
154	20	Senior	M	21	Single	None	White	Medium	Business

(continued)

STUDENT	CRIMES	COLLEGE	SEX	AGE	MAR—STAT	RELIGION	RACE	INCOME	MAJOR
155	24	Senior	M	25	Single	Catholic	White	Medium	Business
156	19	Junior	F	20	Single	Catholic	White	Medium	Business
157	18	Junior	M	37	Married	Catholic	White	Medium	Business
158	17	Junior	F	20	Single	None	White	Medium	Business
159	15	Junior	M	22	Single	None	White	High	Business
160	19	Senior	F	34	Married	Protestant	White	Medium	Business
161	20	Junior	M	33	Married	None	White	Low	Business
162	23	Junior	M	37	Married	Protestant	White	Low	Business
163	18	Junior	M	20	Single	Catholic	White	High	Business
164	18	Senior	F	35	Married	Protestant	Black	Medium	Business
165	19	Senior	M	28	Married	None	White	Medium	Business
166	17	Junior	M	31	Married	Catholic	Other	Medium	Business
167	16	Senior	M	29	Single	Protestant	White	High	Business
168	16	Junior	M	23	Married	Catholic	White	High	Business
169	17	Senior	M	22	Single	Protestant	White	High	Business
170	15	Senior	M	29	Single	None	White	Medium	Business
171	24	Senior	M	22	Single	Protestant	White	High	Business
172	18	Senior	M	27	Single	None	White	High	Business
173	17	Senior	F	20	Single	None	White	High	Business
174	21	Senior	M	23	Single	Other	Other	High	Business
175	17	Senior	M	38	Single	None	White	Low	Business
176	21	Junior	F	24	Single	Catholic	Other	Medium	Business
177	17	Senior	M	23	Single	None	White	Low	Business
178	24	Junior	F	20	Married	Protestant	White	Medium	Business
179	18	Junior	M	27	Married	Catholic	White	Medium	Business
180	18	Junior	F	25	Married	Catholic	White	Medium	Business
181	23	Sophomore	F	22	Married	Protestant	White	High	Business
182	18	Senior	M	24	Married	Catholic	White	Medium	Business
183	15	Senior	M	33	Married	None	White	High	Business
184	15	Junior	F	20	Single	Catholic	White	Medium	Business
185	21	Junior	F	21	Single	Catholic	White	High	Business
186	16	Junior	M	23	Single	Catholic	White	High	Business
187	18	Senior	M	19	Married	Other	White	High	Business
188	16	Junior	M	22	Single	Protestant	White	Medium	Business
189	13	Junior	M	25	Married	None	White	Medium	Business
190	12	Junior	M	33	Married	Jewish	White	High	Business
191	17	Junior	M	36	Married	Catholic	White	High	Business

S T U D E N T	C R I M E S	C O L L E G E	S E X	A G E	M A R — S T A T	R E L I G I O N	R A C E	I N C O M E	M A J O R
192	16	Junior	M	22	Single	Catholic	White	High	Business
193	16	Junior	F	24	Single	Protestant	Other	Medium	Business
194	17	Junior	F	30	Married	Other	White	High	Business
195	17	Junior	M	24	Single	Protestant	White	High	Business
196	16	Senior	M	23	Single	Catholic	White	Medium	Business
197	17	Sophomore	M	18	Single	Protestant	White	High	Business
198	16	Junior	F	20	Single	Catholic	White	High	Business
199	15	Junior	M	20	Single	Protestant	White	High	Business
200	18	Junior	F	36	Single	Protestant	White	High	Business

Patient Treatment Times Database

In order for a health clinic to be capable of handling the desired patient load, designers need to know something about the demand for services in the area in which the clinic is to be located. This includes information on the patient arrival rate and the length of time to treat a patient. Both the arrival rate and the treatment time vary randomly. The arrival rate varies because of the random occurrence of outbreaks of flu and other common illnesses; the time of treatment varies because treatment time depends on a particular patient's illness. Thus the demand for physician and nurse time varies randomly. By studying the frequency distributions of patient arrival rates and treatment times, the designer can specify the numbers of doctors, nurses, technicians, and orderlies, and the amount of physical equipment needed to meet the demand. These numbers affect the length of time a patient must wait in the clinic before receiving attention.

To answer some questions about clinic treatment times, a designer acquired data from an established clinic in a locale that possessed similar characteristics to the proposed new clinic location. The treatment times for 50 patients were randomly selected from the clinic's records.

PATIENT TREATMENT TIMES (IN MINUTES)

21	20	31	24	15	21	24	18	33	8
26	17	27	29	24	14	29	41	15	11
13	28	22	16	12	15	11	16	18	17
29	16	24	21	19	7	16	12	45	24
21	12	10	13	20	35	32	22	12	10

Bonus Level Database

A large firm has a generous but rather complicated policy governing its end-of-year bonuses for lower and middle management. Factors included in the calculation are base salary, grade, performance against preset goals, subjective judgment of contribution to corporate goals, and others. An officer from the Corporate Compensation Group has selected a random sample of 50 males and 50 females from the lower and middle management levels of the firm. Included in the database for each individual are gender, age, and bonus level (expressed as a percentage of base salary).

BONUS LEVEL DATA

FEMALES

Person	Age	Bonus Level	Person	Age	Bonus Level
1	25	14.5	26	36	12.5
2	28	15.0	27	34	12.3
3	26	12.2	28	40	13.7
4	29	13.1	29	44	14.4
5	32	14.0	30	45	13.0
6	30	10.6	31	46	9.3
7	27	9.8	32	47	13.0
8	31	10.0	33	48	12.2
9	33	9.7	34	45	10.8
10	26	7.8	35	47	10.2
11	28	8.3	36	49	11.4
12	30	9.4	37	44	10.6
13	34	8.9	38	46	10.5
14	25	5.2	39	49	11.4
15	26	6.1	40	52	9.7
16	29	6.5	41	53	9.5
17	34	6.8	42	52	8.6
18	37	7.2	43	54	8.0
19	40	8.2	44	50	10.1
20	42	8.8	45	52	13.9
21	43	9.4	46	48	14.6
22	37	9.5	47	47	15.0
23	41	10.0	48	51	15.4
24	41	11.3	49	50	16.2
25	38	11.9	50	49	14.9

BONUS LEVEL DATA

MALES

Person	Age	Bonus Level	Person	Age	Bonus Level
1	25	5.2	26	42	16.3
2	26	6.0	27	37	13.8
3	20	7.1	28	35	12.9
4	26	7.2	29	44	16.9
5	26	9.1	30	46	12.5
6	29	9.1	31	49	12.8
7	25	10.0	32	51	13.2
8	31	10.1	33	52	13.4
9	33	10.3	34	51	12.1
10	28	12.1	35	50	11.2
11	29	14.6	36	50	15.5
12	30	14.8	37	52	16.9
13	33	13.1	38	54	19.3
14	34	10.9	39	55	19.6
15	33	8.5	40	54	18.7
16	37	9.4	41	50	18.6
17	40	9.3	42	47	16.4
18	41	10.2	43	45	16.0
19	42	12.5	44	40	14.7
20	41	12.3	45	41	14.7
21	40	11.7	46	32	10.5
22	36	13.4	47	53	9.8
23	37	14.7	48	51	16.3
24	39	15.0	49	47	14.8
25	41	15.2	50	56	15.2

Annual Returns Database

The Annual Returns Database gives the annual returns (in percent) for large capital stocks and for small capital stocks for the years from 1964 to 1991. For each category of capital stocks, the yearly returns are given for stocks with high book-to-market ratios, medium book-to market ratios and low book-to-market ratios, as well as for an appropriate index. For large capital stocks, the index is the S&P 500 Index; for small capital stocks, the index is a small capital company index.

ANNUAL RETURNS (IN PERCENT)

Book-to-Market Effect in Large and Small Stocks: Annual Returns (%) 1964–1991

	Large Cap				Small Cap			
Year	S&P 500 Index	High Book-to-Market	Medium Book-to-Market	Low Book-to-Market	Small Co. Index	High Book-to-Market	Medium Book-to-Market	Low Book-to-Market
1964	16.5	23.6	15.5	16.8	17.6	21.4	19.6	16.7
1965	12.5	22.4	6.9	16.0	35.6	49.5	34.0	32.4
1966	−10.0	−5.2	−7.4	−11.2	−6.9	−5.5	−4.9	−5.8
1967	24.0	28.4	13.8	30.4	73.7	75.6	74.5	82.4
1968	11.1	22.4	15.8	2.6	38.5	42.6	37.5	29.9
1969	−8.5	−16.0	−14.4	2.3	−25.2	−24.9	−25.2	−23.2
1970	4.0	11.9	8.6	−6.1	−11.7	0.3	−6.7	−20.8
1971	14.3	9.4	7.9	23.0	20.4	14.4	21.7	24.7
1972	19.0	16.0	11.4	22.2	4.2	7.0	6.8	1.0
1973	−14.7	−2.8	−9.5	−20.8	−36.1	−26.0	−32.0	−41.5
1974	−26.5	−22.4	−22.5	−29.5	−28.8	18.2	−24.5	−33.8
1975	37.2	51.9	42.4	33.6	56.7	54.5	58.4	59.1
1976	23.8	45.0	40.1	17.5	48.3	53.6	47.0	37.9
1977	−7.2	0.7	0.2	−9.4	20.1	21.8	18.1	18.0
1978	6.6	6.6	7.6	7.9	20.5	21.8	19.9	18.0
1979	18.4	23.8	24.3	19.6	40.7	38.0	37.8	49.6
1980	32.4	16.5	37.2	34.7	36.8	29.1	30.9	52.1
1981	−4.9	11.2	−8.2	−8.4	2.5	10.5	12.1	−10.1
1982	21.4	27.4	17.8	18.4	25.2	27.7	32.1	19.9
1983	22.5	26.8	25.4	15.1	30.4	44.2	40.1	20.9
1984	6.3	14.1	5.0	−2.0	−4.9	5.1	1.9	−12.9
1985	32.2	29.5	31.0	31.6	30.1	34.7	33.1	28.3
1986	18.5	20.4	18.7	13.8	6.7	16.9	11.8	3.1
1987	5.2	2.0	4.9	7.5	−9.1	−6.3	−4.2	−13.6
1988	16.8	24.7	17.9	11.6	22.4	28.8	28.9	15.3
1989	31.5	28.4	26.0	35.7	15.9	19.6	17.4	18.0
1990	−3.2	−14.0	−6.4	1.4	−20.8	−20.8	−16.4	−20.0
1991	30.6	29.8	22.0	42.0	48.1	39.4	42.9	51.2

Source: FAMA-FRENCH.

Progress of Banks Database

The Federal Deposit Insurance Corporation tracks much data on banks to watch their progress. In this database, all banks in the Kansas City area are ranked by their three-year average return on assets for 1990–1992, one measure of profitability. The following data are included in the database:

- **Profits** show earnings for each of the last three years.

- **Return on assets** shows profits as a percent of total assets. Banks generally target a 1 percent return or higher.

- **Problem loans** are past due 90 days or placed on nonaccrual status. The amount is shown as a percent of total loans.

- **Problem real estate** was repossessed by the bank or classified by regulators as a foreclosure in substance. The amount is shown as a percent of total assets.

- **Problems as a percent of capital** compares the amount of problem loans and real estate to the bank's capital. Capital is the bank's cushion to absorb potential losses from the problems. It consists of the bank's reserve for loan losses and the owners' equity in the bank.

PROGRESS OF BANKS DATA

Bank	Headquarters	Profits			Return on Assets		Problems 1992		Problems as a Percent of Capital		
		1992	1991	1990	1992	3-year average	Loans	Real Estate	1992	1991	1990
Country Club	Kansas City	$2,113	$1,316	$1,045	1.72%	1.48%	0.0%	0.1%	2%	0%	0%
United Missouri*	Kansas City	29,824	30,147	28,520	0.99%	1.05%	0.5%	0.2%	4%	4%	4%
Citizens-Jackson Co.	Warrensburg, Mo.	2,373	1,820	1,700	0.80%	0.92%	0.3%	0.1%	4%	7%	8%
Mark Twain	Kansas City	4,711	3,065	1,661	1.22%	0.88%	1.1%	0.3%	11%	19%	15%
Midland	Kansas City	43	5,980	4,003	0.01%	0.87%	0.2%	5.3%	73%	0%	2%
Blue Ridge	Kansas City	2,586	1,394	1,175	1.13%	0.77%	0.8%	1.0%	19%	27%	31%
Commerce*	Kansas City	19,759	14,739	17,373	0.80%	0.76%	0.3%	0.4%	8%	12%	8%
Boatmen's First National	Kansas City	25,565	24,673	19,452	0.81%	0.75%	2.0%	0.2%	16%	21%	16%
Bannister	Kansas City	1,166	25	564	1.22%	0.65%	0.7%	1.0%	14%	22%	3%
Mercantile*	Kansas City	5,801	6,600	3,961	0.69%	0.61%	1.5%	0.5%	15%	13%	15%
Winterset State	Harrisonville	318	225	-122	0.98%	0.53%	0.0%	0.0%	0%	4%	4%
Missouri	Kansas City	242	164	98	0.60%	0.46%	1.4%	0.9%	28%	39%	16%
First National Platte Co.	Kansas City	200	120	42	0.50%	0.39%	0.2%	0.3%	5%	10%	10%
Central	Kansas City	412	-288	118	0.99%	0.21%	1.7%	0.5%	16%	50%	68%
Superior National	Kansas City	-489	161	485	-2.02%	0.20%	2.4%	5.4%	130%	88%	55%
Hillcrest	Kansas City	200	-181	95	0.47%	0.10%	1.2%	0.5%	16%	29%	29%
Union	Kansas City	335	28	-412	0.50%	-0.02%	0.2%	2.3%	31%	21%	39%
Landmark KCI	Kansas City	-143	-3	6	-0.90%	-0.27%	2.5%	4.8%	71%	84%	58%
Landmark	Kansas City	-286	-1	23	-2.11%	-0.53%	0.4%	6.5%	59%	58%	46%
First Business	Kansas City	-1,561	180	-245	-2.70%	-0.84%	1.6%	0.5%	17%	9%	40%
Merchants/Metro North*	Kansas City	-78,258	-48,464	6,976	-4.57%	-1.95%	7.5%	0.0%	118%	179%	72%

Bank	Headquarters	Profits			Return on Assets		Problems 1992		Problems as a Percent of Capital		
		1992	1991	1990	1992	3-year average	Loans	Real Estate	1992	1991	1990
SOUTH & EAST											
Bank of Grain Valley	Grain Valley	$738	$656	$547	1.55%	1.47%	0.0%	0.0%	0%	5%	0%
Bank of Lee's Summit	Lee's Summit	1,341	947	799	1.41%	1.27%	0.2%	0.9%	10%	2%	1%
Bank of Jacomo	Blue Springs	1,122	770	719	1.36%	1.09%	2.2%	0.1%	16%	21%	31%
Standard	Independence	869	659	654	1.07%	0.93%	0.5%	0.2%	5%	11%	14%
Commercial	Oak Grove	603	412	354	0.96%	0.91%	3.1%	0.6%	18%	21%	24%
Community	Raymore	250	226	226	0.91%	0.89%	0.0%	1.5%	14%	16%	19%
United Missouri Cass Co.	Peculiar	296	270	254	0.93%	0.81%	0.4%	0.0%	1%	1%	0%
Pleasant Hill	Pleasant Hill	421	335	333	0.90%	0.80%	2.1%	0.3%	15%	23%	19%
Bank of Belton	Belton	306	278	230	0.80%	0.74%	0.7%	0.0%	2%	3%	5%
First State of Missouri	Buckner	181	240	220	0.60%	0.71%	0.1%	0.0%	1%	5%	2%
First National	Lee's Summit	398	330	122	0.82%	0.67%	1.3%	1.2%	29%	11%	7%
Sterling National	Sugar Creek	695	314	464	0.81%	0.61%	0.9%	0.3%	11%	14%	5%
Bank 10	Belton	410	376	216	0.63%	0.60%	1.5%	0.7%	18%	16%	6%
First City	Independence	323	132	247	0.70%	0.51%	0.4%	0.2%	5%	8%	36%
Blue Springs Bank	Blue Springs	−181	648	1,231	−0.10%	0.30%	1.8%	2.8%	51%	34%	22%
NORTHLAND											
Truman Bank	Grandview	$150	$268	$60	0.26%	0.27%	5.1%	7.1%	110%	98%	100%
Clayco State	Claycomo	52	218	321	0.29%	1.31%	0.1%	0.0%	0%	0%	0%
Norbank	North Kansas City	421	395	351	1.27%	1.23%	0.0%	2.5%	28%	32%	37%
Wells	Platte City	473	426	286	1.19%	1.03%	0.3%	0.0%	1%	0%	2%
Farmers Exchange	Parkville	469	270	234	1.39%	0.99%	0.0%	0.0%	0%	0%	0%
Kearney Trust	Kearney	460	397	402	0.97%	0.97%	0.4%	0.2%	5%	7%	4%
Bank of Weston	Weston, Mo.	317	371	292	0.92%	0.97%	0.0%	0.3%	3%	2%	4%
Kearney Commercial	Kearney	353	253	219	1.13%	0.95%	0.1%	0.3%	3%	4%	11%
Platte Valley	Platte City	447	307	218	0.95%	0.85%	0.0%	0.0%	0%	0%	0%
Farley State	Parkville	174	186	166	0.78%	0.73%	0.7%	0.3%	7%	2%	6%
Commercial	Liberty	126	596	601	0.13%	0.54%	7.5%	0.1%	34%	4%	5%
Bank of Riverside	Riverside	65	41	25	0.73%	0.54%	0.0%	0.0%	0%	0%	0%
Citizens	Smithville	−189	220	446	−0.53%	0.42%	8.7%	13.1%	151%	96%	58%
First Bank	Gladstone	−4,355	1,830	2,102	−2.05%	−0.06%	5.9%	15.3%	154%	124%	49%
Northland National	Gladstone	−182	101	−30	−0.77%	−0.17%	0.3%	0.6%	13%	21%	50%

*Indicates bank was involved in mergers.

PROGRESS OF BANKS DATA

Bank	Headquarters	Profits			Return on Assets		Problem 1992		Problems as a Percent of Capital		
		1992	1991	1990	1992	3-year average	Loans	Real Estate	1992	1991	1990
Stanley	Stanley	$1,192	$1,081	$837	1.61%	1.44%	0.3%	0.7%	7%	11%	19%
Citizens National	Fort Scott, Kan.	1,210	1,054	1,298	1.31%	1.31%	0.2%	0.0%	1%	0%	12%
Peoples National	Ottawa, Kan.	1,334	819	629	1.37%	1.09%	0.5%	0.4%	10%	8%	15%
Southgate	Prairie Village	431	982	1,024	0.67%	1.06%	0.8%	1.3%	12%	56%	73%
Country Hill	Lenexa	1,764	1,978	885	1.03%	1.06%	0.4%	3.8%	54%	7%	4%
Community	Chapman, Kan.	324	160	151	1.00%	1.02%	3.0%	0.8%	30%	35%	35%
DeSoto State	DeSoto	187	217	219	0.89%	1.02%	1.0%	0.2%	3%	9%	6%
College Boulevard	Overland Park	2,074	2,347	1,391	1.02%	1.01%	0.7%	4.4%	70%	7%	5%
First Kansas	Gardner	547	407	330	1.21%	0.98%	0.5%	1.3%	20%	29%	32%
Metcalf State	Overland Park	1,085	944	920	0.96%	0.97%	0.2%	0.3%	5%	6%	11%
Shawnee State	Shawnee	1,876	1,639	1,677	1.02%	0.97%	0.2%	0.7%	9%	35%	14%
Overland Park State	Overland Park	1,739	1,478	1,906	0.96%	0.94%	0.2%	0.0%	1%	3%	3%
State Bank Spring Hill	Spring Hill	245	216	180	1.02%	0.93%	1.2%	0.0%	4%	2%	3%
Mark Twain Shawnee	Shawnee	749	401	315	1.16%	0.88%	2.0%	1.9%	37%	38%	40%
Bank IV Kansas*	Wichita	53,718	33,774	8,013	1.19%	0.85%	1.5%	0.2%	10%	11%	23%
First National	Olathe	1,874	1,216	1,329	0.95%	0.79%	0.0%	1.3%	16%	20%	24%
United Kansas	Merriam	927	504	468	1.13%	0.75%	1.2%	1.4%	22%	24%	33%
MidAmerican*	Overland Park	2,857	2,698	3,684	0.66%	0.74%	0.4%	0.3%	6%	12%	32%
Bank of Blue Valley	Overland Park	414	225	-37	1.03%	0.63%	0.5%	0.0%	4%	6%	0%
Commerce	Lenexa	0	427	182	0.00%	0.63%	1.6%	0.1%	4%	1%	5%
Oak Park	Overland Park	66	501	509	0.08%	0.48%	2.5%	2.7%	54%	28%	20%
Exchange National	Marysville, Kan.	697	82	81	0.79%	0.46%	0.2%	0.1%	2%	16%	11%
Valley View State	Overland Park	-3,554	4,100	4,008	-0.73%	0.30%	5.6%	9.2%	115%	80%	41%
Parkway Bank	Overland Park	366	159	-295	0.89%	0.26%	0.0%	0.0%	0%	0%	0%
Johnson County	Prairie Village	25	733	837	0.01%	0.25%	2.8%	0.3%	19%	29%	47%
First American	Lenexa	162	89	-102	0.57%	0.22%	1.0%	0.0%	6%	0%	0%
Citizens	Shawnee	24	609	-140	0.03%	0.21%	2.0%	3.3%	49%	46%	64%
Mission	Mission	-3,924	2,134	3,097	-1.02%	0.10%	6.0%	11.8%	120%	111%	98%
First Continental	Overland Park	1,215	652	-1,414	0.65%	0.08%	3.3%	0.8%	22%	26%	155%
Gardner National	Gardner	-135	66	64	-1.63%	-0.02%	6.3%	1.0%	27%	24%	13%
First Shawnee Mission	Fairway	309	-474	63	0.79%	-0.07%	4.5%	2.5%	73%	100%	59%
Olathe Bank	Olathe	230	-310	-437	0.47%	-0.38%	1.7%	4.7%	65%	105%	132%
Midland of Kansas	Mission	-3,696	991	752	-2.97%	-0.61%	2.6%	5.1%	124%	7%	1%
First Colonial	Prairie Village	-39	-1,036	-644	-0.05%	-0.67%	0.4%	0.0%	3%	0%	0%
Heritage	Olathe	-43	-562	-96	-0.17%	-0.85%	5.8%	3.5%	91%	129%	119%

PROGRESS OF BANKS DATA

Bank	Headquarters	Profits			Return on Assets		Problems 1992		Problems as a Percent of Capital		
		1992	1991	1990	1992	3-year average	Loans	Real Estate	1992	1991	1990
Edwardsville	Edwardsville	$416	$394	$347	1.26%	1.31%	0.4%	1.0%	14%	10%	21%
Wyandotte	Wyandotte County	638	693	543	1.08%	1.12%	1.1%	1.1%	21%	14%	31%
Commercial National*	Kansas City, Kan.	4,281	2,635	3,478	1.11%	0.86%	1.6%	0.1%	10%	4%	13%
Twin City State	Kansas City, Kan.	1,263	718	682	1.17%	0.84%	0.6%	0.7%	9%	8%	14%
Commercial State	Bonner Springs	652	564	455	0.93%	0.83%	0.6%	1.2%	19%	21%	20%
Home State	Kansas City, Kan.	953	1,096	519	0.81%	0.75%	1.6%	2.3%	32%	27%	47%
Guaranty	Kansas City, Kan.	805	651	-155	1.19%	0.66%	1.6%	0.7%	14%	23%	29%
Brotherhood	Kansas City, Kan.	1,935	652	375	0.92%	0.44%	5.7%	2.5%	68%	75%	28%
Industrial State	Kansas City, Kan.	-1,751	1,498	2,002	-1.01%	0.32%	1.7%	6.9%	64%	76%	42%
First State	Kansas City, Kan.	687	657	-1,228	1.50%	0.09%	2.3%	0.7%	20%	21%	49%
Citizens	Kansas City, Kan.	24	205	-115	0.06%	0.08%	0.2%	3.6%	44%	63%	47%
Security	Kansas City, Kan.	-3,733	2,073	2,011	-0.81%	0.02%	5.9%	10.7%	141%	122%	-61%
Commerce	Bonner Springs	0	-119	3	0.00%	-0.34%	0.5%	0.4%	8%	21%	3%
Douglass	Kansas City, Kan.	80	-1,535	-340	0.28%	-2.03%	2.4%	0.6%	14%	47%	95%

*Indicates bank was involved in mergers.

2

Useful Statistical Tests and Confidence Intervals

Inferences Concerning the Mean of a Population

A σ known:

1 Statistical test:

Null hypothesis: $\mu = \mu_0$ (μ_0 is specified)

Alternative hypothesis: For a one-tailed test:

    1 $\mu > \mu_0$

    2 $\mu < \mu_0$

    For a two-tailed test:

    3 $\mu \neq \mu_0$

Test statistic: $z = \dfrac{\bar{y} - \mu_0}{\sigma_{\bar{y}}}$ where $\sigma_{\bar{y}} = \dfrac{\sigma}{\sqrt{n}}$

Rejection region: For $\alpha = .05$ (or .01) and for a one-tailed test:

    1 Reject H_0 if $z > 1.645$ (or 2.33).

    2 Reject H_0 if $z < -1.645$ (or -2.33).

    For $\alpha = .05$ (or .01) and for a two-tailed test:

    3 Reject H_0 if $|z| > 1.96$ (or 2.58).

Note: When $n \geq 30$ you may substitute s for σ in the formula for $\sigma_{\bar{y}}$.

2 Confidence interval:

$$\bar{y} \pm z\sigma_{\bar{y}}$$

where $\sigma_{\bar{y}} = \sigma/\sqrt{n}$. *Note:* The values of z for a 90%, a 95%, or a 99% confidence interval for μ are 1.645, 1.96, or 2.58, respectively. When $n \geq 30$, you may substitute s for σ in the formula for $\sigma_{\bar{y}}$.

B σ unknown, and the observations are nearly normally distributed:

1 Statistical test:

Null hypothesis: $\mu = \mu_0$ (μ_0 is specified)

Alternative hypothesis: For a one-tailed test:

$\quad$ 1 $\mu > \mu_0$

$\quad$ 2 $\mu < \mu_0$

$\quad$ For a two-tailed test:

$\quad$ 3 $\mu \neq \mu_0$

Test statistic: $t = \dfrac{\bar{y} - \mu_0}{s/\sqrt{n}}$

Rejection region: For a specified value of α, df $= (n - 1)$, and for a one-tailed test:

$\quad$ 1 Reject H_0 if $t > t_{\alpha}$.

$\quad$ 2 Reject H_0 if $t < -t_{\alpha}$.

$\quad$ For a specified value of α, df $= (n - 1)$, and for a two-tailed test:

$\quad$ 3 Reject H_0 if $|t| > t_{\alpha/2}$.

2 Confidence interval:

$$\bar{y} \pm \frac{ts}{\sqrt{n}}$$

The value of t corresponding to a 90%, a 95%, or a 99% confidence interval is found in Table 3 of Appendix 3 for df $= (n - 1)$ and $a = .05$, .025, or .005, respectively.

Inferences Concerning the Difference Between the Means of Two Populations

A Assumptions:

1 Population 1 is normally distributed with mean equal to μ_1 and variance equal to σ_1^2.

2 Population 2 is normally distributed with mean equal to μ_2 and variance equal to σ_2^2.

B Some results:

1 The sampling distribution of $(\bar{y}_1 - \bar{y}_2)$ is normal.

2 The mean of the sampling distribution, $\mu_{\bar{y}_1 - \bar{y}_2}$ is equal to the difference between the populations means, $(\mu_1 - \mu_2)$.

3 The standard error of the sampling distribution is

$$\sigma_{\bar{y}_1 - \bar{y}_2} = \sqrt{\frac{\sigma_1^2}{n_1} + \frac{\sigma_2^2}{n_2}}$$

C Statistical test:

Null hypothesis: $\mu_1 - \mu_2 = 0$

Alternative hypothesis: For a one-tailed test:

    1 $\mu_1 - \mu_2 > 0$

    2 $\mu_1 - \mu_2 < 0$

For a two-tailed test:

    3 $\mu_1 - \mu_2 \neq 0$

Test statistic:

$$t = \frac{\bar{y}_1 - \bar{y}_2}{s_p \sqrt{\dfrac{1}{n_1} + \dfrac{1}{n_2}}}$$

where

$$s_p = \sqrt{\frac{(n_1 - 1)s_1^2 + (n_2 - 1)s_2^2}{n_1 + n_2 - 2}}$$

Rejection region: For a specified value of α, for df $= (n_1 + n_2 - 2)$, and for a one-tailed test:

    1 Reject H_0 if $t > t_\alpha$.

    2 Reject H_0 if $t < -t_\alpha$.

For a specified value of α, for df $= (n_1 + n_2 - 2)$, and for a two-tailed test:

    3 Reject H_0 if $|t| > t_{\alpha/2}$.

D Confidence interval:

$$\bar{y}_1 - \bar{y}_2 \pm t s_p \sqrt{\frac{1}{n_1} + \frac{1}{n_2}}$$

where

$$s_p = \sqrt{\frac{(n_1 - 1)s_1^2 + (n_2 - 1)s_2^2}{n_1 + n_2 - 2}}$$

The value of t corresponding to a 90%, a 95%, or a 99% confidence interval is found in Table 3 of Appendix 3 for df $= (n_1 + n_2 - 2)$ and $a = .05, .025,$ or .005, respectively.

Inferences About a Population Proportion p

A Assumptions for a binomial experiment:

    1 Experiment consists of n identical trials, each resulting in one of two outcomes, say, success and failure.

2 The probability of success is equal to π and remains the same from trial to trial.

3 The trials are independent of each other.

4 The variable measured is y, the number of successes observed during the n trials.

B Results:

1 The estimator of p is $\hat{p} = y/n$.

2 The mean of $\hat{p}$ is p.

3 The variance of $\hat{p}$ is pq/n, where $q = 1 - p$.

C Statistical test:

Null hypothesis: $p = p_0$ (p_0 is specified)

Alternative hypothesis: For a one-tailed test:

1   $p > p_0$

2   $p < p_0$

For a two-tailed test:

3   $p \neq p_0$

Test statistic: $z = \dfrac{\hat{p} - p_0}{\sigma_{\hat{p}}}$ where $\sigma_{\hat{p}} = \sqrt{\dfrac{p_0 q_0}{n}}$

Rejection region: For $\alpha = .05$ (or .01) and for a one-tailed test:

1   Reject H_0 if $z > 1.645$ (or 2.33).

2   Reject H_0 if $z < -1.645$ (or -2.33).

For $\alpha = .05$ (or .01) and for a two-tailed test:

3   Reject H_0 if $|z| > 1.96$ (or 2.58).

Note: This test is valid when np_0 and nq_0 are both 10 or more.

D Confidence interval:

$$\hat{p} \pm z\sigma_{\hat{p}}$$

where

$$\sigma_{\hat{p}} = \sqrt{\frac{pq}{n}}$$

Note: The z-values corresponding to a 90%, a 95%, or a 99% confidence interval are, respectively, 1.645, 1.96, or 2.58. If $n\hat{p}$ and $n\hat{q}$ are both 10 or greater, we may substitute $\hat{p}$ for p in the formula for $\sigma_{\hat{p}}$.

Inferences Comparing Two Population Proportions p and p_2

A   Assumption: Independent random samples are drawn from each of two binomial populations.

Statistic	Population 1	Population 2
Probability of success	p_1	p_2
Sample size	n_1	n_2
Observed successes	y_1	y_2

B   Results:

1   The estimated difference between p_1 and p_2 is

$$\hat{p}_1 - \hat{p}_2 = \frac{y_1}{n_1} - \frac{y_2}{n_2}$$

2   The mean of $(\hat{p}_1 - \hat{p}_2)$ is $(p_1 - p_2)$

3   The standard error of $(\hat{p}_1 - \hat{p}_2)$ is

$$\sqrt{\frac{p_1 q_1}{n_1} + \frac{p_2 q_2}{n_2}}$$

C   Statistical test:

Null hypothesis: $p_1 - p_2 = 0$

Alternative hypothesis: For a one-tailed test:

   1 $p_1 - p_2 > 0$
   2 $p_1 - p_2 < 0$

   For a two-tailed test:

   3 $p_1 - p_2 \neq 0$

Test statistic: $z = \dfrac{\hat{p}_1 - \hat{p}_2}{\sigma_{\hat{p}_1 - \hat{p}_2}}$

where

$$\sigma_{\hat{p}_1 - \hat{p}_2} = \sqrt{pq\left(\frac{1}{n_1} + \frac{1}{n_2}\right)}$$

and p is approximated by

$$\hat{p} = \frac{y_1 + y_2}{n_1 + n_2}$$

Rejection region: For $\alpha = .05$ (or .01) and for a one-tailed test:

   1   Reject H_0 if $z > 1.645$ (or 2.33).
   2   Reject H_0 if $z < -1.645$ (or -2.33).

   For $\alpha = .05$ (or .01) and for a two-tailed test:

   3   Reject H_0 if $|z| > 1.96$ (or 2.58).

Note: $n\hat{p}$ and $n\hat{q}$ must be greater than or equal to 10 for both populations.

D Confidence interval:

$$\hat{p}_1 - \hat{p}_2 \pm z\sigma_{\hat{p}_1 - \hat{p}_2}$$

where

$$\sigma_{\hat{p}_1 - \hat{p}_2} = \sqrt{\frac{p_1 q_1}{n_1} + \frac{p_2 q_2}{n_2}}$$

Substitute $z = 1.645$, 1.96, or 2.58 for a 90%, a 95%, or a 99% confidence interval, respectively. Also use $\hat{p}_1$ and $\hat{p}_2$ for the unknown parameters p_1 and p_2 in the formula for $\sigma_{\hat{p}_1 - \hat{p}_2}$. Very little error will result provided the sample sizes are large.

Inferences About σ^2 (or σ)

A Assumption: The underlying population of measurements is normal.

B Statistical test:

Null hypothesis:	$\sigma^2 = \sigma_0^2$ (σ_0^2 is specified)
Alternative hypothesis:	For a one-tailed test:

1 $\sigma^2 > \sigma_0^2$
2 $\sigma^2 < \sigma_0^2$

For a two-tailed test:

3 $\sigma^2 \neq \sigma_0^2$

Test statistic:
$$\chi^2 = \frac{(n-1)s^2}{\sigma_0^2}$$

Rejection region: For a specified value of α:

1 Reject H_0 if $\chi^2 > \chi_u^2$, the upper-tail value for $a = \alpha$ and df $= n - 1$.

2 Reject H_0 if $\chi^2 < \chi_l^2$, the lower-tail value for $a = 1 - \alpha$ and df $= n - 1$.

3. Reject H_0 if $\chi^2 > \chi_u^2$, based on $a = \alpha/2$ or $\chi^2 < \chi_l^2$ based on $a = 1 - \alpha/2$; df $= n - 1$.

C Confidence interval:

$$\frac{(n-1)s^2}{\chi_u^2} < \sigma^2 < \frac{(n-1)s^2}{\chi_l^2}$$

Inferences About Two Population Variances

A Assumption:

  1 Population 1 has a normal distribution, with mean μ_1 and variance σ_1^2.

  2 Population 2 has a normal distribution, with mean μ_2 and variance σ_1^2.

  3 Two independent random samples are drawn—n_1 measurements from population 1, and n_2 from population 2.

B Test for comparing two population variances:

Null hypothesis:	$\sigma_1^2 = \sigma_2^2$
Alternative hypothesis:	For a one-tailed test:

        1 $\sigma_1^2 > \sigma_2^2$

        2 $\sigma_1^2 < \sigma_2^2$

        For a two-tailed test:

        3 $\sigma_1^2 \neq \sigma_2^2$

Test statistic: $F = \dfrac{s_1^2}{s_2^2}$

Rejection region: For a given value of α, $\text{df}_1 = (n_1 - 1)$, and $\text{df}_2 = (n_2 - 1)$:

        1   Reject H_0 if $F > F_{\alpha,\,\text{df}_1,\,\text{df}_2}$.

        2   Reject H_0 if $F < 1/F_{\alpha,\,\text{df}_2,\,\text{df}_1}$.

        3   Reject H_0 if $F > F_{\alpha/2,\,\text{df}_1,\,\text{df}_2}$ or if $F < 1/F_{\alpha/2,\,\text{df}_2,\,\text{df}_1}$.

C Confidence interval:

$$\frac{s_1^2}{s_2^2} F_L < \frac{\sigma_1^2}{\sigma_2^2} < \frac{s_1^2}{s_2^2} F_U$$

where $F_L = 1/F_{\alpha/2,\,\text{df}_1,\,\text{df}_2}$ and $F_U = F_{\alpha/2,\,\text{df}_2,\,\text{df}_1}$.

Inference Concerning β_1 in Linear Regression

A Assumptions:

  1 The ϵs in $y = \beta_0 + \beta_1 x + \epsilon$ are normally distributed with mean 0 and variance σ_ϵ^2.

  2 The ϵs are independent.

B Statistical test:

Null hypothesis: $\qquad\qquad\qquad\qquad\qquad$ $\beta_1 \doteq 0$

Alternative hypothesis: $\qquad\qquad\quad$ For a one-tailed test:

$\qquad\qquad\qquad\qquad\qquad\qquad\quad$ 1 $\beta_1 > 0$

$\qquad\qquad\qquad\qquad\qquad\qquad\quad$ 2 $\beta_1 < 0$

$\qquad\qquad\qquad\qquad\qquad\qquad$ For a two-tailed test:

$\qquad\qquad\qquad\qquad\qquad\qquad\quad$ 3 $\beta_1 \neq 0$

Test statistic: $\qquad\qquad\qquad\qquad$ $t = \dfrac{\hat{\beta}_1}{\sqrt{s_\epsilon^2/S_{xx}}}$

Rejection region: $\qquad\qquad\quad$ For a given value of α and df $= n - 2$:

$\qquad\qquad\qquad\qquad\qquad\qquad\quad$ 1 Reject H_0 if $t > t_\alpha$.

$\qquad\qquad\qquad\qquad\qquad\qquad\quad$ 2 Reject H_0 if $t < -t_\alpha$.

$\qquad\qquad\qquad\qquad\qquad\qquad\quad$ 3 Reject H_0 if $|t| > t_{\alpha/2}$.

C Confidence interval:

$$\hat{\beta}_1 \pm t_{\alpha/2} \frac{s_\epsilon}{\sqrt{S_{xx}}}$$

3

Statistical Tables

T A B L E 1   Binomial Probabilities, $P(y)$ for $n \leq 20$

$n = 2$					p						
$y \downarrow$	**0.05**	**0.10**	**0.15**	**0.20**	**0.25**	**0.30**	**0.35**	**0.40**	**0.45**	**0.50**	
0	.9025	.8100	.7225	.6400	.5625	.4900	.4225	.3600	.3025	.2500	2
1	.0950	.1800	.2550	.3200	.3750	.4200	.4550	.4800	.4950	.5000	1
2	.0025	.0100	.0225	.0400	.0625	.0900	.1225	.1600	.2025	.2500	0
	0.95	0.90	0.85	0.80	0.75	0.70	0.65	0.60	0.55	0.50	$y \uparrow$

(continued)

T A B L E 1   continued

n = 3

y ↓	0.05	0.10	0.15	0.20	p 0.25	0.30	0.35	0.40	0.45	0.50	
0	.8574	.7290	.6141	.5120	.4219	.3430	.2746	.2160	.1664	.1250	3
1	.1354	.2430	.3251	.3840	.4219	.4410	.4436	.4320	.4084	.3750	2
2	.0071	.0270	.0574	.0960	.1406	.1890	.2389	.2880	.3341	.3750	1
3	.0001	.0010	.0034	.0080	.0156	.0270	.0429	.0640	.0911	.1250	0
	0.95	0.90	0.85	0.80	0.75	0.70	0.65	0.60	0.55	0.50	y ↑

n = 4

y ↓	0.05	0.10	0.15	0.20	p 0.25	0.30	0.35	0.40	0.45	0.50	
0	.8145	.6561	.5220	.4096	.3164	.2401	.1785	.1296	.0915	.0625	4
1	.1715	.2916	.3685	.4096	.4219	.4116	.3845	.3456	.2995	.2500	3
2	.0135	.0486	.0975	.1536	.2109	.2646	.3105	.3456	.3675	.3750	2
3	.0005	.0036	.0115	.0256	.0469	.0756	.1115	.1536	.2005	.2500	1
4	.0000	.0001	.0005	.0016	.0039	.0081	.0150	.0256	.0410	.0625	0
	0.95	0.90	0.85	0.80	0.75	0.70	0.65	0.60	0.55	0.50	y ↑

n = 5

y ↓	0.05	0.10	0.15	0.20	p 0.25	0.30	0.35	0.40	0.45	0.50	
0	.7738	.5905	.4437	.3277	.2373	.1681	.1160	.0778	.0503	.0313	5
1	.2036	.3281	.3915	.4096	.3955	.3602	.3124	.2592	.2059	.1563	4
2	.0214	.0729	.1382	.2048	.2637	.3087	.3364	.3456	.3369	.3125	3
3	.0011	.0081	.0244	.0512	.0879	.1323	.1811	.2304	.2757	.3125	2
4	.0000	.0005	.0022	.0064	.0146	.0284	.0488	.0768	.1128	.1563	1
5	.0000	.0000	.0001	.0003	.0010	.0024	.0053	.0102	.0185	.0313	0
	0.95	0.90	0.85	0.80	0.75	0.70	0.65	0.60	0.55	0.50	y ↑

n = 6

y ↓	0.05	0.10	0.15	0.20	p 0.25	0.30	0.35	0.40	0.45	0.50	
0	.7351	.5314	.3771	.2621	.1780	.1176	.0754	.0467	.0277	.0156	6
1	.2321	.3543	.3993	.3932	.3560	.3025	.2437	.1866	.1359	.0938	5
2	.0305	.0984	.1762	.2458	.2966	.3241	.3280	.3110	.2780	.2344	4
3	.0021	.0146	.0415	.0819	.1318	.1852	.2355	.2765	.3032	.3125	3
4	.0001	.0012	.0055	.0154	.0330	.0595	.0951	.1382	.1861	.2344	2
5	.0000	.0001	.0004	.0015	.0044	.0102	.0205	.0369	.0609	.0938	1
6	.0000	.0000	.0000	.0001	.0002	.0007	.0018	.0041	.0083	.0156	0
	0.95	0.90	0.85	0.80	0.75	0.70	0.65	0.60	0.55	0.50	y ↑

(continued)

T A B L E   1   Binomial Probabilities, $P(y)$ for $n \leq 20$

n = 7

y↓	0.05	0.10	0.15	0.20	p 0.25	0.30	0.35	0.40	0.45	0.50	
0	.6983	.4783	.3206	.2097	.1335	.0824	.0490	.0280	.0152	.0078	7
1	.2573	.3720	.3960	.3670	.3115	.2471	.1848	.1306	.0872	.0547	6
2	.0406	.1240	.2097	.2753	.3115	.3177	.2985	.2613	.2140	.1641	5
3	.0036	.0230	.0617	.1147	.1730	.2269	.2679	.2903	.2918	.2734	4
4	.0002	.0026	.0109	.0287	.0577	.0972	.1442	.1935	.2388	.2734	3
5	.0000	.0002	.0012	.0043	.0115	.0250	.0466	.0774	.1172	.1641	2
6	.0000	.0000	.0001	.0004	.0013	.0036	.0084	.0172	.0320	.0547	1
7	.0000	.0000	.0000	.0000	.0001	.0002	.0006	.0016	.0037	.0078	0
	0.95	0.90	0.85	0.80	0.75	0.70	0.65	0.60	0.55	0.50	y ↑

n = 8

y↓	0.05	0.10	0.15	0.20	p 0.25	0.30	0.35	0.40	0.45	0.50	
0	.6634	.4305	.2725	.1678	.1001	.0576	.0319	.0168	.0084	.0039	8
1	.2793	.3826	.3847	.3355	.2670	.1977	.1373	.0896	.0548	.0313	7
2	.0515	.1488	.2376	.2936	.3115	.2965	.2587	.2090	.1569	.1094	6
3	.0054	.0331	.0839	.1468	.2076	.2541	.2786	.2787	.2568	.2188	5
4	.0004	.0046	.0185	.0459	.0865	.1361	.1875	.2322	.2627	.2734	4
5	.0000	.0004	.0026	.0092	.0231	.0467	.0808	.1239	.1719	.2188	3
6	.0000	.0000	.0002	.0011	.0038	.0100	.0217	.0413	.0703	.1094	2
7	.0000	.0000	.0000	.0001	.0004	.0012	.0033	.0079	.0164	.0313	1
8	.0000	.0000	.0000	.0000	.0000	.0001	.0002	.0007	.0017	.0039	0
	0.95	0.90	0.85	0.80	0.75	0.70	0.65	0.60	0.55	0.50	y ↑

n = 9

y↓	0.05	0.10	0.15	0.20	p 0.25	0.30	0.35	0.40	0.45	0.50	
0	.6302	.3874	.2316	.1342	.0751	.0404	.0207	.0101	.0046	.0020	9
1	.2985	.3874	.3679	.3020	.2253	.1556	.1004	.0605	.0339	.0176	8
2	.0629	.1722	.2597	.3020	.3003	.2668	.2162	.1612	.1110	.0703	7
3	.0077	.0446	.1069	.1762	.2336	.2668	.2716	.2508	.2119	.1641	6
4	.0006	.0074	.0283	.0661	.1168	.1715	.2194	.2508	.2600	.2461	5
5	.0000	.0008	.0050	.0165	.0389	.0735	.1181	.1672	.2128	.2461	4
6	.0000	.0001	.0006	.0028	.0087	.0210	.0424	.0743	.1160	.1641	3
7	.0000	.0000	.0000	.0003	.0012	.0039	.0098	.0212	.0407	.0703	2
8	.0000	.0000	.0000	.0000	.0001	.0004	.0013	.0035	.0083	.0176	1
9	.0000	.0000	.0000	.0000	.0000	.0000	.0001	.0003	.0008	.0020	0
	0.95	0.90	0.85	0.80	0.75	0.70	0.65	0.60	0.55	0.50	y ↑

(continued)

T A B L E 1   continued

n = 10

p

y↓	0.05	0.10	0.15	0.20	0.25	0.30	0.35	0.40	0.45	0.50	
0	.5987	.3487	.1969	.1074	.0563	.0282	.0135	.0060	.0025	.0010	10
1	.3151	.3874	.3474	.2684	.1877	.1211	.0725	.0403	.0207	.0098	9
2	.0746	.1937	.2759	.3020	.2816	.2335	.1757	.1209	.0763	.0439	8
3	.0105	.0574	.1298	.2013	.2503	.2668	.2522	.2150	.1665	.1172	7
4	.0010	.0112	.0401	.0881	.1460	.2001	.2377	.2508	.2384	.2051	6
5	.0001	.0015	.0085	.0264	.0584	.1029	.1536	.2007	.2340	.2461	5
6	.0000	.0001	.0012	.0055	.0162	.0368	.0689	.1115	.1596	.2051	4
7	.0000	.0000	.0001	.0008	.0031	.0090	.0212	.0425	.0746	.1172	3
8	.0000	.0000	.0000	.0001	.0004	.0014	.0043	.0106	.0229	.0439	2
9	.0000	.0000	.0000	.0000	.0000	.0001	.0005	.0016	.0042	.0098	1
10	.0000	.0000	.0000	.0000	.0000	.0000	.0000	.0001	.0003	.0010	0
	0.95	0.90	0.85	0.80	0.75	0.70	0.65	0.60	0.55	0.50	y↑

n = 12

p

y↓	0.05	0.10	0.15	0.20	0.25	0.30	0.35	0.40	0.45	0.50	
0	.5404	.2824	.1422	.0687	.0317	.0138	.0057	.0022	.0008	.0002	12
1	.3413	.3766	.3012	.2062	.1267	.0712	.0368	.0174	.0075	.0029	11
2	.0988	.2301	.2924	.2835	.2323	.1678	.1088	.0639	.0339	.0161	10
3	.0173	.0852	.1720	.2362	.2581	.2397	.1954	.1419	.0923	.0537	9
4	.0021	.0213	.0683	.1329	.1936	.2311	.2367	.2128	.1700	.1208	8
5	.0002	.0038	.0193	.0532	.1032	.1585	.2039	.2270	.2225	.1934	7
6	.0000	.0005	.0040	.0155	.0401	.0792	.1281	.1766	.2124	.2256	6
7	.0000	.0000	.0006	.0033	.0115	.0291	.0591	.1009	.1489	.1934	5
8	.0000	.0000	.0001	.0005	.0024	.0078	.0199	.0420	.0762	.1208	4
9	.0000	.0000	.0000	.0001	.0004	.0015	.0048	.0125	.0277	.0537	3
10	.0000	.0000	.0000	.0000	.0000	.0002	.0008	.0025	.0068	.0161	2
11	.0000	.0000	.0000	.0000	.0000	.0000	.0001	.0003	.0010	.0029	1
12	.0000	.0000	.0000	.0000	.0000	.0000	.0000	.0000	.0001	.0002	0
	0.95	0.90	0.85	0.80	0.75	0.70	0.65	0.60	0.55	0.50	y↑

(continued)

T A B L E 1   Binomial Probabilities, $P(y)$ for $n \leq 20$

$n = 14$

$y \downarrow$	0.05	0.10	0.15	0.20	p 0.25	0.30	0.35	0.40	0.45	0.50	
0	.4877	.2288	.1028	.0440	.0178	.0068	.0024	.0008	.0002	.0001	14
1	.3593	.3559	.2539	.1539	.0832	.0407	.0181	.0073	.0027	.0009	13
2	.1229	.2570	.2912	.2501	.1802	.1134	.0634	.0317	.0141	.0056	12
3	.0259	.1142	.2056	.2501	.2402	.1943	.1366	.0845	.0462	.0222	11
4	.0037	.0349	.0998	.1720	.2202	.2290	.2022	.1549	.1040	.0611	10
5	.0004	.0078	.0352	.0860	.1468	.1963	.2178	.2066	.1701	.1222	9
6	.0000	.0013	.0093	.0322	.0734	.1262	.1759	.2066	.2088	.1833	8
7	.0000	.0002	.0019	.0092	.0280	.0618	.1082	.1574	.1952	.2095	7
8	.0000	.0000	.0003	.0020	.0082	.0232	.0510	.0918	.1398	.1833	6
9	.0000	.0000	.0000	.0003	.0018	.0066	.0183	.0408	.0762	.1222	5
10	.0000	.0000	.0000	.0000	.0003	.0014	.0049	.0136	.0312	.0611	4
11	.0000	.0000	.0000	.0000	.0000	.0002	.0010	.0033	.0093	.0222	3
12	.0000	.0000	.0000	.0000	.0000	.0000	.0001	.0005	.0019	.0056	2
13	.0000	.0000	.0000	.0000	.0000	.0000	.0000	.0001	.0002	.0009	1
14	.0000	.0000	.0000	.0000	.0000	.0000	.0000	.0000	.0000	.0001	0
	0.95	0.90	0.85	0.80	0.75	0.70	0.65	0.60	0.55	0.50	$y \uparrow$

$n = 16$

$y \downarrow$	0.05	0.10	0.15	0.20	p 0.25	0.30	0.35	0.40	0.45	0.50	
0	.4401	.1853	.0743	.0281	.0100	.0033	.0010	.0003	.0001	.0000	16
1	.3706	.3294	.2097	.1126	.0535	.0228	.0087	.0030	.0009	.0002	15
2	.1463	.2745	.2775	.2111	.1336	.0732	.0353	.0150	.0056	.0018	14
3	.0359	.1423	.2285	.2463	.2079	.1465	.0888	.0468	.0215	.0085	13
4	.0061	.0514	.1311	.2001	.2252	.2040	.1553	.1014	.0572	.0278	12
5	.0008	.0137	.0555	.1201	.1802	.2099	.2008	.1623	.1123	.0667	11
6	.0001	.0028	.0180	.0550	.1101	.1649	.1982	.1983	.1684	.1222	10
7	.0000	.0004	.0045	.0197	.0524	.1010	.1524	.1889	.1969	.1746	9
8	.0000	.0001	.0009	.0055	.0197	.0487	.0923	.1417	.1812	.1964	8
9	.0000	.0000	.0001	.0012	.0058	.0185	.0442	.0840	.1318	.1746	7
10	.0000	.0000	.0000	.0002	.0014	.0056	.0167	.0392	.0755	.1222	6
11	.0000	.0000	.0000	.0000	.0002	.0013	.0049	.0142	.0337	.0667	5
12	.0000	.0000	.0000	.0000	.0000	.0002	.0011	.0040	.0115	.0278	4
13	.0000	.0000	.0000	.0000	.0000	.0000	.0002	.0008	.0029	.0085	3
14	.0000	.0000	.0000	.0000	.0000	.0000	.0000	.0001	.0005	.0018	2
15	.0000	.0000	.0000	.0000	.0000	.0000	.0000	.0000	.0001	.0002	1
	0.95	0.90	0.85	0.80	0.75	0.70	0.65	0.60	0.55	0.50	$y \uparrow$

(continued)

T A B L E 1   continued

n = 18

y↓	0.05	0.10	0.15	0.20	p 0.25	0.30	0.35	0.40	0.45	0.50	
0	.3972	.1501	.0536	.0180	.0056	.0016	.0004	.0001	.0000	.0000	18
1	.3763	.3002	.1704	.0811	.0338	.0126	.0042	.0012	.0003	.0001	17
2	.1683	.2835	.2556	.1723	.0958	.0458	.0190	.0069	.0022	.0006	16
3	.0473	.1680	.2406	.2297	.1704	.1046	.0547	.0246	.0095	.0031	15
4	.0093	.0700	.1592	.2153	.2130	.1681	.1104	.0614	.0291	.0117	14
5	.0014	.0218	.0787	.1507	.1988	.2017	.1664	.1146	.0666	.0327	13
6	.0002	.0052	.0301	.0816	.1436	.1873	.1941	.1655	.1181	.0708	12
7	.0000	.0010	.0091	.0350	.0820	.1376	.1792	.1892	.1657	.1214	11
8	.0000	.0002	.0022	.0120	.0376	.0811	.1327	.1734	.1864	.1669	10
9	.0000	.0000	.0004	.0033	.0139	.0386	.0794	.1284	.1694	.1855	9
10	.0000	.0000	.0001	.0008	.0042	.0149	.0385	.0771	.1248	.1669	8
11	.0000	.0000	.0000	.0001	.0010	.0046	.0151	.0374	.0742	.1214	7
12	.0000	.0000	.0000	.0000	.0002	.0012	.0047	.0145	.0354	.0708	6
13	.0000	.0000	.0000	.0000	.0000	.0002	.0012	.0045	.0134	.0327	5
14	.0000	.0000	.0000	.0000	.0000	.0000	.0002	.0011	.0039	.0117	4
15	.0000	.0000	.0000	.0000	.0000	.0000	.0000	.0002	.0009	.0031	3
16	.0000	.0000	.0000	.0000	.0000	.0000	.0000	.0000	.0001	.0006	2
17	.0000	.0000	.0000	.0000	.0000	.0000	.0000	.0000	.0000	.0001	1
	0.95	0.90	0.85	0.80	0.75	0.70	0.65	0.60	0.55	0.50	y↑

n = 20

y↓	0.05	0.10	0.15	0.20	p 0.25	0.30	0.35	0.40	0.45	0.50	
0	.3585	.1216	.0388	.0115	.0032	.0008	.0002	.0000	.0000	.0000	20
1	.3774	.2702	.1368	.0576	.0211	.0068	.0020	.0005	.0001	.0000	19
2	.1887	.2852	.2293	.1369	.0669	.0278	.0100	.0031	.0008	.0002	18
3	.0596	.1901	.2428	.2054	.1339	.0716	.0323	.0123	.0040	.0011	17
4	.0133	.0898	.1821	.2182	.1897	.1304	.0738	.0350	.0139	.0046	16
5	.0022	.0319	.1028	.1746	.2023	.1789	.1272	.0746	.0365	.0148	15
6	.0003	.0089	.0454	.1091	.1686	.1916	.1712	.1244	.0746	.0370	14
7	.0000	.0020	.0160	.0545	.1124	.1643	.1844	.1659	.1221	.0739	13
8	.0000	.0004	.0046	.0222	.0609	.1144	.1614	.1797	.1623	.1201	12
9	.0000	.0001	.0011	.0074	.0271	.0654	.1158	.1597	.1771	.1602	11
10	.0000	.0000	.0002	.0020	.0099	.0308	.0686	.1171	.1593	.1762	10
11	.0000	.0000	.0000	.0005	.0030	.0120	.0336	.0710	.1185	.1602	9
12	.0000	.0000	.0000	.0001	.0008	.0039	.0136	.0355	.0727	.1201	8
13	.0000	.0000	.0000	.0000	.0002	.0010	.0045	.0146	.0366	.0739	7
14	.0000	.0000	.0000	.0000	.0000	.0002	.0012	.0049	.0150	.0370	6
15	.0000	.0000	.0000	.0000	.0000	.0000	.0003	.0013	.0049	.0148	5
16	.0000	.0000	.0000	.0000	.0000	.0000	.0000	.0003	.0013	.0046	4
17	.0000	.0000	.0000	.0000	.0000	.0000	.0000	.0000	.0002	.0011	3
18	.0000	.0000	.0000	.0000	.0000	.0000	.0000	.0000	.0000	.0002	2
	0.95	0.90	0.85	0.80	0.75	0.70	0.65	0.60	0.55	0.50	y↑

T A B L E 2   Normal Curve Areas

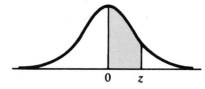

z	.00	.01	.02	.03	.04	.05	.06	.07	.08	.09
0.0	.0000	.0040	.0080	.0120	.0160	.0199	.0239	.0279	.0319	.0359
0.1	.0398	.0438	.0478	.0517	.0557	.0596	.0636	.0675	.0714	.0753
0.2	.0793	.0832	.0871	.0910	.0948	.0987	.1026	.1064	.1103	.1141
0.3	.1179	.1217	.1255	.1293	.1331	.1368	.1406	.1443	.1480	.1517
0.4	.1554	.1591	.1628	.1664	.1700	.1736	.1772	.1808	.1844	.1879
0.5	.1915	.1950	.1985	.2019	.2054	.2088	.2123	.2157	.2190	.2224
0.6	.2257	.2291	.2324	.2357	.2389	.2422	.2454	.2486	.2517	.2549
0.7	.2580	.2611	.2642	.2673	.2704	.2734	.2764	.2794	.2823	.2852
0.8	.2881	.2910	.2939	.2967	.2995	.3023	.3051	.3078	.3106	.3133
0.9	.3159	.3186	.3212	.3238	.3264	.3289	.3315	.3340	.3365	.3389
1.0	.3413	.3438	.3461	.3485	.3508	.3531	.3554	.3577	.3599	.3621
1.1	.3643	.3665	.3686	.3708	.3729	.3749	.3770	.3790	.3810	.3830
1.2	.3849	.3869	.3888	.3907	.3925	.3944	.3962	.3980	.3997	.4015
1.3	.4032	.4049	.4066	.4082	.4099	.4115	.4131	.4147	.4162	.4177
1.4	.4192	.4207	.4222	.4236	.4251	.4265	.4279	.4292	.4306	.4319
1.5	.4332	.4345	.4357	.4370	.4382	.4394	.4406	.4418	.4429	.4441
1.6	.4452	.4463	.4474	.4484	.4495	.4505	.4515	.4525	.4535	.4545
1.7	.4554	.4564	.4573	.4582	.4591	.4599	.4608	.4616	.4625	.4633
1.8	.4641	.4649	.4656	.4664	.4671	.4678	.4686	.4693	.4699	.4706
1.9	.4713	.4719	.4726	.4732	.4738	.4744	.4750	.4756	.4761	.4767
2.0	.4772	.4778	.4783	.4788	.4793	.4798	.4803	.4808	.4812	.4817
2.1	.4821	.4826	.4830	.4834	.4838	.4842	.4846	.4850	.4854	.4857
2.2	.4861	.4864	.4868	.4871	.4875	.4878	.4881	.4884	.4887	.4890
2.3	.4893	.4896	.4898	.4901	.4904	.4906	.4909	.4911	.4913	.4916
2.4	.4918	.4920	.4922	.4925	.4927	.4929	.4931	.4932	.4934	.4936
2.5	.4938	.4940	.4941	.4943	.4945	.4946	.4948	.4949	.4951	.4952
2.6	.4953	.4955	.4956	.4957	.4959	.4960	.4961	.4962	.4963	.4964
2.7	.4965	.4966	.4967	.4968	.4969	.4970	.4971	.4972	.4973	.4974
2.8	.4974	.4975	.4976	.4977	.4977	.4978	.4979	.4979	.4980	.4981
2.9	.4981	.4982	.4982	.4983	.4984	.4984	.4985	.4985	.4986	.4986
3.0	.4987	.4987	.4987	.4988	.4988	.4989	.4989	.4989	.4990	.4990

Source: This table is abridged from Table 1 of *Statistical Tables and Formulas,* by A. Hald (New York: John Wiley & Sons, 1952). Reprinted by permission of A. Hald and the publishers, John Wiley & Sons.

TABLE 3   Critical Values of t

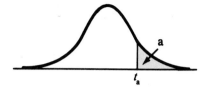

df	a = .10	a = .05	a = .025	a = .010	a = .005
1	3.078	6.314	12.706	31.821	63.657
2	1.886	2.920	4.303	6.965	9.925
3	1.638	2.353	3.182	4.541	5.841
4	1.333	2.132	2.776	3.747	4.604
5	1.476	2.015	2.571	3.365	4.032
6	1.440	1.943	2.447	3.143	3.707
7	1.415	1.895	2.365	2.998	3.499
8	1.397	1.860	2.306	2.896	3.355
9	1.383	1.833	2.262	2.821	3.250
10	1.372	1.812	2.228	2.764	3.169
11	1.363	1.796	2.201	2.718	3.106
12	1.356	1.782	2.179	2.681	3.055
13	1.350	1.771	2.160	2.650	3.012
14	1.345	1.761	2.145	2.624	2.977
15	1.341	1.753	2.131	2.602	2.947
16	1.337	1.746	2.120	2.583	2.921
17	1.333	1.740	2.110	2.567	2.898
18	1.330	1.734	2.101	2.552	2.878
19	1.328	1.729	2.093	2.539	2.861
20	1.325	1.725	2.086	2.528	2.845
21	1.323	1.721	2.080	2.518	2.831
22	1.321	1.717	2.074	2.508	2.819
23	1.319	1.714	2.069	2.500	2.807
24	1.318	1.711	2.064	2.492	2.797
25	1.316	1.708	2.060	2.485	2.787
26	1.315	1.706	2.056	2.479	2.779
27	1.314	1.703	2.052	2.473	2.771
28	1.313	1.701	2.048	2.467	2.763
29	1.311	1.699	2.045	2.462	2.756
30	1.310	1.697	2.042	2.457	2.750
40	1.303	1.684	2.021	2.423	2.704
60	1.296	1.671	2.000	2.390	2.660
120	1.289	1.658	1.980	2.358	2.617
inf.	1.282	1.645	1.960	2.326	2.576

Source: From "Table of Percentage Points of the t-Distribution." Computed by Maxine Merrington, *Biometrika*, Vol. 32 (1941), p.300. Reproduced by permission of the *Biometrika* Trustees.

T A B L E 4   Critical Values of χ^2

CHI

df	a = .995	a = .990	a = .975	a = .950	a = .900
1	0.0000393	0.0001571	0.0009821	0.0039321	0.0157908
2	0.0100251	0.0201007	0.0506356	0.102587	0.210720
3	0.0717212	0.114832	0.215795	0.351846	0.584375
4	0.206990	0.297110	0.484419	0.710721	1.063625
5	0.411740	0.554300	0.831211	1.145476	1.61031
6	0.675727	0.872085	1.237347	1.63539	2.20413
7	0.989265	1.239043	1.68987	2.16735	2.83311
8	1.344419	1.646482	2.17973	2.73264	3.48954
9	1.734926	2.087912	2.70039	3.32511	4.16816
10	2.15585	2.55821	3.24697	3.94030	4.86518
11	2.60321	3.05347	3.81575	4.57481	5.57779
12	3.07382	3.57056	4.40379	5.22603	6.30380
13	3.56503	4.10691	5.00874	5.89186	7.04150
14	4.07468	4.66043	5.62872	6.57063	7.78953
15	4.60094	5.22935	6.26214	7.26094	8.54675
16	5.14224	5.81221	6.90766	7.96164	9.31223
17	5.69724	6.40776	7.56418	8.67176	10.0852
18	6.26481	7.01491	8.23075	9.39046	10.8649
19	6.84398	7.63273	8.90655	10.1170	11.6509
20	7.43386	8.26040	9.59083	10.8508	12.4426
21	8.03366	8.89720	10.28293	11.5913	13.2396
22	8.64272	9.54249	10.9823	12.3380	14.0415
23	9.26042	10.19567	11.6885	13.0905	14.8479
24	9.88623	10.8564	12.4011	13.8484	15.6587
25	10.5197	11.5240	13.1197	14.6114	16.4734
26	11.1603	12.1981	13.8439	15.3791	17.2919
27	11.8076	12.8786	14.5733	16.1513	18.1138
28	12.4613	13.5648	15.3079	16.9279	18.9392
29	13.1211	14.2565	16.0471	17.7083	19.7677
30	13.7867	14.9535	16.7908	18.4926	20.5992
40	20.7065	22.1643	24.4331	26.5093	29.0505
50	27.9907	29.7067	32.3574	34.7642	37.6886
60	35.5346	37.4848	40.4817	43.1879	46.4589
70	43.2752	45.4418	48.7576	51.7393	55.3290
80	51.1720	53.5400	57.1532	60.3915	64.2778
90	59.1963	61.7541	65.6466	69.1260	73.2912
100	67.3276	70.0648	74.2219	77.9295	82.3581

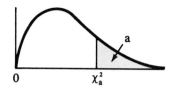

a = .10	a = .05	a = .025	a = .010	a = .005	df
2.70554	3.84146	5.02389	6.63490	7.87944	1
4.60517	5.99147	7.37776	9.21034	10.5966	2
6.25139	7.81473	9.34840	11.3449	12.8381	3
7.77944	9.48773	11.1433	13.2767	14.8602	4
9.23635	11.0705	12.8325	15.0863	16.7496	5
10.6446	12.5916	14.4494	16.8119	18.5476	6
12.0170	14.0671	16.0128	18.4753	20.2777	7
13.3616	15.5073	17.5346	20.0902	21.9550	8
14.6837	16.9190	19.0228	21.6660	23.5893	9
15.9871	18.3070	20.4831	23.2093	25.1882	10
17.2750	19.6751	21.9200	24.7250	26.7569	11
18.5494	21.0261	23.3367	26.2170	28.2995	12
19.8119	22.3621	24.7356	27.6883	29.8194	13
21.0642	23.6848	26.1190	29.1413	31.3193	14
22.3072	24.9958	27.4884	30.5779	32.8013	15
23.5418	26.2962	28.8454	31.9999	34.2672	16
24.7690	27.5871	30.1910	33.4087	35.7185	17
25.9894	28.8693	31.5264	34.8053	37.1564	18
27.2036	30.1435	32.8523	36.1908	38.5822	19
28.4120	31.4104	34.1696	37.5662	39.9968	20
29.6151	32.6705	35.4789	38.9321	41.4010	21
30.8133	33.9244	36.7807	40.2894	42.7956	22
32.0069	35.1725	38.0757	41.6384	44.1813	23
33.1963	36.4151	39.3641	42.9798	45.5585	24
34.3816	37.6525	40.6465	44.3141	46.9278	25
35.5631	38.8852	41.9232	45.6417	48.2899	26
36.7412	40.1133	43.1944	46.9630	49.6449	27
37.9159	41.3372	44.4607	48.2782	50.9933	28
39.0875	42.5569	45.7222	49.5879	52.3356	29
40.2560	43.7729	46.9792	50.8922	53.6720	30
51.8050	55.7585	59.3417	63.6907	66.7659	40
63.1671	67.5048	71.4202	76.1539	79.4900	50
74.3970	79.0819	83.2976	88.3794	91.9517	60
85.5271	90.5312	95.0231	100.425	104.215	70
96.5782	101.879	106.629	112.329	116.321	80
107.565	113.145	118.136	124.116	128.299	90
118.498	124.342	129.561	135.807	140.169	100

Source: From "Tables of the Percentage Points of the χ^2-Distribution," *Biometrika,* Vol. 32 (1941), pp. 188–89, by Catherine M. Thompson. Reproduced by permission of the *Biometrika* Trustees.

T A B L E 5 Upper-tail Values of F, $a = .10$

df_2	df_1 1	2	3	4	5	6	7	8	9
1	39.86	49.50	53.59	55.83	57.24	58.20	58.91	59.44	59.86
2	8.53	9.00	9.16	9.24	9.29	9.33	9.35	9.37	9.38
3	5.54	5.46	5.39	5.34	5.31	5.28	5.27	5.25	5.24
4	4.54	4.32	4.19	4.11	4.05	4.01	3.98	3.95	3.94
5	4.06	3.78	3.62	3.52	3.45	3.40	3.37	3.34	3.32
6	3.78	3.46	3.29	3.18	3.11	3.05	3.01	2.98	2.96
7	3.59	3.26	3.07	2.96	2.88	2.83	2.78	2.75	2.72
8	3.46	3.11	2.92	2.81	2.73	2.67	2.62	2.59	2.56
9	3.36	3.01	2.81	2.69	2.61	2.55	2.51	2.47	2.44
10	3.29	2.92	2.73	2.61	2.52	2.46	2.41	2.38	2.35
11	3.23	2.86	2.66	2.54	2.45	2.39	2.34	2.30	2.27
12	3.18	2.81	2.61	2.48	2.39	2.33	2.28	2.24	2.21
13	3.14	2.76	2.56	2.43	2.35	2.28	2.23	2.20	2.16
14	3.10	2.73	2.52	2.39	2.31	2.24	2.19	2.15	2.12
15	3.07	2.70	2.49	2.36	2.27	2.21	2.16	2.12	2.09
16	3.05	2.67	2.46	2.33	2.24	2.18	2.13	2.09	2.06
17	3.03	2.64	2.44	2.31	2.22	2.15	2.10	2.06	2.03
18	3.01	2.62	2.42	2.29	2.20	2.13	2.08	2.04	2.00
19	2.99	2.61	2.40	2.27	2.18	2.11	2.06	2.02	1.98
20	2.97	2.59	2.38	2.25	2.16	2.09	2.04	2.00	1.96
21	2.96	2.57	2.36	2.23	2.14	2.08	2.02	1.98	1.95
22	2.95	2.56	2.35	2.22	2.13	2.06	2.01	1.97	1.93
23	2.94	2.55	2.34	2.21	2.11	2.05	1.99	1.95	1.92
24	2.93	2.54	2.33	2.19	2.10	2.04	1.98	1.94	1.91
25	2.92	2.53	2.32	2.18	2.09	2.02	1.97	1.93	1.89
26	2.91	2.52	2.31	2.17	2.08	2.01	1.96	1.92	1.88
27	2.90	2.51	2.30	2.17	2.07	2.00	1.95	1.91	1.87
28	2.89	2.50	2.29	2.16	2.06	2.00	1.94	1.90	1.87
29	2.89	2.50	2.28	2.15	2.06	1.99	1.93	1.89	1.86
30	2.88	2.49	2.28	2.14	2.05	1.98	1.93	1.88	1.85
40	2.84	2.44	2.23	2.09	2.00	1.93	1.87	1.83	1.79
60	2.79	2.39	2.18	2.04	1.95	1.87	1.82	1.77	1.74
120	2.75	2.35	2.13	1.99	1.90	1.82	1.77	1.72	1.68
∞	2.71	2.30	2.08	1.94	1.85	1.77	1.72	1.67	1.63

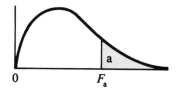

					df_1						
10	12	15	20	24	30	40	60	120	∞	df_2
60.19	60.71	61.22	61.74	62.00	62.26	62.53	62.79	63.06	63.33	1
9.39	9.41	9.42	9.44	9.45	9.46	9.47	9.47	9.48	9.49	2
5.23	5.22	5.20	5.18	5.18	5.17	5.16	5.15	5.14	5.13	3
3.92	3.90	3.87	3.84	3.83	3.82	3.80	3.79	3.78	3.76	4
3.30	3.27	3.24	3.21	3.19	3.17	3.16	3.14	3.12	3.10	5
2.94	2.90	2.87	2.84	2.82	2.80	2.78	2.76	2.74	2.72	6
2.70	2.67	2.63	2.59	2.58	2.56	2.54	2.51	2.49	2.47	7
2.54	2.50	2.46	2.42	2.40	2.38	2.36	2.34	2.32	2.29	8
2.42	2.38	2.34	2.30	2.28	2.25	2.23	2.21	2.18	2.16	9
2.32	2.28	2.24	2.20	2.18	2.16	2.13	2.11	2.08	2.06	10
2.25	2.21	2.17	2.12	2.10	2.08	2.05	2.03	2.00	1.97	11
2.19	2.15	2.10	2.06	2.04	2.01	1.99	1.96	1.93	1.90	12
2.14	2.10	2.05	2.01	1.98	1.96	1.93	1.90	1.88	1.85	13
2.10	2.05	2.01	1.96	1.94	1.91	1.89	1.86	1.83	1.80	14
2.06	2.02	1.97	1.92	1.90	1.87	1.85	1.82	1.79	1.76	15
2.03	1.99	1.94	1.89	1.87	1.84	1.81	1.78	1.75	1.72	16
2.00	1.96	1.91	1.86	1.84	1.81	1.78	1.75	1.72	1.69	17
1.98	1.93	1.89	1.84	1.81	1.78	1.75	1.72	1.69	1.66	18
1.96	1.91	1.86	1.81	1.79	1.76	1.73	1.70	1.67	1.63	19
1.94	1.89	1.84	1.79	1.77	1.74	1.71	1.68	1.64	1.61	20
1.92	1.87	1.83	1.78	1.75	1.72	1.69	1.66	1.62	1.59	21
1.90	1.86	1.81	1.76	1.73	1.70	1.67	1.64	1.60	1.57	22
1.89	1.84	1.80	1.74	1.72	1.69	1.66	1.62	1.59	1.55	23
1.88	1.83	1.78	1.73	1.70	1.67	1.64	1.61	1.57	1.53	24
1.87	1.82	1.77	1.72	1.69	1.66	1.63	1.59	1.56	1.52	25
1.86	1.81	1.76	1.71	1.68	1.65	1.61	1.58	1.54	1.50	26
1.85	1.80	1.75	1.70	1.67	1.64	1.60	1.57	1.53	1.49	27
1.84	1.79	1.74	1.69	1.66	1.63	1.59	1.56	1.52	1.48	28
1.83	1.78	1.73	1.68	1.65	1.62	1.58	1.55	1.51	1.47	29
1.82	1.77	1.72	1.67	1.64	1.61	1.57	1.54	1.50	1.46	30
1.76	1.71	1.66	1.61	1.57	1.54	1.51	1.47	1.42	1.38	40
1.71	1.66	1.60	1.54	1.51	1.48	1.44	1.40	1.35	1.29	60
1.65	1.60	1.55	1.48	1.45	1.41	1.37	1.32	1.26	1.19	120
1.60	1.55	1.49	1.42	1.38	1.34	1.30	1.24	1.17	1.00	∞

Source: From "Tables of Percentage Points of the Inverted Beta (*F*)-Distribution," *Biometrika,* Vol. 33 (1943), pp.73–88, by Maxine Merrington and Catherine M. Thompson. Reproduced by permission of the *Biometrika* Trustees.

T A B L E 6   Upper-tail Values of F, $a = .05$

					df_1				
df_2	1	2	3	4	5	6	7	8	9
1	161.4	199.5	215.7	224.6	230.2	234.0	236.8	238.9	240.5
2	18.51	19.00	19.16	19.25	19.30	19.33	19.35	19.37	19.38
3	10.13	9.55	9.28	9.12	9.01	8.94	8.89	8.85	8.81
4	7.71	6.94	6.59	6.39	6.26	6.16	6.09	6.04	6.00
5	6.61	5.79	5.41	5.19	5.05	4.95	4.88	4.82	4.77
6	5.99	5.14	4.76	4.53	4.39	4.28	4.21	4.15	4.10
7	5.59	4.74	4.35	4.12	3.97	3.87	3.79	3.73	3.68
8	5.32	4.46	4.07	3.84	3.69	3.58	3.50	3.44	3.39
9	5.12	4.26	3.86	3.63	3.48	3.37	3.29	3.23	3.18
10	4.96	4.10	3.71	3.48	3.33	3.22	3.14	3.07	3.02
11	4.84	3.98	3.59	3.36	3.20	3.09	3.01	2.95	2.90
12	4.75	3.89	3.49	3.26	3.11	3.00	2.91	2.85	2.80
13	4.67	3.81	3.41	3.18	3.03	2.92	2.83	2.77	2.71
14	4.60	3.74	3.34	3.11	2.96	2.85	2.76	2.70	2.65
15	4.54	3.68	3.29	3.06	2.90	2.79	2.71	2.64	2.59
16	4.49	3.63	3.24	3.01	2.85	2.74	2.66	2.59	2.54
17	4.45	3.59	3.20	2.96	2.81	2.70	2.61	2.55	2.49
18	4.41	3.55	3.16	2.93	2.77	2.66	2.58	2.51	2.46
19	4.38	3.52	3.13	2.90	2.74	2.63	2.54	2.48	2.42
20	4.35	3.49	3.10	2.87	2.71	2.60	2.51	2.45	2.39
21	4.32	3.47	3.07	2.84	2.68	2.57	2.49	2.42	2.37
22	4.30	3.44	3.05	2.82	2.66	2.55	2.46	2.40	2.34
23	4.28	3.42	3.03	2.80	2.64	2.53	2.44	2.37	2.32
24	4.26	3.40	3.01	2.78	2.62	2.51	2.42	2.36	2.30
25	4.24	3.39	2.99	2.76	2.60	2.49	2.40	2.34	2.28
26	4.23	3.37	2.98	2.74	2.59	2.47	2.39	2.32	2.27
27	4.21	3.35	2.96	2.73	2.57	2.46	2.37	2.31	2.25
28	4.20	3.34	2.95	2.71	2.56	2.45	2.36	2.29	2.24
29	4.18	3.33	2.93	2.70	2.55	2.43	2.35	2.28	2.22
30	4.17	3.32	2.92	2.69	2.53	2.42	2.33	2.27	2.21
40	4.08	3.23	2.84	2.61	2.45	2.34	2.25	2.18	2.12
60	4.00	3.15	2.76	2.53	2.37	2.25	2.17	2.10	2.04
120	3.92	3.07	2.68	2.45	2.29	2.17	2.09	2.02	1.96
∞	3.84	3.00	2.60	2.37	2.21	2.10	2.01	1.94	1.88

					df_1						
10	**12**	**15**	**20**	**24**	**30**	**40**	**60**	**120**	**∞**	**df_2**
241.9	243.9	245.9	248.0	249.1	250.1	251.1	252.2	253.3	254.3	1
19.40	19.41	19.43	19.45	19.45	19.46	19.47	19.48	19.49	19.50	2
8.79	8.74	8.70	8.66	8.64	8.62	8.59	8.57	8.55	8.53	3
5.96	5.91	5.86	5.80	5.77	5.75	5.72	5.69	5.66	5.63	4
4.74	4.68	4.62	4.56	4.53	4.50	4.46	4.43	4.40	4.36	5
4.06	4.00	3.94	3.87	3.84	3.81	3.77	3.74	3.70	3.67	6
3.64	3.57	3.51	3.44	3.41	3.38	3.34	3.30	3.27	3.23	7
3.35	3.28	3.22	3.15	3.12	3.08	3.04	3.01	2.97	2.93	8
3.14	3.07	3.01	2.94	2.90	2.86	2.83	2.79	2.75	2.71	9
2.98	2.91	2.85	2.77	2.74	2.70	2.66	2.62	2.58	2.54	10
2.85	2.79	2.72	2.65	2.61	2.57	2.53	2.49	2.45	2.40	11
2.75	2.69	2.62	2.54	2.51	2.47	2.43	2.38	2.34	2.30	12
2.67	2.60	2.53	2.46	2.42	2.38	2.34	2.30	2.25	2.21	13
2.60	2.53	2.46	2.39	2.35	2.31	2.27	2.22	2.18	2.13	14
2.54	2.48	2.40	2.33	2.29	2.25	2.20	2.16	2.11	2.07	15
2.49	2.42	2.35	2.28	2.24	2.19	2.15	2.11	2.06	2.01	16
2.45	2.38	2.31	2.23	2.19	2.15	2.10	2.06	2.01	1.96	17
2.41	2.34	2.27	2.19	2.15	2.11	2.06	2.02	1.97	1.92	18
2.38	2.31	2.23	2.16	2.11	2.07	2.03	1.98	1.93	1.88	19
2.35	2.28	2.20	2.12	2.08	2.04	1.99	1.95	1.90	1.84	20
2.32	2.25	2.18	2.10	2.05	2.01	1.96	1.92	1.87	1.81	21
2.30	2.23	2.15	2.07	2.03	1.98	1.94	1.89	1.84	1.78	22
2.27	2.20	2.13	2.05	2.01	1.96	1.91	1.86	1.81	1.76	23
2.25	2.18	2.11	2.03	1.98	1.94	1.89	1.84	1.79	1.73	24
2.24	2.16	2.09	2.01	1.96	1.92	1.87	1.82	1.77	1.71	25
2.22	2.15	2.07	1.99	1.95	1.90	1.85	1.80	1.75	1.69	26
2.20	2.13	2.06	1.97	1.93	1.88	1.84	1.79	1.73	1.67	27
2.19	2.12	2.04	1.96	1.91	1.87	1.82	1.77	1.71	1.65	28
2.18	2.10	2.03	1.94	1.90	1.85	1.81	1.75	1.70	1.64	29
2.16	2.09	2.01	1.93	1.89	1.84	1.79	1.74	1.68	1.62	30
2.08	2.00	1.92	1.84	1.79	1.74	1.69	1.64	1.58	1.51	40
1.99	1.92	1.84	1.75	1.70	1.65	1.59	1.53	1.47	1.39	60
1.91	1.83	1.75	1.66	1.61	1.55	1.50	1.43	1.35	1.25	120
1.83	1.75	1.67	1.57	1.52	1.46	1.39	1.32	1.22	1.00	∞

T A B L E 7   Upper-tail Values of F, a = .025

					df_1				
df_2	1	2	3	4	5	6	7	8	9
1	647.8	799.5	864.2	899.6	921.8	937.1	948.2	956.7	963.3
2	38.51	39.00	39.17	39.25	39.30	39.33	39.36	39.37	39.39
3	17.44	16.04	15.44	15.10	14.88	14.73	14.62	14.54	14.47
4	12.22	10.65	9.98	9.60	9.36	9.20	9.07	8.98	8.90
5	10.01	8.43	7.76	7.39	7.15	6.98	6.85	6.76	6.68
6	8.81	7.26	6.60	6.23	5.99	5.82	5.70	5.60	5.52
7	8.07	6.54	5.89	5.52	5.29	5.12	4.99	4.90	4.82
8	7.57	6.06	5.42	5.05	4.82	4.65	4.53	4.43	4.36
9	7.21	5.71	5.08	4.72	4.48	4.32	4.20	4.10	4.03
10	6.94	5.46	4.83	4.47	4.24	4.07	3.95	3.85	3.78
11	6.72	5.26	4.63	4.28	4.04	3.88	3.76	3.66	3.59
12	6.55	5.10	4.47	4.12	3.89	3.73	3.61	3.51	3.44
13	6.41	4.97	4.35	4.00	3.77	3.60	3.48	3.39	3.31
14	6.30	4.86	4.24	3.89	3.66	3.50	3.38	3.29	3.21
15	6.20	4.77	4.15	3.80	3.58	3.41	3.29	3.20	3.12
16	6.12	4.69	4.08	3.73	3.50	3.34	3.22	3.12	3.05
17	6.04	4.62	4.01	3.66	3.44	3.28	3.16	3.06	2.98
18	5.98	4.56	3.95	3,61	3.38	3.22	3.10	3.01	2.93
19	5.92	4.51	3.90	3.56	3.33	3.17	3.05	2.96	2.88
20	5.87	4.46	3.86	3.51	3.29	3.13	3.01	2.91	2.84
21	5.83	4.42	3.82	3.48	3.25	3.09	2.97	2.87	2.80
22	5.79	4.38	3.78	3.44	3.22	3.05	2.93	2.84	2.76
23	5.75	4.35	3.75	3.41	3.18	3.02	2.90	2.81	2.73
24	5.72	4.32	3.72	3.38	3.15	2.99	2.87	2.78	2.70
25	5.69	4.29	3.69	3.35	3.13	2.97	2.85	2.75	2.68
26	5.66	4.27	3.67	3.33	3.10	2.94	2.82	2.73	2.65
27	5.63	4.24	3.65	3.31	3.08	2.92	2.80	2.71	2.63
28	5.61	4.22	3.63	3.29	3.06	2.90	2.78	2.69	2.61
29	5.59	4.20	3.61	3.27	3.04	2.88	2.76	2.67	2.59
30	5.57	4.18	3.59	3.25	3.03	2.87	2.75	2.65	2.57
40	5.42	4.05	3.46	3.13	2.90	2.74	2.62	2.53	2.45
60	5.29	3.93	3.34	3.01	2.79	2.63	2.51	2.41	2.33
120	5.15	3.80	3.23	2.89	2.67	2.52	2.39	2.30	2.22
∞	5.02	3.69	3.12	2.79	2.57	2.41	2.29	2.19	2.11

					df_1						
10	**12**	**15**	**20**	**24**	**30**	**40**	**60**	**120**	**∞**	**df_2**
968.6	976.7	984.9	993.1	997.2	1001	1006	1010	1014	1018	1
39.40	39.41	39.43	39.45	39.46	39.46	39.47	39.48	39.49	39.50	2
14.42	14.34	14.25	14.17	14.12	14.08	14.04	13.99	13.95	13.90	3
8.84	8.75	8.66	8.56	8.51	8.46	8.41	8.36	8.31	8.26	4
6.62	6.52	6.43	6.33	6.28	6.23	6.18	6.12	6.07	6.02	5
5.46	5.37	5.27	5.17	5.12	5.07	5.01	4.96	4.90	4.85	6
4.76	4.67	4.57	4.47	4.42	4.36	4.31	4.25	4.20	4.14	7
4.30	4.20	4.10	4.00	3.95	3.89	3.84	3.78	3.73	3.67	8
3.96	3.87	3.77	3.67	3.61	3.56	3.51	3.45	3.39	3.33	9
3.72	3.62	3.52	3.42	3.37	3.31	3.26	3.20	3.14	3.08	10
3.53	3.43	3.33	3.23	3.17	3.12	3.06	3.00	2.94	2.88	11
3.37	3.28	3.18	3.07	3.02	2.96	2.91	2.85	2.79	2.72	12
3.25	3.15	3.05	2.95	2.89	2.84	2.78	2.72	2.66	2.60	13
3.15	3.05	2.95	2.84	2.79	2.73	2.67	2.61	2.55	2.49	14
3.06	2.96	2.86	2.76	2.70	2.64	2.59	2.52	2.46	2.40	15
2.99	2.89	2.79	2.68	2.63	2.57	2.51	2.45	2.38	2.32	16
2.92	2.82	2.72	2.62	2.56	2.50	2.44	2.38	2.32	2.25	17
2.87	2.77	2.67	2.56	2.50	2.44	2.38	2.32	2.26	2.19	18
2.82	2.72	2.62	2.51	2.45	2.39	2.33	2.27	2.20	2.13	19
2.77	2.68	2.57	2.46	2.41	2.35	2.29	2.22	2.16	2.09	20
2.73	2.64	2.53	2.42	2.37	2.31	2.25	2.18	2.11	2.04	21
2.70	2.60	2.50	2.39	2.33	2.27	2.21	2.14	2.08	2.00	22
2.67	2.57	2.47	2.36	2.30	2.24	2.18	2.11	2.04	1.97	23
2.64	2.54	2.44	2.33	2.27	2.21	2.15	2.08	2.01	1.94	24
2.61	2.51	2.41	2.30	2.24	2.18	2.12	2.05	1.98	1.91	25
2.59	2.49	2.39	2.28	2.22	2.16	2.09	2.03	1.95	1.88	26
2.57	2.47	2.36	2.25	2.19	2.13	2.07	2.00	1.93	1.85	27
2.55	2.45	2.34	2.23	2.17	2.11	2.05	1.98	1.91	1.83	28
2.53	2.43	2.32	2.21	2.15	2.09	2.03	1.96	1.89	1.81	29
2.51	2.41	2.31	2.20	2.14	2.07	2.01	1.94	1.87	1.79	30
2.39	2.29	2.18	2.07	2.01	1.94	1.88	1.80	1.72	1.64	40
2.27	2.17	2.06	1.94	1.88	1.82	1.74	1.67	1.58	1.48	60
2.16	2.05	1.94	1.82	1.76	1.69	1.61	1.53	1.43	1.31	120
2.05	1.94	1.83	1.71	1.64	1.57	1.48	1.39	1.27	1.00	∞

Source: From "Tables of Percentage Points of the Inverted Beta (*F*)-Distribution," *Biometrika,* Vol. 33 (1943), pp. 73–88, by Maxine Merrington and Catherine M. Thompson. Reproduced by permission of the *Biometrika* Trustees.

T A B L E   8   Upper-tail Values of F, $a = .01$

					df_1				
df_2	1	2	3	4	5	6	7	8	9
1	4052	4999.5	5403	5625	5764	5859	5928	5982	6022
2	98.50	99.00	99.17	99.25	99.30	99.33	99.36	99.37	99.39
3	34.12	30.82	29.46	28.71	28.24	27.91	27.67	27.49	27.35
4	21.20	18.00	16.69	15.98	15.52	15.21	14.98	14.80	14.66
5	16.26	13.27	12.06	11.39	10.97	10.67	10.46	10.29	10.16
6	13.75	10.92	9.78	9.15	8.75	8.47	8.26	8.10	7.98
7	12.25	9.55	8.45	7.85	7.46	7.19	6.99	6.84	6.72
8	11.26	8.65	7.59	7.01	6.63	6.37	6.18	6.03	5.91
9	10.56	8.02	6.99	6.42	6.06	5.80	5.61	5.47	5.35
10	10.04	7.56	6.55	5.99	5.64	5.39	5.20	5.06	4.94
11	9.65	7.21	6.22	5.67	5.32	5.07	4.89	4.74	4.63
12	9.33	6.93	5.95	5.41	5.06	4.82	4.64	4.50	4.39
13	9.07	6.70	5.74	5.21	4.86	4.62	4.44	4.30	4.19
14	8.86	6.51	5.56	5.04	4.69	4.46	4.28	4.14	4.03
15	8.68	6.36	5.42	4.89	4.56	4.32	4.14	4.00	3.89
16	8.53	6.23	5.29	4.77	4.44	4.20	4.03	3.89	3.78
17	8.40	6.11	5.18	4.67	4.34	4.10	3.93	3.79	3.68
18	8.29	6.01	5.09	4.58	4.25	4.01	3.84	3.71	3.60
19	8.18	5.93	5.01	4.50	4.17	3.94	3.77	3.63	3.52
20	8.10	5.85	4.94	4.43	4.10	3.87	3.70	3.56	3.46
21	8.02	5.78	4.87	4.37	4.04	3.81	3.64	3.51	3.40
22	7.95	5.72	4.82	4.31	3.99	3.76	3.59	3.45	3.35
23	7.88	5.66	4.76	4.26	3.94	3.71	3.54	3.41	3.30
24	7.82	5.61	4.72	4.22	3.90	3.67	3.50	3.36	3.26
25	7.77	5.57	4.68	4.18	3.85	3.63	3.46	3.32	3.22
26	7.72	5.53	4.64	4.14	3.82	3.59	3.42	3.29	3.18
27	7.68	5.49	4.60	4.11	3.78	3.56	3.39	3.26	3.15
28	7.64	5.45	4.57	4.07	3.75	3.53	3.36	3.23	3.12
29	7.60	5.42	4.54	4.04	3.73	3.50	3.33	3.20	3.09
30	7.56	5.39	4.51	4.02	3.70	3.47	3.30	3.17	3.07
40	7.31	5.18	4.31	3.83	3.51	3.29	3.12	2.99	2.89
60	7.08	4.98	4.13	3.65	3.34	3.12	2.95	2.82	2.72
120	6.85	4.79	3.95	3.48	3.17	2.96	2.79	2.66	2.56
∞	6.63	4.61	3.78	3.32	3.02	2.80	2.64	2.51	2.41

					df_1						
10	12	15	20	24	30	40	60	120	∞	df_2
6056	6106	6157	6209	6235	6261	6287	6313	6339	6366	1
99.40	99.42	99.43	99.45	99.46	99.47	99.47	99.48	99.49	99.50	2
27.23	27.05	26.87	26.69	26.60	26.50	26.41	26.32	26.22	26.13	3
14.55	14.37	14.20	14.02	13.93	13.84	13.75	13.65	13.56	13.46	4
10.05	9.89	9.72	9.55	9.47	9.38	9.29	9.20	9.11	9.02	5
7.87	7.72	7.56	7.40	7.31	7.23	7.14	7.06	6.97	6.88	6
6.62	6.47	6.31	6.16	6.07	5.99	5.91	5.82	5.74	5.65	7
5.81	5.67	5.52	5.36	5.28	5.20	5.12	5.03	4.95	4.86	8
5.26	5.11	4.96	4.81	4.73	4.65	4.57	4.48	4.40	4.31	9
4.85	4.71	4.56	4.41	4.33	4.25	4.17	4.08	4.00	3.91	10
4.54	4.40	4.25	4.10	4.02	3.94	3.86	3.78	3.69	3.60	11
4.30	4.16	4.01	3.86	3.78	3.70	3.62	3.54	3.45	3.36	12
4.10	3.96	3.82	3.66	3.59	3.51	3.43	3.34	3.25	3.17	13
3.94	3.80	3.66	3.51	3.43	3.35	3.27	3.18	3.09	3.00	14
3.80	3.67	3.52	3.37	3.29	3.21	3.13	3.05	2.96	2.87	15
3.69	3.55	3.41	3.26	3.18	3.10	3.02	2.93	2.84	2.75	16
3.59	3.46	3.31	3.16	3.08	3.00	2.92	2.83	2.75	2.65	17
3.51	3.37	3.23	3.08	3.00	2.92	2.84	2.75	2.66	2.57	18
3.43	3.30	3.15	3.00	2.92	2.84	2.76	2.67	2.58	2.49	19
3.37	3.23	3.09	2.94	2.86	2.78	2.69	2.61	2.52	2.42	20
3.31	3.17	3.03	2.88	2.80	2.72	2.64	2.55	2.46	2.36	21
3.26	3.12	2.98	2.83	2.75	2.67	2.58	2.50	2.40	2.31	22
3.21	3.07	2.93	2.78	2.70	2.62	2.54	2.45	2.35	2.26	23
3.17	3.03	2.89	2.74	2.66	2.58	2.49	2.40	2.31	2.21	24
3.13	2.99	2.85	2.70	2.62	2.54	2.45	2.36	2.27	2.17	25
3.09	2.96	2.81	2.66	2.58	2.50	2.42	2.33	2.23	2.13	26
3.06	2.93	2.78	2.63	2.55	2.47	2.38	2.29	2.20	2.10	27
3.03	2.90	2.75	2.60	2.52	2.44	2.35	2.26	2.17	2.06	28
3.00	2.87	2.73	2.57	2.49	2.41	2.33	2.23	2.14	2.03	29
2.98	2.84	2.70	2.55	2.47	2.39	2.30	2.21	2.11	2.01	30
2.80	2.66	2.52	2.37	2.29	2.20	2.11	2.02	1.92	1.80	40
2.63	2.50	2.35	2.20	2.12	2.03	1.94	1.84	1.73	1.60	60
2.47	2.34	2.19	2.03	1.95	1.86	1.76	1.66	1.53	1.38	120
2.32	2.18	2.04	1.88	1.79	1.70	1.59	1.47	1.32	1.00	∞

Source: From "Tables of Percentage Points of the Inverted Beta (*F*)-Distribution," *Biometrika,* Vol. 33 (1943), pp. 73–88, by Maxine Merrington and Catherine M. Thompson. Reproduced by permission of the *Biometrika* Trustees.

TABLE 9   Upper-tail Values of F, $a = .005$

df_2	df_1								
	1	2	3	4	5	6	7	8	9
1	16211	20000	21615	22500	23056	23437	23715	23925	24091
2	198.5	199.0	199.2	199.2	199.3	199.3	199.4	199.4	199.4
3	55.55	49.80	47.47	46.19	45.39	44.84	44.43	44.13	43.88
4	31.33	26.28	24.26	23.15	22.46	21.97	21.62	21.35	21.14
5	22.78	18.31	16.53	15.56	14.94	14.51	14.20	13.96	13.77
6	18.63	14.54	12.92	12.03	11.46	11.07	10.79	10.57	10.39
7	16.24	12.40	10.88	10.05	9.52	9.16	8.89	8.68	8.51
8	14.69	11.04	9.60	8.81	8.30	7.95	7.69	7.50	7.34
9	13.61	10.11	8.72	7.96	7.47	7.13	6.88	6.69	6.54
10	12.83	9.43	8.08	7.34	6.87	6.54	6.30	6.12	5.97
11	12.23	8.91	7.60	6.88	6.42	6.10	5.86	5.68	5.54
12	11.75	8.51	7.23	6.52	6.07	5.76	5.52	5.35	5.20
13	11.37	8.19	6.93	6.23	5.79	5.48	5.25	5.08	4.94
14	11.06	7.92	6.68	6.00	5.56	5.26	5.03	4.86	4.72
15	10.80	7.70	6.48	5.80	5.37	5.07	4.85	4.67	4.54
16	10.58	7.51	6.30	5.64	5.21	4.91	4.69	4.52	4.38
17	10.38	7.35	6.16	5.50	5.07	4.78	4.56	4.39	4.25
18	10.22	7.21	6.03	5.37	4.96	4.66	4.44	4.28	4.14
19	10.07	7.09	5.92	5.27	4.85	4.56	4.34	4.18	4.04
20	9.94	6.99	5.82	5.17	4.76	4.47	4.26	4.09	3.96
21	9.83	6.89	5.73	5.09	4.68	4.39	4.18	4.01	3.88
22	9.73	6.81	5.65	5.02	4.61	4.32	4.11	3.94	3.81
23	9.63	6.73	5.58	4.95	4.54	4.26	4.05	3.88	3.75
24	9.55	6.66	5.52	4.89	4.49	4.20	3.99	3.83	3.69
25	9.48	6.60	5.46	4.84	4.43	4.15	3.94	3.78	3.64
26	9.41	6.54	5.41	4.79	4.38	4.10	3.89	3.73	3.60
27	9.34	6.49	5.36	4.74	4.34	4.06	3.85	3.69	3.56
28	9.28	6.44	5.32	4.70	4.30	4.02	3.81	3.65	3.52
29	9.23	6.40	5.28	4.66	4.26	3.98	3.77	3.61	3.48
30	9.18	6.35	5.24	4.62	4.23	3.95	3.74	3.58	3.45
40	8.83	6.07	4.98	4.37	3.99	3.71	3.51	3.35	3.22
60	8.49	5.79	4.73	4.14	3.76	3.49	3.29	3.13	3.01
120	8.18	5.54	4.50	3.92	3.55	3.28	3.09	2.93	2.81
∞	7.88	5.30	4.28	3.72	3.35	3.09	2.90	2.74	2.62

					df_1						
10	**12**	**15**	**20**	**24**	**30**	**40**	**60**	**120**	**∞**	**df_2**
24224	24426	24630	24836	24940	25044	25148	25253	25359	25465	1
199.4	199.4	199.4	199.4	199.5	199.5	199.5	199.5	199.5	199.5	2
43.69	43.39	43.08	42.78	42.62	42.47	42.31	42.15	41.99	41.83	3
20.97	20.70	20.44	20.17	20.03	19.89	19.75	19.61	19.47	19.32	4
13.62	13.38	13.15	12.90	12.78	12.66	12.53	12.40	12.27	12.14	5
10.25	10.03	9.81	9.59	9.47	9.36	9.24	9.12	9.00	8.88	6
8.38	8.18	7.97	7.75	7.65	7.53	7.42	7.31	7.19	7.08	7
7.21	7.01	6.81	6.61	6.50	6.40	6.29	6.18	6.06	5.95	8
6.42	6.23	6.03	5.83	5.73	5.62	5.52	5.41	5.30	5.19	9
5.85	5.66	5.47	5.27	5.17	5.07	4.97	4.86	4.75	4.64	10
5.42	5.24	5.05	4.86	4.76	4.65	4.55	4.44	4.34	4.23	11
5.09	4.91	4.72	4.53	4.43	4.33	4.23	4.12	4.01	3.90	12
4.82	4.64	4.46	4.27	4.17	4.07	3.97	3.87	3.76	3.65	13
4.60	4.43	4.25	4.06	3.96	3.86	3.76	3.66	3.55	3.44	14
4.42	4.25	4.07	3.88	3.79	3.69	3.58	3.48	3.37	3.26	15
4.27	4.10	3.92	3.73	3.64	3.54	3.44	3.33	3.22	3.11	16
4.14	3.97	3.79	3.61	3.51	3.41	3.31	3.21	3.10	2.98	17
4.03	3.86	3.68	3.50	3.40	3.30	3.20	3.10	2.99	2.87	18
3.93	3.76	3.59	3.40	3.31	3.21	3.11	3.00	2.89	2.78	19
3.85	3.68	3.50	3.32	3.22	3.12	3.02	2.92	2.81	2.69	20
3.77	3.60	3.43	3.24	3.15	3.05	2.95	2.84	2.73	2.61	21
3.70	3.54	3.36	3.18	3.08	2.98	2.88	2.77	2.66	2.55	22
3.64	3.47	3.30	3.12	3.02	2.92	2.82	2.71	2.60	2.48	23
3.59	3.42	3.25	3.06	2.97	2.87	2.77	2.66	2.55	2.43	24
3.54	3.37	3.20	3.01	2.92	2.82	2.72	2.61	2.50	2.38	25
3.49	3.33	3.15	2.97	2.87	2.77	2.67	2.56	2.45	2.33	26
3.45	3.28	3.11	2.93	2.83	2.73	2.63	2.52	2.41	2.29	27
3.41	3.25	3.07	2.89	2.79	2.69	2.59	2.48	2.37	2.25	28
3.38	3.21	3.04	2.86	2.76	2.66	2.56	2.45	2.33	2.21	29
3.34	3.18	3.01	2.82	2.73	2.63	2.52	2.42	2.30	2.18	30
3.12	2.95	2.78	2.60	2.50	2.40	2.30	2.18	2.06	1.93	40
2.90	2.74	2.57	2.39	2.29	2.19	2.08	1.96	1.83	1.69	60
2.71	2.54	2.37	2.19	2.09	1.98	1.87	1.75	1.61	1.43	120
2.52	2.36	2.19	2.00	1.90	1.79	1.67	1.53	1.36	1.00	∞

Source: From "Tables of Percentage Points of the Inverted Beta (*F*)-Distribution," *Biometrika,* Vol. 33 (1943), pp. 73–88, by Maxine Merrington and Catherine M. Thompson. Reproduced by permission of the *Biometrika* Trustees.

T A B L E 10   Critical Values for the Wilcoxon Signed-rank Test
$(n = 5(1)50)$

One-sided	Two-sided	$n = 5$	$n = 6$	$n = 7$	$n = 8$	$n = 9$	$n = 10$	$n = 11$	$n = 12$
$\alpha = .05$	$\alpha = .10$	1	2	4	6	8	11	14	17
$\alpha = .025$	$\alpha = .05$		1	2	4	6	8	11	14
$\alpha = .01$	$\alpha = .02$			0	2	3	5	7	10
$\alpha = .005$	$\alpha = .01$				0	2	3	5	7

One-sided	Two-sided	$n = 13$	$n = 14$	$n = 15$	$n = 16$	$n = 17$	$n = 18$	$n = 19$	$n = 20$
$\alpha = .05$	$\alpha = .10$	21	26	30	36	41	47	54	60
$\alpha = .025$	$\alpha = .05$	17	21	25	30	35	40	46	52
$\alpha = .01$	$\alpha = .02$	13	16	20	24	28	33	38	43
$\alpha = .005$	$\alpha = .01$	10	13	16	19	23	28	32	37

One-sided	Two-sided	$n = 21$	$n = 22$	$n = 23$	$n = 24$	$n = 25$	$n = 26$	$n = 27$	$n = 28$
$\alpha = .05$	$\alpha = .10$	68	75	83	92	101	110	120	130
$\alpha = .025$	$\alpha = .05$	59	66	73	81	90	98	107	117
$\alpha = .01$	$\alpha = .02$	49	56	62	69	77	85	93	102
$\alpha = .005$	$\alpha = .01$	43	49	55	61	68	76	84	92

One-sided	Two-sided	$n = 29$	$n = 30$	$n = 31$	$n = 32$	$n = 33$	$n = 34$	$n = 35$	$n = 36$
$\alpha = .05$	$\alpha = .10$	141	152	163	175	188	201	214	228
$\alpha = .025$	$\alpha = .05$	127	137	148	159	171	183	195	208
$\alpha = .01$	$\alpha = .02$	111	120	130	141	151	162	174	186
$\alpha = .005$	$\alpha = .01$	100	109	118	128	138	149	160	171

One-sided	Two-sided	$n = 37$	$n = 38$	$n = 39$	$n = 40$	$n = 41$	$n = 42$	$n = 43$	$n = 44$
$\alpha = .05$	$\alpha = .10$	242	256	271	287	303	319	336	353
$\alpha = .025$	$\alpha = .05$	222	235	250	264	279	295	311	327
$\alpha = .01$	$\alpha = .02$	198	211	224	238	252	267	281	297
$\alpha = .005$	$\alpha = .01$	183	195	208	221	234	248	262	277

One-sided	Two-sided	$n = 45$	$n = 46$	$n = 47$	$n = 48$	$n = 49$	$n = 50$		
$\alpha = .05$	$\alpha = .10$	371	389	408	427	446	466		
$\alpha = .025$	$\alpha = .05$	344	361	379	397	415	434		
$\alpha = .01$	$\alpha = .02$	313	329	345	362	380	398		
$\alpha = .005$	$\alpha = .01$	292	307	323	339	356	373		

Source: From *Some Rapid Approximate Statistical Procedures* (Revised) by Frank Wilcoxon and Roberta A. Wilcox (Pearl River, N.Y.: Lederle Laboratories, 1964), Table 2. Reproduced by permission of Lederle Laboratories, a division of American Cyanamid Company.

T A B L E   11   Random Numbers

Line	Column									
	1	**2**	**3**	**4**	**5**	**6**	**7**	**8**	**9**	**10**
1	75029	50152	25648	02523	84300	83093	39852	91276	88988	12439
2	73741	30492	19280	41255	74008	72750	70420	67769	72837	27098
3	07049	98408	27011	76385	15212	03806	85928	81312	14514	55277
4	01033	08705	42934	79257	89138	21506	26797	67223	62165	67981
5	48399	78564	35787	07647	23794	73938	29477	11420	03228	16586
6	70459	73480	06740	79124	14078	72352	07410	93292	93057	18715
7	74770	80185	08181	27417	90866	98444	72870	51219	51481	47916
8	24167	13753	65011	66288	12633	79199	61497	56186	83643	96184
9	24316	80240	62592	53393	57028	61626	56508	84407	97873	27571
10	84565	59254	94435	33322	50014	00180	50954	04099	66005	59141
11	60794	32497	47830	94509	36576	68874	84062	84503	50454	42199
12	99104	14833	97062	48867	19645	78069	91602	46991	57523	22219
13	15604	93654	21487	86036	22827	62637	70378	58539	17827	80108
14	20204	00253	19678	15789	17628	63667	23348	67083	92361	50413
15	71233	73676	00958	42662	47344	00104	74530	46238	06655	23791
16	82846	82954	52107	66054	27358	69664	71760	03577	75622	21536
17	48613	97858	49627	17036	55574	80116	80533	62146	48083	29177
18	42313	91287	66900	79817	76803	42462	63542	99089	22655	44130
19	60879	68102	60700	51281	61386	06782	88214	68246	15552	79093
20	34593	95713	62942	16236	30933	39470	58423	95304	46017	18364
21	96033	10917	01205	08978	43021	77321	76736	64527	96534	98457
22	21932	45476	75464	43497	81807	99369	59945	65349	52588	27386
23	91019	99635	78638	75114	42943	81629	03283	85036	80666	18675
24	86053	48238	14952	55565	98821	92843	67663	70387	13356	46650
25	59700	38346	92770	11506	34101	01051	99390	86884	26788	78768

T A B L E 12   Percentage Points of $F_{max} = s^2_{max}/s^2_{min}$

Upper 5% Points

df_2 \ t	2	3	4	5	6	7	8	9	10	11	12
2	39.0	87.5	142	202	266	333	403	475	550	626	704
3	15.4	27.8	39.2	50.7	62.0	72.9	83.5	93.9	104	114	124
4	9.60	15.5	20.6	25.2	29.5	33.6	37.5	41.1	44.6	48.0	51.4
5	7.15	10.8	13.7	16.3	18.7	20.8	22.9	24.7	26.5	28.2	29.9
6	5.82	8.38	10.4	12.1	13.7	15.0	16.3	17.5	18.6	19.7	20.7
7	4.99	6.94	8.44	9.70	10.8	11.8	12.7	13.5	14.3	15.1	15.8
8	4.43	6.00	7.18	8.12	9.03	9.78	10.5	11.1	11.7	12.2	12.7
9	4.03	5.34	6.31	7.11	7.80	8.41	8.95	9.45	9.91	10.3	10.7
10	3.72	4.85	5.67	6.34	6.92	7.42	7.87	8.28	8.66	9.01	9.34
12	3.28	4.16	4.79	5.30	5.72	6.09	6.42	6.72	7.00	7.25	7.48
15	2.86	3.54	4.01	4.37	4.68	4.95	5.19	5.40	5.59	5.77	5.93
20	2.46	2.95	3.29	3.54	3.76	3.94	4.10	4.24	4.37	4.49	4.59
30	2.07	4.20	2.61	2.78	2.91	3.02	3.12	3.21	3.29	3.36	3.39
60	1.67	1.85	1.96	2.04	2.11	2.17	2.22	2.26	2.30	2.33	2.36
∞	1.00	1.00	1.00	1.00	1.00	1.00	1.00	1.00	1.00	1.00	1.00

Upper 1% Points

df_2 \ t	2	3	4	5	6	7	8	9	10	11	12
2	199	448	729	1036	1362	1705	2063	2432	2813	3204	3605
3	47.5	85	120	151	184	21(6)	24(9)	28(1)	31(0)	33(7)	36(1)
4	23.2	37	49	59	69	79	89	97	106	113	120
5	14.9	22	28	33	38	42	46	50	54	57	60
6	11.1	15.5	19.1	22	25	27	30	32	34	36	37
7	8.89	12.1	14.5	16.5	18.4	20	22	23	24	26	27
8	7.50	9.9	11.7	13.2	14.5	15.8	16.6	17.9	18.9	19.8	21
9	6.54	8.5	9.9	11.1	12.1	13.1	13.9	14.7	15.3	16.0	16.6
10	5.85	7.4	8.6	9.6	10.4	11.1	11.8	12.4	12.9	13.4	13.9
12	4.91	6.1	6.9	7.6	8.2	8.7	9.1	9.5	9.9	10.2	10.6
15	4.07	4.9	5.5	6.0	6.4	6.7	7.1	7.3	7.5	7.8	8.0
20	3.32	3.8	4.3	4.6	4.9	5.1	5.3	5.5	5.6	5.8	5.9
30	2.63	3.0	3.3	3.4	3.6	3.7	3.8	3.9	4.0	4.1	4.2
60	1.96	2.2	2.3	2.4	2.4	2.5	2.5	2.6	2.6	2.7	2.7
∞	1.00	1.0	1.0	1.0	1.0	1.0	1.0	1.0	1.0	1.0	1.0

s^2_{max} is the largest and s^2_{min} the smallest in a set of t independent mean squares, each based on $df_2 = n - 1$ df. Values in the column $t = 2$ and in the rows $df_2 = 2$ and ∞ are exact. Elsewhere the third digit may be in error by a few units for the 5% points and several units for the 1% points. The third-digit figures in parentheses for $df_2 = 3$ are the most uncertain.

T A B L E 13   Values of d_n Used in Control Limits for μ

n	d_n
2	1.128
3	1.693
4	2.059
5	2.326
6	2.534
7	2.704
8	2.847
9	2.970
10	3.078
11	3.173
12	3.258

T A B L E 14   Critical Values of T_L and T_U for the Wilcoxon Rank Sum Test: Independent Samples
(Test statistic is rank sum—associated with smaller sample (if equal sample sizes, either rank sum can be used).)

a. $\alpha = .025$ one-tailed; $\alpha = .05$ two-tailed

n_2	n_1 3		4		5		6		7		8		9		10	
	T_L	T_U	T_L	T_U	T_L	T_U	T_L	T_U	T_L	T_U	T_L	T_U	T_L	T_U	T_L	T_U
3	5	16	6	18	6	21	7	23	7	26	8	28	8	31	9	33
4	6	18	11	25	12	28	12	32	13	35	14	38	15	41	16	44
5	6	21	12	28	18	37	19	41	20	45	21	49	22	53	24	56
6	7	23	12	32	19	41	26	52	28	56	29	61	31	65	32	70
7	7	26	13	35	20	45	28	56	37	68	39	73	41	78	43	83
8	8	28	14	38	21	49	29	61	39	73	49	87	51	93	54	98
9	8	31	15	41	22	53	31	65	41	78	51	93	63	108	66	114
10	9	33	16	44	24	56	32	70	43	83	54	98	66	114	79	131

b. $\alpha = .05$ one-tailed; $\alpha = .10$ two-tailed

n_2	n_1 3		4		5		6		7		8		9		10	
	T_L	T_U	T_L	T_U	T_L	T_U	T_L	T_U	T_L	T_U	T_L	T_U	T_L	T_U	T_L	T_U
3	6	15	7	17	7	20	8	22	9	24	9	27	10	29	11	31
4	7	17	12	24	13	27	14	30	15	33	16	36	17	39	18	42
5	7	20	13	27	19	36	20	40	22	43	24	46	25	50	26	54
6	8	22	14	30	20	40	28	50	30	54	32	58	33	63	35	67
7	9	24	15	33	22	43	30	54	39	66	41	71	43	76	46	80
8	9	27	16	36	24	46	32	58	41	71	52	84	54	90	57	95
9	10	29	17	39	25	50	33	63	43	76	54	90	66	105	69	111
10	11	31	18	42	26	54	35	67	46	80	57	95	69	111	83	127

T A B L E   **15**   Values of D_n and D'_n Used in Control Limits for Product Variability

Number of Observations in Sample, n	D_n	D'_n
2	0	3.267
3	0	2.575
4	0	2.282
5	0	2.115
6	0	2.004
7	0.076	1.924
8	0.136	1.864
9	0.184	1.816
10	0.223	1.777
11	0.256	1.744
12	0.284	1.716
13	0.308	1.692
14	0.329	1.671
15	0.348	1.652
16	0.364	1.636
17	0.379	1.621
18	0.392	1.608
19	0.404	1.596
20	0.414	1.586
21	0.425	1.575
22	0.434	1.566
23	0.443	1.557
24	0.452	1.548
25	0.459	1.541
Over 25	$3/\sqrt{n}$	$3/\sqrt{n}$

Source: Reprinted by permission of the American Society for Testing and Materials, copyright 1951.

Glossary of Common Statistical Terms

acceptance region	The set of values of a test statistic that imply acceptance of the null hypothesis.
alternative hypothesis	The hypothesis to be accepted if the null hypothesis is rejected.
analysis of variance	A procedure for comparing more than two population means. There are many applications of analysis of variance beyond those discussed in the text.
arithmetic mean	Average.
bar chart	A graphical method for showing how data fall into a group of categories.
binomial experiment	An experiment involving n identical independent trials. (See Chapter 10 for an exact description of a binomial experiment.)
binomial random variable	A discrete random variable representing the number of successes y in n identical independent trials. (For an exact definition of a binomial experiment, see Chapter 10.)
box-and-whiskers plot	*See* box plot.
box plot	A graphical method for describing data so as to emphasize the symmetry and central tendency of a set of measurements.
Central Limit Theorem	A theorem stating that the sampling distribution of the sample mean (or sum) is approximately normal when certain conditions are satisfied. (See Chapter 6.)
chi-square	A test statistic used to test the null hypothesis of independence for the two classifications of a contingency table. It also has many other statistical applications not discussed in this text.
chi-square distribution	The distribution of a chi-square statistic.
class boundary	The dividing point between two cells in a frequency histogram.
classes	The cells of a frequency histogram.
class frequency	The number of observations falling in a class (referring to a frequency histogram).
classical interpretation of probability	If there are N possible outcomes in an experiment and N_E of these result in event E, then $$P(E) = \frac{N_E}{N}$$
class interval	Each subinterval of a frequency table (or histogram).
class interval width	The width of each class interval in a frequency table (or histogram).
cluster sampling	Sampling so that a simple random sample of groups (clusters) is selected; all items within a selected cluster are then sampled.

coefficient of determination	The square of the sample correlation coefficient, $\hat{\rho}$.
complement	In relation to an event A, the event that A does not occur.
completely randomized design	An experimental design where treatments are assigned to experimental units "at random."
conditional probability (A given B)	The probability that A occurs, given that B has occurred.
confidence coefficient	The probability that an interval estimate (a confidence interval) encloses the parameter of interest.
confidence interval	Two numbers, computed from sample data, that form an interval estimate for some parameter.
contingency table	A two-way table constructed for classifying count data. The entries in the table show the number of observations falling in the cells. The objective of an analysis is to determine whether the two directions of classification are dependent (contingent) on one another.
continuous probability distribution	A smooth curve that gives the theoretical frequency distribution for a continuous random variable. An area under the curve over an interval is proportional to the probability that the random variable will fall in the interval.
continuous random variable	A quantitative random variable that can assume any one of a countless number of values on a line interval.
correlation coefficient	A measure of linear dependence between two random variables.
degrees of freedom	A parameter of Student's t-, the F-, and the chi-square probability distributions, measuring the quantity of information available in normally distributed data for estimating the population variance σ^2.
deviation	The difference between a measurement and the mean of the set of measurements from which the measurement was drawn.
deviation from the mean	The distance between a sample observation and the sample mean $\bar{y}$.
direction observation	A method of data collection whereby the surveyor observes the event(s) of interest.
discrete probability distribution	A listing, a mathematical formula, or a histogram that gives the probability associated with each value of the random variable.
discrete random variable	A random variable that can assume only a countable number of values.
dot diagram	A way to illustrate the variability in a small set of measurements.
Empirical Rule	A rule that describes the variability of data that possess a mound-shaped frequency distribution. (See Chapter 4.)
estimate	A number computed from sample data, used to approximate a population parameter.
estimation	One of two approaches to making inferences about parameters; included in this approach are point estimates and interval estimates (confidence intervals).
event	A collection of outcomes.
expected value of y	For the model $y = \beta_0 + \beta_1 x + \epsilon$, the expected value of y, $E(y)$, is $E(y) = \beta_0 + \beta_1 x$.
expected values of $\hat{\beta}_0$ and $\hat{\beta}_1$	For the linear regression model $y = \beta_0 + \beta_1 x + \epsilon$, the expected values of $\hat{\beta}_0$ and $\hat{\beta}_1$ are, respectively, β_0 and β_1.
experiment	The process by which an observation is obtained.
exploratory data analysis	A relatively new area of applied statistics dealing with data description.
factorial experiment	An experiment where the response is observed at each factor–level combination of the independent variables.
F-distribution	The distribution of an F-statistic.

freehand regression line   A trend line "fit" by eye.

frequency   The number of observations falling in some cell or in some classification category.

frequency polygon   One of several methods for graphing frequency data.

frequency table   A table used to summarize how many measurements in a set fall into each of the subintervals (or classes).

F-statistic   A test statistic used to compare variances from two normal populations; used in the analysis of variance.

hinges   Quantities similar to the upper and lower quartiles of a data set.

histogram   A graphical method for describing a set of data. (See Chapter 3.)

hypothesis testing   *See* statistical test.

independent events   Event A is independent of event B if $P(A \mid B) = P(A)$ (or if $P(B \mid A) = P(B)$).

interaction   The failure of a variable to exert the same effect on a response at different levels of one or more other independent variables.

interval estimate   Two numbers computed from the sample data; the interval formed by the numbers should enclose some parameter of interest. An interval estimate is usually called a *confidence interval*.

Latin square design   An experimental design that employs blocking in two dimensions.

leading digit   The first digit of measurements in a stem-and-leaf plot.

least squares   A method of curve fitting that selects as the best-fitting curve the one that minimizes the sum of the squares of deviations of the data points from the fitted curve. (See Chapter 11.)

level of significance   With regard to the outcome of a specific statistical test of a hypothesis, the probability of drawing a value of the test statistic that is as contradictory (or more contradictory) to the null hypothesis as (or than) the value observed, assuming that the null hypothesis is true.

linear correlation   Dependence between two random variables.

linear equation   An equation of the form $y = \beta_0 + \beta_1 x$.

lower boundary   The lower limit of a class interval.

lower confidence limit   The smaller of the two numbers that form a confidence interval.

lower quartile   The 25th percentile of a set of measurements.

mean   The average of a set of measurements. The symbols $\bar{y}$ and μ denote the means of a sample and a population, respectively.

measures of central tendency   Numerical descriptive measures that locate the center or central value in a set.

median   The middle measurement when a set is ordered according to numerical value. (See Chapter 4.)

method of least squares   A procedure for finding the estimates $\hat{\beta}_0$ and $\hat{\beta}_1$ for the model $y = \beta_0 + \beta_1 x + \epsilon$.

mode   The measurement in a set that occurs with greatest frequency.

mound-shaped frequency distribution   A symmetrical, single-peaked frequency distribution.

mutually exclusive events A and B are mutually exclusive events if the occurrence of one precludes the occurrence of the other.

nonparametric methods   Statistical tests of hypotheses about population probability distributions but not about specific parameters of the distributions.

normal curve   A smooth, bell-shaped curve known as the normal distribution.

normal distribution	A bell-shaped probability distribution. The curve possesses a specific mathematical formula.
null hypothesis	The hypothesis under test in a statistical test of a hypothesis.
numerical descriptive measures	Quantities used to describe a set of measurements.
observational study	A study in which the conditions under which observations are obtained are not fixed (controlled).
one-at-a-time approach	A popular, but inefficient, way to evaluate the effect of each variable in a problem involving more than one independent variable.
outcome	The result of an experiment.
paired-difference test	A statistical test for comparing two population means. The test is based on paired observations, one from each of the two populations.
parameters	Numerical descriptive measures for a population.
parametric methods	Statistical methods for estimating parameters or for testing hypotheses about population parameters.
personal interviews	Interviews conducted in person.
pie chart	A graphical method for describing data. (See Chapter 3.)
point estimate	*See* estimate.
population	The set of measurements, existing or conceptual, of interest to the experimenter. Samples are selected from the population.
predicted value of y	A value of y computed from the prediction equation $\hat{y} = \hat{\beta}_0 + \hat{\beta}_1 x$.
prediction equation	An equation relating a dependent variable y to one or more independent variables.
probability	As a practical matter, a measure of one's belief that a specified event will occur when an experiment is conducted once. The exact definition, giving a quantitative measure of this belief, is subject to debate. The relative frequency concept is the one most widely accepted.
probability distribution	A formula, table, or graph used to display the probabilities for a discrete random variable. *See also* continuous probability distribution, discrete probability distribution.
p-value	The level of significance of a statistical test.
qualitative random variable	A random variable with qualitative outcomes.
qualitative variable	A variable that has qualitative observations.
quantitative random variable	A random variable with quantitative outcomes.
quantitative variable	A variable that has quantitative observations.
randomized block design	An experimental design whereby treatments are assigned to experimental units "at random" within each block.
random sample	A sample of n measurements selected in such a way that every different sample of n elements in the population has an equal probability of being selected.
random variable	A variable associated with an experiment; its values are numerical events that cannot be predicted with certainty.
range of a set of measurements	The difference between the largest and smallest members of the set.
rank correlation coefficient	A coefficient of linear correlation between two random variables that is based on the ranks of the measurements, not on their actual values. *See also* correlation coefficient.
ratio estimation	A way in survey sampling to use information on an auxiliary variable in estimating a population parameter.
regression line	A line fit to data points, using the method of least squares.

rejection region	The set of values of a test statistic that indicates rejection of the null hypothesis.
relative frequency	Class frequency divided by the total number of measurements.
relative frequency concept of probability	If an experiment is repeated n times and n_E of these result in event E, then $$P(E) \approx \frac{n_E}{n}$$
relative frequency histogram	A histogram displaying the fraction of measurements falling in each class.
residual	The difference between a value of y and its predicted value $\hat{y}$: residual $= y - \hat{y}$.
sample	A subset of measurements selected from a population.
sample correlation coefficient	The correlation coefficient computed from a sample of pairs (x, y).
sample mean	Sample average.
sampling distribution	The probability distribution for a sample statistic.
scatterplot	A graphical method for displaying bivariate data.
self-administered questionnaire	A method of collecting data in a sample survey whereby those surveyed complete a questionnaire.
significance level	*See* level of significance.
sign test	A nonparametric statistical test used to compare two populations.
simple random sampling	Sampling so that each sample of size n has an equal chance of being selected.
skewed distribution	An asymmetric distribution of measurements trailing off to the right or to the left.
slope	The term β_1 in the linear equation $y = \beta_0 + \beta_1 x$.
software	Computer programs.
Spearman's rank correlation coefficient	One of several correlation coefficients based on the ranks of the two random variables.
standard deviation	A measure of data variation. (See Chapter 4.)
standardized normal distribution	A normal distribution with mean and standard deviation equal to 0 and 1, respectively. The standardized normal variable is denoted by the symbol z.
standard normal random variable	A random variable having a normal distribution with $\mu = 0$, $\sigma = 1$.
statistical software system	Computer programs used to summarize and statistically analyze data that are integrated into a single system of programs.
statistical test	A procedure for making an inference about one or more population parameters by using information from sample data. The procedure is based on the concept of proof by contradiction.
statistics	Numerical, descriptive measures for a sample.
stem-and-leaf plot	A graphical method for displaying data.
stratified random sampling	Sampling so that a simple random sample is selected within each of the strata.
Student's *t*-distribution	A particular symmetric mound-shaped distribution that possesses more spread than the standard normal probability distribution.
Student's *t*-test	A test statistic used for small-sample tests of means.
subjective probability	One's personal estimate of the probability of an event.

sum of squares about the mean	The sum of the squared differences about the mean, $\Sigma(y - \bar{y})^2$.
sum of squares due to regression	The sum of the squared deviations of the predicted values from the mean, $\Sigma(\hat{y} - \bar{y})^2$.
sum of squares for error	The sum of the squared residuals, $\text{SSE} = \Sigma(y - \hat{y})^2$.
systematic sample	An economical way to sample a population when the elements of the population are arranged in a list.
test statistic	A function of the sample measurements, used as a decision maker in a test of a hypothesis.
trailing digits	The remaining digits (after the leading digits) of measurements in a stem-and-leaf plot.
two-tailed test	A statistical test with a two-sided alternative hypothesis.
type I error	Rejecting the null hypothesis when it is true.
type II error	Accepting the null hypothesis when it is false and the alternative hypothesis is true.
unconditional probability	The probability of an event.
upper confidence limit	The larger of the two numbers that form a confidence interval.
upper quartile	The 75th percentile of a set of measurements.
variable	A phenomenon where observations vary from trial to trial.
variance	A measure of data variation. (See Chapter 4.)
y-intercept	The term β_0 in the linear equation $y = \beta_0 + \beta_1 x$.
z-score	A standardized score formed by subtracting the mean and dividing by the standard deviation.
z-statistic	A standardized normal random variable that is frequently used as a test statistic.

References

Books and Articles

Bergsten, J. W. (1979). *American Statistical Association Proceedings of the Section on Survey Research Methods.* Washington, D. C.: ASA, pp. 239–43. *Biometrika Tables for Statisticians,* vol. I (1953).

Bradbury, et al. (1987). "Answering autobiographical questions: The impact of memory and inference on surveys." *Science,* 157–61.

Bureau of Labor Statistics Handbook of Methods (1982). Washington, D.C.: U.S. Government Printing Office.

Cochran, W. G. (1954). "Some methods for strengthening the common χ^2 test." *Biometrics* 10:417–54.

Conover, W. J. (1980). *Practical Nonparametric Statistics,* 2d ed. New York: John Wiley.

Conover, W. J., and R. L. Iman (1981). "Rank transformations as a bridge between parametric and nonparametric statistics." *American Statistician* 35:124–33.

Deming, W. E. (1982). *Quality, Productivity and Competitive Position.* Cambridge, Mass.: MIJ–CAES.

Dixon, W. J., and M. B. Brown, eds. (1978). *Biomedical Computer Programs,* rev. ed. Los Angeles: University of California Press.

Dixon, W. J., and F. J. Massey (1969). *Introduction to Statistical Analysis,* 3d ed. New York: McGraw-Hill.

Dye, T. R., and L. H. Zeigler (1981). *The Irony of Democracy,* 5th ed. Monterey, Calif.: Duxbury Press.

Hald, A. (1952). *Statistical Tables and Formulas.* New York: John Wiley.

Handbook of Tables for Probability and Statistics, 2d ed. (1968). Cleveland, Ohio: Chemical Rubber Co.

Hammer, M. (1990). "Re-engineering work: Don't automate, obliterate." *Harvard Business Review* (July–August).

Helwig, J. T. (1977). *SAS Supplementary Library User's Guide.* Raleigh, N.C.: SAS Institute.

Hicks, C. R. (1973). *Fundamental Concepts in the Design of Experiments,* 2d ed. New York: Holt, Rinehart & Winston.

Hildebrand, J., and L. Ott (1991). *Statistical Thinking for Managers,* 3d ed. Boston: Duxbury Press.

Hollander, M., and D. A. Wolfe (1973). *Nonparametric Statistical Methods.* New York: John Wiley.

Huntsberger, D. V., and P. Billingsley (1977). *Elements of Statistical Inference,* 4th ed. Boston: Allyn & Bacon.

Mendenhall, W. (1968). *An Introduction to Linear Models and the Design and Analysis of Experiments.* Belmont, Calif.: Wadsworth.

Mendenhall, W. (1987). *Introduction to Probability and Statistics,* 7th ed. Boston: Duxbury Press.

Mendenhall, W., R. Scheaffer, and D. Wackerly (1981). *Mathematical Statistics with Applications,* 2d ed. Boston: Duxbury Press.

Merrington, M. (1941). "Table of percentage points of the *t*-distribution." *Biometrika* 32: 300.

Merrington, M., and C. M. Thompson (1943). "Points of the inverted beta *(F)*-distribution." *Biometrika* 33: 73–88.

National Bureau of Standards (1949). *Tables of the Binomial Probability Distribution.* Washington, D.C.: U.S. Government Printing Office.

Newman, R. W., and R. M. White (1951). *Reference Anthropometry of Army Men.* Report No. 180, Lawrence, Mass.: Environmental Climatic Research Laboratory.

Nie, N., C. H. Hull, J. G. Jenkins, K. Steinbrenner, and D. H. Bent (1975). *Statistical Package for the Social Sciences,* 2d ed. New York: McGraw-Hill.

Omstead, P. S., and J. W. Tukey (1947). "A corner test for association." *Annals of Mathematical Statistics* 18: 495–513.

Ott, L. (1993). *An Introduction to Statistical Methods and Data Analysis,* 4th ed. Boston: Duxbury Press.

Ott, L., R. Larson, and W. Mendenhall (1992). *Statistics: A Tool for the Social Sciences,* 5th ed. Boston: Duxbury Press.

Scheaffer, R. L., W. Mendenhall, and L. Ott (1990). *Elementary Survey Sampling,* 4th ed. Boston: Duxbury Press.

Schuman, H., and S. Presser (1981). *Questions and Answers in Attitude Surveys.* New York: Academic Press.

Siegel, S. (1956). *Nonparametric Statistics for the Behavioral Sciences.* New York: McGraw-Hill.

Tanur, J. M., F. Mosteller, W. H. Kruskal, R. S. Pieters, and G. R. Rising, eds. (1972). *Statistics: A Guide to the Unknown.* San Francisco: Holden-Day.

Thompson, C. M. (1941). "Tables of the percentage points of the χ^2-distribution." *Biometrika* 32: 188–89.

Tukey, J. W. (1977). *Exploratory Data Analysis.* Reading, Mass.: Addison-Wesley.

U.S. Bureau of the Census (1980). *Statistical Abstract of the United States.* Washington, D.C.: U.S. Government Printing Office.

U.S. Department of Justice (1980). *Uniform Crime Reports for the United States.* Washington, D.C.: U.S. Government Printing Office, pp. 60–86.

Statistical Software Systems

BMDP

Dixon, W. S., M. B. Brown, L. Engelman, and R. I. Jennrich, (1990). *BMDP Statistical Software Manual, 1990 Revision.* Berkeley, Calif.: University of California Press.

EXECUSTAT

Strategy Plus, Inc. (1990). *EXECUSTAT.* Boston: PWS-KENT.

MINITAB

Schaefer, R. L., and R. B. Anderson (1989). *The Student Edition of Minitab.* Reading, Mass.: Addison-Wesley.

SAS

SAS Institute (1985). *SAS User's Guide: Basics, Version 5 Edition.* Cary, N.C.: SAS Institute.

SPSS-X

SPSS, Inc. (1988). *SPSS-X User's Guide,* 3d ed. Chicago: SPSS.

SYSTAT

Wilkinson, L. (1987). *SYSTAT: System for Statistics.* Evanston, Ill.: SYSTAT.

Answers to Selected Exercises

Chapter 1

1.1 **a** The population is the set of weights of all shrimp on the diet.

b The sample is the set of weights of the 100 shrimp selected from the pond.

c A single weight that typifies the collection of weights contained in the population, for example, the "average" or mean weight.

d A measure of reliability is needed so that you will know how much faith you can place in your inference.

1.3 **a** The population is the set of numbers of children in all households that receive welfare support in the city.

b The sample is the set of numbers of children corresponding to the 400 households selected from the welfare rolls.

c The characteristic of interest would be a number that typifies the number of children per welfare household—for example, the "average" or mean number per household.

d A measure of reliability is needed so that you will know how much faith you can place in your inference.

Chapter 2

2.1 To estimate the average water consumption per family in a city, an experimenter can choose between many sampling units. If the sampler chooses individual families, the most information of any sampling scheme could be obtained. If the sampler chooses a larger sampling unit, such as dwelling units or city blocks, the expense and difficulty of drawing a representative random sample decreases as the size of the sampling unit increases.

2.3 To estimate the proportion of automobile tires with unsafe tread, a collection of cars would be the easiest to measure. Individual sampling of cars would be difficult and time-consuming. By using a random sample of parking lots, a large representative sample could be collected quickly.

2.5 To obtain an estimate of the number of acres planted in corn within the state, the farms in the state could be stratified according to total farm acreage. A sample could then be taken within each of the strata to determine the percentage of acreage in corn. The strata estimates could be combined together (weighted average) and an estimate of percentage of acreage in corn for the entire state could be obtained. This state percentage could then be multiplied by the total number of plantable acres in the state to obtain the number of acres in the state that are planted in corn.

2.15 **a** Survey

b Simple random sampling

Chapter 3

3.1

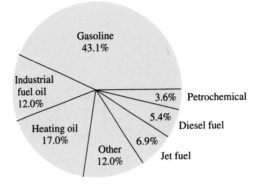

3.3

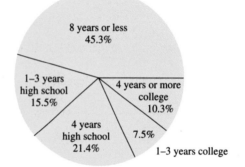

3.5 No

3.7

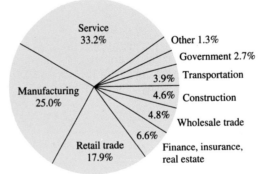

3.9

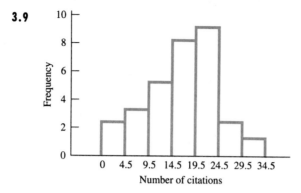

3.11

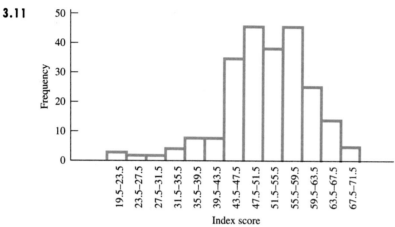

3.13

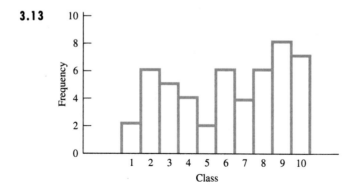

3.15

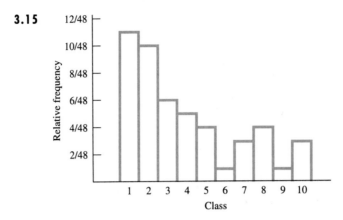

3.17

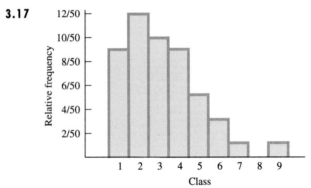

3.19

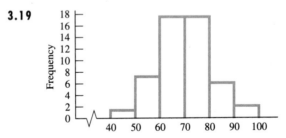

3.21 **a** No

b

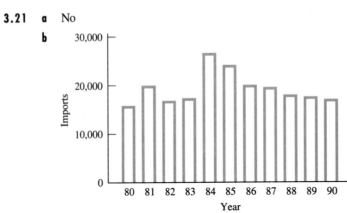

3.23 **a** 0.33

b

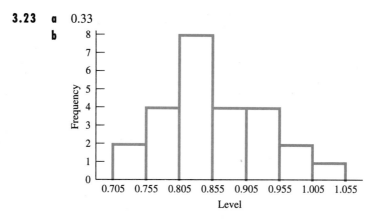

c

Class	Relative frequency
0.705–0.755	2/25
0.755–0.805	4/25
0.805–0.855	8/25
0.855–0.905	4/25
0.905–0.955	4/25
0.955–1.005	2/25
1.005–1.055	1/25

d 7/25; 102/365

3.25 The stem-and-leaf plot is more informative.

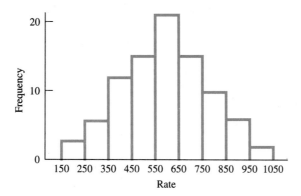

3.27

Class	f
22.5–24.5	1
24.5–26.5	0
26.5–28.5	1
28.5–30.5	10
30.5–32.5	9

3.29

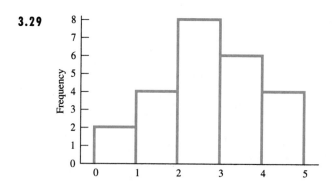

3.31

Freq	Stem	Leaf
4	0	1234
4	0	6678
6	1	023344
5	1	55688
4	2	0144
2	2	79
2	3	12
1	3	6

3.33 Solid line indicates male. Dashed line indicates female. For the years 1967–1990, the difference between males and females has remained relatively constant, whereas there appears to have been a dramatic drop in verbal scores for both sexes from 1970 on, with more of a drop in females' scores.

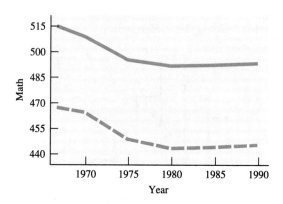

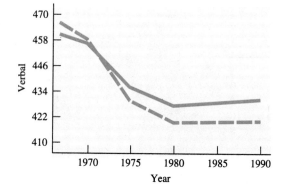

3.35

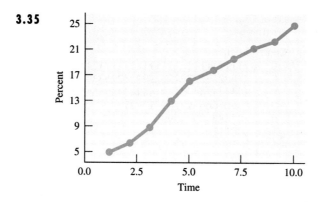

3.41 **a**

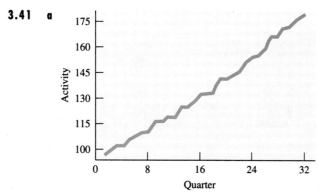

b The trend appears to be linear.

c There appears to be a jump in activity approximately every 3 quarters.

3.43

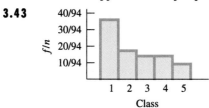

3.45

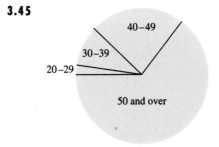

3.47 Range = 44; Largest percentage of deaths is between 56 and 79; Data are slightly mound-shaped

3.49 Stem-and-leaf plot

3.53 **a** Stem-and-leaf of Claims   N = 187
Leaf Unit = 1.0

```
CumFreq  Stem | Leaf
   47      0  | 111111111111111111111111111111111111111111111111111
   89      0  | 2222222222222222222222233333333333333333333
  (37)     0  | 444444444444444444444455555555555555
   61      0  | 66666666677777777777777
   40      0  | 888888889999999
   25      1  | 000111
   19      1  | 233333
   13      1  | 445
   10      1  | 77
    8      1  | 99
    6      2  | 111
    3      2  | 3
    2      2  |
    2      2  | 7
    1      2  |
    1      3  |
    1      3  |
    1      3  | 4
```

b The distribution is very skewed to the right. Most claims are below $10,000. There are a few outliers, two of which are $26,600 and $33,700.

Chapter 4

4.1 mean = 15.19; median = 14.5; mode = 18

4.3 Both trimmed means = 14.83
The extreme observations would affect a 5% trimmed mean.

4.5 mean = 7.6; median = 7.4; mode = 7

4.7 median is 906; mean = 970.1

4.9 width = 2; modal interval midpoints are 49 and 55; median = 54.3, mean = 54.6; A trimmed mean would eliminate some of the extreme values and better represent the center of the distribution.

4.11 mean = 3.17; median and mode remain unchanged

4.13 **a**

	Group 1	Group 2	Group 3
mean	2.923	1.592	.797
median	2.805	1.565	.755
mode(s)	none	1.57, 1.55	.70

b mean = 1.771; median = 1.565; modes = 1.57, 1.55, 0.70

c medians

4.15 4; 3

4.17 range = 26; variance = 79; standard deviation = 8.89

4.19 **a**
```
    .                :               :
+---------+---------+---------+---------+---------+------
0.00      0.60      1.20      1.80      2.40      3.00
```

b 10

c 2.5; 1.58

4.21 **a** $\mu \pm 1\sigma = 520 \pm 110$ or 410 to 630
$\mu \pm 2\sigma = 520 \pm 220$ or 300 to 740
$\mu \pm 3\sigma = 520 \pm 330$ or 190 to 850
The Empirical Rule states that approximately 68% of all test scores will lie between 410 and 630; approximately 95% of all scores will lie between 300 and 740; all or nearly all the measurements will lie between 190 and 850.

b Half of the scores will lie above 510 and half will lie below 510. Further, 98% of all scores will lie below 795 and 2% will lie above.

c The difference between the highest and lowest test scores is 702.

4.23 We expect the distribution to be mound-shaped, centered at 0.013, with a standard deviation of 0.006. The data should conform to the Empirical Rule.

4.25

k	$\bar{y} \pm ks$	Interval boundaries	Freq in interval	Percentage
1	20.32 ± 8.37	11.95 to 28.69	35	70
2	20.32 ± 16.74	3.58 to 37.06	48	96
3	20.32 ± 25.11	−4.79 to 45.43	50	100

Notice that the Empirical Rule gives a very close approximation to the percentages shown in the table.

4.27 **a** Stem-and-leaf of Count   N = 20
Leaf Unit = 10

CumFreq	Stem	Leaf
1	2	5
4	2	677
4	2	
8	3	0011
(8)	3	22233333
4	3	
4	3	677
1	3	9

b The data appear mound-shaped and symmetric about the median.

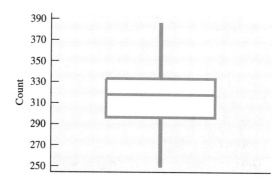

4.29 Resigned 24.0%
Transferred 26.4%
Retired/Fired 49.6%
Approximately an equal amount of employees Resign or Transfer, while almost twice as many employees Retire/Fire as those that Resign or that Transfer.

4.31 **a** mean A = 22; mean B = 17.75; mean C = 26; mean D = 30

b

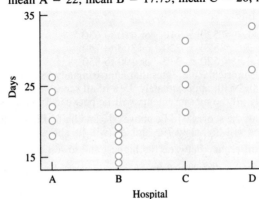

c Hospital stays at B appear to be consistently shorter, while those at hospital D seem the longest. However, the conclusion is questionable due to the small sample sizes.

4.33 M2 and M3 move simultaneously and seem to reflect the same changes in the money supply. The open circle represents M2 and the solid circle represents M3.

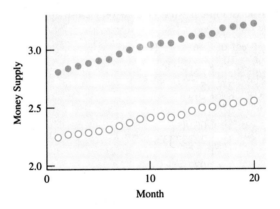

4.35 **a** mode = 5; median = 15; mean = 15.96

b range/4 = 7.5

c $s = 8.5$

d No; they are not mound-shaped

4.37 **a** range = 0.9; $s = 0.28$

b range/4 = 0.225

c

Largest	Range	Std Dev
6.8	6.5	1.97
68.0	67.7	21.32

4.39 $Q_1 \approx 530$, $Q_3 \approx 580$, median = 560. In general, the data are symmetric about the median. There is an outlier at both upper and lower ends of the data with one extreme outlier at the lower end.

4.41 **a** median = 7.04; modal interval is 1.5 − 3.5, midpoint = 2.5

b mean = 8.3

c median

4.43 **a** median = 0.93; mode = 1

b 1.326

c to the right

4.45 **a**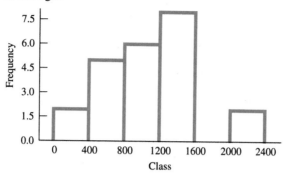

b In the interval (800, 1200)

c median = 1039; mean = 1092

d skewed to the right

4.47 Yes

Stem-and-leaf of Cars   N = 26

Freq	Stem	Leaf
1	54	7
0	55	
0	56	
0	57	
0	58	
0	59	
0	60	
0	61	
1	62	5
1	63	0
0	64	
1	65	6
3	66	477
2	67	79
5	68	88888
4	69	1479
7	70	0123338
1	71	1

4.49

Group	Mean	Median
1-member	93.75	78.5
2-member	98.65	95.0
3-member	113.31	100.0
4-member	124.90	112.5
5-member	131.90	128.5

4.51 New mean = 2.267; Old mean = 4.6
New s = 3.262; Old s = 2.613

4.53 **a** There is such variability in the data, including extreme observations, that the sample mean is not appropriate.

b

	Mean	10% Trim	20% Trim
Plants	677,795	142,328	133,039
Arrests	95	59.7	41.3

A 10% trimmed mean seems appropriate for the number of plants. This trims the two extreme outliers on each end of the data. A 20% trimmed mean seems appropriate for the number of arrests. This trims most of the small data on the lower end of the distribution.

4.55 **a** The open circle represents pharmaceuticals.
The asterisk represents diversified companies.
The solid circle represents chain drugstores.
The open square represents wholesalers.

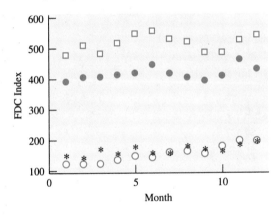

b All components show that the indices increased slightly through the year and generally move together.

4.57 **a** mean = 53.55
b DJIA = 115.74

4.59 The stem-and-leaf plot suggests some extreme observations on both ends of the distribution. 10% trimmed mean = 557.25

4.61 **a** Sixty-two of the 150 residents from coal states encourage a coal national energy policy. Of those who encourage coal, 32.8% are from coal states. From the coal producing states, 41.3% encourage a coal national energy policy. Of all 800 residents surveyed, 7.8% favor a coal national energy policy.

4.63

Treatment	N	Mean	StDev	Range
A	25	12.72	6.75	30.00
B	25	14.04	6.48	22.00
C	25	11.48	4.98	20.00
D	25	19.20	5.74	29.00

4.65 **a** Stem-and-leaf of Obrist   N = 100
Leaf Unit = 1.0

CumFreq	Stem	Leaf
6	3	012244
26	3	5555566777789999999
42	4	0001222222333444
(11)	4	66777788999
47	5	0112344
40	5	566778889
31	6	0333344
24	6	5668888
17	7	144
14	7	59
12	8	0023
8	8	557
5	9	02
3	9	69
1	10	
1	10	
1	11	2

 b mean = 53.67; median = 48.5

4.67 **a** No, because the data is not mound-shaped.

 b The distribution is skewed to the right.

4.71 **a** Discrete

 b Stem-and-leaf of Crimes   N = 200
Leaf Unit = 0.10

CumFreq	Stem	Leaf
2	9	00
2	10	
7	11	00000
15	12	00000000
21	13	000000
29	14	00000000
53	15	000000000000000000000000
85	16	00000000000000000000000000000000
(31)	17	0000000000000000000000000000000
84	18	00000000000000000000000000000000
52	19	000000000000
40	20	00000000000000
26	21	00000000000
15	22	000
12	23	000000
6	24	0000
2	25	00

 c mean = 17.115, median = 17.000, standard deviation = 3.006

Chapter 5

5.1 **a** Subjective probability

 b Relative frequency concept

 c Classical interpretation

d   Relative frequency concept

e   Relative frequency concept

f   Classical interpretation

5.3   {HHH, HHT, HTH, THH, HTT, THT, TTH, TTT}

5.5   **a**   5/8; 1/8; 7/8

b   Not mutually exclusive

5.7   No; No; No

5.9 B is independent of C. None are mutually exclusive.

5.11   No, since $P(A \cap B) = 0.30 \neq 0$.

5.13   **a**   Generator 1 fails to work properly.

b   Generator 2 works properly given that we know generator 1 works properly.

c   Either generator 1 or generator 2 works properly.

5.15   **a**   {GG, GB, BG, BB}

b   3/10

c   6/10

d   1/10

5.17   **a**   No

b $P(B|A) = \dfrac{48}{192}$ does not equal $P(B|\bar{A}) = \dfrac{80}{248}$

5.19   **a**   0.49

b   0.91

5.23   **a**   0.8263

b   0.1737

c   0.9993

d   0.0086

5.25   **a**   0.64

b   0.60

c   0.07

5.27   Yes

5.29   Yes

5.31   **a**   0.0000, 0.0368, 0.0473, 0.0282

b $P(y \leq 100) = \displaystyle\sum_{k=0}^{100} \dfrac{1000!}{k!\,(1000 - k)!}(0.3)^k(0.7)^{1000-k}$

5.33   0.729; 0.972

5.35 $P(0) = \dfrac{3!}{0!3!}\left(\dfrac{1}{2}\right)^3 = \dfrac{1}{8} \qquad P(2) = \dfrac{3!}{2!1!}\left(\dfrac{1}{2}\right)^3 = \dfrac{3}{8}$

$P(1) = \dfrac{3!}{1!2!}\left(\dfrac{1}{2}\right)^3 = \dfrac{3}{8} \qquad P(3) = \dfrac{3!}{3!0!}\left(\dfrac{1}{2}\right)^3 = \dfrac{1}{8}$

5.37 $p = 0.3, n = 10$, binomial

$P(y \leq 100) = P(0) + P(1) + P(2) + \ldots + P(99) + P(100)$

5.39   No; no identical trials.

5.41 Yes

5.43 **a** 0.2580

      b 0.3849

5.45 **a** 0.4947

      b 0.2406

5.47 0.0401

5.49 0

5.51 2.37

5.53 1.96

5.55 **a** 15.87%

      b 2.28%

      c 30.85%

      d 53.28%

5.57 **a** 0.0764

      b 0.0228

      c 0.1949

      d 0.8472

5.59 **a** 11.51

      b 0.44

5.63 (a), (b), (d), (f) are discrete; (c), (e) are continuous

5.65 **a** 95%

      b 2.5%

      c 81.5%

      d 16%

5.67 0.2514

5.69 **a** 1) 0.684   2) 0.707

      b Yes

5.71 **a** 0.24; 0.613; 0.739

      b Yes

5.73 0.0512

5.75 **a** 0.2160

      b 0.2880

      c 0.0640

5.77 **a** 47.72%

      b 0.1587

5.81 If we consider a day to be a trial, then on each day it is possible to have more than one of two outcomes (accident or no accident). In fact we could observe $y = 0, 1, 2, \ldots$ accidents per day. This is not a binomial experiment.

5.83 **a** There are $n = 400$ plants observed on which there either will or will not appear infection by a fungus. However, it is possible that these trials (plants) are not independent. If the plants are planted side by side, contagion could result, so that an infected plant might increase the probability of infection of a plant next to it.

b Assuming that there is no contagion, $p = P$(choosing an infected plant in a single random choice). This is equivalent to the proportion of infected plants in the field from which we are sampling.

c With $n = 400$ and $p = 0.5$, $\mu = np = 200$ and $\sigma = \sqrt{100} = 10$. It is improbable because $y = 242$ lies more than $3\sigma = 30$ away from $\mu = 200$.

5.85 0.59049

5.87 1.645

5.89 -0.5

5.91 0.4649

5.93 70.16 or approximately 70 minutes

5.95 6; 2.19

5.97 **a** 30
    b 0.50
    c 6.68%

5.99 6400; 35.78; Yes

5.101 **a** 0.3069
    b 0.4074
    c 0.4127
    d Not independent

5.103 \$16.00 price for the new book is more unusual; 72.89%

5.105 -1.46; 0.29; 1.56

5.107 0.0594; 0.2514

Chapter 6

6.1 A random sample of n measurements from a population is one in which every different subset of size n from the population has an equal probability of being selected. It is probably not possible to draw a truly random sample.

6.7 Starting where the last group ended; 478, 214, 196, 9, 521

6.9 **a** 10
    b 3.4
    c 1.4

6.11 According to the Central Limit Theorem, the distribution will be approximately normal, with mean 55 and standard error 2.

6.13 The distribution will be approximately normal, with mean 11,520 and standard error 333.08.

6.15 The relative frequency histogram is identical (except for class boundaries) to the histogram shown in Figure 6.3 of the text.

6.17 The sample distribution is approximately normally distributed with mean 60 and standard error 1.25. The 95% interval is (57.55, 62.45).

6.19 **a** 0.7462
    b 0.1587
    c 0.0188

6.21 **a** 0.0548

b To observe a sample mean as large or larger than 7 is highly unlikely, given that the mean oxygen content is 6 ppm. It is more likely that the mean is larger than 6.

6.23 178; 200

6.25 1.00

6.27 **a** 0.2033

b 0.0668

6.29 0.2076

6.31 **a** The sampling distribution will be approximately normal with mean 0.02 and standard error 0.000081.

b Both approximately zero

6.33 **a** 0

b 0

6.37 0.1236

6.39 **a** 62.55%

b 0.9998

6.41 **a** 3

b 2

c 4

6.43 **a**

y	f_i
1	1
3	1
7	1
9	1

b

Sample	$\bar{y}$
{1, 3}	2
{1, 7}	4
{1, 9}	5
{3, 7}	5
{3, 9}	6
{7, 9}	8

c

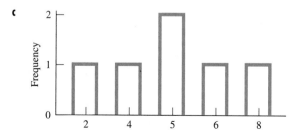

d 5; 2; The Central Limit Theorem applies for large n.

6.47 0.0768

6.49 3500; 32.4

6.51 **a** 0.0099

      b 0.0038

6.53 **a** The distribution would be approximately normal with mean 3 and standard error 0.06.

      b The distribution for $n = 100$ would be approximately normal with mean 3 and standard error 0.03.

      c 0

6.55 The percentage of defective components is greater than 0.05.

6.57 0.0125

6.59 0.3844

6.63 0

6.65 No

6.69 **a** 9.7, 5.1

      b 9.81, 2.16

Chapter 7

7.1 **a** All registered voters in her state

     b Could be selected from voter registration lists

7.3 **a** The lifetimes of all fuses produced by the manufacturer

     b Hypothesis testing

7.5 **a** (101.95, 108.05)

     b (100.99, 109.01)

7.7 **a** Width will be narrower by $1/\sqrt{2}$

     b Width will be cut in half

7.9 (4.04, 6.36)

7.11 (3.02, 3.38)

7.13 (824.7, 875.3)

7.15 (412.88, 447.12)

7.17 (0.164, 0.196)

7.19 246

7.21 125

7.23 6147; 385

7.25 The process is in control because all of these sample means are within the control limits.

7.27 The process may not be in control because 0.647 is above the UCL.

7.29 120

7.31 0.8106; 0.9997; 1

7.33 Yes

7.35 6

7.37 0.2296; 0.5636; 0.8577; 0.9979

7.39 Yes; $H_0: \mu = 525$ versus $H_a: \mu \neq 525$; Reject if $|z| > 2.58$

7.41 **a** $z = 13.25$; Reject H_0

     b Not necessarily

7.43 $z = -2.36$; Reject H_0

7.45 **a** 0.0314

     b 0.0628

7.47 $t = 6.32$; Reject H_0

7.49 **a** Reject H_0 if $t < -1.761$.

     b Reject H_0 if $|t| > 2.074$.

     c Reject H_0 if $t > 2.015$.

7.51 $t = 1.64$; Do not reject H_0

7.53 $t = 1.09$; $0.1 < p\text{-value} < 0.2$; Do not reject H_0

7.55 $t = -0.314$; Do not reject H_0

7.57 $0.001 < p\text{-value} < 0.0025$. There is strong evidence against H_0.

7.59 One-tailed; $t = 4.2133$; Reject H_0

7.61 **a** The mean square feet of coverage from a gallon of the manufacturer's paint

     b Reject H_0 if $z > z_{0.05} = 1.645$.

     c $z = 2.23$; Reject H_0

     d 0.0129.

7.63 **a** (205.1, 244.9)

     b Mean number of passengers (per day, per bus) on all city buses which is unknown

     c Average number of passengers (per day, per bus) on the 35 city buses sampled which is 225

7.65 $t = 2.50$; Reject H_0; $p\text{-value} = 0.0062$

7.67 **a** False; Type II error

     b True

     c False; Type II error

     d False; decrease

7.69 (7.16, 11.24)

7.71 (3.13, 3.27)

7.73 $t = -3.21$; Reject H_0

7.75 $t = -2.71$; Reject H_0

7.77 $t = -2.86$; Reject H_0

7.79 $t = -6.96$; Reject H_0

7.81 0.15

7.83 (21.89, 26.49)

7.85 (1.90, 2.22)

7.87   (c), (e), (g)

7.89   **a**   Normally distributed; to the population mean
         b   Smaller
         c   Wider
         d   Central Limit Theorem
         e   Type II error

7.91   **a** $H_0: \mu = 5.2$ versus $H_a: \mu < 5.2$
         Since σ is unknown, $t = \dfrac{\bar{y} - \mu}{s/\sqrt{n}}$
         Reject H_0 if $t < -1.676$ (50 df)
         b $t = -2.02$; Reject H_0 and conclude that the mean dissolved oxygen count is less than 5.2 ppm.

7.93 $z = -2.36$; p-value 0.0091; Reject H_0

7.95   (407.5, 452.5)

7.97   **a**   (26.4, 34.6)
         b   Wider, (25.1, 35.9)

7.99   (20.96, 23.04)

7.101   (98.35, 98.45)

7.103   **a** $H_a: \mu \neq 29$
         b   0.4716

7.105   (0.718, 0.806)

7.107   (53.93, 62.07)

7.111   **a**   0.762
          b   0.98
          c   (0.7220, 0.8020); We are 98% confident that the population mean proportion of patients per hospital with group medical insurance lies in the interval .722 to .802.

7.113 $t = 2.39$; p-value 0.024; Reject H_0

7.115   **a**   No, because $s > \bar{y}$ so the distribution will be skewed
          b   median

7.117   15.37

7.119   (17.47, 22.53)

7.121   658

7.123   (9.42, 13.54); 99% CI is wider (8.69, 14.27)

7.125   **a**   (4.54, 5.82)
          b   We are 90% confident that this interval captures the population mean damage claim.

7.127   **a** p-value = 0.0002
          b   Reject the null hypothesis and conclude that the mean treatment time is different than 25 minutes.
          c   We could have rejected the null hypothesis when, in fact, it was true which is a Type I error.

Chapter 8

8.1 a   Reject when $|t| > 2.064$
b   Reject when $t > 2.624$
c   Reject when $t < -1.86$
d   Samples are independent; Sampling from normal population; Equal variances

8.3 $0.005 < p$-value < 0.01

8.5 a $(-9.93, 27.93)$
b   90% CI $(-6.7, 24.7)$; 99% CI $(-16.55, 34.55)$
c   The larger the confidence, the wider the interval.

8.7 $t = 1.61$; Do not reject H_0

8.9 $(-3.66, 7.66)$

8.11 a   With this small of a data set, it is difficult to determine the violations to normality. However, the data are not skewed and equal variances seems to be a reasonable assumption.
b $t = -5.85$; p-value < 0.001; Reject H_0

8.13 No, the distribution will be skewed, since $s > \bar{y}$ for the magnesium data.

8.15 $(-6504.4, 5959.6)$

8.17 Age 9 $(-5.2, 41.2)$; Age 13 $(-1.7, 43.7)$; Age 17 $(2.5, 47.5)$

8.19 a $z = 2.918$; since $2.918 > 1.96$, reject H_0
b   A two-sample t-test may be appropriate.

8.21 a $t = -3.28$; p-value $= 0.031$; Reject H_0
b   Still reject H_0

8.23 $t = 4.95$; p-value < 0.001; Reject H_0

8.25 $(48.6, 191.4)$

8.27 Wilcoxon test statistic $= 16$; p-value $= 0.012$; Reject H_0

8.29 a $t = 2.68$; Reject H_0 if $t > 1.782$; Reject H_0
b p-value ≈ 0.01

8.31 $t = 3.66$; Reject H_0 if $|t| > 3.182$; Reject H_0

8.33 $(1.37, 11.03)$

8.35 $(-10.25, -2.35)$

8.37 $(-57.42, 48.26)$

8.39 $t = 6.9$; Reject H_0 if $|t| > 4.604$; Reject H_0

8.41 $t = 2.46$; Reject H_0 if $|t| > 2.776$; Do not reject H_0

8.43 $(2.6, 5.0)$

8.45 $t = -3.07$; Reject H_0 if $t < -1.812$; Reject H_0

8.47 a   Reject H_0; The difference appears to be larger than 0 with a p-value $= 0.0026$.
b   Again, reject H_0 with a p-value $= 0.0023$.

8.49 a $t = 3.57$, p-value $= 0.0008$; There is strong evidence against H_0. Program 1 appears to be slower than Program 2.

b $t = 3.57$, p-value $= 0.0008$
$t' = 3.41$, p-value $= 0.0018$
Wilcoxon test statistic $= 191.0$, p-value $= 0.0034$
All three tests show evidence against H_0.

8.51 (5.4, 10.6); No (5.8, 10.2)

8.53 **a** $s_p = \sqrt{\dfrac{60 + 60}{2}} = 7.75$

b (i) Reject H_0 if $|t| > 2.306$; (ii) No

c (i) Reject H_0 if $t > 1.86$; (ii) Yes

8.55 **a** One-tailed

b (i) H_0: $\mu_w - \mu_m = 0$ versus H_a: $\mu_w - \mu_m < 0$.
(ii) H_0: On the average there is no difference in the spending of women and men candidates.
H_a: Women spend less than men on the average.

8.57 **a** t-test p-value $= 0.0000$; Wilcoxon p-value $= 0.000$
Both tests reject H_0.

b The t-test seems appropriate.

c It does not matter which test was run here. In some situations, where we question the assumptions of the t-test, it is reasonable to perform both tests.

8.59 **a** Yes

b No

c No

d Yes

8.61 $(-0.78, 3.18)$

8.63 $(-\$3252.59, \$2146.59)$
Since the confidence interval includes 0 it neither confirms nor contradicts the officials' claim.

8.65 $t = -1.11$; Reject H_0 if $|t| > 2.12$; Do not reject H_0

8.67 **a** $t = 5.9126$

b $t' = 4.4491$

c Both p-values < 0.001; Reject H_0

8.69 44

8.71 95% CI $= (-35.4, -11.4)$

8.73 **a** There are not enough observations to draw any conclusions concerning normality. There is no suggestion that the population variances are different.

b Results are the same for $\sigma_1 = \sigma_2$.

c $t = -5.8507$, p-value < 0.001

8.75 **a** $t = -5.042$; p-value < 0.001; Reject H_0

b Differences existing at the end of the study may be due to baseline differences. To guard against baseline differences, both treatment groups should have the same type of patients.

8.77 Wilcoxon test statistic $= 4$; Reject H_0

8.79 $t' = -9.375$; p-value < 0.001; Reject H_0

8.83 **a** In general, HMO sources have shorter hospital stays than do non-HMO sources. It would be useful to know the sample sizes, sample means and standard deviations for both groups; all categories.

b $H_0: \mu_1 - \mu_2 = 0$ versus $H_a: \mu_1 - \mu_2 \neq 0$
$t = -14.9$; p-value < 0.001; Reject H_0

8.89 Treatment D versus B: $t = 2.98$; p-value $= 0.0023$; Reject H_0
Treatment D versus C: $t = 5.08$; p-value < 0.001; Reject H_0
Treatment C seems to have the lowest mean HAM-D total score.

8.91 $H_0: \mu_B - \mu_D = 0$ versus $H_a: \mu_B - \mu_D \neq 0$
$t = 1.23$; Reject H_0 if $|t| > 2.021$ (40 df); Do not reject H_0

8.93 $t = -0.17$; p-value $= 0.87$; Do not reject H_0
$t' = -0.17$; p-value $= 0.87$; Do not reject H_0
The results of the t- and t'-test are identical. There is not enough evidence to conclude that the ages of males and females are significantly different.

Chapter 9

9.1 **a** 0.01

b 0.90

c 0.01

9.3 $\chi^2_{.025} = 324.99$
$\chi^2_{.975} = 232.79$

9.5 $H_0: \sigma = 0.15$ versus $H_a: \sigma < 0.15$
$\chi^2 = 46.464$; Reject H_0 if $\chi^2 < 43.19$; Do not reject H_0

9.7 **a** $\bar{y} = 3.999$, $s = 0.016$

b $H_0: \sigma = 0.011$ versus $H_a: \sigma > 0.011$
$\chi^2 = 11.82$; Reject H_0 if $\chi^2 > 37.65$; Do not reject H_0

9.9 The plot reveals the data may be skewed and therefore not from a normal distribution. This violation would affect the validity of the answers in Exercises 9.7 and 9.8.

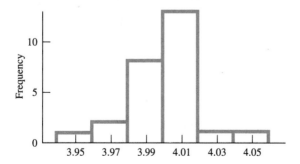

9.11 $0.1 < $ p-value $ < 0.15$

9.13 A $\bar{y}$-chart is used to show the distribution of means in a process. An r-chart is used to show the distribution of ranges (or variability) in a process. Since a $\bar{y}$-chart uses $\bar{r}$ in constructing its limits, the $\bar{y}$-chart should never be used without first constructing the corresponding r-chart.

9.15 LCL $= (0.223)(4.1) = 0.9143$
CL $= 4.1$
UCL $= (1.777)(4.1) = 7.2857$

9.17 **a** 2.22
   b 2.48
   c 1.96
   d 2.70
   e 2.78

9.19 $F = 3.15$; Reject H_0 if $F > 3.79$; Do not reject H_0

9.21 $F = 2.73$; Reject H_0 if $F > 2.32$; Reject H_0

9.23 $F = 3.33$; Reject H_0 if $F > 1.99$; Reject H_0

9.25 **a** (0.728, 185.807)
   b Take square roots throughout (0.85, 13.63)

9.27 $\chi^2 = 11.55$; Reject H_0 if $\chi^2 < 5.407$; p-value < 0.001; Reject H_0

9.29 **a**

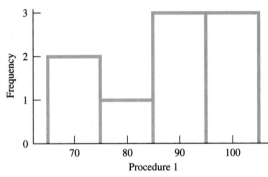

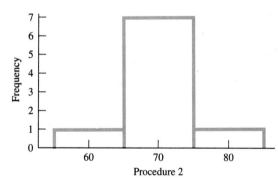

   b Hard to tell with small samples; Procedure 1 may not be; Procedure 2 may be
   c $F = 3.96$; Reject H_0 if $F > 3.44$; Reject H_0

9.31 $F = 2.35$; Reject H_0 if $F > 2.44$; p-value $= 0.12$; Do not reject H_0

9.33 **a** Yes; A solid circle represents technique 1 and an open circle represents technique 2.

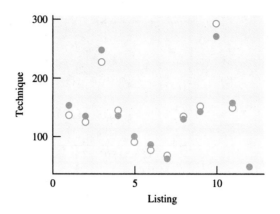

b   mean = 1.67, standard error = 3.31

9.35   Random sample is drawn from a normal population

9.37   Yes

9.39 $F = 3.125$; Reject H_0 if $F > 2.69$; Reject H_0

9.41 $F = 3.62$; Reject H_0 if $F > 3.18$; Reject H_0

9.43   **a**   Unequal variances
          b   No; $F = 5.9$; Reject H_0 if $F > 9.6$; Do not reject H_0

9.45   Treatment D versus A; $F = 1.49$
          Treatment D versus B; $F = 1.39$
          Treatment D versus C; $F = 1.63$
          Reject H_0 if $F > 2.27$; Do not reject any of the null hypotheses

9.47   (0.883, 6.248); (0.377, 2.667)

9.49 $H_0: \sigma^2_{\text{Male}} = \sigma^2_{\text{Female}}$ versus $H_a: \sigma^2_{\text{Male}} \neq \sigma^2_{\text{Female}}$
          $F = 1.61$; Reject H_0 if $F > 1.76$; Do not reject H_0

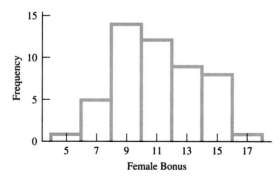

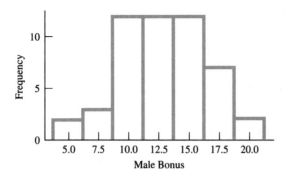

Chapter 10

10.1   1) The experiment consists of n identical trials.
2) Each trial results in one of k outcomes.
3) The probability that a single trial will result in outcome i is p_i, $i = 1, 2, \ldots, k$, and remains constant from trial to trial. (Note $\Sigma_i p_i = 1$.)
4) The trials are independent.
5) We are interested in n_i, the number of trials resulting in outcome i. (Note $\Sigma_i n_i = n$.)

10.3   **a**   7.815
b   21.67
c   24.32

10.5 $\chi^2 = 5.28$; Reject H_0 if $\chi^2 > 9.488$; Do not reject H_0

10.7 $\chi^2 = 49.07$; p-value < 0.001 ; Reject H_0

10.9 $\chi^2 = 181.65$; Reject H_0 if $\chi^2 > 7.815$; Reject H_0

10.11 $\chi^2 = 120.71$; Reject H_0 if $\chi^2 > 7.815$; Reject H_0

10.13 $\chi^2 = 31.15$; p-value < 0.001

10.15 $\chi^2 = 7.608$; Reject H_0 if $\chi^2 > 9.488$; Do not reject H_0

10.17   **a**   Yes; Samples of less than 25 would be suspect.
b $(0.15, 0.25)$

10.19   **a** $(0.78, 0.82)$
b $(0.783, 0.817)$

10.21   **a**   Yes
b   Yes
c $(0.25, 0.51)$; Increase the sample size

10.25   1 $(0.52, 0.60)$
2 $(0.46, 0.54)$
3 $(0.42, 0.50)$

10.27   **a**   A table listed in descending order of proof of illiteracy
b   Who was surveyed? How were they surveyed?

10.29   **a**   Yes; $(800)(0.096) = 76.8$ and $(200)(0.904) = 723.2$
b $z = -5.02$; Reject H_0 if $z < -1.645$; Reject H_0

10.31 $(0.19, 0.61)$

10.33

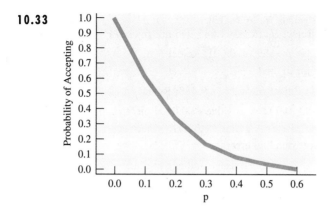

10.35 Yes

10.37 $(-0.145, 0.195)$

10.39 $z = 2.3; 0.01 < p$-value < 0.025

10.41 **a** $z = 3.4; p$-value < 0.001; Reject H_0

 b What was the amount of hair growth? What side effects were observed? What characteristics distinguish responders from nonresponders?

10.43 **a** H_0: The two variables are independent
 H_a: The two variables are not independent

 b $E_{11} = 22.5, E_{12} = 67.5, E_{21} = 27.5, E_{22} = 82.5$
 $\chi^2 = 0.673$; Reject H_0 if $\chi^2 > 3.841$; Do not reject H_0

10.45 $\chi^2 = 7.862; 0.025 < p$-value < 0.01

10.47 **a** Control 10%, Low dose 14%, High dose 19%

 b $\chi^2 = 3.312$; Reject H_0 if $\chi^2 > 5.991$; Do not reject H_0

 c No

10.49 Two; Equal

10.51 $\chi^2 = 12.817$; Reject H_0 if $\chi^2 > 9.488$; Reject H_0

10.53 $\chi^2 = 32.961; p$-value < 0.001; Reject H_0

10.55 **a** $E_{11} = 17.90, E_{12} = 20.97, E_{13} = 23.78, E_{14} = 26.34$
 $E_{21} = 24.14, E_{22} = 28.28, E_{23} = 32.07, E_{24} = 35.52$
 $E_{31} = 16.70, E_{32} = 19.56, E_{33} = 22.18, E_{34} = 24.57$
 $E_{41} = 11.26, E_{42} = 13.20, E_{43} = 14.97, E_{44} = 16.57$

 b $\chi^2 = 57.83$

 c p-value $= 0.0001$

 d Reject H_0

10.57 $(0.64, 0.68)$

10.59 **a** Yes

 b $(-0.094, 0.142)$

10.61 $H_0: p_1 - p_2 = 0$ versus $H_a: p_1 - p_2 > 0$
 $z = 1.99; 0.01 < p$-value < 0.025

10.63 $z = -1.67; 0.025 < p$-value < 0.05; Reject H_0 at $\alpha = 0.05$

10.65 $H_0: p = 0.3$ versus $H_a: p > 0.30$
$z = 0.845$; Reject H_0 if $z > 1.645$; Do not reject H_0

10.67 **a** $z = -4.12$; p-value < 0.001; Reject H_0
b $z^2 = \chi^2$ for df $= 1$

10.69 $\chi^2 = 13.569$; Reject H_0 if $\chi^2 > 9.488$; Reject H_0

10.71 **a** $z = -2.83$; $0.002 < p$-value < 0.01; Reject H_0
b $\chi^2 = 8.00$; Reject H_0 if $\chi^2 > 3.841$; Reject H_0
c Except for rounding errors, $z^2 = \chi^2$

10.73 H_0: Regularity of seat belt usage and age are independent
H_a: Regularity of seat best usage and age are not independent
$\chi^2 = 59.641$; p-value < 0.001; Reject H_0

10.75 **a** Normality of the differences; No
b The t test for $H_0: \mu_d = 0$ versus $H_a: \mu_d > 0$ has $t = 21.39$ and p-value < 0.001. We can conclude that the entry-level blood pressure and the follow-up blood pressure were not the same.

10.77 **a** >1 week $(0.43, 0.61)$; <1 week $(0.04, 0.30)$
b No
c $(0.37, 0.53)$

10.79 **a** No
b $\chi^2 = 21.631$; p-value < 0.001; Reject H_0

10.81 2

10.83 $<\$35,000$: $\chi^2 = 26.105$; Reject H_0 if $\chi^2 > 5.991$; Reject H_0
$>\$35,000$: $\chi^2 = 0.539$; Reject H_0 if $\chi^2 > 5.991$; Do not reject H_0

10.89 Information on the different therapeutic effects

10.91

Religion	Major B	H	NS	SS	Total
P	23	11	7	12	53
C	27	6	7	7	47
J	1	2	0	2	5
O	9	5	8	6	28
N	25	13	15	14	67
Total	85	37	37	41	200

$\chi^2 = 12.397$; There are 2 cells with expected counts less than 1.0 and 4 cells with expected counts that are less than 5. Therefore, this test is probably invalid. One solution would be to combine some of the categories.

Chapter 11

11.1

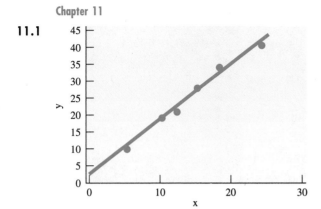

11.3 **a** 7.8

b

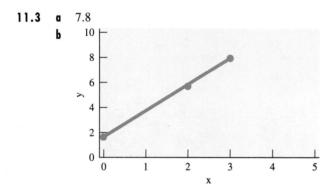

11.5 **a** (i) 5 (ii) 2
b 95

11.7 $\hat{\beta}_0 = -3200$, $\hat{\beta}_1 = 0.25$
a $1425
b $7175

11.9 $\hat{y} = 0.5 + 1.7x$

11.11 $\hat{y} = 2.47 + 1.63x$

11.13 $\hat{y} = -3185.29 + 0.269529x$
a $1800.9965
b $8000.1635

11.15 **a** $\hat{y} = -9.9 + 0.307x$
b $\Sigma(y - \hat{y})^2 = 4.356$

11.17 **a** $\hat{y} = -16.9 + 2.5(x)$
b 108.1

11.19 $\hat{\rho} = 0.993$; This indicates a very strong relationship.

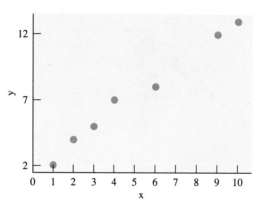

11.21   Correlation coefficient = 0.85
Coefficient of determination = 0.7225
72.25% of the variability of the y values can be explained by the variability of the x's.

11.23   **a**

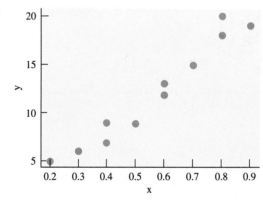

b   The correlation coefficient is 0.971 which indicates a strong positive relationship between basal area and cubic-foot tree volume, i.e. as basal area increases cubic-foot tree volume increases.

11.25 $\rho = 0.890$

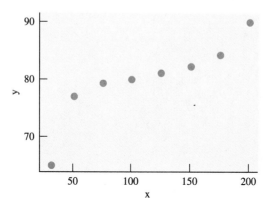

11.27 **b** $\hat{y} = 31.333333 - 7.333333x$

c $x = 2.3$ is observation 7, $\hat{y} = 14.467$

11.29 **a**

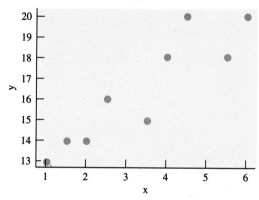

b $\hat{y} = \hat{\beta}_0 + \hat{\beta}_1 x$

c $\hat{y} = 11.82 + 1.36x$

d 18.62

11.31 **a**

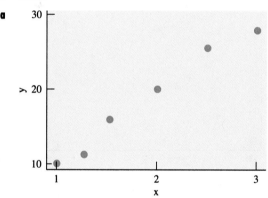

b $\hat{y} = 1.00 + 9.39x$

c 21.358

11.33 **a**

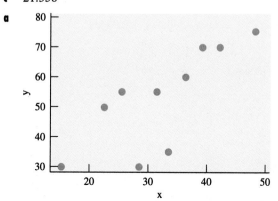

b $\hat{y} = 11.24 + 1.31x$

c Yes

d 0.77

e 50.54

11.35

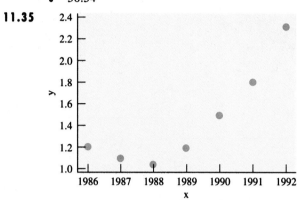

11.37 a

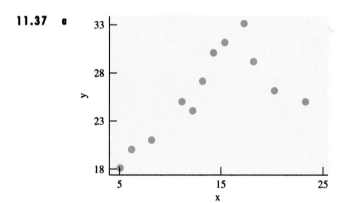

b $y = \beta_0 + \beta_1 x + \beta_2 x^2$

c $\hat{\rho}_s = 0.673$

11.39 a $\hat{y} = -25.6 + 1.84x_1$
$\hat{y} = -21.7 + 1.22x_1 + 1.6x_2$
$\hat{y} = 59.3 + 0.916x_1 + 0.4x_2 - 0.944x_3$

11.41 Englact versus Mathact, $\hat{\rho} = 0.595$
Englact versus Socsact, $\hat{\rho} = 0.668$
Mathact versus Socsact, $\hat{\rho} = 0.567$

11.43 a

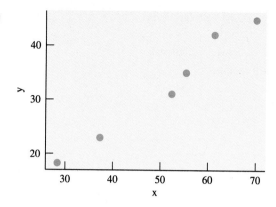

b 0.988

c When a least-square regression line is fit to the data, 97.6% of the variation in the *y*'s can be explained by the variation in the *x*'s.

11.45 a

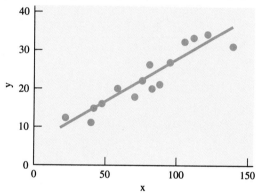

b $\hat{y} = 6.079 + 0.214x$

c Yes

11.47 a

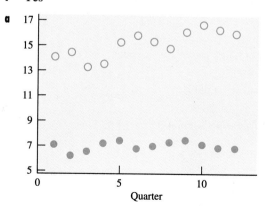

b Slightly

c 0.355

11.49 **a** Strong positive relationship

b Little or no relationship

c Perfect negative relationship

11.51 **a** $y = 2 + x$

b $y = 2 - x$

c $y = 1 + x$

d $y = -2 + x$

11.53 **a**

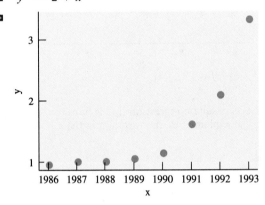

b $\hat{y} = 0.245 + 0.284(\text{Year} - 1985)$

c 2.8; Yes

11.55 $\hat{y} = -430.9 + 0.13791x$

11.57 **a** 0.982

b $\hat{y} = 4.67 + 1.23x$

c 12

11.59 **a** $\hat{y} = -0.40 + 0.18x$

b 46.7

c 0.949

d 90.1%

11.61 **a** False; negative one and one

b False; not necessarily have the same

c True

d False; exactly equal to -1, 0, or 1

e True

f False; the possibility of a relationship

11.63 **a**

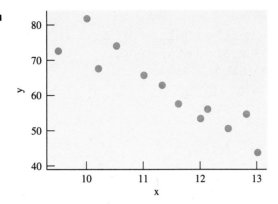

b Yes

c $\hat{y} = 162.57 - 8.84x$

11.65 **a**

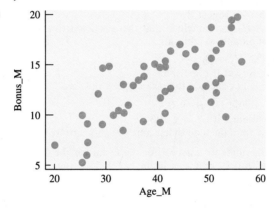

b $\hat{y} = 3.129 + 0.243x$

11.67 **a** $\hat{\rho} = 0.686$, $\hat{\rho}_s = 0.983$

b $\hat{\rho} = 0.286$, $\hat{\rho}_s = 0.971$

11.69 **a** For the large capital stocks:

```
            Low        Medium
Medium      0.811
High        0.739      0.900
```

b There is a fairly strong positive relationship between each of the three categories. The relationship between Medium and High is the strongest.

11.71 0.720

11.73 0.130

Chapter 12

12.1 **a** $\Sigma x = 30$, $\Sigma x^2 = 110$, $\Sigma xy = 374$, $\Sigma y = 117$, $\Sigma y^2 = 1401$

 b $\hat{\beta}_0 = 8.25$, $\hat{\beta}_1 = 1.15$, $s_\epsilon^2 = 0.706$

 c $S_{\hat{\beta}_0} = 0.623$, $S_{\hat{\beta}_1} = 0.188$

12.3 **a** $\hat{y} = 48.93 + 10.33x$

 b 5.7

 c 0.7957

12.5 **a** $\hat{y} = 9.275 + 0.505x$

 b $H_0: \beta_1 = 0$, $H_a: \beta \neq 0$
$t = 7.83$; Reject if $|t| > 3.182$
Reject H_0 and conclude that $\beta_1 \neq 0$.

12.7 **a** 1.07

 b 6.96

 c $H_0: \beta_1 = 0$, $H_a: \beta_1 > 0$
$t = 5.145$; Reject if $t > 2.228$
Reject H_0 and conclude that there is a positive linear relationship between lysine ingestion and weight gain.

12.9 $H_0: \beta_1 = 0$, $H_a = \beta_1 > 0$
$t = 4.23$; Reject if $t > 1.86$
Reject H_0 and conclude that the slope is positive.

12.11 $H_0: \beta_1 = 0$ and $H_a: \beta_1 \neq 0$
$t = 4.68$; Reject if $|t| > 2.306$
Reject H_0 and conclude that x and y are linearly related.

12.13 The 95% confidence interval for the mean weight gain for chickens fed a diet supplemented with 0.19 grams of lysine is (18.55 grams, 20.08 grams).

12.15 (2.595, 2.908)

12.17 **a** $\hat{y} = -1.733 + 1.317x$

 b $t = 6.34$, p-value < 0.001
Reject the H_0 and conclude that $\beta_1 > 0$.

12.19 $\hat{\rho} = -0.972$
$t = -13.08$; Reject if $|t| > 2.228$.
Reject H_0 and conclude $\rho \neq 0$.

12.21 **a** $t = 16.92$; Reject if $|t| > 2.365$.
Reject H_0 and conclude $\rho \neq 0$.

 b This indicates a strong positive linear relationship.

12.23 (0.742, 2.030)

12.25 A value of 40 is well out of the range of the x values used to make this model, hence the confidence interval might be invalid.

12.27 $t = -7.86$; Reject if $t < -1.895$; Reject H_0

12.29 **a** 0.957

 b $t = 8.73$; Reject if $t > 1.895$; Reject H_0

12.31 4.9995 ± 0.064; Narrower; Wider

12.33 9.39368 ± 1.45122

12.35 $t = 5.22$; Reject if $t > 1.943$; Reject H_0

12.37 **b** $\hat{y} = 8.412234 + 5.026596x$

c 28.519

12.39 $\hat{y} = 36.10 + 0.384x$
$H_0: \beta_1 = 0$ and $H_a: \beta_1 \neq 0$
$t = 1.13$; Reject if $|t| > 2.365$. Do not reject H_0.
No, it does not seem to make sense to predict the after-course rating based on the before-course rating.

12.41 $\hat{y} = 738 + 4.7x$
$H_0: \beta_1 = 0$ and $H_a: \beta_1 > 0$
$t = 12.31$; Reject if $t > 1.86$.
Reject H_0 and conclude that the slope is greater than 0.
When $x = 100$ (or 100,000), $\hat{y} = 1208$.

12.43 **a** Yes

b Linear regression model

12.45 **a**

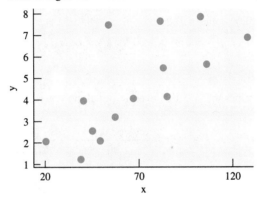

b 0.719

c $H_0: \rho = 0, H_a: \rho \neq 0$
$t = 3.58$; Reject if $|t| > 2.179$.
Reject H_0 and conclude that there is a direct linear relationship between sales and profit before taxes.

12.47 $\hat{\rho}_s = 0.706$; $H_0: \rho = 0, H_a: \rho \neq 0$
$z = 2.34$; Reject if $|z| > 1.96$; Reject H_0

12.49 **a**

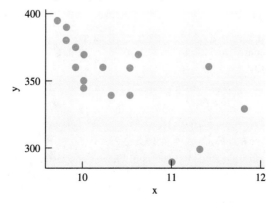

b $\hat{y} = 672.89 - 30.654x$

c Yes, $t = -3.78$, p-value $= 0.002$

12.51 **a** Bonus_M $= 7.55 + 0.0843$ Age_M
$t = 2.06$; p-value $= 0.044$; Reject H_0 at the $\alpha = 0.05$ level.

b Bonus_F $= 3.13 + 0.243$ Age_F
$t = 6.54$; p-value < 0.001; Reject H_0.

12.53 **a** No, because the data points are completely scattered

b Test H_0: $\beta_1 = 0$ and H_a: $\beta_1 \neq 0$.
Do not reject the null hypothesis for each of the three book-to-market ratio categories. The p-values for each of the tests are: High $= 0.555$, Medium $= 0.209$, Low $= 0.201$.

12.55 **a** The correlation coefficient for large and small stocks are:
High 0.769, Medium 0.775, Low 0.792.

b Test H_0: $\rho = 0$ versus H_a: $\rho \neq 0$
Reject the null hypothesis for each of the three book-to-market ratio categories. The p-values for each of the tests are: High p-value < 0.001, Medium p-value < 0.001, Low p-value < 0.001

12.57 $\hat{\rho} = 0.720$; p-value $= 0.0001$

12.59 No, $\hat{\rho} = 0.13$; p-value $= 0.1980$ for a test of H_0: $\rho = 0$ and H_a: $\rho \neq 0$. Do not reject H_0.

Chapter 13

13.1 Two estimates of σ^2, the overall population variance, are available if H_0: $\mu_1 = \mu_2 = \cdots = \mu_k$ is true.

$$s_W^2 = \frac{(n_1 - 1) s_1^2 + \cdots + (n_k - 1) s_k^2}{n_1 + \cdots + n_k - k}$$

and

$$s_B^2 = \frac{\Sigma n_k \bar{y}_i^2 - [(\Sigma n_i \, \bar{y}_i)^2 / n]}{k - 1}$$

should both estimate σ^2. If s_B^2 is too much larger than s_W^2 (as tested by $F = s_B^2 / s_W^2$), then the means are probably not the same.

13.3 (0.86, 2.88)

13.5 $F = 2.51$; Reject H_0 if $F > 2.61$; Do not reject H_0

13.7 **a** $t = -2.424$; Reject if $|t| > 2.179$. Reject H_0

b Yes; Yes; Notice $t^2 = F$.

13.9 $(-5.3, 23.3)$

13.11 **a** Means: A $= 61.0$, B $= 53.0$, C $= 64.8$
Standard Deviations: A $= 3.56$, B $= 3.00$, C $= 5.89$

b 1.732

c H_0: $\mu_A = \mu_B = \mu_C$; $F = 6.055$; p-value $= 0.022$; Reject H_0

d B

13.13 **a** H_0: $\mu_1 = \mu_2 = \mu_3 = \mu_4$

b Reject H_0; $F_{0.01, 3, 60} = 4.13$

c All possible students who could have been tested under these four preconditioning states; If we had been working with non-random samples, the differences we detected using the analysis of variance could not be attributed to differences among the preconditioning states.

d $s_W^2 = 25.5$

e $(-10.03, -3.97)$

13.15 $H_0: \mu_A = \mu_B = \mu_C$; $F = 17.69$; Reject H_0 if $F > 3.89$; Reject H_0

13.17 Reject H_0

Source	Sum of squares	Degrees of freedom	Mean square	F
Between	8.7704	2	4.3852	14.730
Within	2.679	9	0.2977	
Total	11.4494	11		

13.21 **a** $(6.63, 13.81)$

     b $(-1.41, 5.77)$

13.23 $H_0: \mu_1 = \mu_2 = \mu_3 = \mu_4$; $F = 64.5$; Reject H_0 if $F > 3.24$; Reject H_0

13.25 **a** Means: Location 1 = 10.13, 2 = 12.68, 3 = 11.48, 4 = 14.52
Standard Deviations: 1 = 1.49, 2 = 2.84, 3 = 2.72, 4 = 3.72

     b $H_0: \mu_1 = \mu_2 = \mu_3 = \mu_4$; $F = 2.64$; Reject H_0 if $F > 3.10$.
Do not reject H_0.

     c There is no evidence to indicate that the means are different.

     d It appears from part (a) that the means for locations 1 and 4 are different, but this difference was not detected in the analysis of variance.

13.29 $F = 19.5$; Reject H_0 if $F > 6.36$; Reject H_0

13.31 $F = 10.3$; Reject H_0 if $F > 2.87$; Reject H_0

13.33 **b**
```
Line    StDev      Mean
 1     13.8984    280.5
 2     12.9996    265.1
 3     10.2610    264.8
```

     c Reject H_0
```
SOURCE    DF    SUM OF SQUARES    MSE    F-value  p-value
Model      2       1612.467      806.2    5.17     0.0125
Error     27       4207.000      155.8
Total     29       5819.467
```

13.35 $F = 11.046$; p-value < 0.005; Reject H_0

13.37 $F = 10.22$; p-value $= 0.002$; Reject H_0

13.39 Use the Hartley test to check assumptions.
$t = 4$; Do not reject H_0; Analysis of variance seems appropriate.
$F = 0.85$; p-value $= 0.466$; Do not reject H_0

13.41 $F = 0.47$; p-value 0.626; Do not reject H_0

13.43
```
Source       df      SS      MS      F       p
Treatment     3     6298    2099    8.09    0.000
Error        96    24912     260
Total        99    31210
```

Reject the hypothesis of equal mean OBRIST score for each treatment.

13.45
```
Source       df      SS      MS      F       p
Treatment     3      347     116    0.69    0.557
Error        96    15989     167
Total        99    16336
```

Do not reject H_0. Therefore, the four groups are comparable with respect to age.